Bioenergy Resources and Technologies

AF783044

Bioenergy Resources and Technologies

Edited by

Abul Kalam Azad

Mohammad Masud Kamal Khan

School of Engineering and Technology, Central Queensland University, Melbourne, VIC, Australia

ELSEVIER

ACADEMIC PRESS

An imprint of Elsevier

Academic Press is an imprint of Elsevier
125 London Wall, London EC2Y 5AS, United Kingdom
525 B Street, Suite 1650, San Diego, CA 92101, United States
50 Hampshire Street, 5th Floor, Cambridge, MA 02139, United States
The Boulevard, Langford Lane, Kidlington, Oxford OX5 1GB, United Kingdom

Copyright © 2021 Elsevier Inc. All rights reserved.

No part of this publication may be reproduced or transmitted in any form or by any means, electronic or mechanical, including photocopying, recording, or any information storage and retrieval system, without permission in writing from the publisher. Details on how to seek permission, further information about the Publisher's permissions policies and our arrangements with organizations such as the Copyright Clearance Center and the Copyright Licensing Agency, can be found at our website: www.elsevier.com/permissions.

This book and the individual contributions contained in it are protected under copyright by the Publisher (other than as may be noted herein).

Notices
Knowledge and best practice in this field are constantly changing. As new research and experience broaden our understanding, changes in research methods, professional practices, or medical treatment may become necessary.

Practitioners and researchers must always rely on their own experience and knowledge in evaluating and using any information, methods, compounds, or experiments described herein. In using such information or methods they should be mindful of their own safety and the safety of others, including parties for whom they have a professional responsibility.

To the fullest extent of the law, neither the Publisher nor the authors, contributors, or editors, assume any liability for any injury and/or damage to persons or property as a matter of products liability, negligence or otherwise, or from any use or operation of any methods, products, instructions, or ideas contained in the material herein.

Library of Congress Cataloging-in-Publication Data
A catalog record for this book is available from the Library of Congress

British Library Cataloguing-in-Publication Data
A catalogue record for this book is available from the British Library

ISBN 978-0-12-822525-7

For information on all Academic Press publications
visit our website at https://www.elsevier.com/books-and-journals

Publisher: Candice Janco
Acquisitions Editor: Maria Convey
Editorial Project Manager: Timothy Bennett
Production Project Manager: Sojan P. Pazhayattil
Cover Designer: Mark Rogers

Typeset by SPi Global, India

ELSEVIER Book Aid International
Working together to grow libraries in developing countries
www.elsevier.com • www.bookaid.org

Dedication

This book is dedicated to our parents.

Contents

Section B
Biogas energy

4. Biogas: A promising clean energy technology

5. Potential of ionic liquid applications in natural gas/biogas sweetening and liquid fuel cleaning process

Section C
Hydrogen energy

6. Hydrogen production via electrolysis: Mathematical modeling approach

Atta Sojoudi, Ali M. Sefidan, K.C.A. Alam, and Suvash C. Saha

7. Techno-economic evaluation methodology for hydrogen energy systems

Venkatakrishnan Balasubramanian, Nawshad Haque, Suresh Bhargava, Srinivasan Madapusi, and Rajarathinam Parthasarathy

Section E
Socio-economic aspect of bioenergy

12. Social, economic, and environmental aspects of bioenergy resources

Zobaidul Kabir

13. An overview of policy framework and measures promoting bioenergy usage in the EU, the United States, and Canada

Zobaidul Kabir, M.A. Yusuf, and Imran Khan

Contributors

Numbers in parentheses indicate the pages on which the authors' contributions begin.

K.C.A. Alam (157), School of Mechanical and Mechatronics Engineering, University of Technology Sydney, Sydney, NSW, Australia

Abdur Rahman Ansari (91), Jhang-Campus, University of Veterinary and Animal Sciences Lahore, Jhang, Pakistan

M.T. Arif (219), School of Engineering, Deakin University, Waurn Ponds, VIC, Australia

Neha Arora (327), Patel College of Global Sustainability, University of South Florida, Tampa, FL, United States

Muhammad Arshad (91), Jhang-Campus, University of Veterinary and Animal Sciences Lahore, Jhang, Pakistan

Venkatakrishnan Balasubramanian (195), Chemical Engineering, School of Engineering, City Campus, RMIT University, Melbourne; CSIRO Energy, Clayton, VIC, Australia

Dhinesh Balasubramaniyam (291), Department of Mechanical Engineering, Mepco Schlenk Engineering College, Virudhunagar, Tamil Nadu, India

Suresh Bhargava (195), Centre for Advanced Materials and Industrial Chemistry, RMIT University, Melbourne, VIC, Australia

Awais Bokhari (39), Chemical Engineering Department, COMSATS University Islamabad, Lahore, Pakistan

Apinya Chanthakett (219), School of Engineering, Deakin University, Waurn Ponds, VIC, Australia

Arooj Fatima (91), Jhang-Campus, University of Veterinary and Animal Sciences Lahore, Jhang, Pakistan

V. Edwin Geo (291), Department of Automobile Engineering, SRM Institute of Science and Technology, Kattankulathur, Tamil Nadu, India

Pobitra Halder (121), Chemical & Environmental Engineering, School of Engineering, RMIT University, Melbourne, VIC, Australia

Nawshad Haque (195), CSIRO Energy, Clayton, VIC, Australia

Muddasser Inayat (61), Department of Mechanical Engineering, Universiti Teknologi PETRONAS, Seri Iskandar, Perak, Malaysia

Frankowski Jakub (3), Department of Breeding and Agriculture Technology for Fibrous and Energy Plants, Institute of Natural Fibers and Medicinal Plants, Poznan, Poland

Sadia Javed (91), Department of Biochemistry, Government College University Faisalabad, Faisalabad, Pakistan

Zobaidul Kabir (349, 383), School of Environmental and Life Sciences, University of Newcastle, Ourimbah, NSW, Australia

Imran Khan (383), Department of Electrical and Electronic Engineering, Jashore University of Science and Technology, Jashore, Bangladesh

Mohammad Masud Kamal Khan (219), School of Engineering and Technology, CQUniversity, Melbourne, VIC, Australia

Asif Hussain Khoja (39), U.S.-Pakistan Centre for Advanced Studies in Energy (USPCAS-E), National University of Sciences & Technology, Islamabad, Pakistan

Sazal Kundu (121), Chemical & Environmental Engineering, School of Engineering, RMIT University, Melbourne, VIC, Australia

Srinivasan Madapusi (195), Chemical Engineering, School of Engineering, City Campus; Centre for Advanced Materials and Industrial Chemistry, RMIT University, Melbourne, VIC, Australia

M. Taqi Mehran (39), School of Chemical & Materials Engineering, National University of Sciences & Technology, Islamabad, Pakistan

M. Naqvi (39), Department of Engineering and Chemical Sciences, Karlstad University, Karlstad, Sweden

Salman Raza Naqvi (39, 61), School of Chemical & Materials Engineering, National University of Sciences & Technology, Islamabad, Pakistan

Aman M.T. Oo (219), School of Engineering, Deakin University, Waurn Ponds, VIC, Australia

Rajarathinam Parthasarathy (121, 195), Chemical & Environmental Engineering, School of Engineering; Chemical Engineering, School of Engineering, City Campus; Centre for Advanced Materials and Industrial Chemistry, RMIT University, Melbourne, VIC, Australia

Savankumar Patel (121), Chemical & Environmental Engineering, School of Engineering, RMIT University, Melbourne, VIC, Australia

George P. Philippidis (327), Patel College of Global Sustainability, University of South Florida, Tampa, FL, United States

Biplob Pramanik (121), Civil & Infrastructure Engineering, School of Engineering, RMIT University, Melbourne, VIC, Australia

Suvash C. Saha (157), School of Mechanical and Mechatronics Engineering, University of Technology Sydney, Sydney, NSW, Australia

Nor Aishah Saidina Amin (39), Chemical Reaction Engineering Group (CREG), School of Chemical and Energy Engineering, Faculty of Engineering, Universiti Teknologi Malaysia, Johor Bahru, Johor Darul Takzim, Malaysia

Ali M. Sefidan (157), Department of Mechanical Engineering, University of Canterbury, Christchurch, New Zealand

Kalpit Shah (121), Chemical & Environmental Engineering, School of Engineering, RMIT University, Melbourne, VIC, Australia

Muhammad Shahbaz (61), Division of Sustainable Development, College of Science and Engineering, Hamad Bin Khalifa University, Qatar Foundation, Doha, Qatar

Mirza Imran Shahzad (91), Department of Biochemistry and Biotechnology, The Islamia University of Bahawalpur, Bahawalpur, Pakistan

Atta Sojoudi (157), School of Mechanical Engineering College of Engineering, University of Tehran, Tehran, Iran

Ankit Sonthalia (291), Department of Automobile Engineering, SRM Institute of Science and Technology, NCR Campus, Modinagar, Uttar Pradesh, India

Shaharin A. Sulaiman (61), Department of Mechanical Engineering, Universiti Teknologi PETRONAS, Seri Iskandar, Perak, Malaysia

Muhammad Ikhsan Taipabu (251), Department of Chemical Engineering, National Cheng Kung University, Tainan, Taiwan

Syed Ali Ammar Taqvi (39), Department of Chemical Engineering, NED University of Engineering and Technology, Karachi, Pakistan

S. Thiyagarajan (291), Department of Automobile Engineering, SRM Institute of Science and Technology, Kattankulathur, Tamil Nadu, India

Karthickeyan Viswanathan (251), Department of Chemical Engineering, National Cheng Kung University, Tainan, Taiwan

Karthikeyan Viswanathan (291), Department of Mechanical Engineering, Sri Krishna College of Engineering and Technology, Coimbatore, Tamil Nadu, India

Shuang Wang (251), School of Energy and Power Engineering, Jiangsu University, Zhenjiang, Jiangsu, China

Wei Wu (251), Department of Chemical Engineering, National Cheng Kung University, Tainan, Taiwan

M.A. Yusuf (383), Ministry of Finance, Bangladesh Secretariet; Adjunct Faculty, AIUB, Bashundhara, Dhaka, Bangladesh

Editors biography

Dr. Abul Kalam Azad is a Lecturer of Mechanical Engineering in the College of Engineering and Aviation at Central Queensland University, Melbourne Campus, Victoria, Australia. He has a strong record of research publications and achievements in the areas of renewable energy technologies, energy conversion, and thermofluids engineering and in their relevant industrial applications. His research focused on clean energy production and mitigation of environmental pollution. He has an excellent publication track record throughout his career and has 2138 citations and *h-index* 25 (Google Scholar) and 1308 citations and *h*-index 21 (Scopus). He has published about 80 scientific articles including 4 edited refereed books, 11 book chapters, 47 journal articles, and 24 conference papers so far. His first edited book *Clean Energy for Sustainable Development* was published by Academic Press, Elsevier in 2016. His second edited book *Advances in Eco-fuel for a Sustainable Environment* was published as part of Woodhead Publishing Series in Energy, Elsevier in 2018. His third edited book *Advanced Biofuels: Applications, Technologies, and Environmental Sustainability* was published by Woodhead Publishing Series in Energy, Elsevier. His fourth book *Advances in Clean Energy Technologies* and fifth book *Bioenergy Resources and Technologies* are now being published by Academic Press, Elsevier. In addition, he is serving as a member of the editorial board of *Energies* (Q2 journal), section *Sustainable Energy* by MDPI, Switzerland. He has made significant contributions in engineering education, research, and scholarship. He was awarded "On-campus Educator of the Year" and "Student Voice Commendation Award" in 2019 by CQUniversity, Australia. Currently, he is a member of eight national and international professional bodies.

Professor Mohammad Masud Kamal Khan is currently an Adjunct Professor of Mechanical Engineering in the School of Engineering and Technology at the Central Queensland University, Australia. He obtained his PhD in Engineering from the University of Sydney, Australia in 1990. He has a strong academic track record of research and teaching both at national and international levels. His research and teaching areas are in thermofluid mechanics and engineering, energy conversion, and energy-efficient and environmentally sustainable technologies including production and performance assessment of biofuels. He has led various projects in mineral processes and energy areas and provided fundamental solutions to many complex industry-based projects. He has received various research project grants and has published more than 315 research articles in journals and conferences, including 2 edited books and 25 book chapters. The first edited book—*Thermofluid Modeling for Energy Efficiency Applications*, published by Academic Press, Elsevier UK and the second book—*Application of Thermo-fluid Processes in Energy Systems,* published by Springer, Singapore both covered a collection of hot topics on sustainable energy and efficiency. He has delivered many keynote and invited speeches in national and international conferences. He serves in the national and international scientific and advisory committees and on the editorial board of three international journals. He held three visiting professorial positions in the United States (West Virginia State University) and Canada (Ecole Polytechnique, Montreal), and is a member of several national and international bodies.

Preface

Bioenergy generally produced from biological sources is one of the renewable energy sources and is readily available throughout the world. These energy resources are renewable, biodegradable, and eco-friendly, which are the key attributes to meet the increasing energy demand for future energy security and environmental sustainability. Due to severe environmental pollution caused by increased liquid fuel consumption, the world is progressively moving toward renewable, alternative, and low-emitting energy applications. The book *Bioenergy Resources and Technologies* presents research works and articles by experts providing new information to the industries and the academia with advanced approach and applications of bioenergy resources. This book is primarily focused on the recent approaches of new bioenergy feedstocks, efficient energy extraction techniques, and effective applications of those energies covering a range of topics such as low-emission alternative fuels, energy-efficient systems, and environmental sustainability. The themes of the book include, but are not limited to:

- Bioenergy resources and related technologies
- Effective applications of bioenergies
- Prospect, scope, and implementation of bioenergies as an alternative fuel
- Environmental sustainability
- Future energy security
- Bioenergy future, new economic pathway, and community impact.

The book will help in understanding the relevant concept and solution to global issues to achieve bioenergy applications for future energy security as well as environmental sustainability in medium- and large-scale industries. It provides new information and novel techniques, extensive and experimental investigation, in-depth research and analysis and case studies using a wide array of bioenergy resources for sustainable environment. It will be a valuable resource for academics, researchers, practicing engineers, technologists, and students in the field of bioenergy applications.

The contents of the book have been carefully selected and written by experts covering both technical and social aspects from multidisciplinary areas. This book has 14 chapters, divided into five distinct sections such as biomass energy, biogas energy, hydrogen energy, biofuels energy, and socioeconomic aspects of bioenergy. Each section has a specific theme that provides logical continuity of the book.

The first section focuses on biomass energy, which is the primary source of bioenergy. It is one of the most useful forms of energy. This section covers the production of biomass, recently developed advanced approaches such as the catalytic pyrolysis of biomass and biomass gasification to extract energy from biomass.

The second section presents biogas energy production and conversion technologies. A promising approach of biogas production technology and its applications are also explored. In addition, the progressive approach of ionic liquid conversion techniques from biogas is also reported in this section.

The third section describes the hydrogen energy system as one of the clean and zero-emission energy sources. The focus of this section is the recent advancement and approaches of hydrogen energy production and conversion technologies, techno-economic evaluation, and application of hydrogen energy. Finally, the integrated renewable energy system along with hydrogen is also presented in this section.

The fourth section explains advanced biofuels conversion and application in modern transport systems for reducing significant environmental impact. In this section, extensive experimental investigation has been presented on engine performance and emission using different advanced biofuels. In addition, a critical analysis of microalgae biodiesel production and its life cycle assessment is also presented here.

Finally, the fifth section presents the socioeconomic and environmental aspects of bioenergy. It also illustrates the policies and measures to endorse bioenergy for a sustainable environment for the future.

All the chapters have been peer-reviewed by experts in the relevant field, and they have been accepted for publication after addressing the comments and suggestions made by the reviewers and/or editors.

The editors of this book express their sincere thanks to all the authors for their contributions. The successful completion of this book has been the result of the cooperation of many individuals, and the editors express their gratitude to all of them.

The editors also acknowledge the support by Timothy Bennett and Ali Afzal-Khan as an Editorial Project Manager and Maria Convey as an Acquisition Editor at Elsevier, for their help in completing the publication process. They express their profound gratitude and thanks to Maria, Timothy, and Ali for their assistance and guidance for this publication.

Abul Kalam Azad
Mohammad Masud Kamal Khan
College of Engineering and Aviation, Central Queensland University,
Melbourne Campus, Melbourne, VIC, Australia

Section A

Biomass energy

Chapter 1

Usefulness of selected annual plants cultivated for more energy content biomass production purposes in a temperate climate

Frankowski Jakub
Department of Breeding and Agriculture Technology for Fibrous and Energy Plants, Institute of Natural Fibers and Medicinal Plants, Poznan, Poland

1 Introduction

An increasing energy demand as well as the growing public anxiety for the country's energy security related to the dependence on fossil fuel supplies from other countries makes it necessary to seek sustainable solutions to make each country independent of energy imports. Nevertheless, fossil fuels are and will probably remain the main source of energy for many countries for decades. Among the EU countries, in 2015, Cyprus had the largest share of fossil fuels in energy production, followed by the Netherlands (93%), Ireland (92%), and Poland (91%) [1]. While the countries' dependence on coal is still high, in most of them the share of fossil fuels as an energy carrier fell between 1990 and 2015, mostly in Denmark (from 91% in 1990 to 69% in 2015), Latvia (from 83% to 61%), and Romania (from 96% to 74%). Poland also, albeit slightly, managed to reduce its dependence on minerals from 99% in 1990 to 91% in 2015. Due to the EU and national action plans for energy production, it is necessary to take steps to diversify its sources [1–4].

In most EU member countries, there was an increase in the dependence on fossil fuel imports between 1990 and 2015. This was especially visible in the case of Great Britain, where in 1990 the import of minerals for energy production accounted for 2% of all fossil fuels, while in 2015 it was already 43%, while the Netherlands ranked second (from 22% to 56%) and Poland third (from 1% in 1990 to 32% in 2015) [1].

Bioenergy Resources and Technologies. https://doi.org/10.1016/B978-0-12-822525-7.00014-7
Copyright © 2021 Elsevier Inc. All rights reserved.

According to the assumptions adopted by the Council of the European Union, by 2020, for example, Poland should achieve at least a 15% share of energy from renewable sources in the gross final energy production balance [5]. Due to the geographical location and the associated limited possibilities for the development of wind or solar energy compared to countries such as Denmark, Ireland, or Spain, according to researchers' forecasts, one of the pillars for increasing the share of energy from renewable sources in Poland will be biomass [6]. The development of ecological production of biomass as a substrate for the production of bioenergy and bioproducts is currently a priority for the European Union [5]. Therefore, as a result of plant breeding work, as well as through the improvement of technological processes, there is an increase in the efficiency and profitability of crops, and the use of biomass for energy purposes is becoming more and more common [7–9].

The widespread use of obtaining biomass from various plant species should be considered extremely important, as this can have a significant impact on solving the problems of the country's energy security and excessive greenhouse gas emissions [10–13]. Species of plants grown for the conversion of their energy content should have a number of properties that will make cultivation not costly, with a significant yield of biomass. A high dry matter yield per unit of crop area, as well as a high calorific value, primarily determines the quality of the crop and has a significant impact on the profitability of obtaining and using energy crops. The higher the values of these parameters are, the lower the biomass demand for obtaining the desired amount of energy is [14–17].

2 Usefulness of energy crops

Based on the literature on the possibility of using plant raw materials for energy purposes [18], a compilation of possible uses of biomass from energy plantations has been developed (Fig. 1).

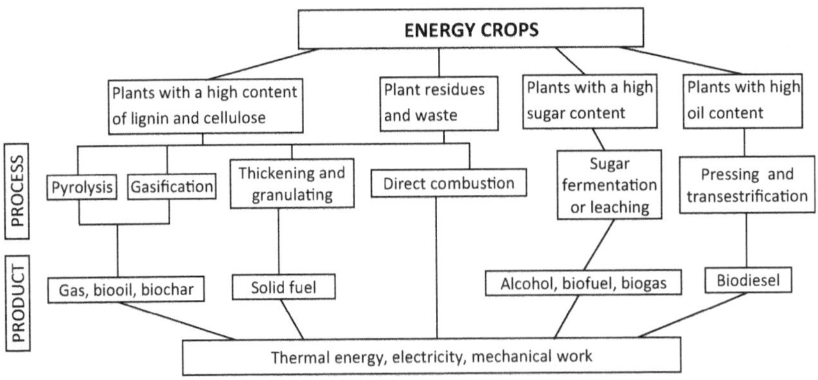

FIG. 1 The possibilities of using energy crops for energetic purposes. *(Source: Own study.)*

Based on Fig. 1, it can be concluded that there is a wide spectrum of possibilities for using plants for energy purposes. Therefore, biomass can have a significant impact on the diversification of energy sources in many countries in which there is an availability of wasteland and land not used for agriculture. In the case of countries in temperate climates, like Poland, Ukraine, Belarus in Central and Eastern Europe, if there is such an area, the climatic conditions (temperatures and, above all, precipitation) are mostly sufficient to obtain biomass for energy purposes. However, the selection of suitable species for energy production has the most significant impact on the quantity and price of fuel produced, and in each country or region the plants should be adapted to the climate change and prevailing microclimate [19–23].

3 Energy crops cultivation potential on the example of Greater Poland

3.1 Soil and climate conditions of the region

In Poland, one of the areas where large quantities of biomass could be produced for energy purposes is Greater Poland (Wielkopolska). A significant part of the region is agricultural areas, where there is a dynamic increase in the area of energy crops.

The geological diversity of the surface formations of Greater Poland is related to the activity of the Scandinavian ice sheet, which is why most of the soil in this area was developed mainly from clays and sands. The most common are brown and white earth soils, which cover 80%–90% of the region. They are rarely rich in clay and calcium, and therefore very fertile, black soils, as well as peat and muck soils, which are post-bog soils. In addition, along the rivers, especially along the Warta, Noteć, and Prosna rivers, sediments arose from river sediments. In areas with high human activity, such as urban agglomerations or opencast brown coal mines, anthropogenically degraded and polluted soils also occur.

Brown ground soils most common in Greater Poland are unevenly distributed. They are predominant in the central and eastern parts of the region. The soils' pH is in a relatively wide range, from neutral to acid. They are usually characterized by high fertility and are considered typical for this part of the country.

White earth soils cover about 33% of the described area and are located primarily in the Notecka Forest. A group of these soils are rusty and mugwort soils. They are distinguished by high permeability and acidic or very acidic reaction, which determines their low fertility [24–26]. According to the map of soils in Poland (Fig. 2), it can be concluded that plantations of energy crops might be located on gray brown soils, leached brown soils, and gley soils formed from loamy sands and light boulder loam as well as on podzol soils, red soils, and podzols formed from sand gravel of various origins.

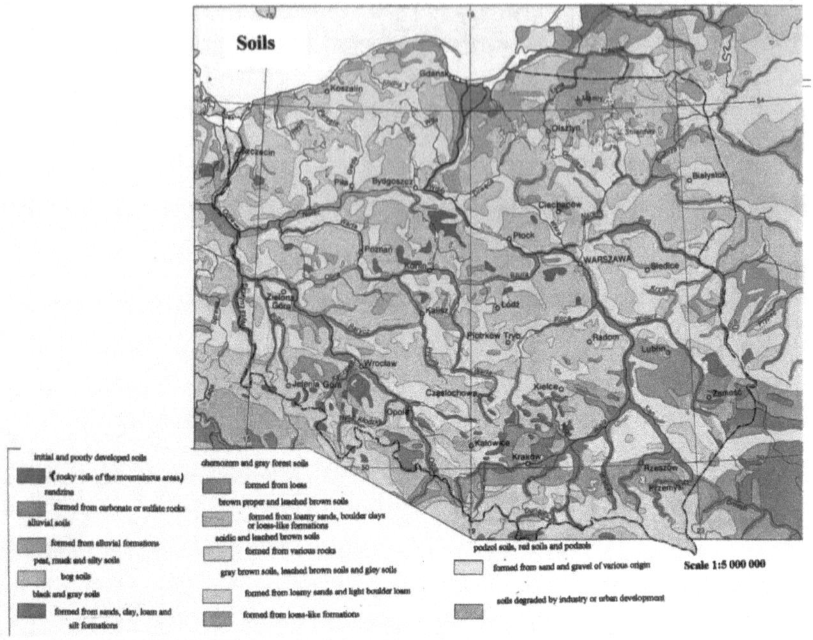

FIG. 2 Soils in Poland. *(Source: http://info-poland.icm.edu.pl/classroom/maps/Polandsoils.jpg.)*

Due to the transitory nature of the climate, annual precipitation may also fluctuate significantly in different years. Usually, the highest rainfall is recorded in the summer months (June–September). The average annual precipitation for western and central Poland is 550 mm, which is insufficient and causes a moisture deficit. It refers especially to the growing season, because the sum of precipitation in Greater Poland in the warmer half of the year is one of the lowest in the country. Since May until October, the average annual precipitation does not exceed 350 mm. Permanent climate changes, which include, among others, anthropogenic processes, cause the phenomenon known as the "Greater Poland steppe," which is progressing [24, 27, 28].

While growing plants for energy purposes, resistance to drought is particularly important because the expenditure incurred for frequent irrigation of plantations would far outweigh the profit from biomass sales. For this reason, in central Poland there are no favorable conditions for growing the most popular energy grasses, since all *Miscanthus* species have large water needs. To obtain reasonable yield, a rainfall of 700 mm is necessary in the period April–October, which is over 100% of the average annual rainfall for Greater Poland [24,29,30].

Usually, the coldest month is January (with the average temperature of about −1°C), and the highest temperatures are recorded in July (18–20°C). However, climate instability means that in individual years the lowest monthly temperatures may be in December or February, and the highest heat may occur in June or August.

Due to the dynamic air circulation, the weather changes frequently. Nevertheless, a constant relationship is observed: the further west, the higher the average annual temperature. In Poznań, capital of Greater Poland, the average equals 8°C. The length of the growing season, which is correlated with air temperature, also increases to the west; for Poznań, it is about 220 days a year in long-term average [24].

However, recent years have seen a significant rise in temperature and precipitation instability. The table below presents the precipitation course and temperature distribution for the last 3 years, in which plot experiments with energy crops were carried out. The results were obtained for the experimental station located in the southern part of Greater Poland (Table 1 and Fig. 3).

Practically in all the analyzed months, the average temperature was above or significantly above the norm (the exception was April). In contrast, precipitation was very uneven. There were weeks without rainfall followed by heavy 1-day rain. However, the shifts of the seasons are observed, the cooler beginning of spring with winter episodes and a very warm autumn. In the summer, however, there are dangerous weather phenomena such as storms and heavy storms. In June, July, and August, the temperatures are often well above average; also, drought periods are more frequent and longer.

3.2 Characteristics of energy plant species dedicated for the region

Species of crops grown for energy purposes should show a number of properties that will make their cultivation not costly, with a significant biomass yield. According to Kuś and Matyka [6], Tworkowski [31], and Grzybek [32], ideal energy crops should be characterized by:

- high calorific value,
- resistance to pests and diseases,
- relatively low soil fertility requirements,
- relatively low water requirements (particularly important drought resistance),
- resistance to low temperatures,
- reasonable growth as compared to irrigated and fertilized crops,
- high propagation and branching ability (does not apply to maize and sorghum, in the cultivation of which these are undesirable features [33]),
- easy biomass drying before processing.
- good parameters related to their combustion.

A high dry matter yield per hectare of crops and a high calorific value determine the quality of the crop and have a considerable impact on the profitability of obtaining plants for energy purposes. The higher these coefficients are, the

TABLE 1 Precipitation and temperature during plant growth season of most species of energy crops in Jutrosin, Greater Poland.

		2016		2018		2019		Three-year average		Long-term monthly average (1960–2000)	
Month	Decade	*Average temp. [°C]*	*Precip. [mm]*	*Average temp. [°C]*	*Precip. [mm]*	*Average temp. [°C]*	*Precip. [mm]*	*Temp. [°C]*	*Precip. [mm]*	*Temp. [°C]*	*Precip. [mm]*
April	I	8.0	22.6	10.0	22.6	8.2	0.8	8.7	15.3	7.3	37.0
	II	7.5	21.0	13.4	4.5	6.6	4.5	9.2	10.0		
	III	5.2	4.0	12.9	10.0	12.7	6.0	10.3	6.7		
May	I	11.1	25.7	14.0	14.7	8.1	11.5	11.1	17.3	12.7	61.0
	II	12.3	3.2	14.4	60.9	10.8	25.0	12.5	29.7		
	III	17.3	48.7	17.1	1.2	16.2	37.4	16.9	29.1		
June	I	18.0	28.1	20.0	13.5	22.6	0.0	20.2	13.9	15.6	53.0
	II	16.6	23.3	18.7	3.3	23.5	7.2	19.6	11.3		
	III	20.1	12.2	14.8	50.5	24.0	10.0	19.6	24.2		
July	I	17.5	15.5	19.8	0.0	18.1	9.0	18.5	8.2	17.4	78.0
	II	17.8	101.0	18.2	92.0	17.9	14.0	18.0	69.0		
	III	21.8	2.0	23.9	0.1	21.7	26.0	22.5	9.37		

August	I	17.7	41.3	25.1	0.0	20.7	17.3	21.2	19.5	16.2	60.0
	II	17.4	0.8	21.4	5.0	20.0	15.3	19.6	7.0		
	III	18.7	11.0	17.9	13.6	22.6	17.5	19.7	14.0		
September	I	19.1	4.0	17.6	3.1	16.9	38.5	17.9	15.2	12.6	49.0
	II	17.6	4.8	17.8	8.5	14.6	6.0	16.7	6.4		
	III	12.4	0.2	12.3	37.0	13.5	16.1	12.7	17.8		

Source: Own study.

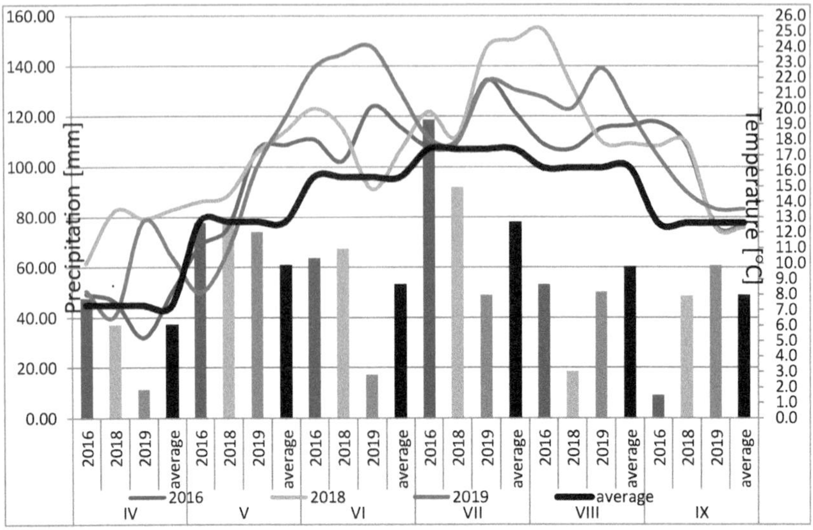

FIG. 3 Precipitation and temperature during the vegetation period of most species of energy crops in Jutrosin, Greater Poland. *((Source: Own study.))*

lower the biomass demand for obtaining the desired amount of energy should be. Research shows that the calorific value is negatively correlated with the ash content, as an increase of its content by 1% in biomass causes a decrease in the calorific value by 0.2 M kg^{-1} [34]. Studies on energy plants from the *Poaceae* family have shown that changing the harvest date has a significant impact on the quality of biomass obtained [30]. The leaves are always characterized by an almost twice as high ash content in comparison to shoots and by 50% higher than in panicles. Delaying harvest until leaves fall results in a lower biomass yield, but its better parameters, which are generally characterized by a lower biomass moisture, which reduces transport costs, facilitates storage and processing. In addition, biomass moisture, which is one of the most important factors determining the calorific value, decreases successively in winter if the plants are not harvested from the field. Based on many studies, it has been found that this happens regardless of the species [32,35].

Resistance to pests and diseases determines the associated savings from fighting pests. Low soil fertility requirements allow the establishment of energy crops on infertile arable land. As a result, they do not compete with crops of other plants that are intended for human food and animal feed [31,32].

Low water requirements are important to reduce or completely eliminate irrigation costs. This is particularly important in Greater Poland, where a significant moisture deficit during the growing season is observed [24]. On the other hand, resistance to freezing is of great importance in the cultivation of perennial energy crops, because the need to cover young plantings implies an increase in the costs of growing.

The rapid growth of biomass and the ease of its harvesting as well as the susceptibility of cultivation to weeds, which with slow initial growth and small plant density raises the need for mechanical and chemical destruction of weeds, have a great impact on the final crop [6].

Easy drying of biomass, which has already finished vegetation, reduces the costs of drying it before further processing. Research conducted in November and March at the University of Warmia and Mazury in Olsztyn showed that the highest humidity among the studied species of energy plants was characterized by energy willow wood, and the lowest by *Miscanthus* sp. The need for additional processing, including drying, may determine the final profitability of energy crops [36].

The biomass parameters that are intended for combustion are also significant, as the effectiveness of this process depends to a large extent on its chemical composition and moisture content [37,38]. In scientific publications, the values of parameters related to biomass burning are different. This is due to the differences in the genotype of plants and environmental conditions in which the plantations studied were located [35]. Low quality of biomass can significantly reduce the amount of energy obtained from energy crops [30], and high chlorine content increases the likelihood of boiler corrosion [6].

An increased sulfur content in biomass is also a threat. According to Stolarski, Szczukowski, and Tworkowski [36], the highest sulfur content, both in biomass extraction in November and March, was determined in prairie spartin (0.162% and 0.107%). In the remaining species tested, the value of this trait did not exceed 0.07% and was the lowest in the Pennsylvanian mallow (0.032%) [36]. Sulfur and chlorine can condense in the form of chemical salts, which, at high temperatures and in the presence of potassium and silicon, can lead to the formation of sticky deposits on the heating parts of boilers [30].

So far, the cultivation of perennial plants was recommended for energy purposes. The most important of them are described below.

3.3 Willow (*Salix* L.)

Of the around 400 species of willows commonly found in temperate climates, only a few are suitable for cultivation for energy purposes and are useful for obtaining wood in short rotations (1-, 3-, and 5-year-old). They belong to the genus *Salix* and are representatives of the subfamily *Vetrix* and *Salix* [39].

The subgenus *Vetrix* includes the most commonly grown willow species for energy conversion. The most valuable specimen for obtaining biomass is the weeping willow (*Salix viminalis* L.). It is characterized by high sucker capacity and a large mass of growing shoots. For this reason, it gives a large annual increase in biomass, which can be up to 15 Mg ha^{-1} dry matter [6,39].

Usually, willow is propagated through shoot cuttings known as cuttings or nurseries. On permanent grasslands, it can be propagated from grasslands in the Eco-Salix system. Although the shrub willow is the most commonly cultivated species among perennial energy plants in Poland, it requires a minimum of

500 mm of precipitation during lush growth and development during the growing season. However, rainfall is lower in most parts of the country from March to October. The occurrence of drought may also cause a decrease in willow yield by up to 50%, which may result in unprofitable plantation in areas with frequent rainfall deficiency, such as Greater Poland [6,36,39,40].

The *Salix* genus includes a number of species as mentioned. One of them is white willow (*Salix alba* L.). Although it does not have high soil requirements and grows quickly, it is not suitable for growing for energy purposes. After cutting, it gives weak shoots from the trunk [39].

Almond-leaved willow or just almond willow (*Salix triandra* L.) is another species included in the *Salix* genius. It gives a satisfactory crop, especially in moisture-rich positions. Although it occurs throughout the lowlands of Poland, it is very sensitive to frost. For this reason, it is estimated that *Salix triandra* L. is a poorer species for biomass production for energy purposes than *Salix viminalis* L. [39].

3.4 Poplar (*Populus* L.)

The genus *Populus* L. belongs to the willow family and includes several dozen species. Poplar, cultivated to obtain biomass in short rotations, has high soil, water, and climate requirements. To achieve high productivity, they require a long growing season and high temperature, especially in summer. For this reason, it is supposed that the cultivation of representatives of the genus *Populus* L. is not effective in soil-climate conditions of Greater Poland, but in another European country (Italy) its cultivation under different crop management systems was checked [41–43].

3.5 Black locust (*Robinia pseudoacacia* L.)

Robinia pseudoacacia, also known as locust beanberry, belongs to the Fabaceae family and lives in symbiosis with the *Rhizobium* bacteria, which have the ability to bind atmospheric nitrogen. It is classified as kenophytes and agriophytes. Comparing biomass yields achieved in Poland, Hungary, Greece, and Turkey, it was found that its cultivation for energy purposes is inefficient and therefore unprofitable in temperate climates [42].

3.6 Multiflora rose (*Rosa multiflora* Thunb.)

This species, known commonly as baby rose or Japanese rose, classified as a wild plant in Poland, is resistant to climatic conditions prevailing in the region. The large-flowered rose creates a well-developed root system, making it resistant to drought. It can be grown on soils of class V and VI with a pH in the range of 5.5–7.5. Establishing a plantation of the thornless Jatar variety for energy purposes is recommended for periods of up to 30 years.

Due to the features of the plant, namely the length of its shoots and strong annual growth, mechanical collection of this species is impossible. The only solution is manual collection using pruning shears or a petrol cutter with a toothed disc. Problematic biomass harvesting, which is time-consuming and inefficient, means that the multifloral rose is not a popular energy plant [30,33].

Owing to the difficulties and costs of the elimination of trees and bushes, a shift to perennial grass varieties was made as their cultivation is deemed easier. The most popular species are described below.

3.7 Giant miscanthus (*Miscanthus × giganteus* J.M. Greef & M. Deuter)

In recent years, the interest in introduced perennial grass species, especially species of the genus Miscanthus, has been increasing. In addition to the cultivation of Amur silvergrass [*Miscanthus sacchariflorus* (Maxim.) Franch.] and maiden silvergrass (*Miscanthus sinensis* Andersson), which are of Asian origin, the giant miscanthus, which is a cross between the two species above, currently has the greatest economic importance. It is a sterile triploid bred in the 1980s in Denmark. As *Miscanthus giganteus* does not produce seeds, only vegetative reproduction is possible. The most effective and economically justified form is reproduction of giant miscanthus by division of rhizomes. It is very important not to overdry the rhizomes. For this reason, rhizomes should be obtained immediately before the planned planting, which is at the beginning of April [42,44].

Miscanthus giganteus belongs to plants with the C4 photosynthesis pathway, which makes it more effective to use water, solar radiation, and fertilizer components. Another advantage is its high yielding potential. However, European research has shown that yields of giant miscanthus can be very diverse and range from 4 Mg DM • ha^{-1} in Germany to even 44 Mg DM ha^{-1} in Greece. It follows that it yields better in countries with warmer climates [6,42].

All *Miscanthus* species have high water needs. To obtain satisfactory cultivation results, a rainfall of 700 mm is necessary in the period from April to October, which is characterized by over 100% of the average annual rainfall for Greater Poland [24,42].

The full yielding of *Miscanthus giganteus* occurs from the third year of cultivation and lasts until the eighth-ninth season, after which it systematically decreases [42]. The dry matter of miscanthus contains, however, a high content of chlorine, which is over 10 times higher than that of shrub willow. Therefore, miscanthus is not very popular, because it can cause corrosion of boilers [6].

Due to low atmospheric precipitation in Greater Poland, it can be assumed that plantations located in this part of the country will have lower productivity than plantations described in the literature. For this reason, it is estimated that cultivating the *Miscanthus × giganteus* species is unprofitable in Greater Poland [30,45].

3.8 Prairie cordgrass (*Spartina pectinata* Bosc)

Spartina pectinata, known by the common names prairie cordgrass, freshwater cordgrass, tall marsh grass, and slough grass, like the miscanthus species discussed above, is a species of perennial grass with the type of C4 photosynthesis. It comes from North America, where it reaches from 1 to 2.5 m high. In the natural environment, it is mainly covered with poor, badly drained ditches and wetlands. It is used to strengthen sandy embankments as an anti-erosion plant. In addition, it occurs on wet meadows, as well as on overgrown dunes because it tolerates soil salinity and has high adaptability. Moreover, it can be cultivated in soil contaminated with heavy metals [39,46,47].

3.9 Reed canary grass (*Phalaris arundinacea* L.)

The Institute of Agrophysics of the Polish Academy of Sciences in Lublin (Poland) conducted research on the use of reed canary grass obtained from several swaths during the year for biogas production. It is perennial grass, resistant to low temperatures, as evidenced by its popularity in Scandinavia where it is very useful for cultivation for energy purposes. By 2016, the Finnish Ministry of Agriculture and Forests has set a priority to achieve its cropping area of 100,000 ha [48]. Its advantage is the low cost of establishing a plantation, which consists in direct sowing of seeds into the ground in an amount of 15–18 kg ha^{-1}. In addition, it is possible to fully mechanize the production of agricultural machinery for growing cereals. It yields best in moist and sandy areas, giving 4–7 Mg of dry matter per hectare. The plantation can be used up to 15 years. To obtain straw, the brain should be cut in early spring, immediately after the snow melts because it starts growing early. Its biomass then has only about 10%–15% humidity, which makes it easy to form cylindrical or perpendicular bales. At that time, it is also characterized by a lower ash content, which is still quite high in March. Depending on the position in which the brain grew, it is 2%–10% [48]. In addition, the biomass of the cane brain contains a lot of nitrogen and chlorine [6], which, with a relatively low yield, is not an alternative source of solid biomass in comparison with much higher *Miscanthus giganteus*, sorghum, or corn. But it can be cultivated in marginal lands [6,48–51].

3.10 Big bluestem (*Andropogon gerardii* Vitman)

Big bluestem, which is known commonly as turkeyfoot or tall bluestem, is a species of prairie grass growing in dense clumps. In a natural environment, it reaches 1–2.5 m in height, depending on soil fertility. It is a thermophilic plant. In Polish conditions, vegetation begins in May. Seeds set up relatively late, which is in October. Andropogon gerardii tolerates salinity and drought well, but in very cold and especially snowless winters, plants are prone to frostbites.

Due to the incomplete resistance to climatic conditions in Poland, this plant is not grown for energy purposes on a larger scale [52].

3.11 Switchgrass (*Panicum virgatum* L.)

This species is the most popular in North America where research on the use of switchgrass for energy purposes has been conducted since the 1990s. In the natural environment, it forms clumps up to 3 m high. It produces strong, short runners, which makes it capable to cover degraded turf and soils which are at risk of erosion. This perennial grass can also be used for planting pioneer species on heaps formed after mining excavations. In relation to big bluestem, the *Panicum virgatum* gives a higher biomass yield. However, it is smaller compared to *Miscanthus giganteus* or sorghum. But it is used for producing solid or liquid or gas biofuels [53–55].

3.12 *Agropyron elongatum* L

Agropyron elongatum L. is a long-term type C3 grass that occurs in dry, saline positions in Southeastern Europe and Asia. It forms compact, 2–3 m-high clumps and does not produce runners. Currently, the Polish variety Bamar grown at the Institute of Plant Breeding and Acclimatization of the National Research Institute in Radzików is used in cultivation mostly for energy purposes.

Agropyron elongatum can be carried out for 8–12 years, which is shorter than other perennial grasses. It can be used in the reclamation of postindustrial areas such as heaps after opencast lignite mining. It is also used to clean soils from heavy metals. Although elongated grassland may soon become an alternative source of biomass, especially as a substrate for biogas plants, currently there are no studies confirming the profitability of its cultivation in soil and climate conditions of Greater Poland [6,56,57].

3.13 Perennial species

Perennial species that can be grown for energy purposes include jerusalem artichoke (*Helianthus tuberosus* L.), Sakhalin knotweed or giant knotweed (*Reynoutria sachalinensis* Nakai), and virginia mallow (*Sida hermaphrodita* Rusby). They are species that tolerate both light rainfall and low winter temperatures well. They do not require good soils and give significant amounts of biomass even in the absence of fertilization. Care for these plants is not complicated and does not require large expenditures. However, the most laborious process is the liquidation of plantations which does not always eliminate all plants in a given field.

Jerusalem artichoke, due to the invasive spread in riverside and ruderal habitats and significantly reduced by biodiversity, as well as ease of regeneration, is a species postulated by the European and Mediterranean Plant Protection Organization for monitoring. In contrast, Sakhalin knotweed is considered an invasive species in Poland, dangerous to native nature. Entering it into the environment or moving it in the natural environment is prohibited. Since 2012, its import, possession or breeding, reproduction, and sale require the permission of the General Director for Environmental Protection. Failure to comply with these restrictions is an offense punishable by detention or even a financial fine by Regulation of September 9, 2011 concerning the list of plants and animals of alien species which, if released into the natural environment, may threaten native species or natural habitats. Nevertheless, they are used for bioenergy production, especially as a raw material to produce solid and liquid biofuel. The tubers of Jerusalem artichoke are a good source of substances for producing bioethanol [58–60].

Virginia mallow can be reproduced both generatively and vegetatively. Planting root sections, however, gives higher yields independently from the substrate used. This species is resistant to drought, because it creates a deep root system. Nevertheless, mallow cultivation can be profitable only when obtaining high biomass yields (over 11 Mg DM ha^{-1}) and its high selling price (320 PLN per 1 Mg of DM). This is due to the need to use mineral fertilizers every year and ensure an intensive chemical protection [6,61,62].

Currently, an increasing opportunity is seen in annual species: the ones cultivated on soils not used for agriculture or on soils which are weak, saline, contaminated with heavy metals, or very weeded. This way, biomass production does not compete with food production.

3.14 Oats (*Avena sativa* L.)

Poland remains at the forefront of producers of oats, although it is not the most commonly grown grain in the country. However, due to its properties (high calorific value, low ash content), it is straw and grain that are mentioned in some publications as a substrate for energy production. Oats have low soil requirements, have phytosanitary properties, and is a good forecrop also for cereals. Grain can be used for the production of bioethanol or for direct combustion in boilers specially adapted for this purpose, having burners for burning grain as starters. However, due to society's unwillingness to use grain as fuel, as it could be a potential source of food, it is estimated that only low quality oats, unfit for consumption, will be used in a few heating installations [63–65].

3.15 Rye (*Secale cereale* L.)

The production of rye in Poland is the largest among the member states of the European Union and, in terms of weight of produced grain, very close to Russia, which is the world leader. In addition to the possibility of using defective grain and straw for the production of heat, rye for energy purposes can be cultivated as a catch crop for biogas production.

While participating in the research carried out at Stary Sielec Experimental Station and Pętkowo Experimental Station, belonging to the Institute of Natural Fibers and Medicinal Plants, a successful experiment was conducted to show the usefulness of common rye cultivated as a catch crop and common maize and sorghum in the main yield as raw materials for agricultural biogas plants. Four years of research have shown that rye harvested at the milk stage-wax grain ripeness and maize grown for silage are both efficient energy raw materials. The total amount of biomass obtained was about 40 Mg ha^{-1} dry matter with high biogas efficiency (17,900 m^3 ha^{-1}). As a result, low energy production costs per unit area were achieved. The cost of bioenergy production from rye as a catch crop and maize as a main crop was 8.11 PLN compared to rye and sorghum which was 8.68 PLN for 1 GJ of energy [45].

3.16 Corn (*Zea mays* L.)

Corn is a plant derived from Central America. However, its wild forms are not known. Various corn varieties that have been bred for centuries are usually classified on the basis of the shape of the grain. Due to the dominance of hybrid forms, they often do not represent the pure initial form, but intermediate types. For biofuel production purposes, hybrids of vitreous maize, also known as flint and horse tooth, are most often grown (also: dent) [33,66].

All varieties of this species are monoecious. They develop strong thick blades, up to 3 m high, finished with panicles—a male inflorescence. Female inflorescence—butt—develops in the middle of the height of the blade at the end of the shortened lateral branching. Corn usually does not spread because in cultivation it is an undesirable trait, hindering agrotechnical operations. As a plant with C4 type photosynthesis, it manages water sparingly; however, due to the large biomass production, it has quite large water needs. The critical moment in this respect is the flowering period of plants. In Poland, the main factor determining the efficiency of maize cultivation is humidity conditions, shaped by the amount of rainfall and its distribution during the growing season. Corn's high thermal requirements can be minimized by choosing varieties with greater tolerance to cold weather and with a shorter vegetation period [45,67,68].

Due to the diversity of environmental conditions in different areas of Poland, there is a regionalization of maize cultivation. The large number of varieties available means that to obtain optimal yields for a given purpose and the place of cultivation, the variety is adapted on the basis of its earliness class, i.e., the FAO number.

The FAO number is a number between 100 and 1000. The smaller the number, the earlier the variety, i.e., its growing season is shorter. In Greater Poland, early, medium, and medium late varieties (FAO 250–290) reach full grain maturity.

In the southern part of the region, even late varieties (FAO 300–350) can be grown for the needs of biogas plants. Based on observations of experiments carried out at Stary Sielec Experimental Station near Rawicz, it was found that also medium-early Opoka variety maize, grown in secondary crops after the ordinary life, is efficient and profitable. Studies have shown that 26 Mg ha^{-1} dry matter maize can be obtained in this way with a biogas capacity of 11,000 m^3 ha^{-1} [45,67,68].

Two production technologies are used in growing plants for energetic purposes: one for silage and the other for grain. The silage technology produces whole plant silage, which is then a substrate for biogas production, while in the grain technology several products can be obtained. They are: dry or pickled grains, silage from grain with the addition of cob cores (CCM—corn-cob-mix), silage from shot flasks, or silage from flasks harvested with cover leaves (LKS—from German: Lieschkolben Schrott). A summary of the possibilities of using maize as an energy raw material is presented in Table 2 [33].

Table 2 shows that maize has various uses. For this reason, it is a good raw material for energy conversion. In the design part of this work, corn cultivation technology for energy purposes is discussed, which is the most profitable according to the author in the soil and climate conditions of Greater Poland.

3.17 Sorghum (*Sorghum bicolor* L.)

Sorghum, belonging to the paniculate plant group, is an annual short-day spring plant with the type of C4 photosynthesis. *Sorghum bicolor* and *Sorghum sudanense* as well as their hybrids are grown on a large scale in the world. They are used as feed or for energetic purposes [69,70].

In Poland, various subspecies and forms of the two-color sorghum variety are mainly used. They form blades from 0.5 to even 4 m in height. The leaves are 0.5–0.8 m long and are covered with a wax layer, which protects them from excessive transpiration. In addition, sorghum has a strongly developed beam root system reaching about 1.9 m deep into the earth. Because of that, it has lower soil requirements than maize and can be grown on very light sandy soils. What is more, it withstands periodic droughts better. Sorghum, especially the Sucrosorgo 506 variety, manages water very sparingly. This is an important feature, especially with a lack of precipitation often occurring in the Greater Poland region. It was observed that in field experiments at Stary Sielec Experimental Station and Pętkowo Experimental Station, belonging to the Institute of Natural Fibers and Medicinal Plants, sweet sorghum best tolerated a small amount of precipitation and gave a higher biomass yield (till 23.4 Mg DM ha^{-1}) in relation to other varieties [45,71,72].

Sorghum is about 30% cheaper to grow than corn because it does not require such intensive care. It is caused by smaller outlays on chemical protection of plantations because sorghum does not yet have natural pests in Poland [45].

TABLE 2 Possibilities of using corn as an energy raw material.

Grain 5–10 Mg ha^{-1}		**Biomass 8–20 Mg s.m. ha^{-1}**	**Straw 3–6 Mg s.m. ha^{-1}**		**Stems 1–2 Mg s.m. ha^{-1}**
Fermentation industry	Combustion: energy and domestic installations	Whole plants or silage	By-product after grain harvest or CCM		By-product after harvesting of whole corn cobs
		Biogas plants	Biogas plants	Combustion in Power plants	Combustion in Power plants
Production volume from 1 Mg of raw material					
370–410 L of ethanol	approx. 19 GJ	600–700 m^3 biogas (350–450 m^3 of methane)	250–300 m^3 of biogas (approx. 150 m^3 of methane)	approx. 15 GJ	approx. 15 GJ

Source: Kołodziej B, Matyka M, editors. Odnawialne źródła energii. Rolnicze surowce energetyczne [Renewable energy sources. Agricultural energy resources]. Poznań: PWRiL; 2020. pp. 314–337, 359–69, 499, 518–21 [in Polish].

Sorghum has very high thermal requirements, so its yield mainly depends on the temperature distribution during the growing season. If sown late (after 15 V), this plant can be grown as a main crop after winter catch crops, and even after early potatoes. The biomass is harvested from the end of September to mid-October before the first autumn frosts occur. They cause a reduction of the sugar content in the green mass, which reduces its quality [33,45,73].

According to Lewandowski and Ryms [74], in the climatic conditions of Poland, sorghum does not give mature seeds. However, according to Burczyk [45], some varieties of two-color sorghum sometimes reach threshing maturity if favorable weather, that is, a long hot summer, is present. Typical tropical sorghum cultivars in a temperate climate prolong the vegetative period and do not produce inflorescences [45].

Sorghum biomass, mainly made of lignocellulose, can be converted for energetic purposes using thermochemical and biochemical methods. Thanks to this, in the process of carbonization, combustion, gasification, pyrolysis, or in the production of bioethanol or biogas, significant amounts of energy (about 10 PLN GJ-1) can be obtained at a low cost [45].

Due to the dynamic development of breeding work and numerous varieties of representatives of the Sorghum genus appearing on the market, it should be assumed that the popularity of this plant will increase. Great progress in the field of sorghum cultivation means that this plant may in a few or several years be an alternative and efficient source for the production of biofuels from maize [73].

In the table below, the characteristics of selected species of energy crops dedicated for the Greater Poland region were summarized (Table 3).

Based on Table 3 and the author's own calculations, it was found that the highest income can be obtained from the cultivation of annual plants and their cultivation is recommended in temperate climates. In order to maintain soil fertility, crop rotation should be applied to prevent soil depletion. The cultivation of perennial energy crops requires the use of higher doses of mineral fertilizers in subsequent years to prevent a decrease in the yield. The high costs of decommissioning the plantation along with the higher mineral fertilization requirement have a negative impact on the economic profitability of the cultivation of perennial species compared to annual plants. It was also found that maize (*Zea mays* L.) and sorghum (*Sorghum bicolor* L.) have the largest production potential in Greater Poland.

4 Possibilities of using biomass for energy purposes

4.1 Biofuel production law

Bioenergy production, as well as other areas of the economy, is based on legislation adopted by the European Union and national law. Studies on the production and use of renewable energy sources are changing quite dynamically, and the long-term perspectives of energy policy are modified

TABLE 3 List the characteristics of selected species of energy crops with the soil and climatic conditions of Greater Poland.

Species		Soil requirements	Water requirements	Fertilizer requirements	Climate requirements (during vegetation period)	Frost resistance	Expansiveness	Susceptibility to diseases and pests	Labor and cost consumption of cultivation	Profit from cultivation
Cereals	*Zea mays* L.	-/+*	++	-	–/+*	++	++	-	-	++
	Sorghum bicolor (L.) Moench	+	++	-	-	++	++	+	-	++
	Avena sativa L.	+	-	+	++	+	++	o	-	+
	Secale cereale L.	+	o	+	+	+	++	o	-	+
Trees and shrubs	*Salix viminalis* L.	-	–	+	+	++	-	-	-	+
	Populus L.	-	+	-	–	o	+	-	o	o
	Robinia pseudoacacia L.	o	+	o	–	o	+	o	o	-
	Rosa multiflora Thumb.	+	+	+	o	+	-	o	-	-
Grasses (*Poaceae*)	*Miscanthus* × *giganteus* J. M. Greef & Deuter	o	–	-	–	-	-	+	–	o
	Spartina pectinata Bosc	++	-	o	-	+	-	+	-	+
	Phalaris arundinacea L.	-	-	o	-	+	-	+	-	+
	Andropogon gerardi Vitman	+	+	o	-	+	-	+	+	o

Continued

TABLE 3 List the characteristics of selected species of energy crops with the soil and climatic conditions of Greater Poland.—cont'd

Species		Soil requirements	Water requirements	Fertilizer requirements	Climate requirements (during vegetation period)	Frost resistance	Expansiveness	Susceptibility to diseases and pests	Labor and cost consumption of cultivation	Profit from cultivation
Perennial plants	*Panicum virgatum* L.	+	+	o	-	+	-	+	+	**o**
	Agropyron elongatum L.	-	+	-	o	+	-	+	-	**o**
	Helianthus tuberosus L.	+	+	-	+	++	–	-	-	**+**
	Reynoutria sachalinensis (F. Schmidt) Nakai	++	–	o	+	++	–	+	-	**o**
	Sida hermaphrodita (L.) Rusby	o	+	-	-	o	–	–	–	**o**

Explanations: "*", depending on the variety; "++", a very positive feature; "+", a positive feature; "O", a neutral feature; "-", a negative feature; "-", a very negative feature. The "Profit from cultivation" is bolded as a summary of characteristics of each species because the profit is also affected by soil and climatic conditions.
Source: Own study.

and updated many times, both in terms of the expected indicators and the deadline for their implementation.

As part of the European Commission's activities in the area of "Energy, climate change and the environment," many activities are undertaken in the field of energy and climate policy. One of the first, to a large extent promoting the use of energy from renewable sources, was the climate and energy package, which is a set of binding rules to ensure that the EU achieves its climate and energy goals by 2020. The package sets out three main goals:

- a 20% reduction in greenhouse gas emissions (compared to 1990),
- a 20% share of energy from renewable sources in total energy consumption in the EU,
- a 20% increase in energy efficiency.

These goals were set by the EU leaders in 2007, and regulations were adopted in 2009. At the same time, these are the main goals of the Europe 2020 strategy for smart, sustainable, and inclusive growth [75].

Then a 2030 climate and energy policy framework was developed containing the EU-wide policy assumptions and goals for 2021–2030. The most important goals for 2030 are:

- a reduction by at least 40% in greenhouse gas emissions (compared to 1990 levels),
- an increase to at least 32% in the share of energy from renewable sources in total energy consumption,
- an increase in energy efficiency by at least 32.5%.

In October 2014, the policy framework was adopted by the Council, and renewable energy and energy efficiency targets were increased in 2018 [76].

One of the latest studies of the European Commission and the European Council is the document of November 28, 2018—COM (2018) 773 entitled "A clean planet for everyone." It is a European long-term strategic vision of a thriving, modern, competitive, and climate-neutral economy. It presents a long-term strategic vision of a thriving, modern, competitive, and climate-neutral economy by 2050.

The strategy illustrates what member states should do to achieve climate neutrality. According to the document, this should be done by investing in realistic technological solutions, empowering citizens, and aligning political action in important areas such as industrial policy, finance, and research. In such a transformation process, it is also important to guarantee social justice.

In line with the wishes of the European Parliament and the European Council, the Commission's vision for a climate-neutral future covers almost all EU policies and is in line with the Paris Agreement's goal of keeping the temperature rise well below 2°C and attempting to reduce it to 1.5°C [77].

The road to an emission-neutral economy is to be based on joint action in line with a set of seven main strategic basic elements:

1. Maximizing the benefits of energy efficiency, including missionless buildings.
2. Maximizing the use of renewable energy sources and electricity to completely decarbonize energy supplies in Europe.
3. Adopting clean, secure, and web-based mobility.
4. A competitive Union industry and a circular economy as a key factor in reducing greenhouse gas emissions.
5. Developing appropriate network infrastructure and interconnections.
6. Reaping the full benefits of bioeconomy and creating the necessary carbon sinks.
7. Elimination of the remaining CO_2 emissions due to carbon capture and storage.

In the second point, the Communication summarizes that currently the energy system in the European Union is based largely on fossil fuels. All assessed scenarios assume that by the middle of this century there will have occurred a radical change due to large-scale electrification of the energy system thanks to the introduction of renewable energy sources, either at the end-user level or for the production of zero-emission fuels and industrial raw materials.

The large-scale use of renewable energy sources will lead to the electrification of our economy and to a high degree of decentralization. By 2050, the share of electricity in final energy demand will at least double, reaching 53%, and electricity production will increase significantly to achieve zero net greenhouse gas emissions, and—depending on the selected energy transformation options—will even reach 2.5 times more than current levels.

4.2 Solid biofuels

For the production of solid biofuels, plant biomass with a high content of lignocellulose compounds, which is waste in various processes coming from energy crops, is most often used. After drying and comminution, briquettes or pellets are usually produced, which are a convenient form of biofuel intended for combustion in domestic furnaces [78,79].

One of the most important parameters determining the suitability of biomass for the production of solid biofuels is its high calorific value. In addition, the dry matter yield per unit of crop area from which the raw material is obtained for energy purposes should also be as high as possible. These factors have a significant impact on the profitability of growing energy crops. The high calorific value of biofuels has the effect of reducing the weight of the raw material needed to produce the desired amount of energy. In addition, it has been shown that the calorific value is correlated with the content of ash remaining after biomass combustion because an increase of its content by 1% results in a decrease in calorific value by 0.2 MJ kg^{-1} [34,80–83].

Studies carried out on the many species have shown that changing the date of harvesting biomass for energy crops has a significant impact on the quality of the obtained raw material. In addition, other studies have shown that grass leaves have a 50% higher ash content than panicles and an almost twice as high content as the mineral substances remaining after burning shoots. Delaying harvest until leaves fall results in less biomass yield, but on the other hand has better physical and chemical parameters. Moreover, the moisture content of biomass, whose content is one of the main factors determining the calorific value of the raw material, gradually decreases in biomass in winter, especially if it is dry and frosty. Based on many analyses, it has been proved that this happens regardless of the species [35,42,74,84].

In the case of biomass obtained during the growing season, easy drying of the biomass reduces expenditure on its drying before further processing. The need for additional drying may determine the final profitability of using biomass for the production of pellets and briquettes [85–87].

Other physical and chemical parameters of solid biofuels produced for combustion are also very important, as the effectiveness of this process depends largely on its properties. In scientific publications, the values of parameters related to biomass combustion of specific species differ significantly. This is primarily due to the environmental conditions in which the experimental plots or plantations were located, from which samples were then taken for laboratory analyses. In addition, different methods of drying and processing the raw material may affect the results obtained. However, a low quality of biomass always significantly reduces the amount of energy obtained from it and, in addition, the high content of chlorine in biofuel increases the likelihood of boiler corrosion. Chlorine and sulfur, under the influence of high temperature, may precipitate in the form of salts, which in the presence of silicon and potassium may lead to the formation of sticky deposits on the heating parts of the installation. For this reason, it is very important to determine the physical and chemical parameters of biomass intended for compaction and pressure agglomeration [6,35,37,88,89].

4.3 Liquid biofuels

Liquid biofuels are used as a replacement or addition to diesel oil. In Poland, as a rule, no biocomponents are used, i.e., bioethanol, biomethanol, esters, dimethyl ethers, pure vegetable oils, or synthetic hydrocarbons, mainly due to the fact that in winter such fuel could freeze in tanks [90–93].

Initially, unprocessed or purified oil pressed from the seeds of plants such as rape, soybean, or sunflower was used as liquid biofuel. Then, when the method of extraction of fatty acid methyl esters—FAME (Fatty Acid Methyl Esters) became popular, both plant and animal, they began to be used as biocomponents in diesel oil or as so-called biodiesel. FAME are produced because of their lower viscosity in relation to the raw material from which they are obtained, which significantly improves their properties as a fuel [94–98].

FAME are usually first-generation biofuels, although they are increasingly produced from waste oils. Nevertheless, mainly as a result of legislative changes aimed at moving away from the production of biofuels from food raw materials, the production of second-generation biofuels from lignocellulosic biomass and third-generation biofuels from algae is rapidly growing [99,100].

The process of converting plant biomass to second-generation biofuels involves several stages. It begins with the appropriate preparation of plant material (so-called effective pretreatment). Undoubtedly, this is the most important stage affecting the effective use of the raw material for the production of second-generation bioethanol. It includes mechanical, chemical, and biological pretreatment of the raw material which is characterized by a compact and complicated structure [101,102].

Pretreatment consists in grinding the solid phase of the raw material, loosening the compact structure of lignocellulose, destroying the crystalline structure of cellulose, and separating lignin from cellulose. Lignin is an undesirable component because it does not produce bioethanol [103,104].

Then enzymatic hydrolysis of biomass is carried out, i.e., the process in which the polysaccharides break down into fermentable sugars (where the selection of effective enzyme preparations is very important), up to ethanol fermentation (using appropriate microorganisms). This fermentation can occur in two ways: by SHF (Separate Hydrolysis and Fermentation) or SSF (Simultaneous Saccharification and Fermentation) methods. Separate Hydrolysis and Fermentation is a two-step process in which hydrolysis is first carried out followed by ethanol fermentation. This allows the reaction conditions to be adjusted, but this inhibits enzyme activity due to an excess of disaccharide cellobiose—$(C_6H_7(OH)_4O)_2O$. It is classified as a reducing sugar and it can be hydrolyzed to glucose enzymatically or with acid.

The shorter and more efficient process is SSF, during which hydrolysis and fermentation occur simultaneously. Biofuel production by this method does not inhibit enzyme activity, but it requires the use of microorganisms that tolerate higher temperatures. Biofuel is created by the breakdown of glucose into ethanol and carbon dioxide by distillery yeast (*Saccharomyces cerevisiae* L.). However, for the ethanol obtained to be used in the transport industry, it is necessary to carry out its distillation and dehydration [105–107].

Undoubtedly, the use of liquid biofuels, instead of diesel and diesel, would have a smaller impact on the natural environment. But, like any product, liquid biofuels have both advantages and disadvantages. The largest of these are energy and cost consumption, as well as the need to use many enzymes, the production and utilization of which entails additional expenditure and financial burden on the environment. For this reason, the production of advanced biofuels is still at an intensive stage of work, including primarily the optimization of production efficiency and the minimization of bioethanol production costs [108–110].

4.4 Gas biofuels

Biogas, as a secondary energy source, is created from organic mass from biomass processing in various biological processes in the absence of oxygen. Its source in the natural environment can be processes occurring at the bottom of oceans and seas, in peat bogs, in liquid manure, or in rumen of ruminants [111,112].

The most common and effective technology for producing biogas under industrial conditions is methane fermentation carried out in special reactors under strict conditions. The gas mixture produced in this way consists of approximately ⅔ of methane and approximately ⅓ of carbon dioxide. In addition, other compounds may be present in different proportions in biogas. These are usually small amounts of hydrogen sulfide, ammonia, hydrogen, and other gases. However, depending on the type of substrate and process step, hydrogen sulfide or ammonia may sometimes predominate among fermentation products. It is usually a natural, short term, but, at the same time, undesirable phenomenon, negatively affecting the efficiency of the process [113,114].

The first stage of the biogas production process is hydrolysis. It undergoes a double exchange reaction process. It runs between water and the organic substance contained in it. Under the influence of enzymes, the breakdown of long molecules—polymers and water-insoluble organic compounds—occurs. Proteins, fats, cellulose, and other complex sugars are hydrolyzed to: amino acids, simple sugars, and polyhydric alcohols and fatty acids, respectively [115,116].

Hydrolysis is followed by an acidogenic phase, i.e., acidogenesis. It distributes the products of the first stage of the process to short-chain organic acids. At this stage, mainly volatile fatty acids (acetic, formic, propionic, butyric, lactic, valerian, and capron) are formed. In addition, alcohols: methanol and ethanol, as well as aldehydes are formed. However, carbon dioxide and hydrogen are released from gaseous products. Some of the compounds formed in acidogenesis are methanogenic. This means that they are directly used by methane bacteria to produce biomethane. They include, among others, methanol and formic and acetic acids. Bacteria from the genera *Bacillus*, *Bifidobacterium*, *Pseudomonas*, and *Clostridium* have a significant share in the proper course of hydrolysis and acidogenesis [117–119].

In the acetogenic phase, also called acetogenesis, there is a breakdown of volatile fatty acids and ethanol into acetates, as well as carbon dioxide and hydrogen, which are also formed in acidogenesis. Bacteria including *Syntrophomonas* sp. and *Syntrophobacter* sp. are responsible for this. Inhibiting their activity increases the pH due to the accumulation of volatile fatty acids. Then, thanks to homoacetogenic bacteria, the production of acetates from carbon dioxide and hydrogen occurs. This enables the redevelopment of acetogenic and, later, methanogenic bacteria. The duration of this phase, as well as the others, depends primarily on the type of substrate and graft used. The last phase is methanogenesis, i.e., the production of biomethane. It is carried out by autotrophic and heterotrophic methane bacteria. These include bacteria of the genus *Methanobacter*, *Methanobrevibacter*, or *Methanococcus* [119–123].

The following are chemical reactions that occur during methanogenesis:

Two-thirds of methane is formed from acetates or alcohols:

$$2\,CH_3CH_3OH + CO_2 \rightarrow 2\,CH_3COOH + CH_4$$

$$CH_3COOH \rightarrow CH_4 + CO_2$$

$$CH_3OH + H_2 \rightarrow CH_4 + H_2O$$

However, one-third is formed in the reaction of carbon dioxide with hydrogen:

$$CO_2 + 4H_2 \rightarrow CH_4 + 2\,H_2O$$

The biogas production process can be carried out at various temperatures due to the reactions occurring during methane fermentation. Nevertheless, it is important to adjust and maintain a constant temperature at such values that are most optimal for the development of the most numerous group of methanogenic microorganisms found in a given vaccine, i.e., inoculum [124,125].

Depending on the optimal temperature range in which methane fermentation is most effective, three types are distinguished: psychrophilic, mesophilic, and thermophilic. The vast majority of bacteria involved in methanogenesis are mesophilic. The optimum of their growth occurs at the temperature of 25–45°C. Thermophilic bacteria tolerate a temperature above 45°C, but usually not higher than 60°C. An appropriate adjustment and keeping the temperature at a constant level favorably affects the physical parameters of the process such as reducing surface tension or viscosity and, above all, the speed of the reaction. Therefore, it also indirectly facilitates mass transport in the reactor [126–129].

In addition to maintaining the temperature at an optimal and constant level, it is very important to ensure anaerobic conditions in the bioreactor, i.e., preventing atmospheric air from entering the reactor, because methane fermentation occurs in anaerobic conditions [130,131].

Another key parameter for efficient fermentation is reaction. The optimal pH for the entire process is 4.5–7.5 and depends primarily on the phase of methane fermentation and, to a lesser extent, on the substrate used. For acidogenesis, the optimal pH oscillates at the level of 5.2–6.3, while for methanogenesis it is higher and ranges from 6.8 to 7.2. If the reaction is too low, which may cause a sudden inhibition of the fermentation process, it can be increased by adding sodium carbonate, calcium carbonate, quicklime, or caustic soda to the bioreactor [118,132,133].

The availability of micro- and macroelements is an important factor conditioning the correct course of the process, resulting from the optimal growth of microorganisms in the fermentation chamber. Organic carbon compounds as well as nitrogen, phosphorus, sulfur, and other elements are necessary for bacterial growth. However, the use of different substrates most often meets the needs of microflora. Nevertheless, it is harder to maintain the right carbon to nitrogen ratio. Its optimal value corresponds to the quotient C: N within 10:1–25:1. When the optimal content of these elements in the fermented mass is disturbed, nitrogen will be used by bacteria in metabolic processes and will reduce the amount of biogas produced. If, on the other hand, the C: N ratio falls below the optimal carbon to nitrogen ratio, then nitrogen will be released in the form of ammonia. This, in turn, will adversely affect the development of methanogenic bacteria, because this compound is toxic to them [118,134–136].

The high content of dry matter is also important for the high efficiency of the fermentation process. It is most commonly assumed that it should be in the range of 12%–15%. Then the mixing process can easily take place (as long as it is homogeneous). Too low water content in the fermentation chamber, as well as lack or too short mixing, can also disturb the optimal course of organic matter decomposition [137–139].

In addition, other factors may also affect the methane fermentation process. Substances that already negatively affect the efficiency of the process in small quantities are called methane fermentation inhibitors. These include, for example, ammonia, heavy metals, detergents, and pesticides. Periodic monitoring of process parameters allows, however, to reduce or even eliminate the adverse effects of undesirable substances. Many problems with biogas production can be avoided by accurately determining the parameters of the substrate and graft, as well as by ensuring the stability of the biogas plant [140,141].

5 Conclusion

To sum up, growing crops for energetic purposes should be based on the analyses of different variants adequate to the prevailing conditions in a given area, be they climatic, soil, and economic ones, as the plantation of perennial species is especially associated with at least a dozen or so years of monoculture and generation of profit can be expected only in a longer perspective. Nevertheless, new plant varieties appearing on the market intended strictly for energetic purposes and an increasing selection of agricultural machinery intended for this field of agriculture, as well as the decreasing demand for biomass, make it worth considering growing this group of plants.

However, based on many years of research and an analysis of data contained in the literature, it was determined that the highest probability to get the greatest income can be obtained from the cultivation of annual plants. In Greater Poland, maize (*Zea mays* L.) and sorghum (*Sorghum bicolor* L.) have the greatest production potential and they may be taken into consideration in other regions with a temperate climate. Their biomass can be used for producing solid, liquids, and gas biofuels. For example, solid and gas biofuels can be used for heating houses, while liquid can be used in vehicles. Increasing the use of biomass for energetic purposes will increase the diversification of energy sources and have a positive impact on the environment.

References

[1] Eurostat. Energy statistics – quantities, https://ec.europa.eu/eurostat/cache/metadata/en/nrg_quant_esms.htm; 2020. [Accessed 3 March 2020].

[2] Umbach F. Global energy security and the implications for the EU. Energy Policy 2010;38 (3):1229–40.

[3] Afifi SN, Hassan MG, Zobaa AF. The impacts of the proposed Nabucco gas pipeline on EU common energy policy. Energ Source B 2013;8(1):14–27.

[4] Lucas JNV, Francés GE, González ESM. Energy security and renewable energy deployment in the EU: liaisons dangereuses or virtuous circle? Renew Sustain Energy Rev 2016;62: 1032–46.

[5] European Union. Directive 2009/28/EC of the European Parliament and of the Council of the European Union of 23 April 2009 on the promotion of the use of energy from renewable sources and amending and subsequently repealing Directives 2001/77/EC and 2003/30/ EC. In: Official Journal of the European Union. Brussels: European Union; 2009. English Version. L 140/16 of 5.6.2009.

[6] Kuś J, Matyka M. Uprawa roślin na cele energetyczne [Growing plants for energy purposes]. Puławy: IUNG-PIB; 2010. p. 64 [in Polish].

[7] Dauber J, Miyake S. To integrate or to segregate food crop and energy crop cultivation at the landscape scale? Perspectives on biodiversity conservation in agriculture in Europe. Energy Sustain Soc 2016;6(1):25.

[8] Zabed H, Sahu JN, Suely A, Boyce AN, Faruq G. Bioethanol production from renewable sources: current perspectives and technological progress. Renew Sustain Energy Rev 2017; 71:475–501.

[9] Hagos K, Zong J, Li D, Liu C, Lu X. Anaerobic co-digestion process for biogas production: progress, challenges and perspectives. Renew Sustain Energy Rev 2017;76:1485–96.

[10] Weldemichael Y, Assefa G. Assessing the energy production and GHG (greenhouse gas) emissions mitigation potential of biomass resources for Alberta. J Clean Prod 2016; 112:4257–64.

[11] Ni Y, Mwabonje ON, Richter GM, Qi A, Yeung K, Patel M, et al. Assessing availability and greenhouse gas emissions of lignocellulosic biomass feedstock supply–case study for a catchment in England. Biofuel Bioprod Biorefin 2019;13(3):568–81.

[12] Kadiyala A, Kommalapati R, Huque Z. Evaluation of the life cycle greenhouse gas emissions from different biomass feedstock electricity generation systems. Sustainability 2016; 8(11):1181.

[13] Murphy F, Sosa A, McDonnell K, Devlin G. Life cycle assessment of biomass-to-energy systems in Ireland modelled with biomass supply chain optimisation based on greenhouse gas emission reduction. Energy 2016;109:1040–55.

[14] Schwerz F, Neto DD, Caron BO, Nardini C, Sgarbossa J, Eloy E, et al. Biomass and potential energy yield of perennial woody energy crops under reduced planting spacing. Renew Energy 2020;153:1238–50.

[15] Nunes LJR, Matias JCO, Catalão JPS. Biomass combustion systems: a review on the physical and chemical properties of the ashes. Renew Sustain Energy Rev 2016;53:235–42.

[16] Pires JR, Souza VG, Fernando AL. Valorization of energy crops as a source for nanocellulose production–current knowledge and future prospects. Ind Crop Prod 2019;140:111642.

[17] Blanco-Canqui H. Growing dedicated energy crops on marginal lands and ecosystem services. Soil Sci Soc Am J 2016;80(4):845–58.

[18] Gradziuk P, Grzybek A. Charakterystyka odnawialnych źródeł energii [Characteristics of renewable energy sources]. In: Grzybek A, Gradziuk P, Kowalczyk K, editors. Słoma. Energetyczne paliwo [The straw. Energy fuel]. Warsaw: Polskie Towarzystwo Biomasy PolBiom; 2001 [in Polish].

[19] Anawar HM, Strezov V. Mutual effects of climate change and energy crops and their controls on biomass and bioenergy production. In: Renewable energy systems from biomass. CRC Press; 2018. p. 75–88.

[20] Tonini D, Vadenbo C, Astrup TF. Priority of domestic biomass resources for energy: importance of national environmental targets in a climate perspective. Energy 2017;124:295–309.

[21] Pandey VC, Bajpai O, Singh N. Energy crops in sustainable phytoremediation. Renew Sustain Energy Rev 2016;54:58–73.

[22] Philippot L, Hallin S. Towards food, feed and energy crops mitigating climate change. Trends Plant Sci 2011;16(9):476–80.

[23] Sims RE, Hastings A, Schlamadinger B, Taylor G, Smith P. Energy crops: current status and future prospects. Glob Chang Biol 2006;12(11):2054–76.

[24] Chmiel J, Kasprowicz M. Flora i roślinność środkowej Wielkopolski. Przewodnik do ćwiczeń terenowych [The flora and vegetation of the central Wielkopolska. A guide to field exercises]. Poznan: UAM; 2004. p. 12–22 [in Polish].

[25] Bieganowski A, Witkowska-Walczak B, Glinski J, Sokolowska Z, Slawinski C, Brzezinska M, et al. Database of Polish arable mineral soils: a review. Int Agrophys 2013;27(3):335–50.

[26] Kabała C, Świtoniak M, Charzyński P. Correlation between the Polish Soil Classification (2011) and international soil classification system World Reference Base for Soil Resources (2015). Soil Sci Annu 2016;67(2):88–100.

[27] Luterbacher J, Xoplaki E, Küttel M, Zorita E, González-Rouco JF, Jones PD, et al. Climate change in Poland in the past centuries and its relationship to European climate: Evidence from reconstructions and coupled climate models. In: The Polish climate in the European context: an historical overview. Dordrecht: Springer; 2010. p. 3–39.

[28] Romanowicz RJ, Bogdanowicz E, Debele SE, Doroszkiewicz J, Hisdal H, Lawrence D, et al. Climate change impact on hydrological extremes: preliminary results from the Polish-Norwegian project. Acta Geophys 2016;64(2):477–509.

[29] Piniewski M, Kardel I, Giełczewski M, Marcinkowski P, Okruszko T. Climate change and agricultural development: adapting Polish agriculture to reduce future nutrient loads in a coastal watershed. Ambio 2014;43(5):644–60.

[30] Wiśniewski G, Podlaski S. Agrotechnika roślin uprawianych na cele energetyczne [Agrotechnics of plants grown for energy purposes]. In: Gradziuk P, editor. Energia odnawialna [Renewable energy]. Płonsk: SGGW; 2008 [in Polish].

[31] Tworkowski J. Ogólna charakterystyka roślin energetycznych [General characteristics of energy plants]. In: Szczukowski S, editor. Perennial energy plants. Warsaw: MULTICO Oficyna Wydawnicza; 2012. p. 6–9 [in Polish].

[32] Grzybek A, editor. Modelowanie energetycznego wykorzystania biomasy [Modeling of biomass energy use]. Warsaw: ITP; 2011. p. 60. 64, 117, 138–39 [in Polish].

[33] Kołodziej B, Matyka M, editors. Odnawialne źródła energii. Rolnicze surowce energetyczne [Renewable energy sources. Agricultural energy resources]. Poznań: PWRiL; 2012. p. 314–37. 359–69, 499, 518–21 [in Polish].

[34] Cassida KA, Muir JP, Hussey MA, Read JC, Venuto BC, Ocumpaugh WR. Biofuel component concentrations and yields of switchgrass in south central US environments. Crop Sci 2005;45(2):682–92.

[35] Monti A, di Virgilio N, Venturi G. Mineral composition and ash content of six major energy crops. Biomass Bioenergy 2008;32:218–22.
[36] Stolarski M, Szczukowski S, Tworkowski J. Produktywność klonów wierzb krzewiastych uprawianych na gruntach ornych w zależności od częstotliwości zbioru i gęstości sadzenia [Productivity of shrub willow clones grown on arable land depending on the frequency of harvesting and planting density]. Fragmenta Agronomica 2002;2:39–51 [in Polish].
[37] Venturi P, Venturi C. Analysis of energy comparision for crops in European agricultural systems. Biomass Bioenergy 2003;25(3):235–55.
[38] Dembiras A. Combustion characteristics of different biomass fuels. Prog Energy Combust Sci 2004;30(2):219–30.
[39] Szczukowski S, editor. Wieloletnie rośliny energetyczne [Perennial energy crops]. Warsaw: MULTICO Oficyna Wydawnicza; 2012. p. 10–38 [in Polish].
[40] Tworkowski J, Szczukowski S, Stolarski M. Yielding and morphological characteristics of willow grown in eco-salix system. Fragmenta Agronomica 2010;27(4):135–46.
[41] Christersson L. Poplar plantations for paper and energy in the south of Sweden. Biomass Bioenergy 2008;32(11):997–1000.
[42] Stolarski M. Topola; Robinia akacjowa; Miskant [Poplar; Black locust; Miscanthus]. In: Szczukowski S, editor. Wieloletnie rośliny energetyczne [Perennial energy crops]. Warsaw: MULTICO Oficyna Wydawnicza; 2012. p. 38–100 [in Polish].
[43] González-García S, Bacenetti J, Murphy RJ, Fiala M. Present and future environmental impact of poplar cultivation in the Po Valley (Italy) under different crop management systems. J Clean Prod 2012;26:56–66.
[44] Vavrova K, Knapek J. Economic assessment of miscanthus cultivation for energy purposes in the Czech Republic. J Jpn Inst Energy 2012;91(6):485–94.
[45] Burczyk H. Przydatność poplonu ozimego oraz kukurydzy i sorgo w plonie wtórym do produkcji biomasy dla biogazowni [Usability of the winter aftercrop as well as the maize and sorghum in the aftercrop to biomass production for a biogas plant]. Problemy Inżynierii Rolniczej 2013;21:87–97 [in Polish].
[46] Helios W, Kozak M, Malarz W, Kotecki A. Effect of sewage sludge application on the growth, yield and chemical composition of prairie cordgrass (Spartina pectinata Link.). J Elem 2014;19(4):1021–36.
[47] Korzeniowska J, Stanislawska-Glubiak E. Phytoremediation potential of Miscanthus × giganteus and Spartina pectinata in soil contaminated with heavy metals. Environ Sci Pollut Res 2015;22(15):11648–57.
[48] Nesheim L, Jørgensen U. Rozwój roślin energetycznych w krajach nordyckich [Development of energy crops in the Nordic countries]. In: Grzybek A, editor. Modelowanie energetycznego wykorzystania biomasy [Modeling of biomass energy use]. Warsaw: ITP; 2011. p. 13–4 [in Polish].
[49] Oleszek M, Król A, Tys J, Matyka M, Kulik M. Comparison of biogas production from wild and cultivated varieties of reed canary grass. Bioresour Technol 2014;156:303–6.
[50] Mander Ü, Järveoja J, Maddison M, Soosaar K, Aavola R, Ostonen I, Salm JO. Reed canary grass cultivation mitigates greenhouse gas emissions from abandoned peat extraction areas. Gcb Bioenergy 2012;4(4):462–74.
[51] Nilsson D, Rosenqvist H, Bernesson S. Profitability of the production of energy grasses on marginal agricultural land in Sweden. Biomass Bioenergy 2015;83:159–68.
[52] Kulig R, Skonecki S, Michałek W, Martyniak D, Piekarski W, Łysiak G, Różyło R. The *Andropogon gerardii* compaction process in terms of ecological solid fuel production. Polish J Environ Stud 2015;24(6):53–7.

[53] Sanderson MA, Reed RL, McLaughlin SB, Wullschleger SD, Conger BV, Parrish DJ, et al. Switchgrass as a sustainable bioenergy crop. Bioresour Technol 1996;56(1):83–93.
[54] Schmer MR, Vogel KP, Mitchell RB, Perrin RK. Net energy of cellulosic ethanol from switchgrass. Proc Natl Acad Sci 2008;105(2):464–9.
[55] Parrish DJ, Fike JH. The biology and agronomy of switchgrass for biofuels. BPTS 2005; 24(5–6):423–59.
[56] Martyniak D, Martyniak J. Nowa energetyczna trawa [A new energy grass]. Farmer 2009;18:28–9 [in Polish].
[57] Lalak J, Kasprzycka A, Martyniak D, Tys J. Effect of biological pretreatment of Agropyron elongatum 'BAMAR' on biogas production by anaerobic digestion. Bioresour Technol 2016; 200:194–200.
[58] Rodrigues MA, Sousa L, Cabanas JE, Arrobas M. Tuber yield and leaf mineral composition of Jerusalem artichoke (*Helianthus tuberosus* L.) grown under different cropping practices. Spanish J Agric Res 2007;5(4):545–53.
[59] Long XH, Shao HB, Liu L, Liu LP, Liu ZP. Jerusalem artichoke: a sustainable biomass feedstock for biorefinery. Renew Sustain Energy Rev 2016;54:1382–8.
[60] Mantovani D, Veste M, Gypser S, Halke C, Koning L, Freese D, Lebzien S. Transpiration and biomass production of the bioenergy crop Giant knotweed Igniscum under various supplies of water and nutrients. J Hydrol Hydromech 2014;62(4):316–23.
[61] Šiaudinis G, Jasinskas A, Šarauskis E, Steponavičius D, Karčauskienė D, Liaudanskienė I. The assessment of Virginia mallow (*Sida hermaphrodita* Rusby) and cup plant (*Silphium perfoliatum* L.) productivity, physico–mechanical properties and energy expenses. Energy 2015; 93:606–12.
[62] Nahm M, Morhart C. Virginia mallow (*Sida hermaphrodita* (L.) Rusby) as perennial multipurpose crop: biomass yields, energetic valorization, utilization potentials, and management perspectives. GCB Bioenergy 2018;10(6):393–404.
[63] Perzon M. Emissions of organic compounds from the combustion of oats–a comparison with softwood pellets. Biomass Bioenergy 2010;34(6):828–37.
[64] Lal R. World crop residues production and implications of its use as a biofuel. Environ Int 2005;31(4):575–84.
[65] Acharya B, Dutta A, Mahmud S, Tushar M, Leon M. Ash analysis of poultry litter, willow and oats for combustion in boilers. J Biomass Biofuel 2014;1:16–26.
[66] Blanco-Canqui H, Lal R. Soil and crop response to harvesting corn residues for biofuel production. Geoderma 2007;141(3–4):355–62.
[67] Kauffman N, Hayes D, Brown R. A life cycle assessment of advanced biofuel production from a hectare of corn. Fuel 2011;90(11):3306–14.
[68] Bavec F, Bavec M. Effects of plant population on leaf area index, cob characteristics and grain yield of early maturing maize cultivars (FAO 100–400). Eur J Agron 2002; 16(2):151–9.
[69] Paterson AH, Bowers JE, Bruggmann R, Dubchak I, Grimwood J, Gundlach H, et al. The Sorghum bicolor genome and the diversification of grasses. Nature 2009;457(7229): 551–6.
[70] Reddy BV, Ramesh S, Reddy PS, Ramaiah B, Salimath M, Kachapur R. Sweet sorghum-a potential alternate raw material for bio-ethanol and bio-energy. Int Sorghum Millets Newslett 2005;46:79–86.
[71] Acevedo E, Fereres E, Hsiao TC, Henderson DW. Diurnal growth trends, water potential, and osmotic adjustment of maize and sorghum leaves in the field. Plant Physiol 1979;64(3): 476–80.

[72] Murray SC, Rooney WL, Hamblin MT, Mitchell SE, Kresovich S. Sweet sorghum genetic diversity and association mapping for brix and height. Plant Genome 2009;2(1):48–62.
[73] Szambelan K, Nowak J, Frankowski J, Szwengiel A, Jeleń H, Burczyk H. The comprehensive analysis of sorghum cultivated in Poland for energy purposes: separate hydrolysis and fermentation and simultaneous saccharification and fermentation methods and their impact on bioethanol effectiveness and volatile by-products from the grain and the energy potential of sorghum straw. Bioresour Technol 2018;250:750–7.
[74] Lewandowski W, Ryms M. Biopaliwa. Proekologiczne odnawialne źródła energii [Biofuels. Pro-ecological renewable energy sources]. Warsaw: WNT; 2013. p. 528 [in Polish].
[75] Anon. Directive 2009/28/EC of the European Parliament and of the Council of 23 April 2009 on the promotion of the use of energy from renewable sources and amending and subsequently repealing Directives 2001/77/EC and 2003/30/EC; 2009.
[76] Anon. Directive (EU) 2018/2001 of the European Parliament and of the Council of 11 December 2018 on the promotion of the use of energy from renewable sources; 2018.
[77] Communication from the Commission to the European Parliament, the European Council, the COUNCIL, the European Economic and Social Committee, the Committee of the Regions and the European Investment Bank. A Clean Planet for all A European strategic long-term vision for a prosperous, modern, competitive and climate neutral economy. COM/2018/773 final; 2018.
[78] Borkowska H, Molas R. Yield comparison of four lignocellulosic perennial energy crop species. Biomass Bioenergy 2013;51:145–53.
[79] Weng JK, Li X, Bonawitz ND, Chapple C. Emerging strategies of lignin engineering and degradation for cellulosic biofuel production. Curr Opin Biotechnol 2008;19(2):166–72.
[80] Koçar G, Civaş N. An overview of biofuels from energy crops: current status and future prospects. Renew Sustain Energy Rev 2013;28:900–16.
[81] Krasuska E, Rosenqvist H. Economics of energy crops in Poland today and in the future. Biomass Bioenergy 2012;38:23–33.
[82] Prade T, Svensson SE, Andersson A, Mattsson JE. Biomass and energy yield of industrial hemp grown for biogas and solid fuel. Biomass Bioenergy 2011;35:3040–9.
[83] Coulson M, Dahl J, Gansekoele E, Bridgwater AV, Obernberger I, Van de Beld L. Ash characteristics of perennial energy crops and their influence on thermal processing. In: Proceedings of the 2nd world biomass conference; 2004. p. 359–62.
[84] Carroll JP, Finnan J. Physical and chemical properties of pellets from energy crops and cereal straws. Biosyst Eng 2012;112:151–9.
[85] Searle SY, Malins CJ. Will energy crop yields meet expectations? Biomass Bioenergy 2014;65:3–12.
[86] Styles D, Thorne F, Jones MB. Energy crops in Ireland: an economic comparison of willow and Miscanthus production with conventional farming systems. Biomass Bioenergy 2008; 32:407–21.
[87] Hauk S, Knoke T, Wittkopf S. Economic evaluation of short rotation coppice systems for energy from biomass – a review. Renew Sustain Energy Rev 2014;29:435–48.
[88] Obernberger I, Brunner T, Bärnthaler G. Chemical properties of solid biofuels—significance and impact. Biomass Bioenergy 2006;30(11):973–82.
[89] Mardoyan A, Braun P. Analysis of Czech subsidies for solid biofuels. Int J Green Energy 2015;12(4):405–8.
[90] Demirbas A. Competitive liquid biofuels from biomass. Appl Energy 2011;88:17–28.
[91] Rathmann R, Szklo A, Schaeffer R. Land use competition for production of food and liquid biofuels: an analysis of the arguments in the current debate. Renew Energy 2010;35:14–22.

[92] Casas A, Ruiz JR, Ramos MJ, Pérez A. Effects of triacetin on biodiesel quality. Energy Fuel 2010;24:4481–9.
[93] Silva PH, Gonçalves VL, Mota CJ. Glycerol acetals as anti-freezing additives for biodiesel. Bioresour Technol 2010;101:6225–9.
[94] Pousa GP, Santo AL, Suarez PA. History and policy of biodiesel in Brazil. Energy Policy 2007;35:5393–8.
[95] Lamers P, Hamelinck C, Junginger M, Faaij A. International bioenergy trade – a review of past developments in the liquid biofuel market. Renew Sustain Energy Rev 2011;15:2655–76.
[96] Ramos MJ, Fernández CM, Casas A, Rodríguez L, Pérez Á. Influence of fatty acid composition of raw materials on biodiesel properties. Bioresour Technol 2009;100:261–8.
[97] Kwon K, Vahdat N, Mbah J. Fatty acid methyl ester biofuels produced from canola oil with honeycomb monolithic catalysts. Fuel 2015;145:116–26.
[98] Lebedevas S, Vaicekauskas A, Lebedeva G, Makareviciene V, Janulis P, Kazancev K. Use of waste fats of animal and vegetable origin for the production of biodiesel fuel: quality, motor properties, and emissions of harmful components. Energy Fuel 2006;20:2274–80.
[99] Rutz D, Janssen R. Biofuel technology handbook. Munich: WIP Renewable Energies; 2007. p. 95.
[100] Lee RA, Lavoie JM. From first-to third-generation biofuels: challenges of producing a commodity from a biomass of increasing complexity. Anim Front 2013;3:6–11.
[101] Nigam PS, Singh A. Production of liquid biofuels from renewable resources. Prog Energy Combust Sci 2011;37:52–68.
[102] Naik SN, Goud VV, Rout PK, Dalai AK. Production of first and second generation biofuels: a comprehensive review. Renew Sustain Energy Rev 2010;14:578–97.
[103] Minu K, Jiby KK, VVN K. Isolation and purification of lignin and silica from the black liquor generated during the production of bioethanol from rice straw. Biomass Bioenergy 2012; 39:210–7.
[104] Cotana F, Cavalaglio G, Nicolini A, Gelosia M, Coccia V, Petrozzi A, et al. Lignin as co-product of second generation bioethanol production from ligno-cellulosic biomass. Energy Procedia 2014;45:52–60.
[105] Alvira P, Tomás-Pejó E, Ballesteros M, Negro MJ. Pretreatment technologies for an efficient bioethanol production process based on enzymatic hydrolysis: a review. Bioresour Technol 2010;101:4851–61.
[106] Ojeda K, Sánchez E, El-Halwagi M, Kafarov V. Exergy analysis and process integration of bioethanol production from acid pre-treated biomass: comparison of SHF, SSF and SSCF pathways. Chem Eng J 2011;176:195–201.
[107] Tan IS, Lee KT. Enzymatic hydrolysis and fermentation of seaweed solid wastes for bioethanol production: an optimization study. Energy 2014;78:53–62.
[108] Zucaro A, Fiorentino G, Ulgiati S. Constraints, impacts and benefits of lignocellulose conversion pathways to liquid biofuels and biochemicals. In: Lignocellulosic biomass to liquid biofuels. Academic Press; 2020. p. 249–82.
[109] Arun N, Dalai AK. Environmental and socioeconomic impact assessment of biofuels from lignocellulosic biomass. In: Lignocellulosic biomass to liquid biofuels. Academic Press; 2020. p. 283–99.
[110] Singh RS, Pandey A, Gnansounou E, editors. Biofuels: production and future perspectives. USA: CRC Press; 2016. p. 578.
[111] Wandera SM, Qiao W, Algapani DE, Bi S, Yin D, Qi X, et al. Searching for possibilities to improve the performance of full scale agricultural biogas plants. Renew Energy 2018; 116:720–7.

[112] Malińska K, Dach J. Biochar as a supplementary material for biogas production. Inżynieria Ekologiczna 2015;41:117–24.
[113] Mursec B, Vindis P, Janzekovic M, Brus M, Cus F. Analysis of different substrates for processing into biogas. J Achiev Mater Manuf Eng 2009;37:652–9.
[114] Rasi S, Veijanen A, Rintala J. Trace compounds of biogas from different biogas production plants. Energy 2007;32:1375–80.
[115] Dandikas V, Heuwinkel H, Lichti F, Drewes JE, Koch K. Correlation between biogas yield and chemical composition of energy crops. Bioresour Technol 2014;174:316–20.
[116] Cieślik M, Dach J, Lewicki A, Smurzyńska A, Janczak D, Pawlicka-Kaczorowska J, et al. Methane fermentation of the maize straw silage under meso-and thermophilic conditions. Energy 2016;115:1495–502.
[117] Shin HS, Youn JH, Kim SH. Hydrogen production from food waste in anaerobic mesophilic and thermophilic acidogenesis. Int J Hydrogen Energy 2004;29:1355–63.
[118] Czerwińska E, Kalinowska K. Warunki prowadzenia procesu fermentacji metanowej w biogazowni [Conditions for conducting the methane fermentation process in a biogas plant]. Technika Rolnicza i Leśna [Agric For Technol] 2014;2:12–4 [in Polish].
[119] Alkaya E, Demirer GN. Anaerobic acidification of sugar-beet processing wastes: effect of operational parameters. Biomass Bioenergy 2011;35:32–3.
[120] Kushkevych I, Kobzová E, Vítězová M, Vítěz T, Dordević D, Bartoš M. Acetogenic microorganisms in operating biogas plants depending on substrate combinations. Biologia 2019; 74:1229–36.
[121] Nie Y, Liu H, Du G, Chen J. Enhancement of acetate production by a novel coupled syntrophic acetogenesis with homoacetogenesis process. Process Biochem 2007;42:599–605.
[122] Christy PM, Gopinath LR, Divya D. A review on anaerobic decomposition and enhancement of biogas production through enzymes and microorganisms. Renew Sustain Energy Rev 2014;34:167–73.
[123] Liu G, Zhang R, El-Mashad HM, Dong R. Effect of feed to inoculum ratios on biogas yields of food and green wastes. Bioresour Technol 2009;100:5103–8.
[124] Im S, Petersen SO, Lee D, Kim DH. Effects of storage temperature on CH4 emissions from cattle manure and subsequent biogas production potential. Waste Manag 2020;101:35–43.
[125] Feng L, Yu Q, Zhen XF, Dong HY, Zheng J, Wang Y. Comparison of methanogenesis between sunflower and corn stalks mixed with pig manure at different temperatures. Appl Ecol Environ Res 2020;18:747–56.
[126] Stolze Y, Bremges A, Rumming M, Henke C, Maus I, Pühler A, Sczyrba A, Schlüter A. Identification and genome reconstruction of abundant distinct taxa in microbiomes from one thermophilic and three mesophilic production-scale biogas plants. Biotechnol Biofuels 2016;9:156.
[127] Zhang L, Loh KC, Sarvanantharajah S, Tong YW, Wang CH, Dai Y. Mesophilic and thermophilic anaerobic digestion of soybean curd residue for methane production: characterizing bacterial and methanogen communities and their correlations with organic loading rate and operating temperature. Bioresour Technol 2019;288:121597.
[128] Chen H, Chang S. Dissecting methanogenesis for temperature-phased anaerobic digestion: impact of temperature on community structure, correlation, and fate of methanogens. Bioresour Technol 2020;123104.
[129] Zhai X, Kariyama ID, Wu B. Investigation of the effect of intermittent minimal mixing intensity on methane production during anaerobic digestion of dairy manure. Comput Electron Agric 2018;155:121–9.

[130] Zhang C, Su H, Baeyens J, Tan T. Reviewing the anaerobic digestion of food waste for biogas production. Renew Sustain Energy Rev 2014;38:383–92.

[131] Lewicki A, Dach J, Czekała W, Janczak D, Cieślik M, Witaszek K, Carmona PCR. Testing the biogas substrate efficiency from the experimental Farms's of Poznan University of Life Sciences in Przybroda biogas plant. Archiv Waste Manage Environ Protect 2014;16:27–30.

[132] Gao WJ, Lin HJ, Leung KT, Liao BQ. Influence of elevated pH shocks on the performance of a submerged anaerobic membrane bioreactor. Process Biochem 2010;45:1279–87.

[133] Jiang D, Ge X, Zhang Q, Zhou X, Chen Z, Keener H, Li Y. Comparison of sodium hydroxide and calcium hydroxide pretreatments of giant reed for enhanced enzymatic digestibility and methane production. Bioresour Technol 2017;244:1150–7.

[134] Ustak S, Munoz J. Cup-plant potential for biogas production compared to reference maize in relation to the balance needs of nutrients and some microelements for their cultivation. J Environ Manage 2018;228:260–6.

[135] Yan Z, Song Z, Li D, Yuan Y, Liu X, Zheng T. The effects of initial substrate concentration, C/N ratio, and temperature on solid-state anaerobic digestion from composting rice straw. Bioresour Technol 2015;177:266–73.

[136] Dioha IJ, Ikeme CH, Nafi'u T, Soba NI, Yusuf MBS. Effect of carbon to nitrogen ratio on biogas production. Int Res J Nat Sci 2015;3:1–10.

[137] Riya S, Suzuki K, Meng L, Zhou S, Terada A, Hosomi M. The influence of the total solid content on the stability of dry-thermophilic anaerobic digestion of rice straw and pig manure. Waste Manag 2018;76:350–6.

[138] Latha K, Velraj R, Shanmugam P, Sivanesan S. Mixing strategies of high solids anaerobic co-digestion using food waste with sewage sludge for enhanced biogas production. J Clean Prod 2019;210:388–400.

[139] Kariyama ID, Zhai X, Wu B. Influence of mixing on anaerobic digestion efficiency in stirred tank digesters: a review. Water Res 2018;143:503–17.

[140] Sun C, Cao W, Banks CJ, Heaven S, Liu R. Biogas production from undiluted chicken manure and maize silage: a study of ammonia inhibition in high solids anaerobic digestion. Bioresour Technol 2016;218:1215–23.

[141] Franco RT, Buffière P, Bayard R. Co-ensiling of cattle manure before biogas production: effects of fermentation stimulants and inhibitors on biomass and methane preservation. Renew Energy 2018;121:315–23.

Chapter 2

Catalytic pyrolysis of biomass using shape-selective zeolites for bio-oil enhancement

Salman Raza Naqvi[a], Syed Ali Ammar Taqvi[b], M. Taqi Mehran[a], Asif Hussain Khoja[c], M. Naqvi[d], Awais Bokhari[e], and Nor Aishah Saidina Amin[f]

[a]*School of Chemical & Materials Engineering, National University of Sciences & Technology, Islamabad, Pakistan,* [b]*Department of Chemical Engineering, NED University of Engineering and Technology, Karachi, Pakistan,* [c]*U.S.-Pakistan Centre for Advanced Studies in Energy (USPCAS-E), National University of Sciences & Technology, Islamabad, Pakistan,* [d]*Department of Engineering and Chemical Sciences, Karlstad University, Karlstad, Sweden,* [e]*Chemical Engineering Department, COMSATS University Islamabad, Lahore, Pakistan,* [f]*Chemical Reaction Engineering Group (CREG), School of Chemical and Energy Engineering, Faculty of Engineering, Universiti Teknologi Malaysia, Johor Bahru, Johor Darul Takzim, Malaysia*

1 Introduction

The rapid urbanization and economic growth require an increase in energy demand day by day. Currently, the energy requirements are fulfilled mostly by fossil fuels, and consequently, it impacts the environment especially greenhouse gas emissions which lead to global warming, forthcoming shortage, and detrimental impact on human health [1]. Higher the consequences, higher the threats to the ecosystem as well. A sharp increase in the fuel consumption in the transportation sector is depleting petroleum reserves and increasing greenhouse gas emissions [2, 3]. This increased petroleum consumption has raised concern over climate and environment and demanded a sustainable, eco-friendly, and economical approach and policies for effective utilization of bio-based transportation fuels [4, 5].

Considering the above facts, the scientific community needs to present a solution to alternative resources of energy. To introduce the concept of bioenergy, renewable resources are identified in comparison to nonrenewable resources to generate their viability. Among renewable resources, plants and living microorganism-based species, called biomass and organic wastes, became popular for providing clean bioenergy [6–8]. Organic wastes including composites

Bioenergy Resources and Technologies. https://doi.org/10.1016/B978-0-12-822525-7.00002-0
Copyright © 2021 Elsevier Inc. All rights reserved.

wastes are gaining much volume because of nonavailable policies and appropriate technologies which alarm the society about fostering global warming issues. These wastes must have to be addressed to clean the environment, efficient life cycle of raw materials to products in a circular economy [9]. Lignocellulosic biomass residues are gained much attention as an alternative resource because of their availability in large amounts, low in price, and can contribute to 10%–15% total world primary energy supply [10–12]. Lignocellulose possesses a higher amount of cellulose, hemicellulose, and then lignin compounds that are responsible to produce a variety of fuels and chemicals upon degradation of their polymeric structure [13, 14]. In a similar pattern, plants and nonedible seeds gained special consideration to produce clean biofuel [15, 16]. The biofuel produced meets the standard criteria of fuel and provides a cheap solution to liquid fuels [17, 18]. To convert these potential feedstocks into fuels and chemicals, thermochemical and biochemical are the two acceptable and widely utilized approaches [19]. Thermochemical conversion processes appear as the most energy efficient and robust in comparison to the biochemical route. Starting from the torrefaction process, onward to the carbonization, pyrolysis, gasification, and lastly combustion are the typical thermochemical conversion processes. Among all thermochemical conversion processes, pyrolysis appears as the most viable strategy to generate high-grade solid, bio-oil, and gaseous fuel from the biomass/organic residues [20, 21]. Pyrolysis process can be operated in three modes: (i) slow, (ii) fast, and (iii) flash with consequential impact on residence time of the vapors. If the pyrolysis process is operated via slow mode (slow heating rate and longer residence time), the process will generate high quantity and quality solid fuel. Fast pyrolysis process is performed usually at higher temperature (450°C–550°C) with faster heating rates (~10,000 °C/min) and a residence time of 10 s. This process will produce more biofuel than gas and solid fuel. Lastly, the flash mode in the pyrolysis process is also adopted widely to produce more bio-oil and gas fuel with heating rates of more than (50,000 °C/min and 1–2 s of residence time). Such observations are reported in our previous publications well in agreement with the literature [22–24]. Pyrolysis process produced much amount of low quality of bio-oil which requires an upgrading/catalytic treatment to make it viable for utilization [25–27]. Higher the efficient catalyst, higher will be the deoxygenation and more aromatics will be produced. However, to design an appropriate catalyst for upgrading process, synthesis techniques and its characterization play an important role [27–29]. Unfortunately, the pyrolysis bio-oil has many technical challenges, i.e., high acidity, high moisture content, low calorific value, corrosiveness, repolymerization, and thermal instability [30]. These properties are promoted in the bio-oil storage and the presence of higher oxygenated compounds [31]. Hence, it is vital in improving the properties of the bio-oil in order to meet the standard specifications recommended to use as transportation fuel.

Catalytic pyrolysis of lignocellulosic biomass using various catalysts is reported comprehensively in the literature [1, 32–34]. However, most of the studies mentioned that zeolites showed much promising deoxygenation of biomass pyrolysis vapors [30, 35]. Among zeolites, microporous zeolites such as ZSM-5 showed high deoxygenation and produced more aromatics [20, 36, 37]. ZSM-5 and its proton form HZSM-5, gained much attention because of suitable pore size and geometry for selective deoxygenation [20]. To achieve higher deoxygenation, a rational design of solid acid catalyst is required. ZSM-5, a medium-pore zeolite possessed straight and zigzag channels of (5.1 Å × 5.5 Å and 5.3 Å × 5.6 Å). This geometry provides enough diffusion path length to generate large species with consequently impact on coke formation at the pore mouth. Hence, the access to active sites at the pore mouth as well as in the channel provides options to crack large oxygenated species. Such a challenge can be addressed through the modification in the zeolite shapes which are suggested through alkaline posttreatment and incorporation of mesoporosity. In consequence, the catalytic effectiveness increased due to enhanced access to active sites and catalytic deoxygenation. Hence, stability is much improved.

Hence, the aim of this study is to synthesis shape-selective small (SAPO-34), medium (MCM-22), and large-pore microporous zeolites (ITQ-2, a delaminated counterpart of MCM-22) and tested for biomass deoxygenation process. These catalysts are compared with commercial small, medium, and large-pore zeolites such as ferrite, ZSM-5, and mordenite. Hence, the catalysts are tested at various pore sizes, shapes at the Si/Al ratio of 20 in a drop-type pyrolyzer. The catalytic upgrading is performed in-bed mode. Catalysts and biomass are mixed physically through a uniform size. The pyrolysis products yield, gas analysis, degree of deoxygenation, and bio-oil composition are reported and discussed in detail.

2 Methodology

For lignocellulosic biomass, rice husk (100–355 μm) is collected from a local rice mill. To prepare the sample, the biomass is dried, sieved, and grinded. To examine the moisture, ash, volatile, and fixed carbon content, the proximate analysis is performed using ASTM standards [38]. For moisture analysis, the sample is heated in a furnace at 105°C for 24 h. Once the sample showed a consistent loss in weight, moisture content is recorded. For ash content, the sample is heated in a furnace at 700°C for 3 h. The volatile matter is recorded by heating the sample at 900°C for 7 min. Through weight (%) formula, the volatiles percentage is recorded. Fixed carbon is estimated by summation of the moisture, ash, and volatiles from hundred. To investigate the carbon, hydrogen, nitrogen, sulfur, and oxygen content, the ultimate analysis was performed using ASTM standards [39]. Elemental analyzer was used to record the ultimate analysis of the sample. The detailed information was published by the author in his previous study [40]. To compare synthesis and commercial zeolites performance, ferrite, ZSM-5, and mordenite zeolites were collected from Zeolyst International.

2.1 Zeolites synthesis

First, a small-medium-pore zeolites, SAPO-34 was prepared on lab scale by following the published protocol [41]. For the source of aluminum, boehmite was used. The precursor of phosphorous was phosphoric acid. Similarly, the precursor of silicon was extracted from a mixture of colloidal silica of 30 wt% of SiO_2 and 0.4 wt% of Na_2O. To incorporate the surface-directing agent, an aqueous solution of TEAOH (Wako Chemicals) was used. Hence, the molar ratio was adjusted using the following pattern: Al_2O_3, 1.0; P_2O_5, 1.0; SiO_2, 0.6; TEAOH, 1.8; H_2O, 77. To prepare the gel solution, the mixture was stirred at 30°C for 24 h with onward drying the sample at 90°C. This process was performed in a stainless-steel Teflon-lined vessel. The crystallization process was performed at 180°C for 1.5–24 h. The product mixture was washed with water. For calcination, the sample was dried at 600°C for 6 h.

Second, layered and its delaminated structure was prepared from MCM-22 precursor. The author already published the protocols [40]. The layered zeolites, MCM-22, were synthesized by preparing the mixture of the sodium aluminate, sodium hydroxide solution, amorphous fumed silica, and hexamethyleneimine. The crystallization process was performed in a stainless-steel vessel at 180°C for 3 days. The mixture was placed in a rotating oven with a rotation speed of 2 rpm. For washing and drying the sample, the mixture was rinsed with water and placed in an oven at 90°C for 24 h. After calcination of the precursor sample at 500°C for 8 h, the sample was named MCM-22. Third, the delaminated counter zeolites, ITQ-2, cetyltrimethylammonium was mixed with 10 wt% of tetrapropylammonium hydroxide solution. The mixtures were refluxed at 80°C for 16 h. An ultrasound force was applied for 1 h to get the separation. TO neutralize the solution, hydrochloric acid was added to the solution to maintain the pH around 2. To separate the mixture, centrifugation was applied. Lastly, the solution was processed for calcination in air at 540°C for 12 h to produce ITQ-2 zeolites.

2.2 Zeolites characterization

The characterization of the synthesized catalysts played a significant role to reveal their surface properties, information about its nature which supports building the discussion of results. To illustrate the physiochemical properties of the synthesized zeolites, the following tools are utilized in this work: XRD, SEM, NH_3-TPD, and BET analysis. To identify the crystallinity and amorphous nature of the synthesized zeolites, X-ray diffraction was performed on a Rigaku Minifle using Cu-Kα radiation. The diffraction pattern was carried out by following under standard operating conditions. The operating conditions are the diffraction angle (2θ) which started at 2° and ended at 80° and a step size of 0.1 step/s. The surface properties of the synthesized zeolites were measured by N_2 adsorption/desorption at −196°C on a Micrometrics ASAP 2010 instrument. This analysis provides the specific surface area (m^2/g) of synthesized zeolites. To find out the micropore volume, T-plot method was incorporated. To estimate the external surface area of the synthesized zeolites, micropore area was subtracted from the BET-specific surface area. These analyses are performed by degassing the samples at 400°C for 24 h. To exploit the surface morphology of prepared zeolites, scanning electron microscopy analysis was performed on Hitachi S-2250 mode. To estimate the total active sites of the prepared zeolites, NH_3-TPD analysis was performed. The procedure of the analysis is as follows: first, pretreatment of zeolites at 550°C for 2 h; second, ammonia adsorption at 120°C; and third, helium flushing for 2 h to desorb ammonia. To ramp the sample through a temperature program method, the sample was processed at a rate of 10°C/min for ammonia desorption for the temperature region of 100°C–800°C. For quantification, TCD was incorporated with the analyzer.

2.3 Catalytic pyrolysis of biomass using zeolites

To perform pyrolysis experiments of no-catalyst and with catalysts, drop-type fixed-bed pyrolyzer was fabricated, installed, and utilized. A novel drop-type pyrolyzer has simple design with a robust facility of providing high heating rates. This facility is very cheap and provides one-step solution for onsite as well as a mobile solution. To upgrade the pyrolysis vapors, the reactor has the facility to catalytic upgrade the pyrolysis vapors in one step (in-bed mode). This research work will test the zeolites as a catalyst in one-step mode. The produced bio-oil showed better pyrolysis oil yield as compared with the two-step mode. Fig. 1 depicts the working mechanism of the drop-type pyrolyzer. The reactor is connected via four main parts: (i) vacuum and nitrogen line, (ii) sample holder, (iii) reaction area, and (iv) collector. Before processing the pyrolysis process, the vacuum and nitrogen lines are connected to ensure the complete inert conditions inside the reaction chamber. Through joint activities of both lines, inside air was replaced with nitrogen. The biomass was placed in the sample holder with the internal diameters of 25 mm. The maximum capacity of the sample holder was 20 g. After placement of the sample, the cover screw is covered tightly to ensure that there will be no chance of gas leakage. The important section is the system of the reaction chamber where biomass with and without catalyst was placed. The dimensions of the reactor are 53-mm internal diameters and 166 mm in height. To supply the heat, the reactor is equipped with an electric heater with perfect insulation so that there will be no loss of heat transfer. To record the temperature inside the reaction chamber, a K-type thermocouple was inserted.

The thermocouple positions are adjusted in a way that temperature was recorded in the center of the reaction vessel. Hence, there are two thermocouples are inserted. The first thermocouple helps to record the heating purpose while the second one controls the pyrolysis temperature. The pyrolysis reaction

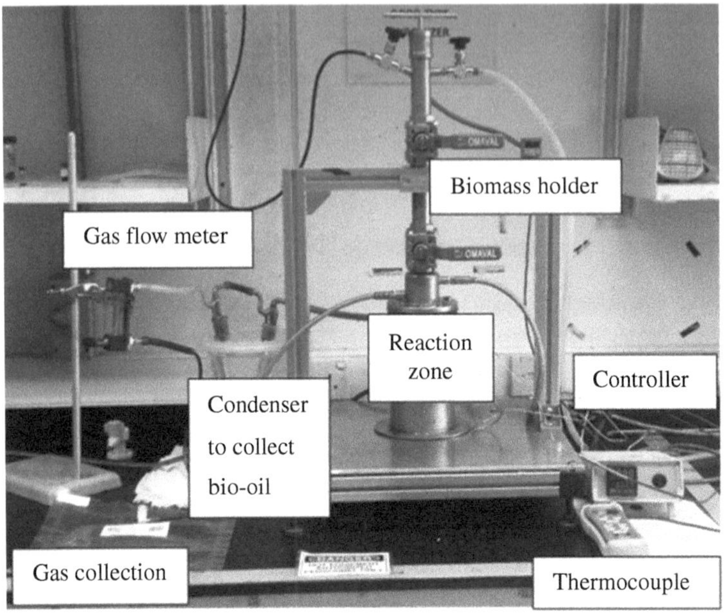

FIG. 1 A working sketch of reactor.

chamber relates to vapor outlet pipe for onward receiving the pyrolysis vapors. The vapors collecting pipe has approximately a length of 10 mm. The option for collecting the leftover solid sample is very easy. Upon cooling, the reaction chamber can be easily de-assembled. To prevent gas leakage, high-temperature silicon gasket is incorporated at the top side of the reactor flange. The pyrolysis experiments were carried out continuously around for 30 min. The condensable vapors are collected in a jointly connected ice bath. This condensing system has the capability to collect all the condensable pyrolysis vapors. On the other hand, to collect the non-condensable gases, the condensing system is well connected with Teflon gas bag. The gas bag has a capacity of 5 L in volume. Before the experimental run, the whole system is connected and run to ensure the smooth running of the process. The bio-oil recovered from the condensing system dividing by the sample mass could provide the yield of the bio-oil. For solid yield, the collected leftover solid sample was divided by the mass of the sample. By summation of the solid and bio-oil yield and subtracting from hundred, the gas yield can be estimated. For catalytic pyrolysis experiments, a definite amount of catalyst is physically mixed with the biomass. The sample processed for catalytic upgrading in the reaction chamber of the pyrolyzer. In the author previous studies, it was observed that the pyrolysis temperature of 450°C appeared as the viable temperature to produce maximum bio-oil [40]. Hence, the catalytic to biomass ratio for catalytic upgrading run was 0.1.

TABLE 1 Design of catalytic experimental runs.

Experimental run	Zeolites	Catalytic loading (ratio)	Temperature (°C)
1	SAPO-34	0.1	450
2	Ferrierite	0.1	450
3	MCM-22	0.1	450
4	ITQ-2	0.1	450
5	ZSM-5	0.1	450
6	Mordenite	0.1	450

The experiments of the catalytic upgrading are listed in Table 1. All the zeolites were tested at the pyrolysis temperature of 450°C at the catalyst to biomass ratio of 0.1. All the experiments were repeated thricely. The catalytic pyrolysis run showed the efficient cracking process. The catalytic loading was enough to crack the pyrolysis vapors at the temperature of 450°C. The catalytic amount was very low, so the coking mechanism was ignored. Hence, it is difficult to recover the catalyst as it was physically mixed with the biomass. The catalytic upgrading was the in-bed mode, so a better performance is expected in a one-step, robust, and cheap reactor. Table 1 shows the experimental design for the catalytic pyrolysis process.

2.4 Products characterization

The bio-oil collected from the condensing system has been processed for testing. This testing was performed using gas chromatography coupled with mass spectroscopy. Bio-oil showed the complex nature of hundreds of compounds. Physically, it was dark brown in color, thick, and not too much viscous. This might be due to the presence of produced water from the reaction. All the collected samples were analyzer within the 48h duration. This physical nature in combination with the analysis elaborates the presence of oxygenated species in the sample. To examine the heating value of the bio-oil, the sample was placed in a bomb calorimeter. As per naked eye observation, it is expected that the sample might reflect very low heating value because of the presence of water and oxygenated molecules. Therefore, bio-oil was characterized by the following mentioned techniques. First, water content by utilizing a Karl Fischer titrator. This analysis was performed by following the ASTM E 203 method. Second, a pH meter was utilized to analyze the acidity of the noncatalytic and catalytic bio-oil. Third, to reveal the chemical composition of the produced bio-oil,

GC-MS with a TCD detector was utilized. The detailed program method was already mentioned by the authors in their previous publications [38].

2.5 Degree of deoxygenation (DOD)

The degree of deoxygenation (DOD) appeared as a viable estimation tool to examine the eliminated oxygen from the bio-oil. This oxygen content represents in the form of oxides or the water phase. In this research work, the degree of deoxygenation is clearly defined as the removal of the oxygen content in the feedstock. This indicator also illustrates the quality of bio-oil. This technique is also utilized by other researchers to identify the deoxygenation capability of oils [42, 43]. This approach will provide how much oxygen has been eliminated from the feedstock and oxygen retained in the bio-oil. The formula to estimate the degree of deoxygenation is mentioned below:

$$\mathrm{DOD} = \left(1 - \frac{(\mathrm{O})_{\text{bio-oil}}}{(\mathrm{O})_{\text{biomass}}}\right) \times 100 \tag{1}$$

where (O) is termed as the mass content of O_2.

3 Results and discussion

3.1 Raw material and catalysts characterization

In the study, the biomass was characterized using proximate and ultimate analysis. In terms of elemental values, the biomass possessed carbon (42.7 wt%) and hydrogen (5.7 wt%). The sample showed 16 MJ/kg heating value. Abu Bakar et al. utilized rice husk for catalytic fast pyrolysis process [44]. The study showed that at dry basis analysis, the sample showed higher oxygen (54 wt %) and carbon (39 wt%) contents. Zhang et al. used pretreated rice husk for catalytic conversion into biofuel over Fe-ZSM-5 catalysts [45]. The pretreated biomass showed higher carbon (43 wt%) and oxygen (39 wt%) content. The as-received biomass in comparison to pretreated biomass showed heating values of 16.2–17.3 MJ/kg. Zhou et al. reported the dry basis analysis of as-received rice husk [46]. The sample showed carbon and oxygen content of 48.5 and 40.4 wt%, respectively, with the empirical formula of $CH_{2.4}N_{0.04}O_{0.6}$. Hence, it can be narrated that the rice husk sample usually contains carbon content (40–50 wt%), hydrogen content (4–5 wt%), and oxygen content (40–50 wt%) with the approximately heating value of 16–17 MJ/kg. Other studies showed the similar trend [11, 23, 47]. Other agricultural residues valorization presented similar observations [48–51]. In terms of proximate analysis, the sample showed 12 wt% of ash content with the volatile matter of 75 wt%. Jeon et al. presented catalytic pyrolysis of rice husk study over mesoporous catalyst [52]. They reported that the rice husk possessed 12 wt% of ash content. Loy et al. used rice husk ash as a catalyst for catalytic upgrading of rice husk pyrolysis vapor using thermogravimetric modeling [53]. As per their results, it was observed that biomass with low moisture content, higher ash content could be utilized for pyrolysis and gasification processes.

To reveal the crystallinity of synthesized zeolites, X-ray diffraction patterns are depicted in Fig. 2. First, the main peaks observed in the SAPO-34 zeolites at the theta of 9.24°, 12.7°, 20.9°, 25.8°, and 31.4°. These peaks are attributed to confirm the presence of the boehmite. The crystallization process induced the chabazite structure in the zeolite. On the other hand, layered MCM-22 and its delaminated counterpart zeolites ITQ-2 showed matching peaks and intensities

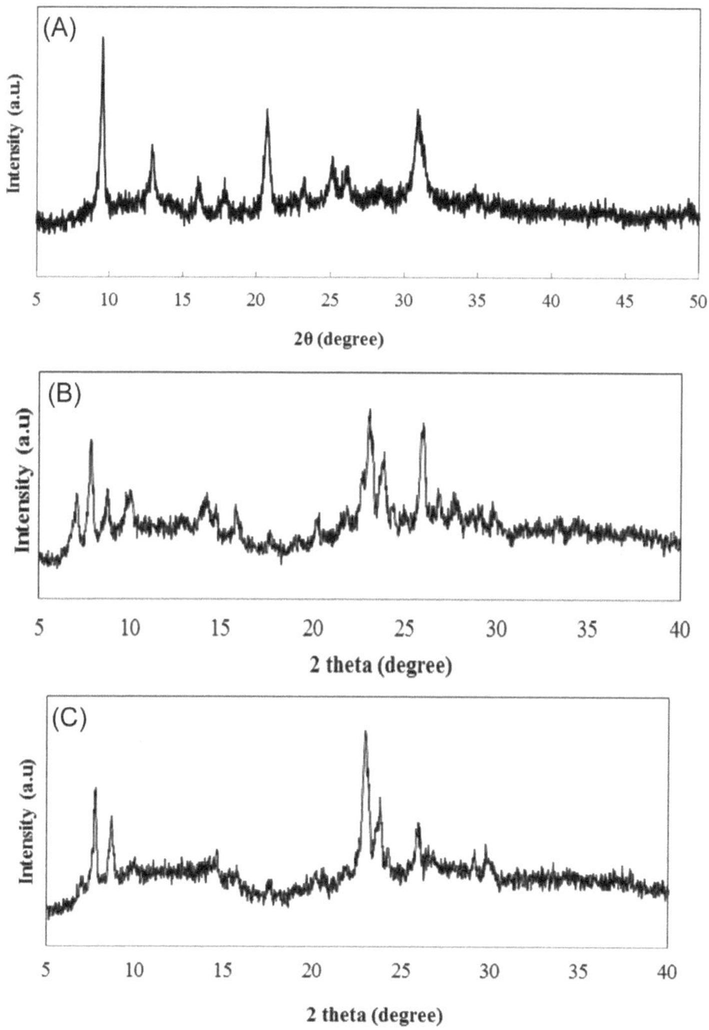

FIG. 2 XRD spectrum of zeolites (A) SAPO-34, (B) MCM-22, and (C) ITQ-2.

which were already presented in the previous literature [54]. This shows a typical structure of the MWW zeolite category. MCM-22 zeolites confirmed the presence of some short peaks with low crystallinity, while ITQ-2 zeolites depicted sharp peaks at 7.5° and 23.5°. MCM-22 and ITQ-2 solid acid catalyst are tested in aqueous-phase dehydration of xylose to the furfural [55]. They also reported that the delaminated structure showed a significant reduction in the peaks over a long range of patterns. Another recent study showed the detailed characterization of MCM-22, MCM-36, and ITQ-2 solid acid catalysts for the dehydration of methanol and ethanol [56]. Another recent study reported the seed-assisted synthesis route for MCM-22 zeolites [57]. All these studies showed matching peaks in comparison to our results.

To identify the surface morphology of the synthesized zeolites, scanning electron microscopy analysis was performed and depicted in Fig. 3A–C. As per the analysis, SAPO-34 zeolites showed cubic crystals in shape. It is also observed that there is no amorphous particle or structure is detected which reflects its high crystalline structure. Some past studies showed similar observations for the morphological analysis of the SAPO-34 zeolites [58, 59]. On the other hand, MCM-22 zeolites showed a layered structure. SEM analysis also

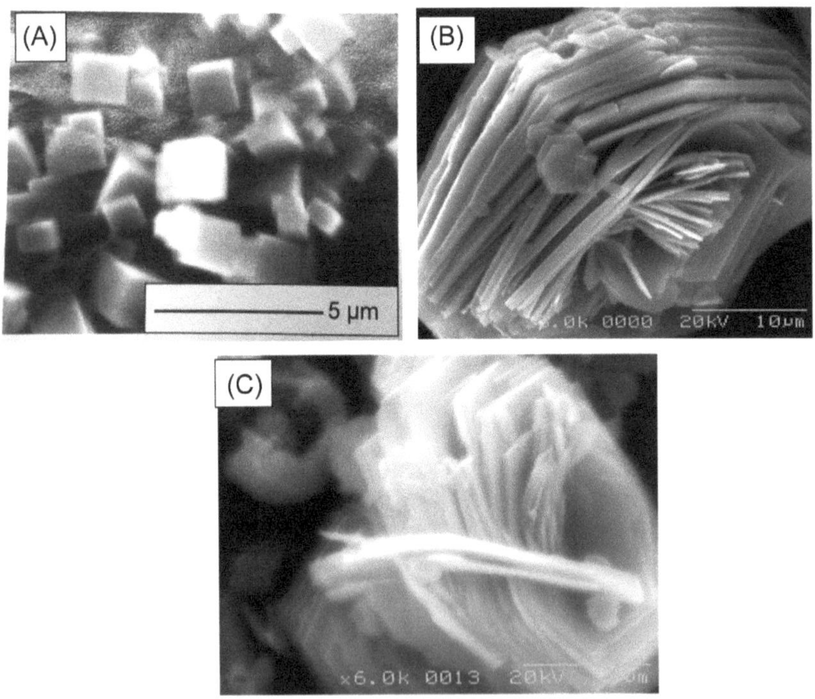

FIG. 3 Morphological analysis of synthesized zeolites; (A) SAPO-34, (B) MCM-22, and (C) ITQ-2.

confirmed the particle diameter with a range of 6–7 μm (Fig. 3B). Lastly, its delaminated counter particles showed irregular, nonuniform, and thinner sheet structure are existing as thinner sheets than MCM-22. Upon swelling, this layered structure was destroyed into irregular sheets. These sheets provide more accessible area as compared with layered sheets. These results are comparable with the existing literature [60].

Table 2 shows the textural properties and active site information. In comparison to other microporous zeolites, ITQ-2 showed a more specific and external surface area in comparison to MCM-22 and SAPO-34. SAPO-34 is ranked in a small-pore zeolite. This small-pore zeolite possessed a ring size of 8 Å with pore mouth of 4.3 Å. SAPOS-34 showed a larger surface area than layered zeolites. In comparison to SAPO-34, ferrite zeolite is also ranked as small-pore zeolite with the surface area of 302 m^2/g and pore size of 4.8 and 4.2 and 5.4 Å. On the other hand, layered zeolites showed a thin plate structure with pore structure of 10 Å. This structure is equipped with twin internal cages structure of 5.5 × 5.1 Å. In comparison to layered structure, ITQ-2 showed irregular layered structure. This provides an opportunity to access the internal as well as the external surface of the catalyst. This facility accumulated into the enhanced BET surface of the catalyst, hence ITQ-2 has a higher surface area (m^2/g) than MCM-22. Few studies claim that this nature is attributed due to exfoliation of the sheets [61, 62]. Another study reported the presence of shorter pores on the external surface in comparison to MCM-22. The thin structure of ITQ-2 provides external surface area and shorter pores than MCM-22 [40]. ZSM-5 and mordenite zeolites ranked low external surface area with 10 MR and 12 MR pore shape and 5.5 and 7 Å pore size, respectively. For measuring a total number of active sites,

TABLE 2 Properties of the synthesized and commercial zeolites.

Zeolites	BET surface area (m^2/g)	External surface area (m^2/g)	Pore shape	Pore size (Å)	Micropore volume (cm^3/g)	Active sites (mmol/g)
SAPO-34	537	40	8 MR	4.3	0.19	5087
Ferrrite	302	53	8 MR	4.8	0.12	1578
MCM-22	326	100	10 MR	5.1	0.12	724
ITQ-2	598	442	10 MR, 12 MR	5.5 and 7.1	0.07	5082
ZSM-5	390	141	10 MR	5.5	0.12	6024
Mordenite	412	50	12 MR	7	0.18	100

NH_3-TPD experiments were conducted. The available acid sites decrease in the following rank: ZSM-5 > SAPO-34 > ITQ-2 > ferrite > MCM-22 > mordenite.

3.2 Pyrolysis products yield

Fig. 4 shows the yield (wt%) of the pyrolysis products from the no-catalyst and catalytic upgrading process. The bio-oil is further divided into water and organics oil. In the absence of catalyst, the bio-oil (aqueous + organics) showed the yield of 15 and 20 wt%, respectively. In comparison to the no-catalytic system, the yield of the bio-oil was found to be decreased for catalytic upgrading process. This might be due to the secondary cracking of the volatiles over zeolites [63, 64]. Xu et al. reported the activity of MFI nanosheet zeolite during the upgrading of biomass pyrolysis vapors [65]. The catalytic activity was enhanced by generating mesoporosity on the catalytic surface. This mesoporosity increased the aromatic and olefins production in comparison to conventional ZSM-5 zeolites. This will prompt the upgrading process and reduced the coking tendency. Kumar et al. investigated the catalytic upgrading process by employing Ni and Cu-supported zeolites using in situ and ex situ modes [66]. Through this combined approach, a high degree of deoxygenation was observed which favored higher aromatic production. Mohamed et al. investigated the catalytic biomass pyrolysis using catalyst mixtures in a microwave system [67]. However, author observed a significant reduction in catalytic activity due to oxygenated coke formation on the catalyst surface. Jia et al. investigated the catalytic upgrading performance in a micro-fluidized-bed reactor at 500°C with various catalyst to biomass ratio [68]. The coke-trapped mechanism inside and outside the catalyst pore was revealed using microporous and hierarchical zeolites. For catalytic pyrolysis, the water yield increases in the following order:

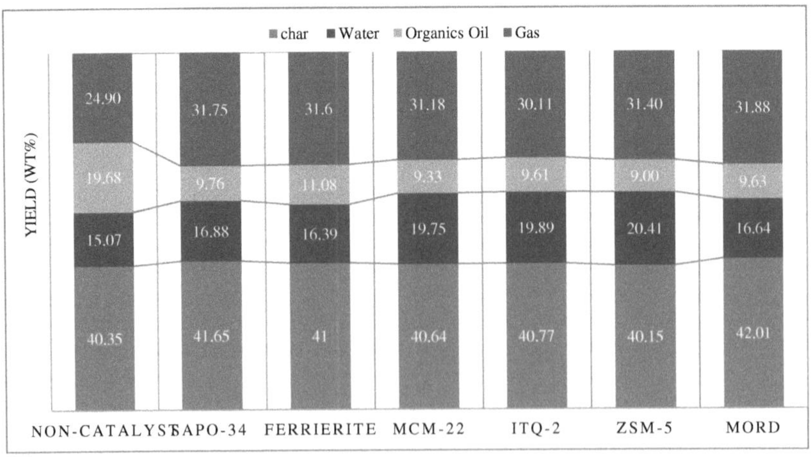

FIG. 4 Noncatalytic and catalytic pyrolysis products yield (wt%).

ferrite > mordenite > SAPO-34 > MCM-22 > ITQ-2 > ZSM-5. As per results, the organics oil yield increases in the following order: ZSM-5 > MCM-22 > ITQ-2 > mordenite > SAPO-34 > ferrite. The small-pore zeolites showed enough accessible active sites to crack much-oxygenated species and produce more organics. Medium-pore zeolites such as MCM-22, ITQ-2, and ZSM-5 produced more water as compared with large- and small-pore zeolites. Dehydration, decarbonylation, and decarboxylation might be the responsible mechanism for deoxygenation [63]. For gas yield, ZSM-5 showed higher gas amount as compared with other zeolites due to its strong acidity. Hertzog et al. studied the fast pyrolysis process over mesoporous and hierarchical zeolites [69]. As per their results, deoxygenation and aromatization were the two main modifications caused by the zeolites in heavy oil. The gas yield increased as per incremental catalyst to biomass ratio. Another study by Hernando et al. reported similar observation and identified the secondary cracking for gas yield using ZSM-5 and beta zeolites through modification of Mg and Zn [70]. The incorporation of the Mg and Zn produced more active sites on external surface of ZSM-5 and beta zeolites. This activity cracks the large molecules and produced aromatic hydrocarbons. A very recent study investigated the zeolite-encapsulated metal catalyst for biomass conversion process [71]. This study identifies the secret of employing metal encapsulation in terms of controlling shape selectivity and mass transfer of the species. To maintain the catalyst stability and catalytic active sites, metal encapsulation into the zeolite channel provides a viable option to accelerate the activity and reduce coking tendency.

3.3 Gas analysis and DOD

The results of noncatalytic and catalytic pyrolysis gas composition are presented in Fig. 5. The gas yield was observed higher in catalytic experiments as compared to no-catalyst process. In gas analysis, four main gaseous components are identified such as hydrogen, methane, carbon dioxide, and carbon monoxide. With the addition of zeolites, there is significant increase in the formation of CO and CO_2 was observed. This formation is attributed to the decarbonylation and decarboxylation mechanism reactions. Firstly, SAPO-34 produced more CO and CO_2 as compared with ferrite zeolites. This might be due to the internal pore structure, higher surface area, and more available active sites. All these factors might be responsible for the decarbonylation and decarboxylation reactions. Chen et al. studied the catalytic conversion of cellulose using SAPO zeolites [72]. As per their results, the conversion was increased, and more gaseous components were observed. Secondly, in comparison of medium-pore zeolites, ZSM-5 appeared the most viable catalyst for promoting decarbonylation and decarboxylation reactions. However, in comparison of MCM-22 and ITQ-2, irregular and nonuniform layered sheets (ITQ-2 zeolites) provide accessible active sites for pyrolysis vapors cracking and showed higher deoxygenation, Therefore, ITQ-2 showed more deoxygenation by following

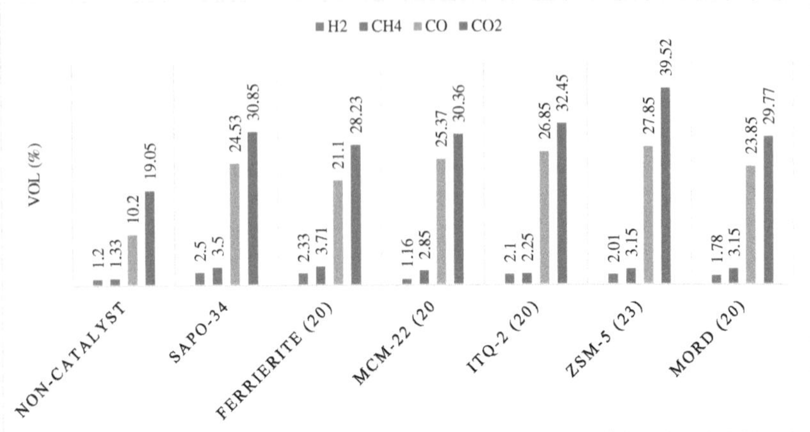

FIG. 5 Gas compositions.

dehydration, decarbonylation, and decarboxylation mechanisms to produce high-quality bio-oil. Che et al. studied the catalytic upgrading process of wood sawdust using metal oxide and ZSM-5 [73]. Shape-selective zeolites such as ZSM-5 showed higher aromatics formation because of unique pore structure and properties. Du et al. investigated the performance of ZSM-5 as support for catalytic biomass pyrolysis process [74]. The study examined the performance in terms of liquid yield and aromatics production. By utilization ZSM-5, the liquid yield was decreased, and aromatic hydrocarbon content was increased. The Lewis acid active sites are responsible for this mechanism. One of the major hindrances in effective utilization of ZSM-5 in the biomass conversion process was his tendency to produce coke and challenges associated with its regeneration. Heracleous et al. reported the characteristics nature of deactivated and regenerated ZSM-5 for biomass conversion process [75]. Coking was the main reason to deactivate the catalyst in cracking process and reduces the deoxygenation compared with MCM-22 and small-pore zeolites. Our results are in line with the existing studies. As per agreement, ZSM-5 was the most suitable catalyst for higher decarbonylation and decarboxylation reactions.

One of the main objectives of this research work was to examine the performance of various zeolites in terms of removal of oxygen content from the produced bio-oil. To understand this process, an indicator was established by a few researchers, called the degree of deoxygenation [34, 76, 77]. The degree of deoxygenation of noncatalytic and catalytic pyrolysis is shown in Fig. 6. The oxygen removal is less in noncatalytic pyrolysis as compared with catalytic pyrolysis. The catalyst helped reasonably to reject more oxygen from the original biomass. The DOD increases in the following order: ferrite > SAPO-34 > mordenite > MCM-22 > ITQ-2 > ZSM-5. Among all zeolites, ZSM-5

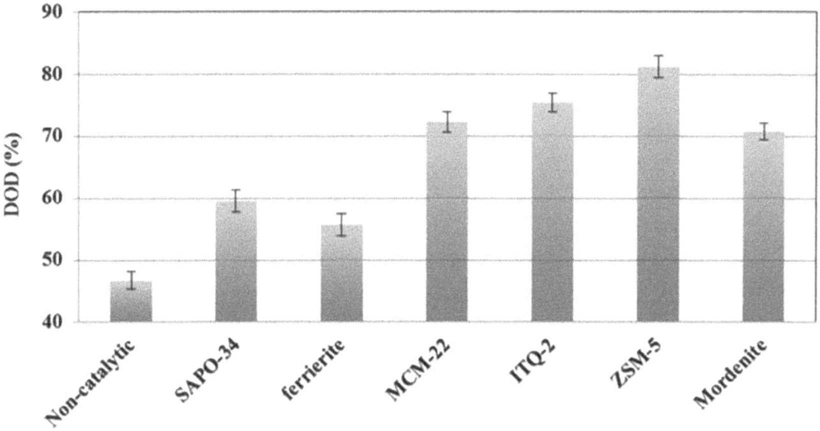

FIG. 6 Degree of deoxygenation of bio-oils.

appeared as the best suitable catalysts for oxygen rejection in the biomass pyrolysis. Saraeian et al. published the impact of deoxygenation of biomass pyrolysis molecules [78]. This tool will help to identify the performance in terms of removal of oxygen content from the raw material toward the product. This indicator also provides a concept to develop a rational catalyst for biomass conversion processes. Kalogiannis et al. presented the pilot-scale study of biomass pyrolysis process using acidic catalysts [79]. According to their results, dehydration was the main reason to promote deoxygenation following by decarbonylation and decarboxylation.

3.4 Pore size and shape impact

Fig. 7 compares the product distribution of noncatalytic and catalytic pyrolysis bio-oil using zeolites. As per GC-MS analysis, the produced bio-oil consisted of oxygenated species, aromatic, and polyaromatic hydrocarbons (PAHs). Using small-pore zeolites, moderate catalytic activity was observed and generated oxygenated species such as acids, ketones, furans, char, CO, and CO_2. On the other hand, medium-pore zeolites possessed much stronger acidity to crack these oxygenated species and produced phenols, aromatic, and polyaromatic hydrocarbons. The large-pore zeolites showed slight oxygenated species and reasonable aromatic hydrocarbons. From these results, it can be narrated that small pore could not facilitate the diffusion and negatively impact aromatic formation. Fig. 8 provides comparison information among small-, medium-, and large-pore zeolites for the formation of oxygenated species (area%) and aromatic and PAHs. SAPO-34 being a small pore zeolite with chabazite structure lacked strong Bronsted active sites, therefore, very low aromatics formation was observed. Among medium- and small-pore zeolites, ZSM-5 showed less

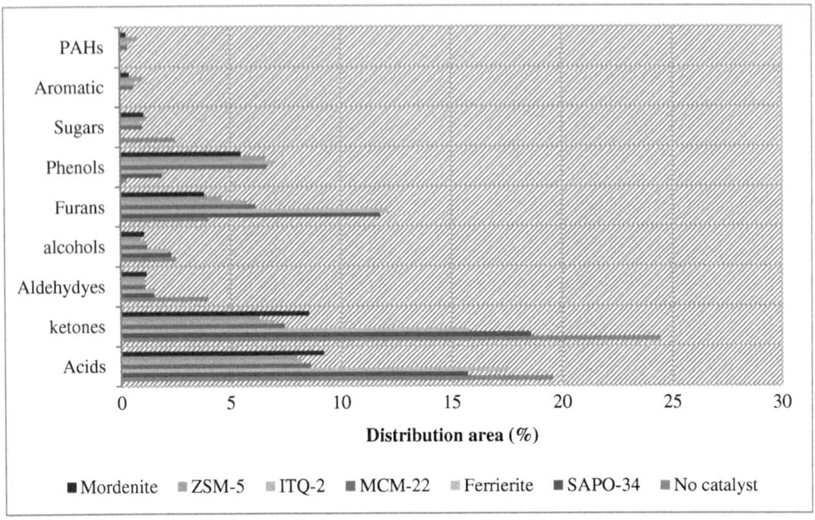

FIG. 7 Bio-oil composition from noncatalytic and catalytic pyrolysis process.

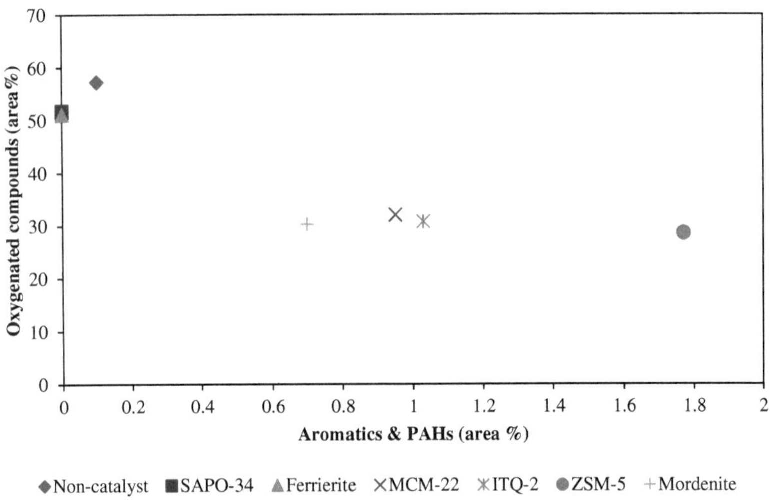

FIG. 8 Oxygenated species and aromatic formation for various zeolites catalyst.

oxygenated species with more aromatics following by ITQ-2 zeolite. Few recent studies also mentioned that ZSM-5 showed promising results in terms of biomass pyrolysis vapors deoxygenation and produced much aromatics [80–82]. However, the disadvantage of employing ZSM-5 in biomass pyrolysis process that there is coke formation of the catalyst pore mouth which decreases its cracking ability [75]. Biomass conversion over zeolites showed shape-selective routes for aromatics production. The diffusion across the pore mouth, length, internal cage with mass transfer hindrance must be identified to develop a rational catalyst. The zeolites showed to have a higher specific surface area, the possibility to generate more external active sites, and capability to crack molecules with less coking tendency is favored for large aromatics production.

Fig. 8 shows the performance of zeolites for oxygenated species and aromatics formation based on their pore size and shape. Catalysts play an important role to reduce the formation of undesired products and to improve aromatics hydrocarbon in the bio-oil during the pyrolysis process. As per our results, medium-pore zeolites and delaminated structure are appeared up as effective cracking catalyst because of their structure and strong acidity. The author already described the kinetics of catalysis pyrolysis process [47]. Addition of catalyst reduces the activation energy in the first region and then increases it in the second region for all chemical reaction orders. This study also implies that plugging of zeolites pore mouth is observed due to limitation in diffusion and catalyst decay. The results are well in agreement with the published studies [1, 83–85].

4 Conclusion

This study provides state-of-the-art information about using pore size and shape-selective zeolites in the biomass pyrolysis process. Microporous zeolites (synthesized and commercial) have unique structures and acidity, which heavily impact on deoxygenation of biomass pyrolysis vapors. For oxygenated species, small-pore zeolites showed much better results in terms of their formation but could not successfully crack them into aromatics. Layered and its delaminated counterpart zeolites showed the higher formation of aromatics in comparison to small-pore zeolites and with a higher degree of deoxygenation. Due to appropriate pore structure and acidity, ZSM-5 was observed as the best catalyst for effective deoxygenation and higher aromatics formation. From the results, it can be narrated that large micropore diameter, large surface, and external surface area with higher acidity are the promising features required for a catalyst to be used effectively in the biomass pyrolysis process. Pore size and shape-selective microporous zeolites can adjust the intermediate species in their pore, provide access to external surface area, and elucidate well the deoxygenation mechanism.

Acknowledgment

Authors would like to acknowledge the National University of Sciences and Technology for the support.

References

[1] Bhoi PR, Ouedraogo AS, Soloiu V, Quirino R. Recent advances on catalysts for improving hydrocarbon compounds in bio-oil of biomass catalytic pyrolysis. Renew Sust Energ Rev 2020;121:109676.

[2] Aguilera RF, Eggert RG, Gustavo Lagos CC, Tilton JE. Depletion and the future availability of petroleum resources. Energy J 2009;30(1).

[3] Keller F, Lee RP, Meyer B. Life cycle assessment of global warming potential, resource depletion and acidification potential of fossil, renewable and secondary feedstock for olefin production in Germany. J Clean Prod 2020;250:119484.

[4] Lap T, Benders R, Köberle A, van der Hilst F, Nogueira L, Szklo A, Schaeffer R, Faaij A. Pathways for a Brazilian biobased economy: towards optimal utilization of biomass. Biofuels Bioprod Biorefin 2019;13(3):673–89.

[5] Nda-Umar UI, Ramli I, Taufiq-Yap YH, Muhamad EN. An overview of recent research in the conversion of glycerol into biofuels, fuel additives and other bio-based chemicals. Catalysts 2019;9(1):15.

[6] Halder P, Azad AK. Recent trends and challenges of algal biofuel conversion technologies. In: Advanced Biofuels. Woodhead Publishing; 2019. p. 167–79.

[7] Naqvi SR, Ali I, Nasir S, Taqvi SAA, Atabani AE, Chen WH. Assessment of agro-industrial residues for bioenergy potential by investigating thermo-kinetic behavior in a slow pyrolysis process. Fuel 2020;278:118259.

[8] Malik S, Fatima F, Imran A, Chuah LF, Klemeš JJ, Khaliq IH, Asif S, Aslam M, Jamil F, Durrani AK, Akbar MM. Improved project control for sustainable development of construction sector to reduce environment risks. J Clean Prod 2019;240:118214.

[9] Naqvi SR, Prabhakara HM, Bramer EA, Dierkes W, Akkerman R, Brem G. A critical review on recycling of end-of-life carbon fibre/glass fibre reinforced composites waste using pyrolysis towards a circular economy. Resour Conserv Recycl 2018;136:118–29.

[10] Mutsengerere S, Chihobo CH, Musademba D, Nhapi I. A review of operating parameters affecting bio-oil yield in microwave pyrolysis of lignocellulosic biomass. Renew Sust Energ Rev 2019;104:328–36.

[11] Naqvi SR, Jamshaid S, Naqvi M, Farooq W, Niazi MBK, Aman Z, Zubair M, Ali M, Shahbaz M, Inayat A, Afzal W. Potential of biomass for bioenergy in Pakistan based on present case and future perspectives. Renew Sust Energ Rev 2018;81:1247–58.

[12] Bentsen NS. Biomass for biorefineries: availability and costs. In: Biorefinery. Cham: Springer; 2019. p. 37–48.

[13] Ayoub M, Wei WJ, Ahmad M, Mathialagan R, Farrukh S, Danish M, Ullah S, Naqvi SR. Glycerol conversion to diglycerol via etherification under microwave irradiation. In: Triglycerides and Cholesterol. IntechOpen; 2020.

[14] Khan Z, Yusup S, Aslam M, Inayat A, Shahbaz M, Naqvi SR, Farooq R, Watson I. NO and SO2 emissions in palm kernel shell catalytic steam gasification with in-situ CO2 adsorption for hydrogen production in a pilot-scale fluidized bed gasification system. J Clean Prod 2019;236:117636.

[15] Azad AK, Adhikari J, Halder P, Rasul MG, Hassan N, Khan MM, Naqvi SR, Viswanathan K. Performance, emission and combustion characteristics of a diesel engine powered by Macadamia and grapeseed biodiesels. Energies 2020;13(11):2748.

[16] Jamil U, Khoja AH, Liaquat R, Naqvi SR, Omar WNNW, Amin NAS. Copper and calcium-based metal organic framework (MOF) catalyst for biodiesel production from waste cooking oil: a process optimization study. Energy Convers Manag 2020;215:112934.

[17] Wang S, Viswanathan K, Esakkimuthu S, Azad K. Experimental investigation of high alcohol low viscous renewable fuel in DI diesel engine. Environ Sci Pollut Res Int 2020. https://doi.org/10.1007/s11356-020-08298-y.
[18] Shakeel K, Javaid M, Muazzam Y, Naqvi SR, Taqvi SAA, Uddin F, Mehran MT, Sikander U, Niazi M. Performance comparison of industrially produced formaldehyde using two different catalysts. Processes 2020;8(5):571.
[19] Adams P, Bridgwater T, Lea-Langton A, Ross A, Watson I. Biomass conversion technologies. In: Greenhouse Gas Balances of Bioenergy Systems. Academic Press; 2018. p. 107–39.
[20] Naqvi SR, Uemura Y, Yusup SB. Catalytic pyrolysis of paddy husk in a drop type pyrolyzer for bio-oil production: the role of temperature and catalyst. J Anal Appl Pyrolysis 2014; 106:57–62.
[21] Naqvi SR, Uemura Y, Osman N, Yusup S. Production and evaluation of physicochemical characteristics of paddy husk bio-char for its C sequestration applications. Bioenergy Res 2015; 8(4):1800–9.
[22] Ubando AT, Chen WH, Tan RR, Naqvi SR. Optimal integration of a biomass-based polygeneration system in an iron production plant for negative carbon emissions. Int J Energy Res 2019. https://doi.org/10.1002/er.4902.
[23] Naqvi SR, Uemura Y, Osman NB, Yusup S, Nuruddin MF. Physiochemical properties of pyrolysis oil derived from fast pyrolysis of wet and dried rice husk in a free fall reactor. In: Applied Mechanics and Materials, Vol. 625. Trans Tech Publications Ltd; 2014. p. 604–7.
[24] Shahbaz M, Al-Ansari T, Aslam M, Khan Z, Inayat A, Athar M, et al. A state of the art review on biomass processing and conversion technologies to produce hydrogen and its recovery via membrane separation. Int J Hydrog Energy 2020;45(30):15166–95.
[25] Uemura Y, Tran NT, Naqvi SR, Nishiyama N. Nano-catalysts for upgrading bio-oil: catalytic decarboxylation and hydrodeoxygenation. In: AIP Conference Proceedings, Vol. 1877, No. 1. AIP Publishing LLC; 2017. p. 020002.
[26] Bokhari A, Chuah LF, Michelle LZY, Asif S, Shahbaz M, Akbar MM, Inayat A, Jamil F, Naqvi SR, Yusup S. Microwave enhanced catalytic conversion of canola-based methyl ester: optimization and parametric study. In: Advanced Biofuels. Woodhead Publishing; 2019. p. 153–66.
[27] Manzoor N, Sadiq M, Naqvi M, Sikandar U, Naqvi SR. Experimental study of CO2 conversion into methanol by synthesized photocatalyst (ZnFe2O4/TiO2) using visible light as an energy source. Catalysts 2020;10(2):163.
[28] Naqvi SR, Naqvi M, Noor T, Hussain A, Iqbal N, Uemura Y, Nishiyama N. Catalytic pyrolysis of Botryococcus braunii (microalgae) over layered and delaminated zeolites for aromatic hydrocarbon production. Energy Procedia 2017;142:381–5.
[29] Naqvi SR, Bibi A, Naqvi M, Noor T, Nizami AS, Rehan M, Ayoub M. New trends in improving gasoline quality and octane through naphtha isomerization: a short review. Appl Petrochem Res 2018;8(3):131–9.
[30] Rahman MM, Liu R, Cai J. Catalytic fast pyrolysis of biomass over zeolites for high quality bio-oil–a review. Fuel Process Technol 2018;180:32–46.
[31] Imran A, Bramer EA, Seshan K, Brem G. High quality bio-oil from catalytic flash pyrolysis of lignocellulosic biomass over alumina-supported sodium carbonate. Fuel Process Technol 2014;127:72–9.
[32] Kabir G, Hameed BH. Recent progress on catalytic pyrolysis of lignocellulosic biomass to high-grade bio-oil and bio-chemicals. Renew Sust Energ Rev 2017;70:945–67.
[33] Iliopoulou EF, Triantafyllidis KS, Lappas AA. Overview of catalytic upgrading of biomass pyrolysis vapors toward the production of fuels and high-value chemicals. Wiley Interdiscipl Rev: Energy Environ 2019;8(1):e322.

[34] Nolte MW, Shanks BH. A perspective on catalytic strategies for deoxygenation in biomass pyrolysis. Energ Technol 2017;5(1):7–18.
[35] Liu R, Sarker M, Rahman MM, Li C, Chai M, Cotillon R, Scott NR. Multi-scale complexities of solid acid catalysts in the catalytic fast pyrolysis of biomass for bio-oil production—a review. Prog Energy Combust Sci 2020;80:100852.
[36] Lok CM, Van Doorn J, Almansa GA. Promoted ZSM-5 catalysts for the production of bio-aromatics, a review. Renew Sust Energ Rev 2019;113:109248.
[37] Liu R, Rahman MM, Sarker M, Chai M, Li C, Cai J. A review on the catalytic pyrolysis of biomass for the bio-oil production with ZSM-5: focus on structure. Fuel Process Technol 2020;199:106301.
[38] Naqvi SR, Naqvi M. Catalytic fast pyrolysis of rice husk: influence of commercial and synthesized microporous zeolites on deoxygenation of biomass pyrolysis vapors. Int J Energy Res 2018;42(3):1352–62.
[39] Gao N, Jia X, Gao G, Ma Z, Quan C, Naqvi SR. Modeling and simulation of coupled pyrolysis and gasification of oily sludge in a rotary kiln. Fuel 2020;279:118152.
[40] Naqvi SR, Uemura Y, Yusup S, Sugiura Y, Nishiyama N. In situ catalytic fast pyrolysis of paddy husk pyrolysis vapors over MCM-22 and ITQ-2 zeolites. J Anal Appl Pyrolysis 2015;114:32–9.
[41] Hirota Y, Murata K, Tanaka S, Nishiyama N, Egashira Y, Ueyama K. Dry gel conversion synthesis of SAPO-34 nanocrystals. Mater Chem Phys 2010;123(2–3):507–9.
[42] Kubička D, Horáček J, Setnička M, Bulánek R, Zukal A, Kubičková I. Effect of support-active phase interactions on the catalyst activity and selectivity in deoxygenation of triglycerides. Appl Catal B Environ 2014;145:101–7.
[43] Choi IH, Lee JS, Kim CU, Kim TW, Lee KY, Hwang KR. Production of bio-jet fuel range alkanes from catalytic deoxygenation of Jatropha fatty acids on a WOx/Pt/TiO2 catalyst. Fuel 2018;215:675–85.
[44] Bakar MSA, Titiloye JO. Catalytic pyrolysis of rice husk for bio-oil production. J Anal Appl Pyrolysis 2013;103:362–8.
[45] Zhang S, Zhang H, Liu X, Zhu S, Hu L, Zhang Q. Upgrading of bio-oil from catalytic pyrolysis of pretreated rice husk over Fe-modified ZSM-5 zeolite catalyst. Fuel Process Technol 2018;175:17–25.
[46] Zhou L, Yang H, Wu H, Wang M, Cheng D. Catalytic pyrolysis of rice husk by mixing with zinc oxide: characterization of bio-oil and its rheological behavior. Fuel Process Technol 2013;106:385–91.
[47] Naqvi SR, Uemura Y, Osman N, Yusup S. Kinetic study of the catalytic pyrolysis of paddy husk by use of thermogravimetric data and the Coats–Redfern model. Res Chem Intermed 2015;41(12):9743–55.
[48] Shahbaz M, Yusup S, Inayat A, Patrick DO, Ammar M, Pratama A. Cleaner production of hydrogen and syngas from catalytic steam palm kernel shell gasification using CaO sorbent and coal bottom ash as a catalyst. Energy Fuel 2017;31(12):13824–33.
[49] Shahbaz M, Yusup S, Pratama A, Inayat A, Patrick DO, Ammar M. Parametric study and optimization of methane production in biomass gasification in the presence of coal bottom ash. Procedia Eng 2016;148:409–16.
[50] AlNouss A, Parthasarathy P, Shahbaz M, Al-Ansari T, Mackey H, McKay G. Techno-economic and sensitivity analysis of coconut coir pith-biomass gasification using ASPEN PLUS. Appl Energy 2020;261:114350.
[51] How BS, Ngan SL, Hong BH, Lam HL, Ng WPQ, Yusup S, Ghani WAWAK, Kansha Y, Chan YH, Cheah KW, Shahbaz M. An outlook of Malaysian biomass industry commercialisation: perspectives and challenges. Renew Sust Energ Rev 2019;113:109277.

[52] Jeon MJ, Kim SS, Jeon JK, Park SH, Kim JM, Sohn JM, Lee SH, Park YK. Catalytic pyrolysis of waste rice husk over mesoporous materials. Nanoscale Res Lett 2012;7(1):1–5.

[53] Loy ACM, Gan DKW, Yusup S, Chin BLF, Lam MK, Shahbaz M, Unrean P, Acda MN, Rianawati E. Thermogravimetric kinetic modelling of in-situ catalytic pyrolytic conversion of rice husk to bioenergy using rice hull ash catalyst. Bioresour Technol 2018;261:213–22.

[54] Corma A, Diaz U, Fornés V, Guil JM, Martinez-Triguero J, Creyghton EJ. Characterization and catalytic activity of MCM-22 and MCM-56 compared with ITQ-2. J Catal 2000; 191(1):218–24.

[55] Antunes MM, Lima S, Fernandes A, Pillinger M, Ribeiro MF, Valente AA. Aqueous-phase dehydration of xylose to furfural in the presence of MCM-22 and ITQ-2 solid acid catalysts. Appl Catal A Gen 2012;417:243–52.

[56] Marosz M, Samojeden B, Kowalczyk A, Rutkowska M, Motak M, Díaz U, Palomares AE, Chmielarz L. MCM-22, MCM-36, and ITQ-2 zeolites with different Si/Al molar ratios as effective catalysts of methanol and ethanol dehydration. Materials 2020;13(10):2399.

[57] Rodrigues MV, Vieira LH, Campos GP, Martins L. Effect of different seed sources on the hydrothermal crystallization of MCM-22 zeolite catalysts. Cryst Eng Comm 2018;20(25): 3467–75.

[58] Sun Q, Wang N, Bai R, Chen X, Yu J. Seeding induced nano-sized hierarchical SAPO-34 zeolites: cost-effective synthesis and superior MTO performance. J Mater Chem A 2016;4 (39):14978–82.

[59] Guo G, Sun Q, Wang N, Bai R, Yu J. Cost-effective synthesis of hierarchical SAPO-34 zeolites with abundant intracrystalline mesopores and excellent MTO performance. Chem Commun 2018;54(30):3697–700.

[60] Juttu GG, Lobo RF. Characterization and catalytic properties of MCM-56 and MCM-22 zeolites. Microporous Mesoporous Mater 2000;40(1–3):9–23.

[61] Hernando H, Fermoso J, Ochoa-Hernández C, Opanasenko M, Pizarro P, Coronado JM, Čejka J, Serrano DP. Performance of MCM-22 zeolite for the catalytic fast-pyrolysis of acid-washed wheat straw. Catal Today 2018;304:30–8.

[62] Min HK, Park MB, Hong SB. Methanol-to-olefin conversion over H-MCM-22 and H-ITQ-2 zeolites. J Catal 2010;271(2):186–94.

[63] Khan W, Jia X, Wu Z, Choi J, Yip AC. Incorporating hierarchy into conventional zeolites for catalytic biomass conversions: a review. Catalysts 2019;9(2):127.

[64] Palizdar A, Sadrameli SM. Catalytic upgrading of biomass pyrolysis oil over tailored hierarchical MFI zeolite: effect of porosity enhancement and porosity-acidity interaction on deoxygenation reactions. Renew Energy 2020;148:674–88.

[65] Xu M, Mukarakate C, Iisa K, Budhi S, Menart M, Davidson M, Robichaud DJ, Nimlos MR, Trewyn BG, Richards RM. Deactivation of multilayered MFI nanosheet zeolite during upgrading of biomass pyrolysis vapors. ACS Sustain Chem Eng 2017;5(6):5477–84.

[66] Kumar R, Strezov V, Lovell E, Kan T, Weldekidan H, He J, Jahan S, Dastjerdi B, Scott J. Enhanced bio-oil deoxygenation activity by Cu/zeolite and Ni/zeolite catalysts in combined in-situ and ex-situ biomass pyrolysis. J Anal Appl Pyrolysis 2019;140:148–60.

[67] Mohamed BA, Ellis N, Kim CS, Bi X. Microwave-assisted catalytic biomass pyrolysis: effects of catalyst mixtures. Appl Catal B Environ 2019;253:226–34.

[68] Jia LY, Raad M, Hamieh S, Toufaily J, Hamieh T, Bettahar MM, Mauviel G, Tarrighi M, Pinard L, Dufour A. Catalytic fast pyrolysis of biomass: superior selectivity of hierarchical zeolites to aromatics. Green Chem 2017;19(22):5442–59.

[69] Hertzog J, Carré V, Jia L, Mackay CL, Pinard L, Dufour A, Mašek O, Aubriet F. Catalytic fast pyrolysis of biomass over microporous and hierarchical zeolites: characterization of heavy products. ACS Sustain Chem Eng 2018;6(4):4717–28.

[70] Hernando H, Moreno I, Fermoso J, Ochoa-Hernández C, Pizarro P, Coronado JM, Čejka J, Serrano DP. Biomass catalytic fast pyrolysis over hierarchical ZSM-5 and Beta zeolites modified with Mg and Zn oxides. Biomass Convers Biorefin 2017;7(3):289–304.
[71] Limlamthong M, Yip AC. Recent advances in zeolite-encapsulated metal catalysts: a suitable catalyst design for catalytic biomass conversion. Bioresour Technol 2020;297:122488.
[72] Chen X, Chen Y, Chen Z, Zhu D, Yang H, Liu P, Li T, Chen H. Catalytic fast pyrolysis of cellulose to produce furan compounds with SAPO type catalysts. J Anal Appl Pyrolysis 2018;129:53–60.
[73] Che Q, Yang M, Wang X, Chen X, Chen W, Yang Q, Yang H, Chen H. Aromatics production with metal oxides and ZSM-5 as catalysts in catalytic pyrolysis of wood sawdust. Fuel Process Technol 2019;188:146–52.
[74] Du S, Gamliel DP, Valla JA, Bollas GM. The effect of ZSM-5 catalyst support in catalytic pyrolysis of biomass and compounds abundant in pyrolysis bio-oils. J Anal Appl Pyrolysis 2016;122:7–12.
[75] Heracleous E, Pachatouridou E, Hernández-Giménez AM, Hernando H, Fakin T, Paioni AL, Baldus M, Serrano DP, Bruijnincx PCA, Weckhuysen BM, Lappas AA. Characterization of deactivated and regenerated zeolite ZSM-5-based catalyst extrudates used in catalytic pyrolysis of biomass. J Catal 2019;380:108–22.
[76] Chen W, Li K, Xia M, Yang H, Chen Y, Chen X, Che Q, Chen H. Catalytic deoxygenation co-pyrolysis of bamboo wastes and microalgae with biochar catalyst. Energy 2018;157:472–82.
[77] Naqvi SR, Uemura Y, Yusup S, Sugiur Y, Nishiyama N, Naqvi M. The role of zeolite structure and acidity in catalytic deoxygenation of biomass pyrolysis vapors. Energy Procedia 2015;75:793–800.
[78] Saraeian A, Nolte MW, Shanks BH. Deoxygenation of biomass pyrolysis vapors: improving clarity on the fate of carbon. Renew Sust Energ Rev 2019;104:262–80.
[79] Kalogiannis KG, Stefanidis SD, Karakoulia SA, Triantafyllidis KS, Yiannoulakis H, Michailof C, Lappas AA. First pilot scale study of basic vs acidic catalysts in biomass pyrolysis: deoxygenation mechanisms and catalyst deactivation. Appl Catal B Environ 2018;238:346–57.
[80] Hernando H, Ochoa-Hernández C, Shamzhy M, Moreno I, Fermoso J, Pizarro P, Coronado JM, Čejka J, Serrano DP. The crucial role of clay binders in the performance of ZSM-5 based materials for biomass catalytic pyrolysis. Cat Sci Technol 2019;9(3):789–802.
[81] Yung MM, Starace AK, Griffin MB, Wells JD, Patalano RE, Smith KR, Schaidle JA. Restoring ZSM-5 performance for catalytic fast pyrolysis of biomass: effect of regeneration temperature. Catal Today 2019;323:76–85.
[82] Che Q, Yang M, Wang X, Yang Q, Williams LR, Yang H, Zou J, Zeng K, Zhu Y, Chen Y, Chen H. Influence of physicochemical properties of metal modified ZSM-5 catalyst on benzene, toluene and xylene production from biomass catalytic pyrolysis. Bioresour Technol 2019; 278:248–54.
[83] Soongprasit K, Vichaphund S, Sricharoenchaikul V, Atong D. Activity of fly ash-derived ZSM-5 and zeolite X on fast pyrolysis of Millettia (Pongamia) Pinnata waste. Waste Biomass Valoriz 2020;11(2):715–24.
[84] Hu C, Zhang H, Wu S, Xiao R. Molecular shape selectivity of HZSM-5 in catalytic conversion of biomass pyrolysis vapors: the effective pore size. Energy Convers Manag 2020;210:112678.
[85] Liao Y, Zhong R, d'Halluin M, Verboekend D, Sels BF. Aromatics production from lignocellulosic biomass: Shape selective dealkylation of lignin-derived phenolics over hierarchical ZSM-5. ACS Sustain Chem Eng 2020;8:8713–22.

Chapter 3

Advance strategies for tar elimination from biomass gasification techniques

Muddasser Inayat[a], Muhammad Shahbaz[b], Salman Raza Naqvi[c], and Shaharin A. Sulaiman[a]
[a]*Department of Mechanical Engineering, Universiti Teknologi PETRONAS, Seri Iskandar, Perak, Malaysia,* [b]*Division of Sustainable Development, College of Science and Engineering, Hamad Bin Khalifa University, Qatar Foundation, Doha, Qatar,* [c]*School of Chemical & Materials Engineering, National University of Sciences & Technology, Islamabad, Pakistan*

1 Introduction

1.1 Bioenergy

Energy is the main driving force for the development and economic growth of any country [1–3]. Currently, the primary source of the world's energy is still majorly covered by oil, gas, and coal. Although the conventional source of energy fulfils the major proportion of the world energy as shown in Fig. 1, overuse of these fuels causes several environmental issues, such as environmental pollution, global warming, and imbalance ecosystem [5–7]. Numerous researchers try to cop these issues via clean renewable energy sources. The well-known renewable energy sources are wind, solar, hydro, geothermal, and ocean energies. Among many renewable energy sources, biomass is cheap, carbon-neutral, and widely available source of energy via thermochemical conversion techniques, such as combustion, pyrolysis, and gasification [8–10].

The word biomass is used for organic materials, which are originally derived from living organisms. These materials are derivative of plants and animals, such as wood from forests, crops residual from the agricultural sector, seaweed, and organic, industrial, human, and animal wastes. Biomass is a general term used for the phytomass or plant biomass [11]. Biomass is one of the primary and reasonable sources of energy, particularly in rural areas, where it is abundantly and readily available [12]. Biomass is considered a renewable source of energy that has the highest potential to fulfill the energy requirements of the modern world for both the developed and developing countries. Currently, it

Bioenergy Resources and Technologies. https://doi.org/10.1016/B978-0-12-822525-7.00010-X
Copyright © 2021 Elsevier Inc. All rights reserved.

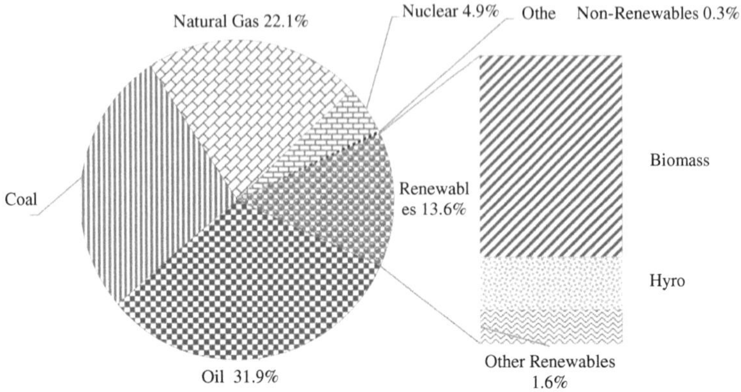

FIG. 1 World primary energy in 2018 [4].

contributes around 14% of the world's primary energy that makes it in the top four largest energy source after oil, gas, and coal [13].

1.2 Biomass thermochemical conversion techniques

The thermochemical platforms are available, such as combustion, pyrolysis, and gasification for biomass conversion into syngas [14,15]. Thermochemical conversion of biomass has some advantages like the use of multiple feedstock with a larger generation rate and no effect of climate condition due to operation at high temperature. Among all thermochemical conversion process, the use of gasification process for syngas production is advocated by the advantages like high efficiency, the limited formation of oxides of nitrogen and sulfur, high yield of gaseous than liquid products, reduction in the volume of gases for application in a small scale, and transportation [16,17]. Fig. 2 shows the flowchart of the different thermochemical conversion process and the products obtained [18].

1.2.1 Biomass combustion

The process of burning any fuel in the presence of air to release its energy in the form of heat of reaction is termed as combustion. In this process, the biomass burns directly in the presence of excess air in a boiler or furnace, where steam produced converts this heat energy into mechanical energy to run turbine or steam-driven pumps as per the demand of the industry [19]. This process is an energy-inefficient process as thermal power plant has only approximately 40% efficiency [20]. Biomass combustion is usually not preferred for syngas production because the combustion products are carbon dioxide and water. In practice, however, many other gases are emitted depending on the source and composition of the biomass. SO_x as a result of sulfur in biomass, carbon

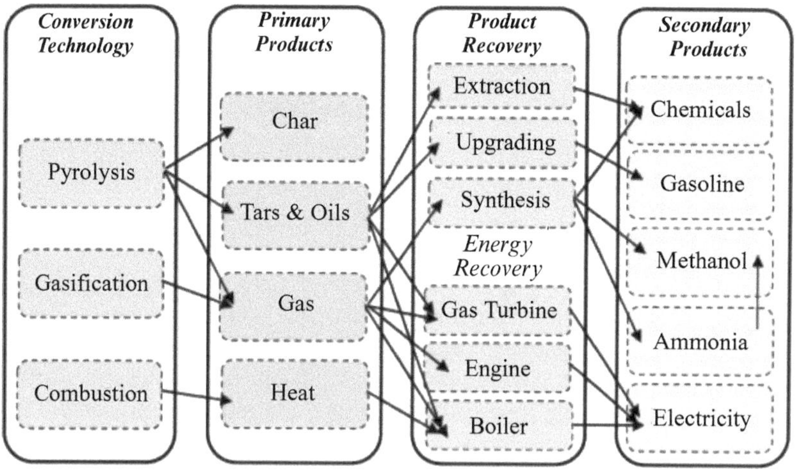

FIG. 2 Thermochemical conversion process of biomass [18].

monoxide as a result of incomplete combustion and NO_x are due to the presence of nitrogen in air and biomass [21]. Environmental laws restrict the level of these pollutants to a specific level to avoid any health problems. Remedial treatment for these unwanted gases cost very much and increases the operating cost [19]. As a result, the cost per unit of energy is increased. This is an inefficient energy process due to the wastage of a large amount of energy in the form of flue gases enthalpy [22]. Although wastage of energy can be minimized by installing waste heat recovery boiler/economizer, there are still limitations which restrict the heat recovery to a certain level [23]. These limitations are acid dew point (ADP) and water dew point (WDP) [24]. Due to the reason mentioned above, researchers have focused on further exploration of the energy-efficient technologies.

1.2.2 Biomass pyrolysis

Pyrolysis is the thermal decomposition occurring in the absence of oxygen. The composition of product yields such as gas, liquid bio-oil, and char depend on the type of pyrolysis method and operating parameters. The pyrolysis process is classified into slow pyrolysis and fast pyrolysis based on the heating rate [25,26]. The slow pyrolysis normally used for charcoal production. In this process, biomass is heated up to the pyrolysis temperature with a lower heating rate of 10°C/s which takes a longer time to reach desired temperature [27]. Whereas in fast pyrolysis the heating rate is high about 10^3–30^3°C/s and it takes the smaller time to reach the pyrolysis temperature [28,29]. In pyrolysis, the process has been taking place in the temperature range of 400–500°C, and the pyrolysis vapors are condensed [30]. Products formed during the pyrolysis of lignocellulosic biomass consist of a solid residue called char, a liquid called bio-oil, and uncondensed gases and tar [31].

1.2.3 Biomass gasification and its principle

In the early 17th century, a scientist named Thomas Shirley attempted to produce CH4 from coal gasification process [32]. Initially, the gasification process was developed to provide gas to the town for cooking and power for street lighting. Hence, the town gas gained popularity as a gaseous fuel. Among thermochemical conversion, the gasification process involved multiple complex reactions. In this gasification process, the feedstock is converted into a product, such as chemicals and gases in various stages of the gasifier [32]. By taking a case study of downdraft gasifier, the biomass undergoes into several gasification steps such as drying, pyrolysis, oxidation, and reduction processes. This occurred at various zones in the gasifier. Fig. 3 showed the occurrence of multiple steps of gasification temperature zones of biomass [33].

1.3 Gasification mechanism

Gasification is a higher temperature (700–1000°C) partial oxidation thermal conversion process to convert carbon feedstock into gaseous fuel. In gaseous products, the following are the major product gases; carbon monoxide, hydrogen, methane, and carbon dioxide. The by-products of the gasification process are char, ash, and tarry materials. Table 1 listed the common gasification reactions which are observed in the biomass gasification process.

1.4 Types of gasifiers

Gasifiers design and operation are comprehensively studied over the last two centuries [38]. The scales of the gasifiers are small, medium, and large which are available and discussed in the literature. The gasifiers are categorized into three main types: (i) fixed bed, (ii) moving bed (bubbling and circulating fluidized), and (iii) entrained gasifiers [39].

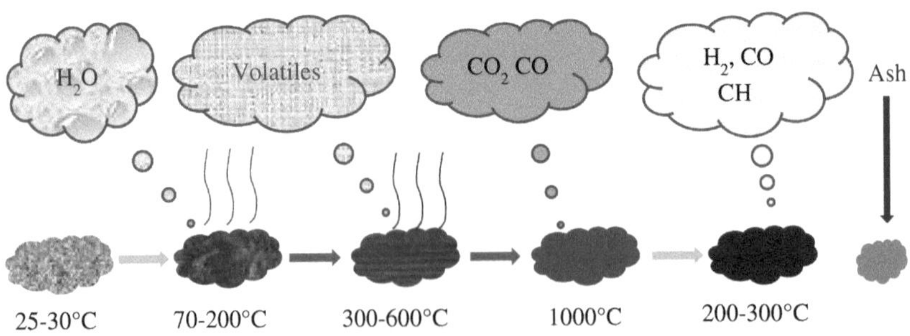

FIG. 3 Various temperature zones in the gasifier [32].

TABLE 1 Common gasification reactions [34–37].

Reaction	Reaction type
Volatile matter = CH_4 + C Mildly exothermic	Oxidation of char
$C + 1/2\ O_2 \rightarrow CO$ $\Delta H^* = -110.6$ kJ/mol	
$C + O_2 \rightarrow CO_2$ $\Delta H^* = -393.6$ kJ/mol	
$CO + 1/2\ O_2 \rightarrow CO_2$ $\Delta H^* = -283$ kJ/mol	Oxidation of volatiles
$H_2 + 1/2\ O_2 \rightarrow H_2O$ $\Delta H^* = -241.9$ kJ/mol	
$C + CO_2 \rightarrow 2CO$ $\Delta H^* = 172.5$ kJ/mol	Boudouard
$C + H_2O \leftrightarrow CO + H_2$ $\Delta H^* = 131.3$ kJ/mol	Water gas primary
$C + 2H_2O \leftrightarrow CO_2 + 2H_2$ $\Delta H^* = 90.2$ kJ/mol	Water gas secondary
$C + 2H_2 \leftrightarrow CH_4$ $\Delta H^* = -74.9$ kJ/mol	Methanation
$CO_2 + 4H_2 \leftrightarrow CH_4 + 2H_2O$ $\Delta H^* = -165$ kJ/mol	
$C + 3H_2 \leftrightarrow CH_4 + H_2O$ $\Delta H^* = -206$ kJ/mol	
$CO + H_2O \leftrightarrow CO_2 + H_2$ $\Delta H^* = -41.2$ kJ/mol	Water-gas shift

ΔH^*; Amount of energy required for reaction in kJ/mol.

1.4.1 Fixed-bed gasifier

The simplest apparatus initially used in the gasification process is the fixed bed. The fixed-bed mode can be operated in updraft, downdraft, and cross-draft mode. In this mode, the reactor is larger in length and velocity of the gas flow is low [39].

1.4.1.1 Updraft gasifier

Fig. 4A showed the schematic diagram of the updraft fixed-bed gasifier. The feedstock entered from the top of the gasifier while the oxidizing agent entered from the bottom of the gasifier. After gasification process, the product gas moved upward and collected from the side [32].

1.4.1.2 Downdraft gasifier

Fig. 4B showed the sketch of the downdraft gasifier. In this configuration, the biomass and product gas moved toward down of the gasifier. Tar formation was observed from 200°C to 500°C. The oxidizing agent assisted for tar cracking into noncondensable gases. Due to this facility, downdraft gasifier has greater computability for low tar formation in the product gas ($<1\,g/Nm^3$) [32].

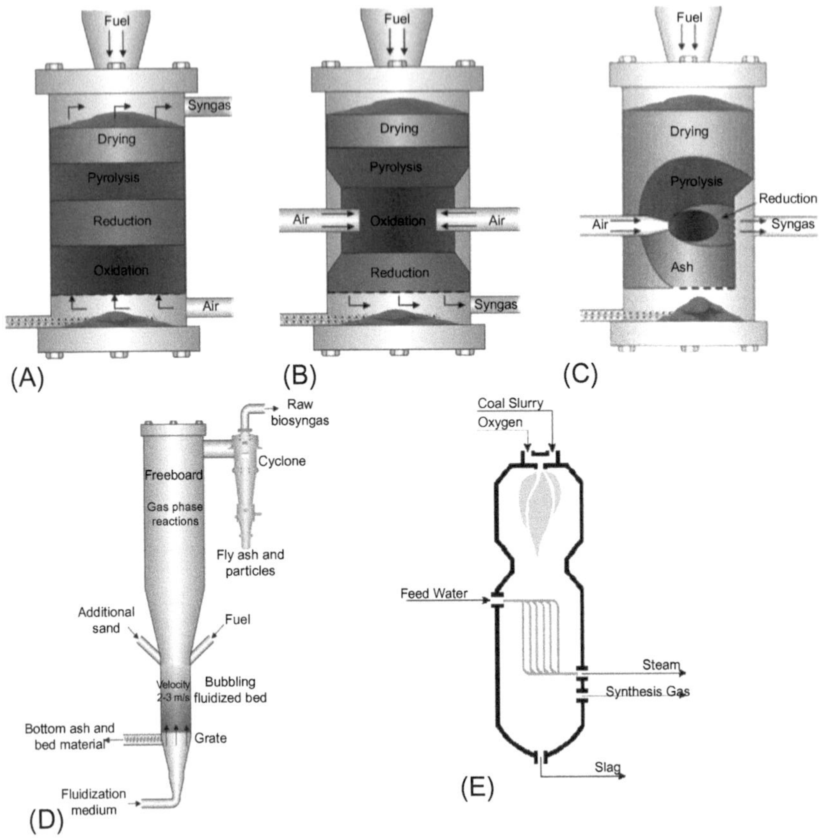

FIG. 4 Schematic diagram of gasifiers: (A) updraft fixed-bed gasifier, (B) downdraft fixed-bed gasifier, (C) cross-draft gasifier, (D) fluidized-bed gasifier, and (E) entrained flow gasifier [35].

1.4.1.3 Cross-draft gasifier

Thirdly, another aspect of fixed-bed gasifier is cross-draft gasifier. In this facility, oxidizing agent entered from the side, while the biomass is injected from the top of the gasifier. The product gas is collected from the opposite side (Fig. 4C). In comparison to updraft and downdraft, cross-draft gasifier took much less starting time and depicted high-temperature profile. This facility could be applied where low tar content is required with the utilization of small particle size [32].

1.4.2 Fluidized-bed gasifier

Moving-bed gasification system (bubbling and circulating fluidized bed) has gained much attention because it provides well-homogenized mixing by handling even bigger particle size. In fluidized-bed gasification system, the gasification agent blows the bed of solid particles. The feedstock is injected from the bottom of the gasifier. This facility is famous because of lower tar production and is applied in many commercial applications. The major challenge is to control the temperature in the reactor and equipped to provide large-scale quality syngas production [32,39,40].

1.4.3 Entrained flow gasifier

The most successful design with the capability to operate gasifier at a large scale is entrained gasifier. This gasifier can be operated using updraft and downdraft mode by handling the gasification process at a higher temperature (1400°C) with very low residence time. Multiple companies such as Shell, GE energy, and Conoco Philips build entrained gasifier for commercial-scale syngas production. Low-ranking coal feeds provide challenges to entrained gasification system which leads to slagging and lower gasification efficiency [32].

1.5 What is tar

Tar has appeared as an undesired side-product from the gasification process. It has been introduced and reported in many ways, by considering the syngas quality specification for end use application, tar collection, and analysis method [41]. It is identified as a mixture of complex organics condensed hydrocarbon material with aromatics that was produced during thermal or partial oxidation of carbonaceous materials like gasification of coal and biomass [42]. In appearance, tar is a viscous and thick liquid. Tar consists typically of one ring to five rings complex aromatic compounds along with other complex polycyclic aromatic hydrocarbon and oxygen-containing hydrocarbons called tar [43,44]. Another definition decided in International Energy Agency meeting on Bioenergy Gasification Task that "*hydrocarbons with molecular weight higher than benzene defined as tar*" in which the protocol for its measuring and analysis has also been decided in Brussels in 1998 [45].

1.5.1 Formation for tar

Usually, tar is formed during the thermochemical conversion process from the decomposition of lignocellulosic biomass (carbonaceous materials) at the pyrolysis stage [46]. The components which involve in pyrolysis are mainly oxygenated hydrocarbons. The tar formation in the thermal degradation is resultant of complex reactions mechanism occurring simultaneously [47]. The tar production extremely depends on the reaction mechanism and process conditions [45,47], such as equivalence ratio (φ) (actual air supplied/stoichiometric air needed for complete combustion), temperature [48], feedstock and blending ratio (H/C and O/C ratio), feedstock chemical composition [49], and catalyst application, etc. The temperature and equivalence ratio are important operating factors in the gasification process, that not only strongly affect the reaction mechanisms but also tar formation and composition during the process [47].

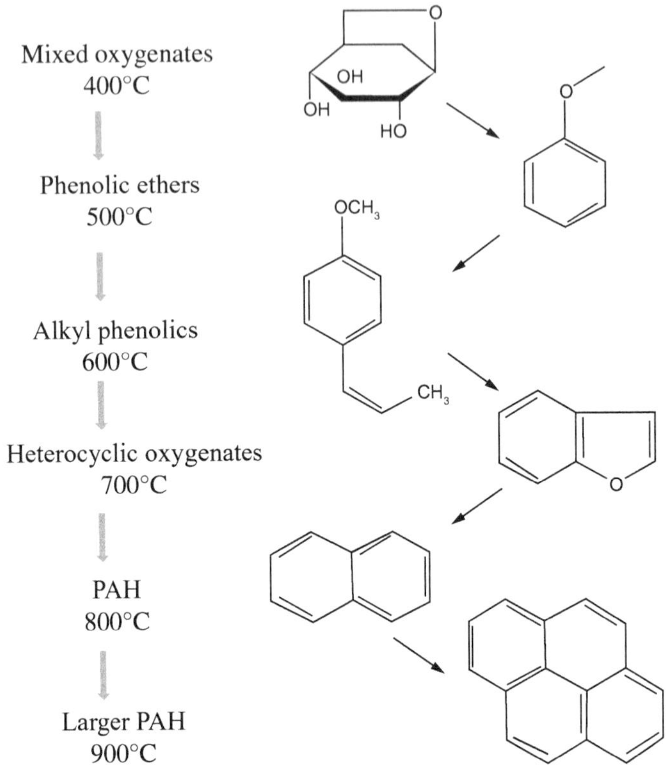

FIG. 5 Tar formation process as a function of reaction temperature [50].

The reaction temperature plays a vital role in tar formation as the increase in temperature that increases the oxygenated hydrocarbons converted into light hydrocarbons, such as aromatics and olefins [47]. Further, these light hydrocarbons convert into higher hydrocarbons and longer PAH (polycyclic aromatic hydrocarbons). A typical process of tar formation as a function of the reaction temperature is presented in Fig. 5. Typically organic compounds become more stable at increasing reaction temperature. Typically, at all stages, the process temperature affects the thermochemical decomposition including solid to gas conversion, fuel devolatilization, and char gasification reaction [48].

1.5.2 Chemical composition of tar

The chemical composition of tar and its quantity produced from gasification depends on several parameters, such as gasification temperature, type of feedstock, gasification medium, and type of gasifier and other operating factors like heating rate, equivalence ratio, and pressure. An increase in temperature can decompose class 1 and 2 tars that leads to the production of more class 3

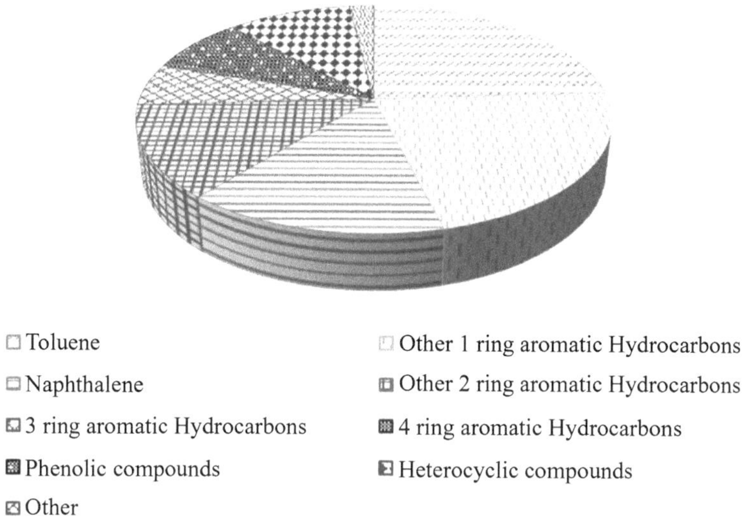

FIG. 6 Typical composition of biomass tars [50].

and 5 tars. Therefore temperature affects the tar yield as well as tar type [50]. The typical tar composition is presented in Fig. 6.

Tar composition present in syngas produced from birch, straw, and miscanthus at temperatures of range of 700–900°C as presented in Table 2 [51]. The primary pyrolyzed tar from biomass is slowly changed its composition at a higher temperature of 700–900°C. A gradual increase in temperature progressively reduced or completely eliminated tar that consists of PAHs and heteroatom compounds. As temperature increases, specific tar components breakdown into C_1–C_5 hydrocarbons and small molecular gases and promote the aromatic rings of tar. The reaction mechanisms that are responsible for such processes are the addition of C_2H_2, abstraction of H, and cyclodehydrogenation. The dehydrogenation of aromatic components causes the polymerized aromatic compounds. The higher temperature of 900°C generates the PAHs contents such as naphthalene, anthracene, and phenanthrene temperature increase higher the conversion pathway resulted in very stable aromatic hydrocarbons [52].

1.5.3 Classification of tar and their properties

In appearance, tar is a viscous and thick liquid, based on the process conditions of gasifier Milne et al. [53] divided the tar into four classes; first is primary tar, secondary tar, alkyl-tertiary tar, and tertiary tar. Primary tar produced during the pyrolysis of biomass building blocks such as sugars, acids, alcohols, aldehydes, ketones, phenols, guaiacols, catechols, furans syringols, and oxygenates. At a temperature higher than 500°C the primary tar transforms into bulky molecules,

TABLE 2 Chemical composition from birch, straw, and miscanthus thermal cracking tar (mg/g dry biomass feedstock) [51].

	Birch			Straw			Miscanthus		
Compound	*700°C*	*850°C*	*900°C*	*700°C*	*850°C*	*900°C*	*700°C*	*850°C*	*900°C*
Naphthalene	1.0	2.5	2.3	2.2	3.3	3.9	1.6	3.3	3.7
Toluene	1.5	1.5	1.7	1.1	1.1	0.7	3.2	2.7	2.2
Benzene	0.6	1.1	2.4	0.7	0.8	0.9	3.3	3.1	5.4
o-Xylene	1.2	1.2	0.7	1.3	1.0	0.6	1.5	1.2	0.7
p-Xylene	0.6	0.4	0.1	0.4	0.2	0.1	0.6	0.4	0.2
Indene	0	0.1	0	0.2	1.0	0.4	1.3	1.4	0.6
1-Methylnaphthalene	0.1	0	0	0.3	0.2	0.1	0.1	0.2	0.1
2-Methylnaphthalene	0.1	0	0	0.3	0.3	0.1	0.5	0.5	0.3
Acenaphthylene	0	0	0	0.5	0.5	0.5	0.4	0.8	0.6
Biphenyl	0	0	0	0.2	0.2	0.1	0	0.1	0.1
Phenanthrene	0	0	0	0.3	0.3	0.6	0.2	0.5	0.5
Fluorene	0.1	0	0	0.1	0.2	0.2	0	0	0
Fluoranthene	0	0.1	0.1	0.1	0.1	0.2	0.1	0.1	0.1
Pyrene	0	0	0	0.1	0.1	0.3	0	0.1	0.1
Anthracene	0	0.3	0.3	0.1	0.1	0.2	0.1	0.2	0.2
All light aromatics	4.6	6.1	5.2	7.2	8.6	8.0	9.6	11.5	9.4

i.e., olefins and phenols. Alkyl tertiary consists of aromatics with methyl derivatives, i.e., methyl acenaphthylene, indene, and toluene.

Tar is divided into five types based on tar properties such as solubility, condensability, and chemical and molecular weight. The information was reported by the Energy Research Center of Netherlands (ECN) in collaboration with the Applied Scientific Research Center of Netherlands and Twente University, the Netherlands, as shown in Table 3 [50,54].

TABLE 3 Classification and properties of tar based on appearance, type, solubility and condensability, and chemical composition [50,54].

Tar-class	Class name	Properties	Tar compounds
1	GC-undetectable	Very heavy tar compounds that cannot be detected using GC	This tar can be measured; (total gravimetric tar—GC detectable tar fraction)
2	Heterocyclic	• Heteroatoms aromatic compounds • Highly soluble in water	Pyridine, phenol, cresols, quinoline, dibenzophenol, and isoquinoline
3	Light-aromatic (single ring)	• Light hydrocarbons with one ring • Do not pose a problem regarding solubility and condensability	Ethylbenzene, toluene, styrene, xylenes
4	Light-PAH compounds (2–3 rings)	• Two and three ring compounds • Condense at very low temperature even at very low concentration	Methylnaphthalene, indene, naphthalene, fluorene, phenanthrene, biphenyl, acenaphthalene, and anthracene,
5	Heavy-PAH compounds (4–7 rings)	• Contain more than three rings, • Condensable at high temperature and low concentration	Coronene, perylene, pyrene, fluoranthene, coronene, and chrysene
Primary tar	–	• Low molecular weight • Oxygenated hydrocarbon compounds • Formed between 400°C and 700°C	Levoglucosan, furfural, and hydroxyacetaldehyde

Continued

TABLE 3 Classification and properties of tar based on appearance, type, solubility and condensability, and chemical composition—cont'd

Tar-class	Class name	Properties	Tar compounds
Secondary tar	–	• Phenolic and olefin compounds. • Formed between 700°C and 850°C	Phenol, cresol, and xylene
Tertiary tar	–	• Complex aromatic compounds • Formed between 850°C and 1000°C	Benzene, naphthalene, pyrene, and toluene

The condensation behavior of tar is the effect of all components of the tar in the gas and their individual effects on the total vapor pressure of the tar are critical. When the vapor pressure crosses over the saturation pressure of tar, the gas is saturated according to Raoult's law. Thermodynamically, this condition results in the condensation of saturated vapor. The tar dew point is the temperature at which the actual partial pressure of the tar is equal to the saturation pressure of the tar. Therefore, in case of condensation problems, the dew point of tar is an effective parameter for evaluating the performance of gas cleaning systems. It is believed that when the dew point of the tar is lowered below the lowest expected temperature, contamination problems are solved by condensation or tar aerosols. To use this approach in design problems, a calculator was developed to predict the dew point of tar as a function of the concentration of individual tar components in the gas produced [45].

1.5.4 *Tar reactivity*

Reactivity of tar produced from biomass gasification is very important for the sophisticated implication and development of biomass gasification. The removal of tar from the product gas is difficult therefore, removal methods are based on condensation, filtration, and chemical removal by a catalyst. Obtaining the very least quantity of tar in product gas needs advance research and development. The equipment is likely to be more complex and costly as compared to the conventional gas cleaning equipment. Hence, decomposition of tar is a more attractive alternative that breakdown tar into more valuable cleaner gas content that needs further study. Tar breakdown can be done by thermal cracking, catalyst cracking, and plasma. Tar undergoes the decomposition reaction at high temperature and resident time. Reactions may occur between liquid tar or gaseous tar with carbonaceous ash/char and reaction between tar and liquid content. These kinds of reactions are called tar polymerization. The decomposition reactions may occur between gas phases of tar. It may also react with gaseous tar and gaseous reactants that also breakdown the tar compounds. The radical formation reaction governs the craking reaction rate for tar compounds. Radical formation reaction plays a very important role in thermal cracking which included radical formation due to chemical bound breaking, accelerate reactions to the formation of new chemical bonds, transfer of hydrogen, isomerization reaction, and finally stop the reaction once two radicals react [55].

1.5.5 Acceptable limits for tar

The tar stays in vapor state in gas till gas is cooled or condensed on cold surfaces. When this occurs in heat exchangers, ducts, or in filters, tar deposits on its surface that cause the blockages which reduce the efficiency of the process, increase emissions and thus increase the operating costs of the system [56]. Tar present in the syngas can define its utility; either it can be used directly in IC engines, combustion system, fuel cell gas turbines, or raw material for synthesis methanol. The limits in the application of tar are presented in Table 4. In direct combustion systems, the syngas is burnt directly that is still hot, so there is little chance of tar condensation, there are no restrictions on the amount of tar and particulates, but the flue gases produced follow the emission standard of this technology. In IC engines, tar can condense in fuel injection systems; while these engines are more pollution resistant compared to the gas turbines, whereas there are restrictions in gas turbine and its blades are more sensitive to tar deposited on the surface when hot gases passing through its blades after combustion [28,54,57].

TABLE 4 Acceptable limits for tar in syngas for different applications [54,57–59].

Syngas application	Tar concentration (mg/Nm^3)
Direct combustion	No specified limit
Compressors	50–500
Internal combustion engine	<100
Gas turbine	<5.0
Fuel cells	<1.0
Synthesis of methanol	<0.1
Fischer Tropsch synthesis	<1.0μL/L (class 2, BTX)

1.5.6 *Relative issues of tar in downstream gasification process*

The formation of tar is the major issue in biomass gasification that affects and causes the following complications in downstream equipment [46]:

i. Condensation of tar and successive pipe choking of downstream equipment.
ii. Generation of tar aerosol compounds.
iii. Formation of complex structures polymerization. High tar formation in the gas produced from gasification causes undesirable maintenance level for engines and turbines that increases the maintenance cost [50]. The tar treatment influence the type of tar and different types of tar affects the equipment differently.

1.6 Tar abatement techniques

There are several approaches that have been reported in the literature to reduce the amount of tar in the syngas [37,54,60]. The choice of an appropriate technique is based on tar removal efficiency, economic feasibility, and minimal impact on syngas composition. The available tar removal methods can be classified into primary and secondary methods.

1.6.1 *Primary tar abatement techniques*

In primary methods, operating parameters are controlled to reduce the tar formation during the gasification operation. Operating conditions such as gasifier design and bed additives (catalyst) were key factors in controlling the quality of the syngas. Temperature, pressure, residence time, and medium of gasification seem to be the most important parameters that affect the quality of the gas produced [19].

1.6.1.1 Operating condition

It was reported by Yu et al. [20,61] that the pyrolysis of birch wood was carried out in a free-fall paralyzer to study the effect of varied temperature on the pyrolysis. It was found that higher temperature favor the formation of gas contents. It has been found that tar can be removed by maintaining high temperatures in a fluidized-bed gasifier. The tar yield decreased by up to 40% by raising the temperature from 700°C to 900°C [61]. The concentration of aromatic compounds and oxidized hydrocarbons (one and two-rings) decreased significantly with increasing temperature. However, the production of three rings and four rings aromatics concentration increased with increasing temperature. Brag et al. [62] reported almost complete elimination of phenol and 50% reduction of toluene during high-temperature pyrolysis of the birch wood. However, an increase in benzene and naphthalene was noted when the temperature rose from 700°C to 900°C [62]. Kinoshita et al. [63] reported the tar concentration decreased as temperature increased during gasification of sawdust in fixed-bed gasifier. The effect of pressure on the gasification of biomass has been reported in the literature. Knight [64] reported the effect of pressure on the gasification of woodchips. It was found that increasing the pressure up to 21 bar increased the concentration of polycyclic aromatic hydrocarbons. In another study, it was found that tar concentration with lower hydrocarbons than naphthalene decreased with increasing pressure during gasification with 100% carbon conversion [65]. Equivalence ratio (ER) is an important parameter of gasification. It was reported that the concentration of tar decreased rapidly to 2 g/Nm^3 as the equivalence ratio increased, which also affects the composition of the syngas during the air used as gasification medium [66]. The CO_2 content increased, while the H_2 and CO content decreased at high equivalence ratio values that eventually reduce the calorific value of the gas produced. O_2 availability has been reported to increase with a high equivalence ratio that increased the oxidation of products. The tar concentration decreased 30% by increasing the equivalence ratio from 0.22 to 0.32 at 700°C [63,66]. However, gas quality and lower calorific value were decreased at a significant increase in the equivalence ratio.

1.6.2 Secondary tar abatement techniques

These methods are used to treat the syngas from a gasifier. Secondary tar removal methods can be divided into physical and chemical treatments. Mechanical/physical technique used scrubbers, cyclones, electrostatic precipitators, fabric filters, and ceramic filters [67]. Mechanical or physical tar removal method further can be classified as hot or dry cleaning and cold or wet cleaning [50,54,68]. Normally dry gas cleaning is done without gas cooling when the gas temperature is higher than 500°C and after gas cooling temperature slightly lower than 200°C whereas wet gas cleaning is done once the gas has cooled to a typical temperature of 20–60°C [54]. The mechanical tar removal method only physically removes tar from the gaseous phase to condensable form, whereas catalytic cracking is the most attractive option which converts the unwanted tar into valuable fuel gas and enhances the overall gasification performance [69,70].

The chemical tar removal technique allows cracking of tar compounds using high temperature (heat) or catalysts (catalytic) at the exit end of the gasifier. Secondary methods have been found to be very useful in tar elimination from syngas. In addition, these methods have also been shown to be effective in reducing ammonia [71,72]. High-temperature thermal cracking and catalyst cracking are the methods widely utilized to decompose tar into syngas. The thermal reforming of tar at an elevated temperature converts the large complex organic molecules of tar into smaller simple gas molecules. Without catalyst application, it is not an attractive technique as it needs high energy, and it also causes soot formation [37]. The thermal decomposition of tar involved a series

TABLE 5 Typical catalytic gasification reaction mechanisms [73].

Reaction mechanism	Reaction type
$pC_nH_x \rightarrow qC_mH_y + rH_2$	Cracking
$C_nH_x + nH_2O \rightarrow (n + x/2)H_2 + nCO$	Steam reforming
$C_nH_x + nCO_2 \rightarrow (x/2)H_2 + 2nCO$	Dry reforming
$C_nH_x \rightarrow nC + (x/2)H_2$	Carbon formation
$C_nH_x + (n/2)O_2 \rightarrow nCO + (x/2)H_2$	Partial oxidation

of tar decomposition mechanisms like thermal cracking, dry reforming, steam reforming, carbon, partial oxidation, and formation [47] as presented in Table 5 [45,74]. In these reactions, C_nH_x and C_mH_y represented tar and hydrocarbon, respectively. C_nH_x is the combination of various organic compounds, whereas C_mH_y presents the lower carbon number while C_nH_x representing higher carbon number [45,74].

1.6.2.1 Mechanical tar removal methods

Mechanical tar removal methods are mainly applied to remove particles from the syngas. These methods have been shown to be effective in physically capturing both tar and particulates from syngas [75]. Bakers et al. [76] reported that tar is present in syngas in two forms based on the syngas exit temperature from the gasifier. At higher temperatures, tar could be detected as a vapor, while it could be present as droplets of liquid in syngas at low temperatures. Thus, at higher temperatures, the removal of tar and particles could be separated. When tar and oil condense at lower temperatures, their removal cannot be apart from particulate removal [54]. Several mechanical or physical gas cleaning systems are available to remove both tar and particulates from syngas. Based on operating conditions and the desired level of gas cleaning, mechanical methods are classified into two groups: hot or dry and cold or wet tar removal. Dry tar removal is usually performed before the gas cooling typically temperature above 500°C, and is reduced to 200°C after the gas has cooled. Normally for dry gas cleaning carried out by using a cyclone, electrostatic precipitators (ESP), rotating particle separators (RPS), bag filters, sand bed filters, ceramic filters, baffle filters, fabric/tube filters, and adsorbers [77]. The wet tar removal is normally performed at a low temperature of 20–60°C because it is done after the gas had cooled. Typically, equipment used for wet cleaning is spray towers, impingement scrubbers, wet electrostatic precipitators, packed column scrubber (wash tower), venturi scrubbers, and wet cyclones. The prime purpose of dry cleaning and wet cleaning units in downstream of the gasification system is to remove the tar and particles from the produced syngas [54]. A typical

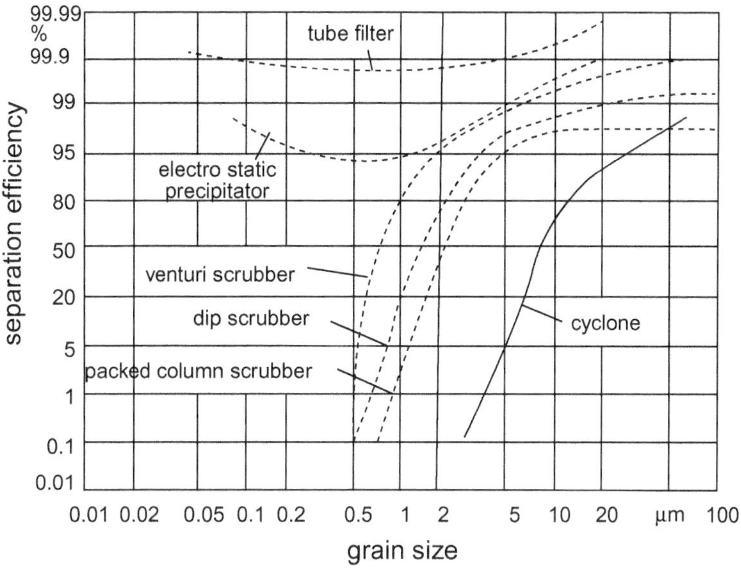

FIG. 7 Typical separation efficiencies of mechanical gas cleaning systems [54,78].

performance of different particles capturing units of mechanical or physical gas cleaning systems is shown in Fig. 7 [78]. This figure shows the general separation efficiency of mechanical gas cleaning systems at different droplet and particle sizes based on their collection technique.

Nevertheless, tar is present in syngas in the form of vapor and can be removed by condensation as liquid droplets. In fixed-bed updraft gasifier normally tar is present in liquid droplets due to low exit temperature. Whereas, in fluidized-bed and downdraft gasifiers, it is in the form of vapor because of the high exit temperature of syngas. Merely at high temperature tar and particulates can be divorced from each other. A temperature at oil and tar condense into liquids, oil, and tar cannot be separated from the particulates [54].

Dry tar removal Mainly dry tar removal systems consist of a cyclone, electrostatic precipitators (ESP), rotating particle separators (RPS), bag filters, sand bed filters, ceramic filters, baffle filters, fabric/tube filters, and adsorbers to remove tar and particulates [54]. Typically RPS and fabric filters alone cannot remove enough tar, it needs extra tar removal devices. The sand bed filter is also used to remove tar from syngas. Activated carbon is very suitable for tar removal due to its adsorption characteristics and reasonable prices. Currently, another newly developed tar removal method is the catalytic filter. This technique combines both catalytic cracking of tar and filtration for particles from syngas in one go. The ceramic candle filter consists of nickel-based tar cracking

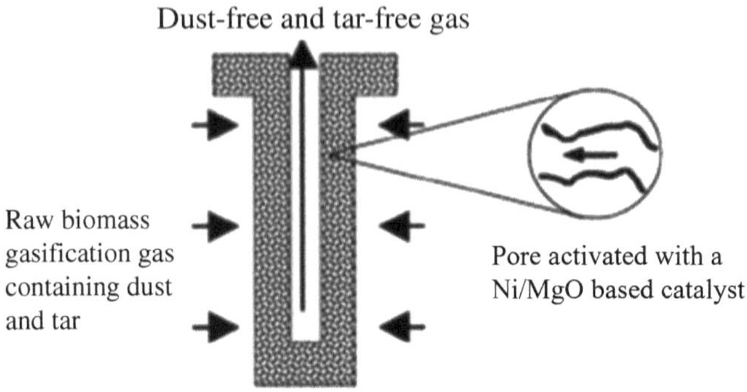

FIG. 8 Schematic representation and operating mechanism of the candle filter.

catalyst, which supports catalyst cracking activities, and operating mechanism of candle filter is presented in Fig. 8 [54,78]. However, ceramic filters are not an attractive option due to the high cost and complex process. Problems related to the RPS, fabric filters, and ceramic filters can be addressed through the activated carbon tar adsorbers [54]. On the other hand, tar capture by the filter is difficult to remove by any means eventually deposition of tar blocks the filter pores, which causes plugging. Usually, barrier filters are not appropriate for tar removal although in some cases barrier filters are successfully attested [67]. The tar removal efficiency of dry mechanical cleaning system for syngas is given in Table 6.

TABLE 6 Tar removal efficiency of a dry mechanical cleaning system for syngas.

Cleaning devices	Temperature (°C)	Tar removal efficiency (%)	Remarks	Reference
Sand bed filter	10–20	50–97	Choking and not easy to clean	[18]
Cyclone with humidifier		50–90	Supplementary tar elimination needed	[76]
Ceramic filter (glass)	650–700	75–97	Costly and complex	[79]
Activated carbon filter	25	20–80	Choking and not easy to clean	[78]

TABLE 6 Tar removal efficiency of a dry mechanical cleaning system for syngas—cont'd

Cleaning devices	Temperature (°C)	Tar removal efficiency (%)	Remarks	Reference
Ceramic filters (quartz)	650–700	75.6–94	Costly and complex	[79]
Fabric filter	130	0–50	Supplementary tar elimination needed	[58]
Rotational particle separator	130	30–70	Supplementary tar elimination needed	[58]
Fixed-bed tar adsorber	80	50	Choking and not easy to clean	[78]
Catalytic tar cracker	900	>95	Costly and complex	[58]

Wet tar removal Wet tar removal methods consist of spray towers, impingement scrubbers, wet electrostatic precipitators, packed column scrubber (wash tower), venturi scrubbers, and wet cyclones. These techniques have been extensively used for the gasification plant. In general, wet tar removal cleaning is an efficient technique to remove tar and particles from syngas. Wet scrubbing is a promising technique in wet syngas cleaning. A normally wet scrubber uses water to scrub the syngas which removes both tar and particulates at the same time. Verities of scrubbing devices are used in wet tar removal, the most common types are packed columns, packed bed, venturi, and impingement scrubbers. A wet scrubber is capable of decreasing the tar and particulates 20–40 mg/m^3 and 10–20 mg/m^3, respectively [54]. Tar and particulates removal efficiencies of a various wet mechanical cleaning system for syngas is presented in Table 7.

Among wet scrubber, spray towers are the simplest in design, low-cost, and have the lowest efficiency; whereas impingement scrubber has an overall 70% efficiency. Higher tar removal efficiency can be achieved by using a series of wet impinger scrubber. Three impingers can attain above 95% efficiency. Besides being efficient, the most important necessary characteristic, it is simple in structure [80]. Venturi scrubber is also a highly efficient wet gas cleaning device to remove tar and particulates. Usually, its tar removal efficiency is about 50%–90% [78].

TABLE 7 Tar removal efficiency of a various wet mechanical cleaning system for syngas.

Cleaning devices	Temperature (°C)	Tar reduction (%)	Reference
Wash tower	50–60	10–25	[58]
Venturi scrubber		50–90	[58]
Rotational atomizer	<100	85–90	[58]
Wet electrostatic precipitator	40–50	0–60	[58]

Usually, wet scrubber uses water to scrub the syngas. The use of water shows unacceptable conditions regarding continuous operation behavior and regeneration efficiency. Critical issues with using water as a washing agent involve operational problems such as saponification, surface tension effects, clogging of apparatus, the low solubility of hydrocarbon compounds, comparatively expensive wastewater treatment [81]. Furthermore, after the wet cleaning process, the heating value of the syngas and net energy efficiency of the gasification process is lower due to the loss of syngas enthalpy.

1.7 Chemical tar removal technique

Chemical tar removal technique is also secondary tar removal method which preforms out the primary reactor normally in the secondary reactor. Chemical tar removal method improved the gaseous such H2 and CO by the decomposition of tar present in syngas that eventually increased the calorific value of gas. The chemical technique can be classified into three groups namely, catalytic tar cracking, thermal tar decomposition, and tar removal by thermal plasma.

1.7.1 Catalytic tar cracking

Tar decomposition via catalyst cracking is one of the promising methods in gasification. Catalytic gasification has more advantages in terms of energy efficiency, and it improves the syngas composition to reform higher aromatic ring of tar into valuable gas fraction [60]. Catalytic biomass gasification gets the attention of researcher since mid-1980s. The major motivation behind the advancements in the catalytic biomass gasification area is tar-free syngas from biomass gasification. The catalysts reduce the tar formation and improve the syngas composition. In general, catalysts improve the char gasification, syngas composition, and decrease the tar yield during biomass gasification [82].

Tar removal using catalytic gasification increased the economic feasibility of the biomass gasification process.

Application of catalysts can be divided into two distinct groups based on the position of the catalyst in the gasification process. The first application of the catalyst called as primary catalyst application in which catalyst is directly mixed with the feedstock and fed to the gasifier, reactions involved in catalytic gasification are presented in Table 5. In the second application, the catalysts are placed into the second reactor located downstream from the gasifier. Being an independent catalyst in a secondary reactor, the catalyst may behave differently under the different operating conditions than those in the gasification unit. The catalyst actively reforms the hydrocarbons and methane as follows [37]:

$$C_nH_m + nH_2O \leftrightarrow nCO + \left(n + \frac{m}{2}\right) H_2 \tag{1}$$

$$C_nH_m + nCO_2 \leftrightarrow 2CO + \left(\frac{m}{2}\right) H_2 \tag{2}$$

There are numerous types of catalysts considered as potential tar cracking catalysts. Catalyst can be divided as the nickel-based catalyst, non-nickel transition metal catalyst, alkali metal catalyst, natural catalyst, zeolite catalyst, and activated carbon catalyst [50,54]. Nickel-based catalysts have been widely used for tar craking because of their high activity; however, a major problem with the nickel-based catalyst is quick deactivation due to carbon fouling on the catalyst [50,82]. Even though noble metal-based catalysts are very stable, high catalytic activity, and high carbon fouling resistance, however, these are very costly catalysts. The transition metal catalysts such as Cu, Fe, and Co have also been shown excellent performance for the tar cracking. However, these are also having the problem of carbon deposition and easily deactivate like nickel-based catalyst under heavy tar conditions [50,54]. Alkali and alkaline earth metal (AAEMs) catalysts are very useful for tar reforming due to the high catalytic activity. Mainly, alkali metals present in biomass always perform the catalytic role in thermal conversion and may act as a tar reforming agent in gasification. However, the main problem with alkali metal catalysts is their evaporation during the reaction and difficulty in recovering. Natural catalysts such as dolomite olivine and shells have been extensively used for tar reforming as it is low price, abundantly available, and easily disposable. The main disadvantage of these catalysts is very fragile after calculations which are not suitable for fluidized-bed reactors due to loss of bed material and low catalytic activities as compared to man-made catalysts [50]. Zeolite is also a good catalyst for tar cracking. It has a high surface area and adsorption capacity. It is thermally very stable and has high resistance to sulfur compounds and easily regenerates. Activated carbon such as biomass-derived char also used as a catalyst tar cracking. Biomass-derived char have advantages like high porous textural structures, low-cost, and the by-product of biomass gasification. Furthermore, the minerals contained in biomass can also improve the catalytic activity of biomass char [50,54].

Although many catalysts are used as a potential for tar reforming, however, the catalyst can be selected according to fundamental criteria.

i. The catalysts must be effective in tar cracking.
ii. Catalysts must be able to provide the desired tar removal level in the syngas for the deliberate process.
iii. If syngas is the desired product, then the catalysts must be capable of reforming methane and tar.
iv. The catalysts should be inexpensive and easily available.
v. The catalysts should be stable and reusable.
vi. The catalysts should be resistant to deactivation as a result of coking, carbon fouling, and sintering.

1.7.2 Thermal decomposition

The thermal decomposition can define as tar compounds breakdown into lighter gases under the significant effect of temperature. High temperature affects the stability composition of the tar and breakdown it into lighter components. Morf et al. [83] reported that 88% of tar reduction during the pyrolysis of wood. Fassinou et al. [84] carried out the *Pinus pinaster* pyrolysis in two-stage gasifier. It was observed that an increase in temperature and resident time increased the gas yield and reduced the tar formation. In another study, 6.3% phenol conversion was observed at a temperature of 700°C and conversion increased up to 98% at a temperature of 900°C [85]. Phuphuakrat et al. [86] reported 78% tar decomposition at the temperature of 800°C. An addition of air/oxygen in gasification gasifier, decomposed tar in syngas via partial oxidation. The partial oxidation of gaseous components has lowered the calorific value of the syngas, which can lead to additional problems with the use of gases in conventional gas turbines or engines [87]. It is reported that an effective tar elimination required a significantly higher temperature range from 700°C to 900°C [88].

1.7.3 Tar removal by thermal plasma

Plasma can be defined as totally or partially ionized gas. Plasma can be formed in many ways such as combustion, shocks (electrically, chemically, and magnetically driven) combustion, flames, and electric discharges of matter (annealing, corona, microwave, arc, beam of electrons, etc.) [89]. Plasma is a highly reactive form of the matter and is conductive due to the existence of atoms, ions, radicals, molecules, and neutral particles. In thermal plasmas, background gas temperature is very high whereas, other charged and neutral particles remain in the thermal equilibrium state. It has been found that chemical reactions increase dramatically in the hot environment due to the existence of reactive particles like ions, electrons, and radicals. However, plasma requires high energy consumption to maintain the reaction [90]. Thermal plasma is generally has been used to treat solid or municipal waste. It is identical to thermal decomposition techniques in that the waste decomposes at a high temperature of the plasma. Plasma gasification is a beneficial process to obtain energy from such waste.

H_2O/Ar plasma torches were used to covert plastic and wood waste into useful fuel gas syngas. High concentrations of H_2 and CO up to 90% was obtained with low tar concentration in syngas however, higher energy consumption made plasma uneconomic [91]. Materazzi et al. [92] introduced plasma in two-stage fluidized-bed gasifier plasma to convert the solid wastes in to syngas. It was found that two-stage fluidized-bed gasifier efficiently decomposed the tar content. Plasma is useful to decompose the condensable tar in gasification. The synergic effect was prominent at higher temperature that converts residual carbon into the CO with the help of plasma. In another study, H_2 was observed to be higher in plasma gasification compared to the normal gasification [93]. Utilization of plasma in gasification for the decomposition of biomass tar is an attractive approach. Its application for tar elimination is increasing with the passage of time because it is capable to produce clean free of tar syngas. Plasma setup has advantages of handling of a vast range of hazardous and heterogeneous materials with a smart and small size of setup [94]. Thermal plasma used in gasification increased significant yield of H_2 and CO and improve the quality of syngas. However, substantial energy input is required to produce the large volume of the clean syngas that somehow make this process uneconomical.

2 Conclusion

This chapter summarized the potential of biomass as an alternative source of energy and thermochemical conversion techniques in detail. Tar formation in the thermochemical process, its classification, and tar reactivity are explained in detailed. Furthermore, acceptable limits of different technologies and their relative issue to tar are described. Finally, tar removal techniques such as primary and secondary tar reduction techniques are summarized. Primary methods included tar reduction by controlling operating parameters, whereas secondary methods are mechanical removal of tar, chemical tar removal, catalyst application for tar decomposition, thermal degradation of tar, and application of thermal plasma for tar decomposition.

Acknowledgments

The authors would like to acknowledge and thank Universiti Teknologi PETRONAS Malaysia and Hamad Bin khalifa University Doha, Qatar for financial and technical assistance.

References

[1] Fouquet R. Path dependence in energy systems and economic development. Nat Energy 2016;1(8):1–5.

[2] Shahbaz M, et al. A state of the art review on biomass processing and conversion technologies to produce hydrogen and its recovery via membrane separation. Int J Hydrogen Energy 2020;45(30):15166–95.

[3] Inayat M, Sulaiman SA, Naz MY. Thermochemical characterization of oil palm fronds, coconut shells, and wood as a fuel for heat and power generation. In: MATEC Web of Conferences. EDP Sciences; 2018.
[4] Agency IE. Renewables information: overview. In: Renewables information. International Energy Agency; 2018. p. 1–12.
[5] Jie X, et al. The decarbonisation of petroleum and other fossil hydrocarbon fuels for the facile production and safe storage of hydrogen. Energ Environ Sci 2019;12(1):238–49.
[6] Shahbaz M, et al. Cleaner production of hydrogen and syngas from catalytic steam palm kernel shell gasification using CaO sorbent and coal bottom ash as a catalyst. Energy Fuel 2017;31(12):13824–33.
[7] Inayat M, et al. Catalytic and noncatalytic gasification of wood–coconut shell blend under different operating conditions. Environ Prog Sustain Energy 2019;38(2):688–98.
[8] Tahir MH, et al. Demonstrating the suitability of canola residue biomass to biofuel conversion via pyrolysis through reaction kinetics, thermodynamics and evolved gas analyses. Bioresour Technol 2019;279:67–73.
[9] Shahbaz M, et al. Fluidization of palm kernel shell, palm oil fronds, and empty fruit bunches in a swirling fluidized bed gasifier. Sci Technol 2017;35(2):150–7.
[10] Shahbaz M, et al. Parametric study and optimization of methane production in biomass gasification in the presence of coal bottom ash. Procedia Eng 2016;148:409–16.
[11] Naqvi M, et al. Waste biomass gasification based off-grid electricity generation: a case study in Pakistan. Energy Procedia 2016;103(103):406–12.
[12] Naqvi SR, et al. Production and evaluation of physicochemical characteristics of paddy husk bio-char for its C sequestration applications. Bioenergy Res 2015;8(4):1800–9.
[13] Yusup S, et al. Emerging technologies for biofuels production. In: Biofuels: alternative feedstocks and conversion processes for the production of liquid and gaseous biofuels. Elsevier; 2019. p. 45–76.
[14] Salman RN, et al. Fast pyrolysis of rice husk in a drop type pyrolyzer for bio-oil and bio-char production. J Basic Appl Sci 2014;8(5):294–8.
[15] Inayat M, et al. Application of response surface methodology in catalytic co-gasification of palm wastes for bioenergy conversion using mineral catalysts. Biomass Bioenergy 2020;132:105418.
[16] Lahijani P, Zainal ZA. Gasification of palm empty fruit bunch in a bubbling fluidized bed: a performance and agglomeration study. Bioresour Technol 2011;102(2):2068–76.
[17] Iovane P, Donatelli A, Molino A. Influence of feeding ratio on steam gasification of palm shells in a rotary kiln pilot plant. Experimental and numerical investigations. Biomass Bioenergy 2013;56:423–31.
[18] Bridgwater A. Catalysis in thermal biomass conversion. Appl Catal Gen 1994;116(1):5–47.
[19] Demirbas A. Potential applications of renewable energy sources, biomass combustion problems in boiler power systems and combustion related environmental issues. Prog Energy Combust Sci 2005;31(2):171–92.
[20] Sengupta S, Datta A, Duttagupta S. Exergy analysis of a coal-based 210 MW thermal power plant. Int J Energy Res 2007;31(1):14–28.
[21] Finkelman RB. Potential health impacts of burning coal beds and waste banks. Int J Coal Geol 2004;59(1):19–24.
[22] Saidur R, Ahamed J, Masjuki H. Energy, exergy and economic analysis of industrial boilers. Energy Policy 2010;38(5):2188–97.

[23] Huijbregts W, Leferink R. Latest advances in the understanding of acid dewpoint corrosion: corrosion and stress corrosion cracking in combustion gas condensates. Anti-Corrosion Methods Mater 2004;51(3):173–88.
[24] Blanco J, Peña F. Increase in the boiler's performance in terms of the acid dew point temperature: environmental advantages of replacing fuels. Appl Therm Eng 2008;28(7):777–84.
[25] Naqvi SR, Naqvi M. Catalytic fast pyrolysis of rice husk: Influence of commercial and synthesized microporous zeolites on deoxygenation of biomass pyrolysis vapors. Int J Energy Res 2018;42(3):1352–62.
[26] Naqvi SR. Recent developments on biomass utilization for bioenergy production in Pakistan. Sci Proc Ser 2020;2(2):156–60.
[27] Xu Q. Investigation of co-gasification characteristics of biomass and coal in fluidized bed gasifiers; 2013.
[28] Basu P. Biomass gasification and pyrolysis: practical design and theory. Academic Press; 2010.
[29] Naqvi SR, et al. Catalytic consequences of micropore topology on biomass pyrolysis vapors over shape selective zeolites. Energy Procedia 2017;105:557–61.
[30] Butler E, et al. A review of recent laboratory research and commercial developments in fast pyrolysis and upgrading. Renew Sustain Energy Rev 2011;15(8):4171–86.
[31] Lin B-J, et al. Oxidative reaction interaction and synergistic index of emulsified pyrolysis bio-oil/diesel fuels. Renew Energy 2019;136:223–34.
[32] Basu P. Biomass gasification, pyrolysis and torrefaction: practical design and theory. Academic Press; 2013.
[33] Bhavanam A, Sastry RC. Biomass gasification processes in downdraft fixed bed reactors: a review. Int J Chem Eng Appl 2011;2(6):425–33.
[34] Taba EL, et al. Fuel blending effects on the co-gasification of coal and biomass – a review. Biomass Bioenergy 2013;57:249–63.
[35] Sharma A. Assessing the suitability of various feedstocks for biomass gasification. The Department of Biological and Agricultural Engineering, Louisiana State University; 2011. p. 94.
[36] Sulaiman SA, et al. On gasification of different tropical plant-based biomass materials. Asian J Sci Res 2013;6(2):245–53.
[37] Sutton D, Kelleher B, Ross JRH. Review of literature on catalysts for biomass gasification. Fuel Process Technol 2001;73(3):155–73.
[38] Ammar M, et al. Influence of effective parameters on product gas ratios in sorption enhanced gasification. Procedia Eng 2016;148:735–41.
[39] Reed T, Das A. Handbook of biomass downdraft gasifier engine systems. Biomass Energy Foundation; 1988.
[40] Puig-Arnavat M, Bruno JC, Coronas A. Review and analysis of biomass gasification models. Renew Sustain Energy Rev 2010;14(9):2841–51.
[41] Inayat M, et al. Effect of various blended fuels on syngas quality and performance in catalytic co-gasification: a review. Renew Sustain Energy Rev 2019;105:252–67.
[42] Rios MLV, et al. Reduction of tar generated during biomass gasification: a review. Biomass Bioenergy 2018;108:345–70.
[43] Arena U. Fluidized bed gasification. In: Scala F, editor. Fluidized bed technologies for near-zero emission combustion and gasification. Woodhead Publishing; 2013. p. 765–812 [Chapter 17].

[44] Gredinger A, Spörl R, Scheffknecht G. Comparison measurements of tar content in gasification systems between an online method and the tar protocol. Biomass Bioenergy 2018;111:301–7.
[45] Li C, Suzuki K. Tar property, analysis, reforming mechanism and model for biomass gasification—an overview. Renew Sustain Energy Rev 2009;13(3):594–604.
[46] Adnan MA, et al. Iron oxide over silica-doped alumina catalyst for catalytic steam reforming of toluene as a surrogate tar biomass species. Energy Fuel 2017;31(7):7471–81.
[47] Inayat M, et al. Catalytic and noncatalytic gasification of wood–coconut shell blend under different operating conditions. Environ Prog Sustain Energy 2019;38(2):688–98.
[48] Hernández JJ, Ballesteros R, Aranda G. Characterisation of tars from biomass gasification: effect of the operating conditions. Energy 2013;50:333–42.
[49] Lapuerta M, et al. Gasification and co-gasification of biomass wastes: effect of the biomass origin and the gasifier operating conditions. Fuel Process Technol 2008;89(9):828–37.
[50] Guan G, et al. Catalytic steam reforming of biomass tar: prospects and challenges. Renew Sustain Energy Rev 2016;58:450–61.
[51] Myrén C, et al. Catalytic tar decomposition of biomass pyrolysis gas with a combination of dolomite and silica. Biomass Bioenergy 2002;23(3):217–27.
[52] Feng D, et al. Effects of H2O and CO_2 on the homogeneous conversion and heterogeneous reforming of biomass tar over biochar. Int J Hydrogen Energy 2017;42(18):13070–84.
[53] Milne TA, Evans RJ, Abatzaglou N. Biomass gasifier "'Tars'": their nature, formation, and conversion. Golden, CO, US: National Renewable Energy Laboratory; 1998.
[54] Anis S, Zainal ZA. Tar reduction in biomass producer gas via mechanical, catalytic and thermal methods: a review. Renew Sustain Energy Rev 2011;15(5):2355–77.
[55] Vreugdenhil BJ, Zwart R, Neeft JPA. Tar formation in pyrolysis and gasification. ECN; 2009.
[56] Burhenne L, et al. Technical demonstration of the novel Fraunhofer ISE biomass gasification process for the production of a tar-free synthesis gas. Fuel Process Technol 2013;106:751–60.
[57] Milne TA, Evans RJ, Abatzoglou N. Biomass gasifier "Tars": their nature, formation, destruction, and tolerance limits in energy conversion devices; 1997.
[58] Hasler PH, Nussbaumer T. Gas cleaning for IC engine applications from fixed bed biomass gasification. Biomass Bioenergy 1999;16(6):385–95.
[59] Woolcock PJ, Brown RC. A review of cleaning technologies for biomass-derived syngas. Biomass Bioenergy 2013;52:54–84.
[60] Asadullah M. Biomass gasification gas cleaning for downstream applications: a comparative critical review. Renew Sustain Energy Rev 2014;40:118–32.
[61] Yu Q, et al. Temperature impact on the formation of tar from biomass pyrolysis in a free-fall reactor. J Anal Appl Pyrolysis 1997;40:481–9.
[62] Brage C, et al. Tar evolution profiles obtained from gasification of biomass and coal. Biomass Bioenergy 2000;18(1):87–91.
[63] Kinoshita CM, Wang Y, Zhou J. Tar formation under different biomass gasification conditions. J Anal Appl Pyrolysis 1994;29(2):169–81.
[64] Knight RA. Experience with raw gas analysis from pressurized gasification of biomass. Biomass Bioenergy 2000;18(1):67–77.
[65] Wang W, et al. Catalytic hot gas cleaning of fuel gas from an air-blown pressurized fluidized-bed gasifier. Ind Eng Chem Res 2000;39(11):4075–81.
[66] Narvaez I, et al. Biomass gasification with air in an atmospheric bubbling fluidized bed. Effect of six operational variables on the quality of the produced raw gas. Ind Eng Chem Res 1996;35 (7):2110–20.

[67] Han J, Kim H. The reduction and control technology of tar during biomass gasification/pyrolysis: an overview. Renew Sustain Energy Rev 2008;12(2):397–416.
[68] Baruah D, Baruah DC. Modeling of biomass gasification: a review. Renew Sustain Energy Rev 2014;39:806–15.
[69] Noichi H, Uddin A, Sasaoka E. Steam reforming of naphthalene as model biomass tar over iron–aluminum and iron–zirconium oxide catalyst catalysts. Fuel Process Technol 2010;91 (11):1609–16.
[70] Inayat M, Sulaiman SA, Sanaullah K. Effect of blending ratio on co-gasification performance of tropical plant-based biomass; 2016.
[71] Simell PA, Hepola JO, Krause AOI. Effects of gasification gas components on tar and ammonia decomposition over hot gas cleanup catalysts. Fuel 1997;76(12):1117–27.
[72] Simell PA, et al. Catalytic decomposition of gasification gas tar with benzene as the model compound. Ind Eng Chem Res 1997;36(1):42–51.
[73] Stark AK, et al. Prediction and validation of major gas and tar species from a reactor network model of air-blown fluidized bed biomass gasification. Energy Fuel 2015;29(4):2437–52.
[74] Devi L, Ptasinski KJ, Janssen FJJG. Pretreated olivine as tar removal catalyst for biomass gasifiers: investigation using naphthalene as model biomass tar. Fuel Process Technol 2005;86 (6):707–30.
[75] Thambimuthu KV. Gas cleaning for advanced coal-based power generation. 53. London: IEA Coal Research; 1993.
[76] Baker EG, et al. Engineering analysis of biomass gasifier product gas cleaning technology. Richland, WA, USA: Pacific Northwest Lab.; 1986.
[77] Bhoi PR. Wet scrubbing of biomass producer gas tars using vegetable oil; 2014.
[78] Hasler P, Nussbaumer T. Gas cleaning for IC engine applications from fixed bed biomass gasification. Biomass Bioenergy 1999;16(6):385–95.
[79] de Jong W, et al. Biomass and fossil fuel conversion by pressurised fluidised bed gasification using hot gas ceramic filters as gas cleaning. Biomass Bioenergy 2003;25(1):59–83.
[80] Khummongkol D, Tangsathitkulchai C. A model for tar-removal efficiency from biomass-produced gas impinging on a water surface. Energy 1989;14(3):113–21.
[81] Lettner F, Timmerer H, Haselbacher P. Biomass gasification -state of the art description. Guideline for safe and eco-friendly biomass gasification Intelligent Energy-Europe (IEE). Austria: Graz University of Technology-Institute of Thermal Engineering; 2007.
[82] Bocci E, et al. State of art of small scale biomass gasification power systems: a review of the different typologies. Energy Procedia 2014;45:247–56.
[83] Morf P, Hasler P, Nussbaumer T. Mechanisms and kinetics of homogeneous secondary reactions of tar from continuous pyrolysis of wood chips. Fuel 2002;81(7):843–53.
[84] Fassinou WF, et al. Pyrolysis of Pinus pinaster in a two-stage gasifier: influence of processing parameters and thermal cracking of tar. Fuel Process Technol 2009;90(1):75–90.
[85] Abu El-Rub Z, Bramer EA, Brem G. Review of catalysts for tar elimination in biomass gasification processes. Ind Eng Chem Res 2004;43(22):6911–9.
[86] Phuphuakrat T, et al. Characterization of tar content in the syngas produced in a downdraft type fixed bed gasification system from dried sewage sludge. Fuel 2010;89(9):2278–84.
[87] Fjellerup J, et al. Formation, decomposition and cracking of biomass tars in gasification. Technical University of Denmark; 2005.
[88] Fagbemi L, Khezami L, Capart R. Pyrolysis products from different biomasses: application to the thermal cracking of tar. Appl Energy 2001;69(4):293–306.
[89] Liu C-J, Xu G-H, Wang T. Non-thermal plasma approaches in CO2 utilization. Fuel Process Technol 1999;58(2–3):119–34.

[90] Petitpas G, et al. A comparative study of non-thermal plasma assisted reforming technologies. Int J Hydrogen Energy 2007;32(14):2848–67.
[91] Hlina M, et al. Production of high quality syngas from argon/water plasma gasification of biomass and waste. Waste Manag 2014;34(1):63–6.
[92] Materazzi M, et al. Performance analysis of RDF gasification in a two stage fluidized bed–plasma process. Waste Manag 2016;47:256–66.
[93] Diaz G, et al. Enhanced hydrogen production using steam plasma processing of biomass: experimental apparatus and procedure. Int J Hydrogen Energy 2015;40(5):2091–8.
[94] Kim H-H. Nonthermal plasma processing for air-pollution control: a historical review, current issues, and future prospects. Plasma Process Polym 2004;1(2):91–110.

Section B

Biogas energy

Chapter 4

Biogas: A promising clean energy technology

Muhammad Arshad[a], Sadia Javed[b], Abdur Rahman Ansari[a], Arooj Fatima[a], and Mirza Imran Shahzad[c]

[a]*Jhang-Campus, University of Veterinary and Animal Sciences Lahore, Jhang, Pakistan,* [b]*Department of Biochemistry, Government College University Faisalabad, Faisalabad, Pakistan,* [c]*Department of Biochemistry and Biotechnology, The Islamia University of Bahawalpur, Bahawalpur, Pakistan*

1 Introduction

Biogas is the blend of various gases formed as a result of the microbial activities on organic matter present in many kinds of wastes including agricultural, industrial, and municipal wastes [1]. It consists of mostly methane (CH_4) in the range of 50%–80%. Some other gases, CO_2, H_2, N_2, and O_2, are present in low proportions [2]. If the methane present in the mixture is above 50%, then the mixture is combustible. Anaerobic digestion is the biotechnology used worldwide for biogas production through the treatment of organic waste. Biomass wastes such as manure are utilized by anaerobic digestion to produce biogas for the reason that it is a renewable energy source as well as a help in solving ecological and agrochemical problems [3]. The anaerobic fermentation of animal sludge for the generation of biogas does not decrease its significance as a fertilizer supplement source. The essential nitrogenous and other compounds remain present in the treated manure but also make it available for the plants [4]. In order to enhance the generation of biogas through anaerobic digestion, various studies have taken place all over the world. Two continuously stirred tank reactors were used to improve biogas yield and its conversion efficiency [5]. In developing nations, biogas technology is used on a vast scale for the generation of power and heat. It is also upheld by government agencies and quite a lot of international organizations, including the UN and the EU [6], because it decreases the discharge of greenhouse gas from manure and helps in the production of sustainable energy [7,8]. Its increased production for the rural households will help in reducing the random cutting of trees used for burning of fire, thereby decreasing the adverse effect of soil erosion, greenhouse gas emanation, and

Bioenergy Resources and Technologies. https://doi.org/10.1016/B978-0-12-822525-7.00011-1
Copyright © 2021 Elsevier Inc. All rights reserved.

much more. Biogas that is generated from waste products of animals is used as a source of sustainable biofuel. This source of energy has been observed and declared as clean and low in cost and is also able to generate a residue with a high fertilizer value for crop production [8,9]. Anaerobic bacteria are widely found in the rumen of the ruminant obtained from the slaughterhouse and are used in the production of biogas.

The global desire to shrink greenhouse gas (GHG) release for the control of adverse climate changes and to fulfill worldwide energy requirements has improved and enthused biogas production from many sources. Biological production of methane is a proven environment-friendly source of energy that can take part in the reduction of carbon cost. Biogas composition and its properties depend on the temperature, pH, and retention time during the anaerobic process. Moreover, energy generation from biogas depends on the biomaterials used as feedstock, the pretreatment procedures, and the biological digestion process. Policies to control and enhance the generation and utilization of biogas are vital for the promotion of such environment-friendly sources. The biogas can be produced from a massive range of resources with very diverse nature. Easily available sources of raw materials must be preferred, and the type of process must be optimized for several substrates [10]. Metabolic profiles of the employed microbes also impact the fluctuating nature of biogas.

From the efficiency and yield point of view of biogas production, it was concluded that methane % (presence of methane in biogas) has a significant importance in the valorization and application of biogas. Moreover, several impurities such as H_2S (causing corrosion), uncontrolled emissions have many negative effects on the utilization system. By the way, promising biogas upgrading and cleaning technologies have gained much attention of the bio-renewable community [10,11]. By the utilization of thermophilic upflow reactors, a higher content of 23%–96% of methane was obtained with H_2 and CO_2 that we provide externally and are finally converted into CH_4 [11]. Methane contents were enhanced up to 98% by the process of hydrogenotrophic methanogenesis in the two upflow reactors [12]. In 2018, methane yield was increased with the decrease in lag phase almost by 150 through a biogas upgrading technique. The technique was also applicable under thermophilic conditions [13]. In the selection of low-cost technology, circumstance, site-specificity, and case-sensitivity must be considered. The financial and environmental costs are the other factors that must be thought over before adoption of biogas upgrading technology with the feasibility of each strategy employed [10].

According to Amon et al. [14] production of biogas from numerous organic resources is based on the concentration of the constituents convertible to volatile acids that may be turned into biogas. Chemical nature and ability to biodegradation are highly substantial parameters for the biogas production with better methane concentration. To achieve better production of biogas, most appropriate indigenous materials must be preferred. Some substrates may be deficient in some nutritive factors necessarily required for microbial growth and other operational parameters [15]. Pretreatment of the biomass is essential for better availability of substrate to microbes. The co-digestion of organic matter from different resources can fulfill the nutritional deficiencies to enhance the production of biogas through biological digestion of biomass [16]. It is worth mentioning that the cost of raw material impacts the cost-competitiveness of the biological anaerobic digestion process. The cost of raw material, its availability, the distance at which the raw material is available (the distance from the anaerobic digestion site) majorly affects the economy of the process.

2 Anaerobic digestion

Anaerobic digestion is a biological process to produce biogas by the decomposition of organic matter, which is an excellent source of renewable energy. The nutrients present in the organic matter are released and biodegraded. The process is environmentally friendly. Biogas is a proven form of sustainable energy source. It can compensate for different forms of energy. The reliance on fossil fuels is thereby reduced, and climate change is mitigated with the reduced emission of greenhouse gases. The energy security problems can be resolved through diversification of different energy forms.

Biogas production and information as to its advantages are well known to humans from many centuries. Currently, one of the significant challenges is to increase the harvest of renewable energy from the substrates that are not easy to digest. The biological digestion process is highly efficient in green energy production with the environment-friendly disposal of wastewater from the industrial sector. The process can also be environment friendly in disposing of the urban wastes and sewage sludge. The animal manure and slaughterhouse waste also can be digested through the process for the provision of green energy.

The biological process of anaerobic digestion comprises four phases of biochemical transformation. Fig. 1 represents the process of biochemical transformations in the order of their happening.

From these four biochemical transformations, hydrolysis is the rate-limiting stage, particularly for those substrates that are difficult to degrade. In anaerobic digestion, easily biodegradable substrates' digestion is limited by methanogenesis. In most of the cases, decreased degradation of recalcitrant substrates forces the anaerobic digestion process to operate below the ideal performance level. The application of pretreatment techniques, especially in the case of lignocellulose substances, can improve the decomposition. Agricultural biomass is mostly comprised of lignocellulosic substances, but their relative amount is different in the biomass. As the materials available for anaerobic digestion vary in their properties, so the use of a specific pretreatment is essential and can enhance the digestibility and production of biogas. The pretreatments can be used together or in sequence to enhance the conversion of the organic matter into bioenergy.

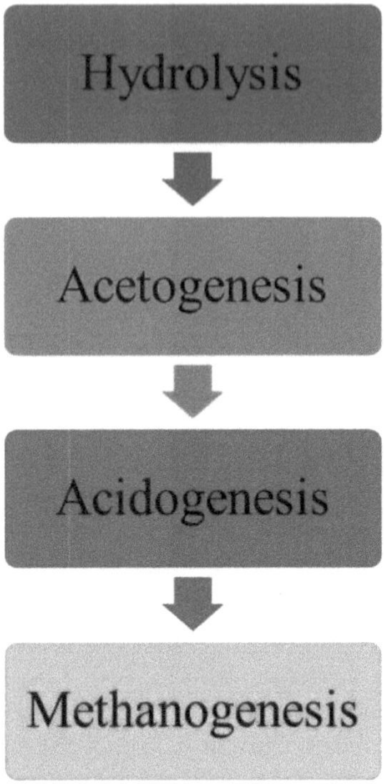

FIG. 1 Four phases of anaerobic digestion.

2.1 Paybacks of the anaerobic digestion

Following are the paybacks of the process:

1. Different wastes and biomass are biodegraded and transformed into biogas by the process of anaerobic digestion, so the wastes are disposed of [17].
2. Due to the high organic loading rate, a very limited area is required.
3. No oxygen is required for the process, as it takes place in anaerobic condition.
4. The process of anaerobic digestion helps in the reduction of odor arising from the wastewater. In the methane production process from waste, a sense of balance is maintained among the formation of volatile acids and their conversion to biogas. The odor is decreased with the conversion of volatile intermediates into methane [18].
5. Methane and carbon dioxide are the end products of this biological process; both these gases are the leaders for greenhouse gas phenomena. So such harmful gases are locked in, and their open disposal in the atmosphere is prohibited because the anaerobic digestion process occurs in the fully sealed

off chambers. In this way, the damaging impact of these gases is avoided [19]. Hence, global warming and climate change can be prevented [20].

6. Pollution of wastewater resources as a result of agricultural, industrial, and human activities is prevented by anaerobic digestion. Risk of pathogens present in the sewage and wastewater is reduced due to efficient anaerobic degradation, which are otherwise disposed of into water bodies that serve as a reservoir for freshwater [21]. Organic matter in the high COD and BOD is present in wastewater and industrial effluent is reduced.
7. Animal manure can be properly managed in an economical way by decomposing it through anaerobic digestion before its application on agricultural fields [22]. Otherwise, the animal manure is spread into nearby water bodies in the monsoon rainfall [23]. The chemical oxygen demand (COD) and biological oxygen demand (BOD) of the lagoons are increased, thus increasing the pollution. So the decomposition of animal manure in a limited place can limit the animal manure from being permeated by rainfall [24]. The major benefits of the process are presented in Fig. 2.
8. The decomposition process of anaerobic digestion conserves the nutrients during biological breakdown. Only carbon is the element removed from the substrate, while other elements as well as phosphorus and nitrogen are well maintained in the form of their compounds and further converted into their active biological formations. In this way, the solubility and bioavailability of desirable nutrients are increased for the plants so that they can uptake and utilize directly [25].
9. The occurrence of the biological anaerobic digestion process for the treatment of waste is an economical and inexpensive process. The procedure is very straightforward and requires no specialized and refined equipment. The input of the energy for the occurrence of the process is much lower as compared to the orthodox aerobic treatment processes [26]. The net

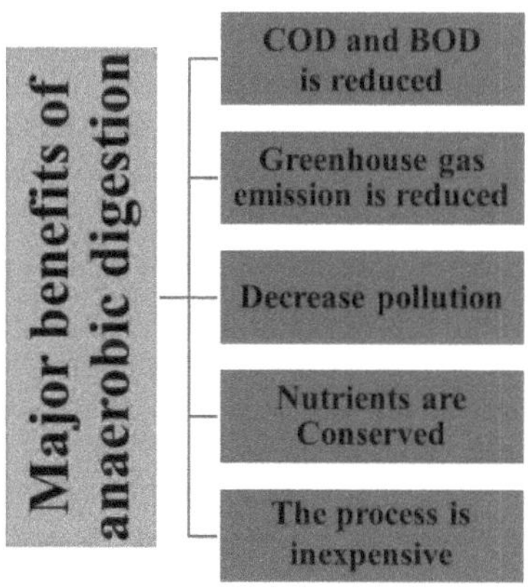

FIG. 2 Benefits of anaerobic digestion.

energy of the process is positively available for anaerobic digestion, whereas the net energy produced by the biogas surpasses the energy required for the operation of the digester [27]. In some cases, positive energy balance becomes hard to achieve when the extra input energy is desired for heating purposes [28]. The extra energy needed for temperature maintenance of the digester within range for, in effect, performance depends on the type of digester, the inlet temperature of the substrate used, the materials employed for fabrication, position of the digester underground or above the ground, the use of insulation, and finally use of heat exchangers. In cold areas, an insulated and underground digester can help in the maintenance of internal temperature by prevention of heat loss from the body of the digester [29]. The utilization of heat exchangers can also help in the maintenance and optimization of internal temperature. Heat exchangers can use the heat recovered from the degassed sludge, thus obtaining the positive energy balance [27].

3 Principles of anaerobic digestion

Different waste to bioenergy processes have been explored for the energy generation from biomass and waste substances and can be divided into biochemical and thermochemical techniques [30–32]. Biochemical technologies include two major processes, anaerobic and aerobic fermentation. Combustion, gasification, and pyrolysis are the major players under thermochemical technologies [30–34]. The breakdown process of organic matter through consortia of microbes in the absence of oxygen produces biogas that was comprised majorly of CH_4 and CO_2. The digestate obtained keeps a high nutritional content. This digestate can be used as a bio-fertilizer in the agriculture field [35–37]. Hydrolysis, acidogenesis, acetogenesis, and methanogenesis are four major processes [38]. Before the anaerobic digestion process, lignocellulosic wastes and agricultural biomass are pretreated to soften the texture and increase the surface area for better accessibility of microbes for hydrolysis [39,40].

The presence of total solids, concentration of carbon and nitrogen, the pH of the media, and the temperature of the media are as well tuned and maintained to enhance the biological process of anaerobic digestion [40]. After the treatment phase, the biomass is entered into the digesters with a predetermined amount of inoculums to get a faster digestion process [41]. During the first phase of anaerobic digestion, complex molecules are decomposed into water-soluble monomers [36–38]. The small monomers are then processed into short-chain volatile fatty acids during acidogenesis [36–38]. These volatile fatty acids are broken down into acetic acids, H_2 and CO_2, in the third phase. In the fourth phase of methanogenesis, CH_4 and CO_2 are produced [36–38]. The biogas produced can be used in thermal and electrical energies in combined heat and power stations while the digestate may be employed as a bio-fertilizer to improve soil fertility [36–40]. The process has been used for a variety of feedstocks together with agricultural waste, wastewater, sewage sludge, municipal solid wastes, and animal manure high organic matter and moisture content [42–44].

4 Co-digestion

Recently, co-digestion has been adopted as a procedure to enhance biodigester efficiency in the production of biogas [45]. Co-digestion can utilize municipal waste, waste of food industry, animal waste, slaughterhouse waste with crop residue and animal manure, which are major stakeholders in biogas production [46]. The process is not simple, but it needs to estimate the biogas production capability of each component. Knowledge about the presence of necessary nutrients is vital before the ratio of each substrate is set for an enhanced yield of biomethane. In a study, six feedstock mixtures of different origins were tested for biomethane production potential, and a difference up to 100% was set up [47].

Lignocellulosic materials are hard to digest and need pretreatment. Considering the point, before co-digestion with a lignocellulosic biomass, pretreatment of the feedstock is vital to optimize the procedure. Many researchers have explored keeping the process economically favorable [48]. Currently, lime (CaO) was used as the pretreatment material of miscanthus biomass and it was noticed that the biochemical methane potential was raised up to 37% [49]. With the co-digestion strategy, cornstalks were digested with swine manure; as a resultant, the final volume of methane increased up to 22%. The time reduction was up to 60% when hydrogen peroxide is pretreated with a lignocellulosic biomass [50].

Hydrothermal carbonization is another essential and inexpensive pretreatment material. Addition of hydrochar in the food waste increased the methane specific yield up to 2.5-folds [51], while biochar as a substrate in swine manure anaerobic fermentation increased 39% in methane production [52].

Still focused on raw materials for pretreatment, many studies put emphasis on the readily ease use of materials, for example, availability of sugars for conversion to biogas through various pretreatment procedures. These procedures are favorable for release of glucose from polymers like cellulose [53,54]. The major inhibiting reason for effective hydrolysis of the lignocellulosic biomass is the crystalline nature of the polysaccharide and the hindrance in conversion to biogas. Breakdown of hydrogen bonds and the disruption of crystals is essential for enzyme action. When enzymes and catalysts access the structure of cellulose, then the hydrolysis of the glycosidic bonds is possible to release the monomers of glucose. Glucose yield is increased twofold by the utilization of moderate acids, for example, phosphoric acid and trifluoroacetic acid, during the breakdown of cellulose with a commercial enzyme [51].

Several examples of pretreatment procedures that use a single substrate are present. Organic-bonding metals from a sewage sludge are removed by using EDTA, and by this pretreatment, the metal in the substrate is reduced from 5.1% to 1.4% [55]. Besides, this biomethane production increased up to 48%. Co-digestion of the various substrates has been presented in Fig. 3.

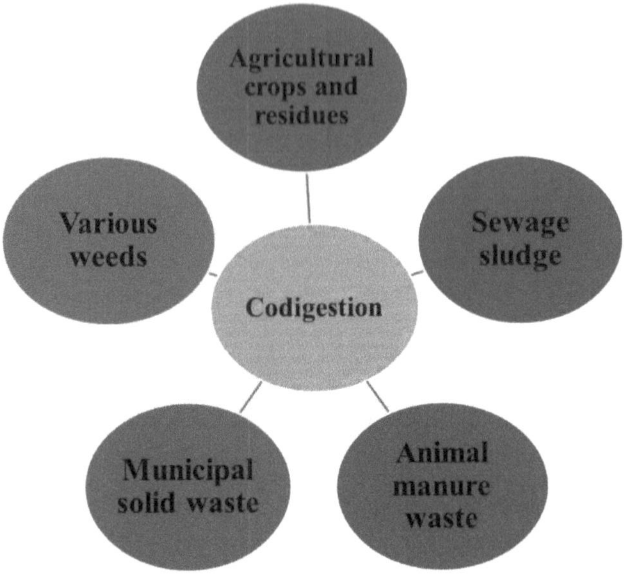

FIG. 3 Co-digestion of various substrates.

The feedstocks for the production of biogas are so easily available that even feathers can also be used as feedstock. Many studies in the literature are presented; besides, some strategies are already present to enhance the production of methane by pretreatment procedures such as alkaline, thermal, enzymatic, and biological pretreatments. These procedures and strategies are applied alone or in combination to enhance up to 20 times the production of CH_4 [56–58]. It was observed that utilization of liquid nitrogen with the decrease in particle size and increased solubility could enhance the bioavailability, and thus the methane production can be enhanced from wool by more than 80% [59].

The pretreatment procedures can be a burden considering the aspect of economics, energy, and environment. Even though biogas production is enhanced, the utilization of many pretreatment procedures can enhance energy consumption and cost by utilizing hydrolytic enzymes and even the release of carbon [60]. The use of inoculums containing consortia of microbes that are previously adapted and selected can be a sustainable strategy. In many other bioprocesses, the strategy is being employed. It will favor the production of biogas, with cost-effectiveness and in an environment-friendly manner.

5 Feedstock for the generation of biogas

Biogas can be produced from any raw materials containing polymers of biomolecules, especially carbohydrates. The raw materials include but are not limited to agricultural waste (e.g., crop residue), animal waste (such as animal manure),

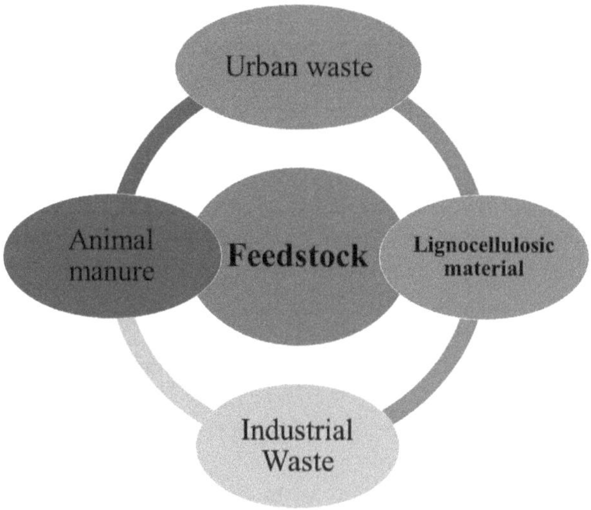

FIG. 4 Major sources of feedstock.

landfill, waste of the food industry, aquatic biomass, and lignocellulosic raw materials. These are all abundantly and widely available. Most of these feedstocks are comprised of complex molecules and break down very slowly. Long retention time is needed for these substrates. Some raw materials are either formed from harmful intermediates or keep hazardous compounds, which limit the production of biogas. But, due to their great quantity and cheaper availability, there is a need to explore novel procedures for their better usage [61]. Before the raw material is utilized, its biochemical nature must be known. The raw materials are composed of carbohydrates or protein. Based on the biochemical composition, the operational parameters are temperature, and pH must be characterized. Biodegradability, bio-accessibility, and bioavailability are the other significant factors that impact optimization of biogas production [62]. Fig. 4 shows the sources of various kinds of feedstock.

Although, any material keeping organic matter can be used for the production of biogas, but it can be produced easily from those raw materials that can be easily accessed and decomposed by the microbial community. Digestible raw materials, for example, residue from crops, waste from municipal activities, waste of food, and industrial wastewater with high organic content, are plentiful and easily available for the production of biogas [63].

5.1 Urban waste

High trade and industrial activity, and increased population have boosted the production of urban solid waste or municipal waste [64]. Such aspects are in control for producing more amount of waste, especially in large urban areas, apartments, and small commercial establishments, for example, restaurants and bakeries [65]. After the separation, urban solid waste can be used again for other purposes, but it is essential to maximize the value of the organic matter acquired from this source. Municipal solid waste is commonly comprised of waste from kitchen and food, garden waste that can be easily treated by anaerobic digestion. Organic matter presence can be up to 46% that can be transformed into biogas [66].

For waste treatment, the first step is its transport from the urban site to the biodigester. Then is the separation of the waste, so the efficiency in separation affects the overall efficiency of the digestion process. Additionally, separation of the residue at the source allows the appropriate use of the digestible organic fraction and decreases the effect related to the use of conventional energy in the process of mechanical separation [67]. Producing a flow that has no costs of raw material and only needs transportation to anaerobic digesters is also essential if food waste is taken to landfills [68], where approximately 97% of food wastes end up [69].

The maximum value of organic waste is procured when bioenergy is produced for various applications. So anaerobic digestion is well thought out as a cost-effective and environmentally friendly solution for the disposal of organic waste [70]. Each ton of organic content originated from municipal solid waste yields up to 200 m^3 of biogas and methane yield of up to 330 L/kg of total volatile solids [71,72].

The construction and development of biogas plants are increasing day by day in order to treat food waste in urban areas and attained much better results in biogas production [73], improving the management of food and food components whose waste is unavoidable. Such wastes primarily contain lignocellulosic materials and are rich in carbohydrate materials [74], substrates potentially appropriate for biogas production. The success of the anaerobic digestion of a carbohydrate-rich substrate, though, relies upon equilibrium between acidogenesis and methanogenesis [66].

Chongqing, a city of China, has constructed biogas plants in 2009 for the bio-digestion of waste based on food. Anaerobic fermentation technology was employed to bioprocess 1000 tons of waste from food on a daily basis and manufactured 28 million m^3 of biogas per year, producing 33 million kWh of electricity. In Poland, Lubelskie Voivodeship city utilized food waste for the production of biogas and generation of electricity. The electricity generated accounts for 1.5% of total electricity consumption in the city and 0.18% of electricity generated from sustainable resources countrywide in 2012 [75,76].

The plan for solid waste management with the anaerobic digestion process was executed to minimize the cost of waste transportation and maximize the benefits of the final product to the local community. Contribution of the community in waste management can be ensured by providing an incentive for households to separate their own waste. In developed countries, domestic waste is the biggest contributor to food waste [68]. Decentralization of anaerobic digestion needs a space of <3000 t/a [76–78], which can be accomplished through multiple units or smaller digester cells for batch feeding [76]. Changes in the waste composition create asymmetry in the system by the use of insufficient carbon and nitrogen rates, which are essential for the optimization of the process.

Fluctuation in the production of methane due to variation in the composition of municipal solid waste was studied. Urban organic waste is composed of domestic waste urban and rural, restaurant, and food processing waste bakery and cheese, with almost 10kg from a global sample of each waste source. The study revealed that restaurant waste samples keep a higher value of biochemical methane potential (491–535mL CH_4g/VS in 30days) and a more significant part of organic matter as compared to waste samples from houses (274–368mL CH_4g/VS in 30days) [65]. Garden waste significantly decreased the production of methane when added to domestic waste. The samples of commercial food waste were not favorable because ammoniacal nitrogen is present in excess, causing a low carbon and nitrogen ratio (C/N). C/N ratios of 25–30 are ideal for anaerobic digestion of the organic substances of urban solid waste [79].

A pilot-scale landfill site was studied to measure the biogas produced by urban solid waste where municipal waste was landfill from 101.60kg of waste. The temperature had a direct connection with biogas production; the higher temperature produces more methane content (0.799m^3/t day) as microbial activity was enhanced [64].

For anaerobic digestion of urban waste, raw material needs pretreatment steps, starting with the mechanical separation of inorganic components, for example, plastics and metals [80–82]. In the next step, particles must be ground to decrease the particle size [69] or even liquefied to enhance the anaerobic digestibility of food particles [83,84]. Pretreatment needs further analysis to conclude their economic viability because their goal is to stimulate hydrolysis and enhance the solubility of the particles. Mathematical modifications can help in the selection of suitable combinations of raw materials and pretreatment procedures [66].

Waste materials from restaurants and food processing industries are comprised of fats, oils, and greases. Such constituent parts are identified as high lipid content with readily degradable organic components. Lipids can produce approximately double the amount of biogas compared to carbohydrates and proteins [85,86]. A sample of these lipid particles, along with the inoculum of anaerobic sludge, was examined [87], and it was verified that those lipid particles have the ability to be an excellent source for biomethane production.

In sanitary sewage treatment plants, biogas can be produced from the treatment of the effluent in anaerobic reactors and from anaerobic digestion of sludge obtained from the treatment [88]. The utilization of sludge as auxiliaries in co-digestion needs a more precise study of its aspects because sludge may have an adverse effect on the environment and it also affects the survival and activities of microbes responsible for anaerobic degradation. Sludge might also have harmful substances, for example, surfactants and detergents that may reduce the production of biogas [64].

Municipal organic wastes are used as an energy source to solve waste disposal problems and fewer effects on the environment and lower costs as compared to landfill disposal [67,89]. In the resolution of ratios of low macro and micronutrients, harmful substances dilution, the stability of pH, temperature, moisture content, various substrates can be used as an anaerobic co-digestion alternative. By this, control of the procedure is challenged and is directly related to the physical-chemical parameters of the raw material, which can expand across regional, seasonal, and socioeconomic circumstances [52,66,67].

5.2 Animal manure waste

From agricultural waste, the key raw material for biogas production is animal manure obtainable from swine, cattle, and poultry. The utilization of biodigesters takes part in the incorporation of agricultural activities, transforming manure, which normally has a minute or no commercial value, into energy. Animal manure is considered as a high ideal substrate for biogas production because it better integrates energy production and nutrient recycling and decreases CH_4 and NO_2 emanations more than conventional manure management [18,90].

The presence of high water content in animal manure makes it economically harder to operate biogas plants only with animal manure. There have been attempts to enhance the efficiency by the addition of lignocellulosic material; as a result, the dry matter and C/N ratio are increased. Biogas yield was improved by the addition of green vegetation from 18% to 40% when co-digested with animal manure, while a 10%–80% increase was noticed with the agricultural residues of wheat straw, rice straw, and corn stems, relying upon the pretreatment conditions and dry matter contents [91].

The optimal percentage for biogas production changes greatly with co-digestion. Methane yield of 298.2 $mL\,g^{-1}$ VS is produced by mixing chicken manure and corn stover in the ratio of 1:3, which is higher as compared to chicken manure alone (291.1 $mL\,g^{-1}$ VS). In another batch test, the co-digestion of chicken manure and corn stover is seen with a 1:1 ratio [92–94,87].

The anaerobic digestion process of poultry manure is more complicated than those of other animal manures as poultry waste contains a high content of nitrogen; as a result, inhibition of ammonia occurs [88]. Ammonia is produced due to the presence of a high content of uric acid and undigested proteins in the anaerobic digestion of chicken waste [95,96]. In addition to toxicity, chicken manure consists of a fraction of wood chips used as bedding materials, so such high NH_3 and lignocellulose particles are less affected by anaerobic destruction for biogas production [97–99]. Various methods to prevent inhibition effects during anaerobic digestion of ammonia-rich organic substances have been explored. Co-digestion of chicken manure with other substrates is an excellent approach to control the inhibition caused by ammonia [100].

5.3 Industrial effluents and waste

The industrial effluents and wastewater generated from industrial processes are utilized for biogas production through anaerobic digestion as its treatment strategy and resultant energy was harvested [101]. The use of industrial wastewater for biogas production has greatly increased during the past 10 years, with the number of articles published increasing from 53 during 2000–2004 to 90 during 2010–2014. Many articles on organic wastewater arising from industrial processes are most cited [102–104]. A substitute for energy production from utilization of wastewater from inorganic sources, such as chromium, has also been studied and is a more environment-friendly approach that prevents landfill disposal of these wastes [105,106].

High organic load in industrial waste and wastewater has made it an ideal candidate for its significant use in biogas production [107]. Most of the times, landfill methods are used to dispose of the industrial effluents that produce a bad smell. The process has environmental concerns and high operational costs. Significant benefits of the biological anaerobic digestion process are the low cost of the energy utilized [108].

In addition to generating electricity, the biogas produced can be used in vehicles, domestic heating, and chemical industries. Brazil, in particular, had 15 biogas plants with a total production of 114.7 MW of energy, equivalent to 0.83% of the total national biomass capacity, as of January 2017 [109].

Wastewater from the leather industry keeps chromium and vegetable tannin that is used in the production of biogas. An increased amount of biogas is produced because the enzymes and coenzymes rely upon the metals for their activity, and the concentration of sludge was optimum for production. When chromium was used in the experiment, an increased amount of methane was produced. When the particles containing 7.79% chromium were used in experiments, 21.5 mL of biogas/VSS and 10.7 mL of methane/g VSS were produced. It was also stated that a low concentration of chromium in the sludge had the ability to produce biogas and methane and was anaerobically biodegradable [110].

Wastewater from cassava processing digestion and several types of waste types (fruit bagasse and leaves) in various ratios of raw material: inoculum (2:1, 3:1, 4:1, 5:1) over 45 days of cultivation were used [111]. When the inoculum ratio of 2:1, C/N ratio of 40:1, and initial concentration of 16% total solids in the fruit bagasse particles were maintained, the highest methane production was achieved [112]. The study shows that the thermophilic phase in the production of biogas is more effective as compared to the mesophilic phase because of the breakdown of substrate increase in the thermophilic phase [101]. The objective of tapioca production is biogas production. Over 45 days of production, 1% tapioca residue, 0.08% yeast, 0.04% urea, and 10% bacteria were added, evaluated, and compared with production of biogas in a tank without adding yeast. In the presence of yeast, a higher amount of biogas production was noted as compared to biogas production without yeast [112]. The supplementation of the yeast *Saccharomyces cerevisiae* enhanced hydrolysis of the effluent. The polysaccharides were hydrolyzed to monosaccharides and were used by the methanogenic bacteria as a source of the substrate, and, as a result, production of biogas was increased [78].

Biogas was also produced from the anaerobic digestion of solid waste from the palm oil industry. Solid-state anaerobic digestion and several types of waste types in various ratios of raw material: inoculum (2:1, 3:1, 4:1, 5:1) over 45 days of cultivation were used [112]. When the inoculum ratio of 2:1, C/N ratio of 40:1, and initial concentration of 16% total solids in the residue of fruit bagasse particles were maintained, the highest methane production was achieved. The study also shows that the thermophilic phase in the production of biogas is more effective as compared to the mesophilic phase because of the breakdown of the substrate increase in the thermophilic phase [113].

Sludge from the pulp industry was used for the production of biogas through anaerobic digestion in the thermophilic phase. When water passed through a press filter system, three types of sludge were achieved. Higher accumulated methane production was achieved during the second type of sludge during an assay over 30 days. Cumulative methane yields of 3.5 and 3.3 mLN CH_4/g VS were achieved by the primary sludge and mixture of both primary and secondary sludge in a 2.5:1 ratio, respectively [114].

In Brazil, the ability of organic residues, including vinasse to produce biogas, was analyzed. Distilleries and the sugarcane industries generated this waste and had the ability to produce biogas of 1142,614.106 m^3/year, which also prevents the emission of 0.725 Mt CO_2/year. This waste also had the ability to generate electricity of 254,675 MW and is viable to utilize because Brazil is the world's largest sugarcane producer [88].

Waste obtained from wine had the ability to produce biogas. Various kinds of waste are produced from wineries such as grape bagasse, must, primary and secondary sludge. A mixture of these wastes containing bagasse and primary and secondary sludge and must was used in the experiment. Must produce the highest yield of biogas, at 1151.71 m^3 ton VS^{-1}, and the mixture of all residues, at 289.13 m^3 ton VS^{-1} [115]. Together, these studies explain that the production of biomethane from industrial waste is a renewable choice for energy production in both developed and developing countries [116].

5.4 Lignocellulosic material

Lignocellulosic biomass is a sustainable raw material and has the ability to produce methane because highly fermentable monomers were present in it [117]. This raw material is comprised of glucose, mannose, xylose, arabinose, and other organic compounds that have high productivity and are easily broken through the process of anaerobic digestion [118,119]. Production of methane through the biomass fluctuates due to the presence of the energy ratio of each and every biomaterial. On the other hand, the utilization of lignocellulosic biomaterials in these production techniques has a drawback as compared to other sources of biomass [120].

The presence of lignin in these materials prevents the action of microbes on the substrate during anaerobic digestion; due to this, the yield of the procedure is directly affected. The transformation ratio of the biomaterials into biogas is increased by the pretreatment procedures.

After the pretreatment procedure is completed, the sugars obtained are transferred to the next phase. The polymers such as cellulose and hemicellulose are decomposed into their simple monomeric structures glucose, xylose, mannose, and arabinose by the process of hydrolysis and degraded by anaerobic digestion [48,110]. Variations in inoculums are used to attain a better transformation ratio of such materials into biogas; such inoculum variations impact when animal waste is used in anaerobic reactors in wastewater treatment plants [111,112]. Various researchers have reported that many sources of lignocellulosic feedstock are capable of producing biogas and these substrates vary in their origin and composition, with most lignocellulosic feedstocks obtained from cereal.

Biogas is produced from cornstalks in the presence of hydrogen peroxide (H_2O_2) and sulfuric acid (H_2SO_4) in various concentrations, using various pretreatments and orbital shaker agitation [50]. An inoculum of mesophilic microbes and pretreated cornstalks was used to test the biochemical potential of biogas production. Biogas production when H_2O_2 is used as a pretreatment in the generation of biogas removes 71.6% of lignin, decreases hemicellulose by 19.3%, and increases cellulose content by 73.4%. By this pretreatment, the final volume of biogas is increased up to 22% and the time to obtain this volume is decreased.

Biomethane is produced from biomasses with high lignin content (corn stover, wheat straw, flax, hemp, miscanthus, and willow). These biomasses undergo enzymatic pretreatments with the laccase enzymes obtained from Trametes Versicolor and peroxidase obtained from *Bjerkandera adusta*. An inoculum from a bovine-manure-effluent treatment plant was added to test the yield of biomethane. As a result, biomethane yield is increased in all the biomasses after the action of enzymes [113].

The yield of methane production is fluctuating and directly associated with physical and chemical characteristics of the substrates used. However, the measurements of the biogas produced, and the pretreatment and types of inoculum used are important factors in yield.

6 Various pretreatment techniques used for biogas production

Most of the raw materials are not appropriate for biogas production due to the following problems:

Microorganisms cannot digest such materials, or the digestion process is very slow; either of those substances is there which can retard the process. Such materials may result from the decomposition of the waste. The major task of pretreatment measures is to ease and speed up the anaerobic digestion through the removal of physical, structural, or biological hindrances. By removing such blockades, the organic matter is fully released and easy to get to the microbial population for their growth.

The complicated structures of agricultural waste hinder the use of agricultural residues for biological digestion. The crystalline nature of agricultural waste reduces the available surface for the action of microbes, and the presence of lignin restricts the digestion process [17]. In particular, the hydrolysis rate is reduced [18]. The hydrolysis process requires monomers that are released from the polysaccharides thus readily available for microbial biocatalysts during the digestion process in anaerobic conditions [19,20]. The pretreatment procedures can be categorized into biological, physical, chemical and the arrangements based on these [21,121]. Well-matched raw materials, enzymes, and the microbial community guide in choosing pretreatment methods.

6.1 Physical pretreatments

In the physical pretreatment procedures, physical parameters such as temperature and pressure are carried out, rather than the application of chemicals or microbes [23]. Physical pretreatment procedures enlarge the area available for microbial action; the penetration ability in terms of pore size is increased, with the destruction of the structure, in the reduction of crystalline properties.

The major issue of the physical pretreatment procedures is extraordinary utilization of energy. The other economic challenge is the everyday repair of the equipment used for the pretreatment procedures [24]. Steam explosion, mechanical milling, pyrolysis, microwave and ultrasound irradiation, and extrusion are the common physical procedures used to reduce the size and increase the surface area.

From the pretreatment procedures, milling is the mechanical process used for the reduction of biomass size [25], normally practiced before the next procedures are applied to ease the process and make it more effective [26]. Mechanical procedures take account of crushing (wet and dry) and many of the milling techniques. The moisture content of raw material decides the mechanical procedure [23].

The structure of agricultural waste is altered, and the particle size is decreased by the application of chemical or biological treatment [21,22]. The reduction of polymerization and crystalline nature enhances the rate of biological digestion [27]. The size of the finished product is dependent on the technique applied [25]. Extra reduction in the particle size of agri-biomass may reduce the biological digestion due to overproduction of volatile acids, thus disrupting the process [28]. Aggregation of decreased size procedures accompanied by other pretreatment techniques is additionally advantageous.

Microwave irradiation and ultrasound technology is one of the physical pretreatments commonly used [23]. It is the best procedure from orthodox techniques that utilize the disturbance in chemical bonds of cell walls. The procedure is very simple to operate and upright in energy efficiency [25]. The irradiations bring a thermal effect in the probable breakage of hydrogen bonds and improve the ease of access. Alteration in the biomass's cell wall structure increased the reactivity of cellulose. The crystalline nature of the cellulose has been decreased. It is set up that after microwave pretreatment, enzyme access to lignocellulose feedstocks has been enhanced [121]. The acid or alkaline pretreatment techniques are easily applied after microwave pretreatment procedures [29]. The only disadvantage of this pretreatment procedure is the extraordinary equipment cost.

To reduce the degree of polymerization and the crystalline region, ultrasound treatment is very effective. The lignin molecules are degraded easily and better accessed for the chemical reactions. Thus, the treated biomass leads to an improvement in the better bacteria's action [23]. The microbubbles are created by the noise, produced by ultrasound waves, and generate homogenous cavities, which are the major cause of physical and chemical effects in liquid solutions.

The coalescing impact of physical and chemical disturbance causes the destruction of the cell wall and its structure, but the structure of cellulose remains intact to some extent. Pretreatment by ultrasound waves cannot disturb the hemicellulose structure, reduce the ratio of the area of the fibers to the surface area, and negatively affects subsequent enzymatic hydrolysis. Some researchers have reported the cracking of cellulose [30]. Biogas production performance has been increased from sludge after pretreatment with ultrasound waves.

The pretreatment method of extrusion is a physic-thermal process because it involves mixing, heating, and cutting materials, which produce physical and chemical alterations [31]. The agricultural residue is fed to one end of the extruder and then transported along the barrel using a drive screw [122]. As the material moves, it is subjected to strong friction, mixing, and cutting after the release of pressure at the end of finishing. The method is applicable to biomass materials containing more than 15%–20% moisture [25]. This method is useful because it requires less energy than mechanical friction, and a higher mechanical cut breaks the biomass structure, resulting in defibrillation and shortening of fibers.

The polymers of cellulose, hemicellulose, lignin, and proteins are removed [33]. The thermal decomposition of sugars and amino acids in the extrusion process may lead to slow degradation or even results in the production of nondegradable substances. In certain conditions, inhibiting substances can also be produced, such as furfural [34]. Caution must be exercised to escape from such issues.

Pyrolysis is another heat-absorbing technique which utilizes very low energy, because the biomass is pretreated at a temperature above 300°C [25]. During the course of pyrolysis, the cellulose is swiftly degraded, resulting in the discharge of various products of gaseous nature with coal-like materials, which are further cured with weak acid and filtered with water. The major portion of the leachate is glucose, which can act as a carbon source for biogas production [21].

The freezing of the biomass is a recently developed physical pretreatment. It has the ability to increase the enzyme capability to digest biomass and lignocellulosic material on a large scale. The technique is presented with its specific properties, such as environmentally friendly, better biogas production, and minimum use of toxic chemicals, but the energy cost is very high, so it is not applicable on an industrial scale [25].

The most common pretreatment technique applied to lignocellulosic material is the steam explosion. The technique is based on heating the biomass with saturated steam at very high pressure (a short time). Afterward, the pressure is rapidly decreased, which leads to the biomass explosion [35]. With the technique, hemicellulose is decomposed into its monomers, and the lignin is also degraded [23]. The method is highly viable to pretreat the biomass for experimental and commercial uses [36].

Based on the benefits and difficulties of various pretreatment procedures, successful physical pretreatment must keep the following properties [18]:

(1) the technique must be capable of enhancing the digestion process;
(2) carbohydrates are not decomposed;
(3) the inhibitors are not produced;
(4) use of chemicals must be very low or not all;
(5) the expensive treatment of bioreactors not to be used;
(6) volume reduction should be at the minimum level;
(7) energy use and input must be very low;
(8) waste must not be produced;
(9) the technique must be profitable and environment friendly;
(10) waste produced can be easily disposable;
(11) use of water must be avoided.

The type of biomass or agricultural residue must be taken into account before choosing pretreatment techniques.

6.2 Chemical treatment

The pretreatment methods relying on the chemicals are extensively tested for lignocellulosic materials and biomass. The chemical pretreatment methods are usually considered satisfactory. Chemical-based pretreatment has received more attention as compared to physical and biological approaches, because it is low in cost, gives very swift results, and is effectual in increased degradation of complex organic materials [37,38]. On the other hand, requirements of distinct equipment and creation of pollution are the disadvantages [21].

The chemical pretreatment procedures are better to tackle the fat-rich biomass that cannot be digested otherwise [39]. The most commonly used chemical treatments are based on acidic, basic, oxidizing, ionic, and organic solvent processes.

Chemical pretreatment with acids consists of organic and inorganic acids. Organic acid procedures include formic, acetic, and propionic acids, while inorganic acid procedures use sulfuric, nitric, hydrochloric, and phosphoric acids [21,40–43]. The major role of chemical treatment is to separate and remove lignin, plant fibers, and enhanced hydrolysis. Hydrolysis of cellulose and hemicellulose is mostly managed with this technique. Even though cellulose is based on crystals that are highly resistant to acids, the acids can decompose. For hemicellulose, solutions are used that are most effective in dissolving lignin in biomass. The enzyme can easily attack after acid pretreatment of hemicellulose materials [23].

Concentrated acids in the range of 30%–70% are used at low temperatures up to 40°C, while dilute acids are employed at high temperatures above 200°C. The concentrated acids are more in effect for hydrolysis of cellulosic materials. On the other hand, the procedure becomes very toxic, corrosive, thus becoming dangerous. Costly equipment made up of special alloys and nonmetallic materials in the form of reactors are required. Moreover, the acids used for the pretreatment must be recovered after biomass treatment due to economic reasons. The process is also very highly energy consumptive. Consequently, the dilute acid technique is preferred as compared to concentrated acid for pretreatment of agricultural biomass. So the dilute acid procedure has become most widely applied and a widely used chemical treatment method. Sugar monomers such as xylose, arabinose, and galactose present in the hemicellulose are obtained up to 100% by dilute acid pretreatment procedures. The structure of lignin is significantly disturbed, but in some cases it remains ineffective. The major role of dilute acid pretreatment is to significantly enhance the vulnerability of cellulosic materials to microbial decomposition [21,39].

More lignin-containing biomass sources are better susceptible to alkaline treatment as compared to any acid pretreatment. For alkaline pretreatment, sodium hydroxide, potassium hydroxides along with calcium, and ammonium hydroxide are usually employed. The most commonly used bases remove entanglement and increase the porosity of the agricultural residue. The surface area is enhanced by a swelling of the structure. Through the reduction of the degree of polymerization and decreased crystals, the structure of the lignin is broken along with other polymers [44,123]. The most common performance enhancer is the sodium hydroxide base in alkaline treatment. In many studies, improved decomposition of the lignocellulosic biomass is reported [45]. The presence of the remaining base in the pretreated biomaterials avoids the lowering of pH during acid formation in anaerobic digestion, which is very helpful [45,46].

Wet oxidation pretreatment is applied when the water is added with other oxidation factors like air, oxygen, and hydrogen peroxide (H_2O_2). The temperature is kept high from 125°C to 300°C along with high pressure from 0.5 to 20 M Pascal. The most critical factor is reaction time which varies from a few minutes to hours. Production of reactive oxygen and free radical increases the reaction rates. The use of wholesome oxygen gas can enhance the rate of reaction, but the application of pure oxygen gas leads to an increase in the cost of operation. Normally, air is applied for oxidization in the pretreatment of wet oxidation. The energy inputs are reduced due to the exothermic nature of the process, and the heat produced in this reaction is utilized to maintain the desired temperature. The moment pretreatment starts, heat becomes sufficient to maintain the temperatures so that the process may be initiated at a comparatively lower temperature. The presence of water is critical to this process if the pretreated material is dry [124]. After the pretreatment with wet oxidation, biological access of microorganisms and enzymes to the cellulosic portion by the removal of lignin and hemicellulose is increased [17].

To improve enzymatic digestion, pretreatment with ionic fluids such as *N*-methyl morpholine, 1-*N*-butyl-3 methylimidazolium, and benzyldimethyl (tetradecyl) ammonium chloride is very effective [47]. In the pretreatment of biomass with ionic liquids, the oxygen and hydrogen atoms from the hydroxyl groups act as a donor or receptor in the complexes to interact [48]. The hydrogen bonds are broken as groups of cellulose hydroxyl and ionic fluids interact. After the breakage of the hydrogen bond, the structure of cellulose is broken [49], and soluble cellulose is precipitated speedily.

Pretreatment of lignocellulose biomaterials with organic solvents, especially with low boiling alcohols, comes under the organosolv method. In the procedure, the lignin fraction is removed through the cleavage of the ether bond, and then it is dissolved [50]. In the organosolv method, the lignin portion is removed by breakage of the link between lignin and carbohydrates. So enzymatic attack becomes easy. The major advantage of the process is that the microbial population can also use these low molecular weight alcohols and organic acids [51,125].

7 Production of biogas in different countries

Biogas is being produced worldwide from organic matter as a sustainable energy source. Leading nations promoting biogas business are the United States, the United Kingdom, Italy, and Brazil. In the European Union, Italy is the major producer of biogas keeping almost 1300 biogas production facilities with the generation of eight GWh of electricity [126]. The biogas production facilities in Italy are incentivized by a high feed-in tariff of 0.28/KWh for the facilities of capacity below 1 MW for a period of 20 years [127].

In the UK, biogas has also been considered as a renewable energy source. Green bioelectricity generation has been reached in the UK up to 6 GW till

2017, from the use of biomass and municipal solid waste through a biological digestion process [128]. Denmark, Austria, the Czech Republic, and Bulgaria are among the major players of biogas production from different wastes in Europe. Denmark has researched the production of biogas on an extensive scale and become highly knowledgeable in the utilization of biogas for green electricity generation. The country has developed technologies for the pretreatment, co-digestion of compost, source-separated municipal solid waste (MSW), and clean biological industrial wastes for many years. Denmark utilizes "green pricing," an instrument that gives opportunities for companies using biogas to supply electricity to encourage the utilization of biogas [129,130]. The Czech Republic has created the subsidy system for biogas production facilities from agricultural waste [131]. In Bulgaria, biogas production facilities have been promoted for the utilization of the "energy efficiency and renewable energy sources for Bulgaria" during which the plan may accept up to twenty grants [132].

Due to the feed-in tariff and green electricity act, in Austria, the installed capacity of biogas has been raised from 15 MW in 2002 to about 80 MW by the top of 2015 [133]. The feed-in tariff differs depending on the plant's technology and capacity, also with the source of biogas like landfill gas or sewage gas [134,135].

Sweden is the leader in the use of biogas for transport fuel. Its biogas production reached 1.7 TW in 2013. Major biogas production facilities include sources as a co-digestion plant (34%), waste treatment plant (40%), landfill gas valorization (14%), and farm-scale plants (5%) also as industrial waste treatment plants (7%). More than 50% of used vehicles use biogas, making Sweden the leader [136].

Including India, China, and other Asian countries, Thailand, Vietnam, Nepal, and Bangladesh are encouraging biogas systems. Biogas production and electricity generation from the biogas has been presented in Table 1. In Vietnam, more than one million biological production plants of biogas have been installed. These anaerobic digestion facilities include 0.15 million production units being operated on an industrial scale, while there were approximately 0.5 million production units of medium size till 2013 for the safe disposal of industrial effluents and municipal wastewater [137]. In Thailand, electricity generation from indigenous sources means from biogas has been raised from 120 to 600 MW till 2011 [138]. The Japanese government is also trying to improve the production of biogas through the installation of biogas production facilities, and efforts are being made to market industrial biogas plants [139].

Brazil and the United States are major players in the production of biogas worldwide. According to the American Biogas Council, more than 2100 biogas production facilities are in operation in the United States [140,141]. The US government had incentivized and offered subsidy in federal taxes for the installation of biogas production facilities. The biogas production projects get performance-based incentives and soft loans. Iowa state provides tax credits

TABLE 1 Electricity generation from biogas in different countries.

Sr.#	Countries	Biogas/electricity/energy production (GWh/year)
1	Australia	1587
2	Austria	564.52
3	Brazil	5219
4	Denmark	3723
5	Finland	692
6	France	1833
7	Germany	33,345
8	Korea	2815
9	Sweden	2044
10	Switzerland	1454
11	Netherland	1153
12	United Kingdom	8317

value of $ 0.015 per kWh for 10 years. In Pennsylvania and Oregon, 25%–50% of the installation cost of biogas production facilities from agricultural residues is subsidized; in Massachusetts, almost 75% of the cost is off and 25% on the construction phase [142].

In Brazil, more than 127 biogas production facilities are in operation and making profits till 2017, being run on livestock manure and municipal wastes. Almost one hundred million m^3 of biogas is being produced from Brazil's livestock wastes, agricultural residue, wastewater from different sources, and municipal wastes on an annual basis [143].

8 Conclusion

Anaerobic digestion to produce biogas production is a well-established technology for sustainable energy generation. Recently, increasing trends of biogas production have opened new horizons for taking advantage in the form of sustainable energy, getting its higher potential applications. The biogas market is rising and developing procedures to control the process for better utilization of available feedstock. A better understanding of the microbial community is going to play a more critical role in tailoring the biogas production process. It can be easily predicted that in future the biogas production facilities are intended to transform into innovative bioenergy industrial units for a safe and sound operation.

References

[1] Ezeonu S, Dioha I, Eboatu A. Daily biogas production from different wastes and identification of methanogenic bacteria involved. Niger J Solar Energy 2005;15:80–5.

[2] Kalia V, Sonakya V, Raizada N. Anaerobic digestion of banana stem waste. Bioresour Technol 2000;73(2):191–3.

[3] Budiyono B, Widiasa IN, Johari S, Sunarso S. Increasing biogas production rate from cattle manure using rumen fluid as inoculums. Int J Sci Eng 2014;6(1):31–8.

[4] Alvarez R, Lidén G. The effect of temperature variation on biomethanation at high altitude. Bioresour Technol 2008;99(15):7278–84.

[5] Boe K, Angelidaki I. Serial CSTR digester configuration for improving biogas production from manure. Water Res 2009;43(1):166–72.

[6] Rabiu A, Yaakub H, Liang J, Samsudin A. Enhancing biogas production rate of cattle manure using rumen fluid of ruminants. J Agric Vet Sci 2014;7:25–8.

[7] Mueller RL, Macey JR, Jaekel M, Wake DB, Boore JL. Morphological homoplasy, life history evolution, and historical biogeography of plethodontid salamanders inferred from complete mitochondrial genomes. Proc Natl Acad Sci 2004;101(38):13820–5.

[8] Sommer RJ, Hume AJ, Ciak JM, VanNostrand JJ, Friggens M, Walker MK. Early developmental 2, 3, 7, 8-tetrachlorodibenzo-p-dioxin exposure decreases chick embryo heart chronotropic response to isoproterenol but not to agents affecting signals downstream of the beta-adrenergic receptor. Toxicol Sci 2005;83(2):363–71.

[9] Albihn A, Vinnerås B. Biosecurity and arable use of manure and biowaste—treatment alternatives. Livest Sci 2007;112(3):232–9.

[10] Sun Q, Li H, Yan J, Liu L, Yu Z, Yu X. Selection of appropriate biogas upgrading technology-a review of biogas cleaning, upgrading and utilisation. Renew Sustain Energy Rev 2015;51:521–32.

[11] Bassani I, Kougias PG, Treu L, Porte H, Campanaro S, Angelidaki I. Optimization of hydrogen dispersion in thermophilic up-flow reactors for ex situ biogas upgrading. Bioresour Technol 2017;234:310–9.

[12] Kougias PG, Treu L, Benavente DP, Boe K, Campanaro S, Angelidaki I. Ex-situ biogas upgrading and enhancement in different reactor systems. Bioresour Technol 2017;225:429–37.

[13] Bu F, Dong N, Khanal SK, Xie L, Zhou Q. Effects of CO on hydrogenotrophic methanogenesis under thermophilic and extreme-thermophilic conditions: microbial community and biomethanation pathways. Bioresour Technol 2018;266:364–73.

[14] Amon T, Amon B, Kryvoruchko V, Zollitsch W, Mayer K, Gruber L. Biogas production from maise and dairy cattle manure—influence of biomass composition on the methane yield. Agric Ecosyst Environ 2007;118(1–4):173–82.

[15] Achinas S, Achinas V, Euverink G. A technological overview of biogas production from biowaste. Engineering 2017;3(3):299–307. https://doi.org/10.1016/J.ENG.2017.03.002.

[16] Nielfa A, Cano R, Fdz-Polanco M. Theoretical methane production generated by the co-digestion of organic fraction municipal solid waste and biological sludge. Biotechnol Rep 2015;5:14–21.

[17] Nasir IM, Ghazi TIM, Omar R. Production of biogas from solid organic wastes through anaerobic digestion: a review. Appl Microbiol Biotechnol 2012;95(2):321–9.

[18] Holm-Nielsen JB, Al Seadi T, Oleskowicz-Popiel P. The future of anaerobic digestion and biogas utilisation. Bioresour Technol 2009;100(22):5478–84.

[19] Koneswaran G, Nierenberg D. Global farm animal production and global warming: impacting and mitigating climate change. Environ Health Perspect 2008;116(5):578–82.

[20] Carlsson M, Naroznova I, Møller J, Scheutz C, Lagerkvist A. Importance of food waste pretreatment efficiency for global warming potential in life cycle assessment of anaerobic digestion systems. Resour Conserv Recycl 2015;102:58–66.
[21] Jaiyeola AT, Bwapwa JK. Treatment technology for brewery wastewater in a water-scarce country: a review. S Afr J Sci 2016;112(3–4):1–8.
[22] Mathias JFCM. Manure as a resource: livestock waste management from anaerobic digestion, opportunities and challenges for Brazil. Int Food Agribusiness Manag Rev 2014;17 (4):87–110.
[23] Pollard AT, Morra MJ. Estrogens: properties, behaviors, and fate in dairy manure-amended soils. Environ Rev 2017;25(4):452–62.
[24] Muralidharan A. Feasibility, health and economic impact of generating biogas from human excreta for the state of Tamil Nadu, India. Renew Sustain Energy Rev 2017;69:59–64.
[25] Möller K, Müller T. Effects of anaerobic digestion on digestate nutrient availability and crop growth: a review. Eng Life Sci 2012;12(3):242–57.
[26] Angenent LT, Karim K, Al-Dahhan MH, Wrenn BA, Domíguez-Espinosa R. Production of bioenergy and biochemicals from industrial and agricultural wastewater. Trends Biotechnol 2004;22(9):477–85.
[27] Nielsen M, Holst-Fischer C, Malmgren-Hansen B, Bjerg-Nielsen M, Kragelund C, Møller HB, et al. Small temperature differences can improve the performance of mesophilic sludge-based digesters. Biotechnol Lett 2017;39(11):1689–98.
[28] Wahid R, Hjorth M, Kristensen S, Møller HB. Extrusion as pretreatment for boosting methane production: effect of screw configurations. Energy Fuel 2015;29(7):4030–7.
[29] Pham C, Vu C, Sommer SG, Bruun S. Factors affecting process temperature and biogas production in small-scale rural biogas digesters in winter in northern Vietnam. Asian Australas J Anim Sci 2014;27(7):1050.
[30] Turkenburg WC, Faaij A. Renewable energy technologies. In: UNDP/UNDESA/WEC: energy and the challenge of sustainability. World Energy; 2000.
[31] McKendry P. Energy production from biomass (part 2): conversion technologies. Bioresour Technol 2002;83(1):47–54.
[32] Demirbas A. Biofuels sources, biofuel policy, biofuel economy and global biofuel projections. Energ Conver Manage 2008;49(8):2106–16.
[33] Sharholy M, Ahmad K. Municipal solid waste management in Indian cities—a review. Waste Manag 2008;28:459–67.
[34] Demirbas A. Waste management, waste resource facilities and waste conversion processes. Energ Convers Manage 2011;52(2):1280–7.
[35] Rahman KM, Woodard R, Manzanares E, Harder MK. An assessment of anaerobic digestion capacity in Bangladesh. Renew Sustain Energy Rev 2014;32:762–9.
[36] Deublein D, Steinhauser A. Biogas from waste and renewable resources: an introduction. John Wiley & Sons; 2011.
[37] Kranert M, Kusch S, Huang J, Fischer K. Anaerobic digestion of waste. Waste to energy. Springer; 2012. p. 107–35.
[38] Appels L, Baevens J, Degrève J, Dewil R. Principios y potencial de la digestión anaerobia de lodos activados por residuos. Prog Combust Energía Sci 2008;34:755–81.
[39] Zhang L-L, Wang S-T, Yang G-H, Song D, Wang J-Q, Lin J-S. The chemical mechanism of heavy oil viscosity reduction in carbon dioxide flooding. J Petrochem Univ 2011;24:1–5.
[40] Zupančič GD, Grilc V. Anaerobic treatment and biogas production from organic waste. In: Management of organic waste. IntechOpen; 2012. p. 1–28.
[41] Vickers NJ. Animal communication: when i'm calling you, will you answer too? Curr Biol 2017;27(14):R713–5.

[42] Ward A, Hobbs P, Holliman P, Jones D. Optimisation of the anaerobic digestion of agricultural resources. Bioresour Technol 2008;99(17):7928–40.
[43] Khalid A, Arshad M, Anjum M, Mahmood T, Dawson L. The anaerobic digestion of solid organic waste. Waste Manag 2011;31(8):1737–44.
[44] Appels L, Lauwers J, Degrève J, Helsen L, Lievens B, Willems K, et al. Anaerobic digestion in global bio-energy production: potential and research challenges. Renew Sustain Energy Rev 2011;15(9):4295–301.
[45] Adelard L, Poulsen TG, Rakotoniaina V. Biogas and methane yield in response to co-and separate digestion of biomass wastes. Waste Manag Res 2015;33(1):55–62.
[46] Wangliang L, Zhikai Z, Guangwen X. Enhancement of biogas yield of poplar leaf by high-solid codigestion with swine manure. Appl Biochem Biotechnol 2016;179(2):270–82.
[47] Valenti F, Porto SM, Selvaggi R, Pecorino B. Evaluation of biomethane potential from by-products and agricultural residues co-digestion in southern Italy. J Environ Manage 2018;223:834–40.
[48] Paudel SR, Banjara SP, Choi OK, Park KY, Kim YM, Lee JW. Pretreatment of agricultural biomass for anaerobic digestion: current state and challenges. Bioresour Technol 2017;245:1194–205.
[49] Thomas HL, Seira J, Escudié R, Carrère H. Lime pretreatment of miscanthus: impact on bmp and batch dry co-digestion with cattle manure. Molecules 2018;23(7):1608.
[50] Venturin B, Camargo AF, Scapini T, Mulinari J, Bonatto C, Bazoti S, et al. Effect of pretreatments on corn stalk chemical properties for biogas production purposes. Bioresour Technol 2018;266:116–24.
[51] Zhao K, Li Y, Zhou Y, Guo W, Jiang H, Xu Q. Characterisation of hydrothermal carbonisation products (hydrochars and spent liquor) and their biomethane production performance. Bioresour Technol 2018;267:9–16.
[52] Gómez X, Meredith W, Fernández C, Sánchez-García M, Díez-Antolínez R, Garzón-Santos J, et al. Evaluating the effect of biochar addition on the anaerobic digestion of swine manure: application of Py-GC/MS. Environ Sci Pollut Res 2018;25(25):25600–11.
[53] Hewetson BB, Zhang X, Mosier NS. Enhanced acid-catalysed biomass conversion to hydroxymethylfurfural following cellulose solvent-and organic solvent-based lignocellulosic fractionation pretreatment. Energy Fuel 2016;30(11):9975–7.
[54] Shiga TM, Xiao W, Yang H, Zhang X, Olek AT, Donohoe BS, et al. Enhanced rates of enzymatic saccharification and catalytic synthesis of biofuel substrates in gelatinised cellulose generated by trifluoroacetic acid. Biotechnol Biofuels 2017;10(1):310.
[55] Lu Y, Xu Y, Dong B, Dai X. Enhancement of anaerobic methane production by removal of organic-bonding metals from sewage sludge. Huanjing Kexue 2018;39(1):284–91.
[56] Forgács G, Lundin M, Taherzadeh MJ, Horváth IS. Pretreatment of chicken feather waste for improved biogas production. Appl Biochem Biotechnol 2013;169(7):2016–28.
[57] Kabir MM, Forgács G, Horváth IS. Enhanced methane production from wool textile residues by thermal and enzymatic pretreatment. Process Biochem 2013;48(4):575–80.
[58] Patinvoh RJ, Feuk-Lagerstedt E, Lundin M, Horváth IS, Taherzadeh MJ. Biological pretreatment of chicken feather and biogas production from total broth. Appl Biochem Biotechnol 2016;180(7):1401–15.
[59] Kuzmanova E, Zhelev N, Akunna JC. Effect of liquid nitrogen pretreatment on various types of wool waste fibres for biogas production. Heliyon 2018;4(5), e00619.
[60] Van Fan Y, Klemeš JJ, Lee CT, Perry S. Anaerobic digestion of municipal solid waste: energy and carbon emission footprint. J Environ Manage 2018;223:888–97.
[61] Taherzadeh MJ, Karimi K. Pretreatment of lignocellulosic wastes to improve ethanol and biogas production: a review. Int J Mol Sci 2008;9(9):1621–51.

[62] Hagos K, Zong J, Li D, Liu C, Lu X. Anaerobic co-digestion process for biogas production: progress, challenges and perspectives. Renew Sustain Energy Rev 2017;76:1485–96.

[63] Patinvoh RJ, Osadolor OA, Chandolias K, Horváth IS, Taherzadeh MJ. Innovative pretreatment strategies for biogas production. Bioresour Technol 2017;224:13–24.

[64] Barros RM, Tiago Filho GL, Santos AHM, Ferreira CH, Pieroni MF, Moura JS, et al. A potential of the biogas generating and energy recovering from municipal solid waste. Renew Energy Focus 2018;25:4–16.

[65] Browne JD, Allen E, Murphy JD. Assessing the variability in biomethane production from the organic fraction of municipal solid waste in batch and continuous operation. Appl Energy 2014;128:307–14.

[66] Tyagi VK, Fdez-Güelfo L, Zhou Y, Álvarez-Gallego C, Garcia LR, Ng WJ. Anaerobic co-digestion of organic fraction of municipal solid waste (OFMSW): progress and challenges. Renew Sustain Energy Rev 2018;93:380–99.

[67] Morero B, Vicentin R, Campanella EA. Assessment of biogas production in Argentina from co-digestion of sludge and municipal solid waste. Waste Manag 2017;61:195–205.

[68] Florkowski WJ, Us A, Klepacka AM. Food waste in rural households support for local biogas production in Lubelskie voivodship (Poland). Resour Conserv Recycl 2018;136:46–52.

[69] Levis JW, Barlaz MA, Themelis NJ, Ulloa P. Assessment of the state of food waste treatment in the United States and Canada. Waste Manag 2010;30(8–9):1486–94.

[70] De Clercq D, Wen Z, Fan F, Caicedo L. Biomethane production potential from restaurant food waste in megacities and project level-bottlenecks: a case study in Beijing. Renew Sustain Energy Rev 2016;59:1676–85.

[71] Bolzonella D, Pavan P, Mace S, Cecchi F. Dry anaerobic digestion of differently sorted organic municipal solid waste: a full-scale experience. Water Sci Technol 2006;53(8):23–32.

[72] Hartmann H, Angelidaki I, Ahring BK. Co-digestion of the organic fraction of municipal waste with other waste types. In: Biomethanization of the organic fraction of municipal solid wastes. IWA Publishing; 2002. p. 181–200.

[73] Deng L, Liu Y, Zheng D, Wang L, Pu X, Song L, et al. Application and development of biogas technology for the treatment of waste in China. Renew Sustain Energy Rev 2017;70:845–51.

[74] Karimi S, Karimi K. Efficient ethanol production from kitchen and garden wastes and biogas from the residues. J Clean Prod 2018;187:37–45.

[75] Lubelskiego ZW. Regionalny Program Operacyjny Województwa Lubelskiego na lata 2014–2020. Projekt [Regional operational Programme of Lublin Voivodeship for 2014-2020 draft-in Polish]. Lublin; 2013.

[76] Anyaoku CC, Baroutian S. Decentralized anaerobic digestion systems for increased utilisation of biogas from municipal solid waste. Renew Sustain Energy Rev 2018;90:982–91.

[77] Tahir A, Arshad M, Anum F, Abbas M, Javed S, Shahzad MI, Ansari AR, Bano I, Shah FA. Ecofuel future prospect and community impact. In: Advances in eco-fuels for a sustainable environment. Woodhead Publishing; 2019. p. 459–79.

[78] Righi S, Oliviero L, Pedrini M, Buscaroli A, Della Casa C. Life cycle assessment of management systems for sewage sludge and food waste: centralised and decentralised approaches. J Clean Prod 2013;44:8–17.

[79] Arshad M. Clean and sustainable energy technologies. In: Rasul M, editor. Clean energy for sustainable development. Academic Press; 2017. p. 73–89.

[80] Kayhanian M, Hardy S. The impact of four design parameters on the performance of a high-solids anaerobic digestion of municipal solid waste for fuel gas production. Environ Technol 1994;15(6):557–67.

[81] Arshad M, Abbas M. Future biofuel production and water usage. In: Arshad M, editor. Perspectives on water usage for biofuels production. Cham: Springer; 2018.

[82] Gutiérrez EC, Wall DM, O'Shea R, Novelo RM, Gómez MM, Murphy JD. An economic and carbon analysis of biomethane production from food waste to be used as a transport fuel in Mexico. J Clean Prod 2018;196:852–62.

[83] Kavitha S, Banu JR, Priya AA, Yeom IT. Liquefaction of food waste and its impacts on anaerobic biodegradability, energy ratio and economic feasibility. Appl Energy 2017;208:228–38.

[84] Arshad M, Bano I, Younus M, Khan A, Rahman A. Health concerns associated with biofuel production. In: Arshad M, editor. Perspectives on water usage for biofuels production. Cham: Springer; 2018.

[85] Gallert C, Winter J. Bacterial metabolism in wastewater treatment systems. Wiley Online Library; 2005.

[86] Arshad M, Abbas M. Water sustainability issues in biofuel production. In: Arshad M, editor. Perspectives on water usage for biofuels production. Cham: Springer; 2018.

[87] Kobayashi T, Kuramochi H, Xu K-Q. Variable oil properties and biomethane production of grease trap waste derived from different resources. Int Biodeter Biodegr 2017;119:273–81.

[88] dos Santos IFS, Vieira NDB, de Nóbrega LGB, Barros RM, Tiago Filho GL. Assessment of potential biogas production from multiple organic wastes in Brazil: impact on energy generation, use, and emissions abatement. Resour Conserv Recycl 2018;131:54–63.

[89] Arshad M, Zia MA, Shah FA, Ahmad M. An overview of biofuel. In: Arshad M, editor. Perspectives on water usage for biofuels production. Cham: Springer; 2018.

[90] Hijazi O, Munro S, Zerhusen B, Effenberger M. Review of life cycle assessment for biogas production in Europe. Renew Sustain Energy Rev 2016;54:1291–300.

[91] Sreekrishnan T, Kohli S, Rana V. Enhancement of biogas production from solid substrates using different techniques—a review. Bioresour Technol 2004;95(1):1–10.

[92] Li Y, Zhang R, Chen C, Liu G, He Y, Liu X. Biogas production from co-digestion of corn stover and chicken manure under anaerobic wet, hemi-solid, and solid state conditions. Bioresour Technol 2013;149:406–12.

[93] Li Y, Zhang R, Liu X, Chen C, Xiao X, Feng L, et al. Evaluating methane production from anaerobic mono-and co-digestion of kitchen waste, corn stover, and chicken manure. Energy Fuel 2013;27(4):2085–91.

[94] Arshad M, Bano I, Khan N, Shahzad MI, Younus M, Abbas M, Iqbal M. Electricity generation from biogas of poultry waste: an assessment of potential and feasibility in Pakistan. Renew Sustain Energy Rev 2018;81:1241–6.

[95] Belostotskiy DE, Ziganshina EE, Siniagina M, Boulygina EA, Miluykov VA, Ziganshin AM. Impact of the substrate loading regime and phosphoric acid supplementation on performance of biogas reactors and microbial community dynamics during anaerobic digestion of chicken wastes. Bioresour Technol 2015;193:42–52.

[96] Abouelenien F, Fujiwara W, Namba Y, Kosseva M, Nishio N, Nakashimada Y. Improved methane fermentation of chicken manure via ammonia removal by biogas recycle. Bioresour Technol 2010;101(16):6368–73.

[97] Costa J, Barbosa S, Alves M, Sousa D. Thermochemical pre-and biological co-treatments to improve hydrolysis and methane production from poultry litter. Bioresour Technol 2012;111:141–7.

[98] Ziganshina EE, Belostotskiy DE, Shushlyaev RV, Miluykov VA, Vankov PY, Ziganshin AM. Microbial community diversity in anaerobic reactors digesting Turkey, chicken, and swine wastes. J Microbiol Biotechnol 2014;24:1464–72.

[99] Arshad M, Ahmed S. Cogeneration through bagasse: a renewable strategy to meet the future energy needs. Renew Sustain Energy Rev 2016;54:732–7.

[100] Wang X, Yang G, Feng Y, Ren G, Han X. Optimising feeding composition and carbon–nitrogen ratios for improved methane yield during anaerobic co-digestion of dairy, chicken manure and wheat straw. Bioresour Technol 2012;120:78–83.

[101] Hultberg M, Lind O, Birgersson G, Asp H. Use of the effluent from biogas production for cultivation of Spirulina. Bioprocess Biosyst Eng 2017;40(4):625–31.

[102] Ng CA, Wong LY, Chai HY, Bashir MJ, Ho C-D, Nisar H, et al. Investigation on the performance of hybrid anaerobic membrane bioreactors for fouling control and biogas production in palm oil mill effluent treatment. Water Sci Technol 2017;76(6):1389–98.

[103] Mohamad IN, Rohani R, Nor MTM, Claassen P, Muhammad M, Masdar MM, et al. An overview of gas-upgrading technologies for biohydrogen produced from treatment of palm oil mill effluent. J Eng Sci Technol 2017;12:725–55.

[104] Agustini CB, Spier F, da Costa M, Gutterres M. Biogas production for anaerobic co-digestion of tannery solid wastes under presence and absence of the tanning agent. Resour Conserv Recycl 2018;130:51–9.

[105] Priebe G, Kipper E, Gusmão A, Marcilio N, Gutterres M. Anaerobic digestion of chrome-tanned leather waste for biogas production. J Clean Prod 2016;129:410–6.

[106] Mannucci A, Munz G, Mori G, Lubello C. Anaerobic treatment of vegetable tannery wastewaters: a review. Desalination 2010;264(1–2):1–8.

[107] Appels L, Baeyens J, Degreve J, Dewil R. Principles and potential of the anaerobic digestion of waste-activated sludge. Prog Energy Combust Sci 2008;34(6):755–81.

[108] Bradshaw A. Regulatory change and innovation in Latin America: the case of renewable energy in Brazil. Util Policy 2017;49:156–64.

[109] Agustini C, da Costa M, Gutterres M. Biogas production from tannery solid wastes–scale-up and cost saving analysis. J Clean Prod 2018;187:158–64.

[110] Primaloka AD, Ardhannari L, Matin HH, Sumardiono S, editors. Study of biogas production from cassava industrial waste by anaerobic process. MATEC web of conferences. EDP Sciences; 2018.

[111] Arshad M, Abbas M, Iqbal M. Ethanol production from molasses: environmental and socio-economic prospects in Pakistan: feasibility and economic analysis. Environ Technol Innov 2019;14:100317.

[112] Suksong W, Jehlee A, Singkhala A, Kongjan P, Prasertsan P, Imai T, et al. Thermophilic solid-state anaerobic digestion of solid waste residues from palm oil mill industry for biogas production. Ind Crops Prod 2017;95:502–11.

[113] Arshad M, Khan ZM, Khalil-ur-Rehman, Shah FA, Rajoka MI. Optimisation of process variables for minimisation of byproduct formation during fermentation of blackstrap molasses to ethanol at industrial scale. Lett Appl Microbiol 2008.

[114] Lopes ACP, Silva CM, Rosa AP, de Ávila Rodrigues F. Biogas production from thermophilic anaerobic digestion of Kraft pulp mill sludge. Renew Energy 2018;124:40–9.

[115] Guerini Filho M, Lumi M, Hasan C, Marder M, Leite LC, Konrad O. Energy recovery from wine sector wastes: a study about the biogas generation potential in a vineyard from Rio Grande do Sul, Brazil. Sustain Energy Technol Assess 2018;29:44–9.

[116] Prabakar D, Manimudi VT, Mathimani T, Kumar G, Rene ER, Pugazhendhi A. Pretreatment technologies for industrial effluents: critical review on bioenergy production and environmental concerns. J Environ Manage 2018;218:165–80.

[117] Solarte-Toro JC, Chacón-Pérez Y, Cardona-Alzate CA. Evaluation of biogas and syngas as energy vectors for heat and power generation using lignocellulosic biomass as raw material. Electron J Biotechnol 2018;33:52–62.

[118] Surendra K, Takara D, Hashimoto AG, Khanal SK. Biogas as a sustainable energy source for developing countries: opportunities and challenges. Renew Sustain Energy Rev 2014;31:846–59.

[119] Arshad M, Zia MA, Bajwa MA, Bhatti HN. Improving bio-ethanol yield; using Virginiamycin and sodium Flouride at Pakistani distillery. Afr J Biotechnol 2011;10(53):11071–4.

[120] Zheng Y, Zhao J, Xu F, Li Y. Pretreatment of lignocellulosic biomass for enhanced biogas production. Prog Energy Combust Sci 2014;42:35–53.

[121] Zhao X, Luo K, Zhang Y, Zheng Z, Cai Y, Wen B, et al. Improving the methane yield of maise straw: focus on the effects of pretreatment with fungi and their secreted enzymes combined with sodium hydroxide. Bioresour Technol 2018;250:204–13.

[122] Pilli S, Yan S, Tyagi RD, Surampalli RY. Thermal pretreatment of sewage sludge to enhance anaerobic digestion: a review. Crit Rev Environ Sci Technol 2015;45(6):669–702.

[123] Li J, Zhang R, Siddhu MAH, He Y, Wang W, Li Y, et al. Enhancing methane production of corn Stover through a novel way: sequent pretreatment of potassium hydroxide and steam explosion. Bioresour Technol 2015;181:345–50.

[124] Schroyen M, Vervaeren H, Vandepitte H, Van Hulle SW, Raes K. Effect of enzymatic pretreatment of various lignocellulosic substrates on production of phenolic compounds and biomethane potential. Bioresour Technol 2015;192:696–702.

[125] Valenti F, Zhong Y, Sun M, Porto SM, Toscano A, Dale BE, et al. Anaerobic co-digestion of multiple agricultural residues to enhance biogas production in southern Italy. Waste Manag 2018;78:151–7.

[126] Torrijos M. State of development of biogas production in Europe. Procedia Environ Sci 2016;35:881–9.

[127] Arshad M, Hussain T, Chaudhry N, Sadia H, Aslam B, Tahir U, Abbas M, Qureshi N, Nazir A, Rajoka MI, Iqba M. Enhancing profitability of ethanol fermentation through gamma ray mutagenesis of *Saccharomyces cerevisiae*. Pol J Environ Stud 2019;28(1):35–41.

[128] Murdock HE, Gibb D, André T, Appavou F, Brown A, Epp B, et al. Renewables 2019 global status report; 2019.

[129] Aryal N, Kvist T. Alternative of biogas injection into the Danish gas grid system—a study from demand perspective. ChemEngineering 2018;2(3):43.

[130] Vochozka M, Maroušková A, Šuleř P. Economic, environmental and moral acceptance of renewable energy: a case study—the agricultural biogas plant at Pěčín. Sci Eng Ethics 2018;24(1):299–305.

[131] Kolchakov V, Petrova V, Mitova T, Ivanov P, Marinova S. Possibilities for biogas production from waste—potential, barriers, and legal notices. In: Energy solutions to combat global warming. Springer; 2017. p. 181–91.

[132] Arshad M, Hussain T, Iqbal M, Abbas M. Enhanced ethanol production at commercial scale from molasses using high gravity technology by mutant *S. cerevisiae*. Braz J Microbiol 2017;48:403–9.

[133] Stürmer B. Biogas–part of Austria's future energy supply or political experiment? Renew Sustain Energy Rev 2017;79:525–32.

[134] Pablo-Romero MP, Sánchez-Braza A, Salvador-Ponce J, Sánchez-Labrador N. An overview of feed-in tariffs, premiums and tenders to promote electricity from biogas in the EU-28. Renew Sustain Energy Rev 2017;73:1366–79.

[135] Larsson M, Grönkvist S, Alvfors P. Upgraded biogas for transport in Sweden–effects of policy instruments on production, infrastructure deployment and vehicle sales. J Clean Prod 2016;112:3774–84.
[136] Arshad M, Ahmed S, Zia MA, Rajoka MI. Kinetics and thermodynamics of ethanol production by *Saccharomyces cerevisiae* MLD10 using molasses. Appl Biochem Biotechnol 2014;172(5):2455–64.
[137] Sehgal K. Current state and future prospects of global biogas industry. In: Biogas. Springer; 2018. p. 449–72.
[138] Chaiyapong P, Chavalparit O. Enhancement of biogas production potential from Acacia leaf waste using alkaline pretreatment and co-digestion. J Mater Cycles Waste Manag 2016;18 (3):427–36.
[139] Takeuchi Y, Andriamanohiarisoamanana FJ, Yasui S, Iwasaki M, Nishida T, Ihara I, et al. Feasibility study of a centralised biogas plant performance in a dairy farming area. J Mater Cycles Waste Manag 2018;20(1):314–22.
[140] Kemausuor F, Adaramola MS, Morken J. A review of commercial biogas systems and lessons for Africa. Energies 2018;11(11):2984.
[141] Pasqual JC, Bollmann HA, Scott CA, Edwiges T, Baptista TC. Assessment of collective production of biomethane from livestock waste for urban transportation mobility in Brazil and the United States. Energies 2018;11(4):997.
[142] Sam A, Bi X, Farnsworth D. How incentives affect the adoption of anaerobic digesters in the United States. Sustainability 2017;9(7):1221.
[143] Langeveld JW, Peterson EC. Feedstocks for biogas production: biogas and electricity generation potentials. In: Biogas. Springer; 2018. p. 35–49.

Chapter 5

Potential of ionic liquid applications in natural gas/biogas sweetening and liquid fuel cleaning process

Pobitra Halder[a], Savankumar Patel[a], Sazal Kundu[a], Biplob Pramanik[b], Rajarathinam Parthasarathy[a], and Kalpit Shah[a]
[a]*Chemical & Environmental Engineering, School of Engineering, RMIT University, Melbourne, VIC, Australia,* [b]*Civil & Infrastructure Engineering, School of Engineering, RMIT University, Melbourne, VIC, Australia*

1 Introduction

The need for energy is increasing rapidly with the increase in population as well as technological advancement. Natural gas, the primary source of fossil fuel all over the world, has versatile energy applications in industry, homes, electricity generation, and transportation. At the end of 2017, the world has total proved natural gas reserves of 193.5 trillion $m^3$and the consumption was about 3670.4 billion m^3 [1]. However, the foremost issue relating to its use is the presence of CO_2. Natural gas obtained from gas wells contains about 4%–50% of CO_2 and burning of natural gas is responsible for almost 20% CO_2 emission of total human-produced CO_2 emissions [2,3]. This acid gas is highly corrosive and capable of destroying the pipeline very fast; it also has a very low heat content, which lowers the natural gas heating performance [4]. It is necessary to remove CO_2 from natural gas before transporting through the pipeline to increase the heating value, to reduce corrosion, and to maintain a typical pipeline specification of 2%–5% CO_2.

The traditional liquid fuels such as crude oil, gasoline, diesel, and kerosene contain a high content of sulfur compounds including disulfides, dibenzothiophene, benzothiophene, and hydrogen sulfide [5]. In 2015, the consumption of diesel, gasoline, and crude oil was approximately 1245.1 million tone oil equivalent (Mtoe), 999.59 Mtoe, and 17.87 Mtoe, respectively. Typically crude oil contains about 85%–89% carbon, 10%–13% hydrogen, and 0.25%–4% sulfur

Bioenergy Resources and Technologies. https://doi.org/10.1016/B978-0-12-822525-7.00001-9
Copyright © 2021 Elsevier Inc. All rights reserved.

with a trace amount of oxygen and nitrogen [6]. In some cases, crude oil contains more than 7% sulfur content [7]. The combustion of fuels with high sulfur content produces SO_x, which leads to acid rain. Therefore, different countries have imposed stringent standards on sulfur content in diesel and gasoline [8]. For instance, the United States adopted a national standard of less than 10ppm; Mexico adopted less than 15ppm and Columbia, Chile, and Uruguay along with East Asia (i.e., Malaysia and Singapore) and South Asia (i.e., China and Hong Kong) have set a requirement of 50ppm sulfur in diesel.

Currently, global energy policy has been changed and emphasis has been given to explore sustainable alternative resources such as bio-oil/biodiesel, biogas/bioCNG, and bio-alcohol. The thermochemical conversion of biomass at elevated temperature in the absence of oxygen produces liquid fuel termed as bio-oil. In 2014, almost 14% of total global energy consumption came from biomass energy. However, only 0.8% of the biomass was converted into biofuel for the transportation sector [9]. In 2015, global biofuel production reached 133 billion liters containing 74% ethanol, 22% biodiesel, and 4% hydrogenated vegetable oils. Bio-oil is considered one of the promising candidates in the replacement of liquid fuel for transport and power generation [10]. Biogas is a combustible gas similar to the natural gas produced by anaerobic digestion of renewable feedstocks, including agricultural residues, plant residues, livestock manure, municipal, and industrial wastes [11,12].

However, the main concern for bio-oil utilization is its high concentrations of oxygenates such as aldehydes, alcohols, ketones, and carboxylic acids compared to that of crude petroleum, which reduces its performance as engine fuel. Bio-oil produced from fast pyrolysis contains approximately 35%–40% of oxygen, which is very high compared to the heavy oil (only 1%) [13]. This high content of oxygen leads to low heating value (16–19MJ/kg), instability, high viscosity, and high corrosiveness, which limit the direct use in diesel engines [14]. However, the oxygenates have interesting applications such as a meat browning agent feedstock for fermentation and chemical reagent [15]. Therefore, the removal of the oxygenates from bio-oil can provide a solution for using bio-oil in diesel engines as a replacement for petroleum fuel as well as various commercial opportunities of separated oxygenates. Similar to natural gas, biogas is a methane-rich fuel which contains about 50%–70% methane and 25%–50% CO_2 [16]. The huge amount of CO_2 reduces the heating value of biogas and limits its wider application.

A number of processes, for example, physical absorption, chemical absorption, water scrubbing absorption, membrane separation, pressure swing adsorption, and cryogenic/low-temperature separation methods have been used for upgrading biogas/natural gas mainly for separating CO_2 [17,18]. On the other hand, various acid, base, and metal-based catalysts have been explored for producing low oxygen hydrocarbon fuel. Hydrodesulfurization is a widely used technology for the production of low sulfur petroleum fuels. However, these gas sweetening methods, hydrodesulfurization, and catalytic deoxygenation methods are energy intensive, health, and environmental hazardous as well as associated with significant solvent and catalyst degradation and losses [19,20]. In addition, these solvents and catalysts have limited regeneration efficiency, stability to high pressure, and sorption capability.

Currently, ionic liquids (ILs) have been considered as emerging green solvents because of their nonflammability, negligible vapor pressure, chemical and thermal stability, and no emission of volatile organic compounds [21–23]. ILs are the low-temperature organic salts, mainly consisting of an organic cation and inorganic anion can exist in liquid form at room temperature [24,25]. There are numerous categories of cations and anions, which can be combined for synthesizing the IL of desired properties. In spite of the number of methods available for gas absorption from liquid fuel and gaseous fuel, the use of ILs can be an alternative option because of their unique properties. This chapter presents the future role of ILs for natural gas/biogas sweetening and liquid fuel cleaning process.

2 Absorption of CO_2 from methane-rich gas mixture

2.1 Treatment with traditional methods

The methods for CO_2 capture from the gas mixture are mostly classified as precombustion and post-combustion methods. The gas sweetening methods are mainly precombustion type and quite different from the post-combustion methods [26]. Among the gas sweetening technologies as mentioned earlier, the absorption method is a widely used commercial technology; for instance, 40% of the total biogas plants (more than 200) in the European region implemented water scrubbing absorption and 25% of them executed chemical absorption method [27,28]. Physical solvent absorption is suitable when the CO_2 concentration in the gas stream is very high and feed gas pressure is more than about 20 bar [29]. In the case of the chemical absorption method, various solvents such as aqueous solutions of primary, secondary, tertiary amines, and their mixture are widely used, which react with the CO_2 through the formation of covalent bonds. In contrast, the physical solvent, for example, the mixture of polyethylene glycol and dimethyl ethers absorb CO_2 according to Henry's law [26]. Pressure swing adsorption uses zeolites, activated carbons, metal oxides, and silica gel as adsorbents because of their high surface area and pore volume [30]. The adsorbents interact with CO_2 through covalent bonding and Van der Waals force. Recently, the regeneration of zeolites was investigated at 280°C and atmospheric pressure with an N_2 velocity of 0.46 cm/s [31]. In the case of cryogenic technology, CO_2 is separated in liquid phase from biogas at a very low temperature (−78.5°C) [32]. Recently, membrane processes have gained attention for offshore separation processes because of the relatively few disadvantages over the other methods. Some researchers studied the membrane separation technology and developed different models such as the Taguchi model, 2D model, and Aspen simulation for estimating CO_2 separation from natural gas [33–35]. Table 1 shows the highlights of the various CO_2 absorption methods.

2.2 CO_2 separation using ionic liquid treatment

2.2.1 Conventional ionic liquid treatment

The use of ILs as an absorbent for CO_2 separation from natural gas/biogas stream can greatly suppress some issues of the traditional natural gas sweetening methods. The current researches have focused on the CO_2 solubility in ILs and different solubility parameters. Table 2 depicts the available literature for

TABLE 1 Characteristics of various methods for CO_2 separation [32,36,37].

Biogas cleaning Methods	CH_4 purity (Vol%)	Features of the CO_2 absorption method
Water scrubbing absorption	96.1	• Simple technology • High water consumption. Development of microbial clogging and low flexibility toward input gas variation
Pressure swing adsorption	95.8	• Low emission and energy requirement • Relatively more expensive • Additional steps and adsorbent regeneration requirement
Chemical absorption	94.6	• High CO_2 absorption rate • High energy consumption and high cost • Loss of the absorbent
Membrane permeation	90.3	• Higher efficiency because of the high surface area • No moving parts and simple in operation • No foam formation High cost of membrane because of their finite life • Produces additional barrier to mass transfer
Physical absorption	93	• Low energy requirement • Large quantity of absorbent requirement • Unsuitable for low concentration of CO_2
Cryogenic	88	• Suitable for high concentrated gas • Requirement of additional equipment (i.e., turbines, compressors, and heat exchangers) causes high investment cost • High energy requirement

TABLE 2 Carbon dioxide solubility data in different ILs.

ILs name	Temperature (K)	Pressure (MPa)	CO_2 dissolved in IL (mole fraction)	Method of analysis	Reference
[Hmim][Tf_2N]	293–413	10	4.7 (mol/kg)	Gravimetric	[38]
[Emim][Tf_2N]	310–450	0–15	60 mol%	Bubble point pressures measuring technique	[39]
[Bmim][PF_6]	298.15	0.667	0.118	Vapor adsorption apparatus	[40]
[C_6mim][PF_6]		0.927	0.167		
[Emim][BF_4]		0.875	0.106		
[C_6mim][BF_4]		0.899	0.163		
[Emim][Tf_2N]		0.903	0.209		
[C_6mim][Tf_2N]		0.859	0.236		
[C_1C_2Im][Tf_2N]	303.63–344.23	47.917–57.097[a]	$(13.68–26.14) \times 10^{-3}$	Isochoric saturation technique	[41]
[C_1C_4pyrr][Tf_2N]	303.78–344.15	48.678–57.264[a]	$(14.70–27.15) \times 10^{-3}$		
[$N_{1132\text{-}OH}$][Tf_2N]	304.16–344.84	46.70–54.386[a]	$(12.21–23.84) \times 10^{-3}$		
[Emim][PF_6]	322.80–366.03	1.49–9.40	0.0511		[42]
[Bmim][PF_6]	293.15–393.15	0.105–9.685	0.0170		[43]
[Bmim][PF_6]	283.15–323.15	0.050–1.300	0.0079		[44]
[C_6mim][PF_6]	298.31–363.58	0.64–9.20	0.0163		[45]
[Bmim][BF_4]	303.7–344.5	0.018–0.084	0.0002		[46]

[a] *Pressure in kPa.*

CO_2 solubility data in different ILs. Mortazavi-Manesh et al. and Althuluth et al. reported the higher selectivity of CO_2 in [Bmim][Tf_2N] and [Bmim][BF_4] compared to CH_4 which indicated the applicability of ILs for the separation of CO_2 from CO_2/CH_4 gas mixture [47,48]. The understanding of the CO_2 separation from CO_2/H_2S/[Bmim][PF_6] system was characterized by a ternary equation at room temperature [49]. However, the result of this analysis was found unsatisfactory. Baltus et al. studied the CO_2 separation performance of [C_8F_{13}mim][Tf_2N] from gas stream generated in coal-fired power plants in terms of permeance measurement of CO_2 [50]. Xu et al. investigated biogas upgradation by removing CO_2 utilizing [Bmim][Tf_2N] [19]. They observed that [Bmim][Tf_2N] performed better compared to pressured water scrubbing and monoethanolamine aqueous scrubbing in terms of purity of CO_2 recovered, energy consumption, and environmental aspect. So far, [$C_n$mim] cation-based ILs have been widely synthesized and investigated for CO_2 separation [51]. However, Bara et al. observed better performance on the CO_2 separation when the alkyl chains in the imidazolium cation were substituted by ethylene glycol or nitrile groups [52]. Fig. 1 illustrates the process flow of CO_2 separation from a gas mixture using IL.

Yokozeki et al. developed an equation-of-state model to correlate the CO_2 solubility data of 18 room-temperature ionic liquids, including 9 commercially available ILs and 9 newly synthesized ILs at 298 K under less than 2 MPa [54]. The solubility mechanism of ILs indicated that 10 of them showed physical absorption, while 8 of them revealed chemical absorption. Therefore, it can be seen from the literature that ILs are capable of absorbing CO_2 selectively from the gas mixture.

2.2.2 *Treatment with ionic liquid-solvent-functional group mixture*

Although ILs has sufficient affinity to dissolve CO_2 effectively, the introduction of the different functional groups along with ILs improves the CO_2 solubility

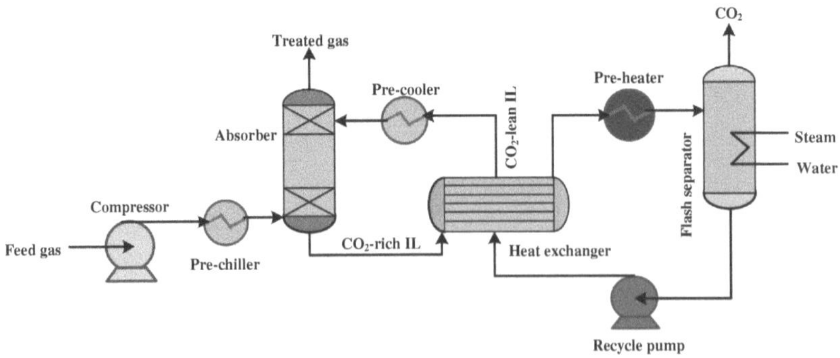

FIG. 1 Schematic representation of CO_2 absorption from a gas mixture with IL [53].

and task-specific functionality. So far, few researchers have studied the reaction kinetics of CO_2 solubility in functionalized ILs. The comparative study of various amine-functionalized ILs through ab initio calculations indicated that the nature and the position of the functional group in ILs substantially affect the reaction enthalpy [55]. Furthermore, the reaction mechanism of CO_2 absorption in an aqueous solution of tetramethylammonium glycinate was investigated in a stirred cell absorber with a planar gas-liquid interface and demonstrated as the fast pseudo-first-order reaction [56]. Another study presented both first-order and second-order reaction kinetics of CO_2 absorption in phosphonium-based anion-functionalized ILs. The second-order reaction rate constants followed Arrhenius behavior and limited by the mass transfer rate of CO_2. The aprotic heterocyclic anions had low solubility because of its highly aromatic structure and electron-withdrawing group [57]. The physical absorption mechanism of CO_2 in amine-functionalized ILs and in [Bmim][BF_4] was compared to a great extent. The results were interpreted according to the thin-film model proposed by Van Krevelen and Hoftijzer and the Danckwerts surface renewal model [58]. The single-step reaction kinetics between CO_2 and primary amine forms carbamate in a 2:1 stoichiometric ratio of amine-functionalized IL and CO_2 [59]. Fig. 2 demonstrates the chemical reaction behavior of CO_2 in amine-functionalized cation and amine-functionalized anion containing ILs.

Sánchez et al. analyzed the impacts of functionalized chains on the CO_2 absorption behavior in ILs using [Bmim][BF_4], [Bmim][DCA], primary

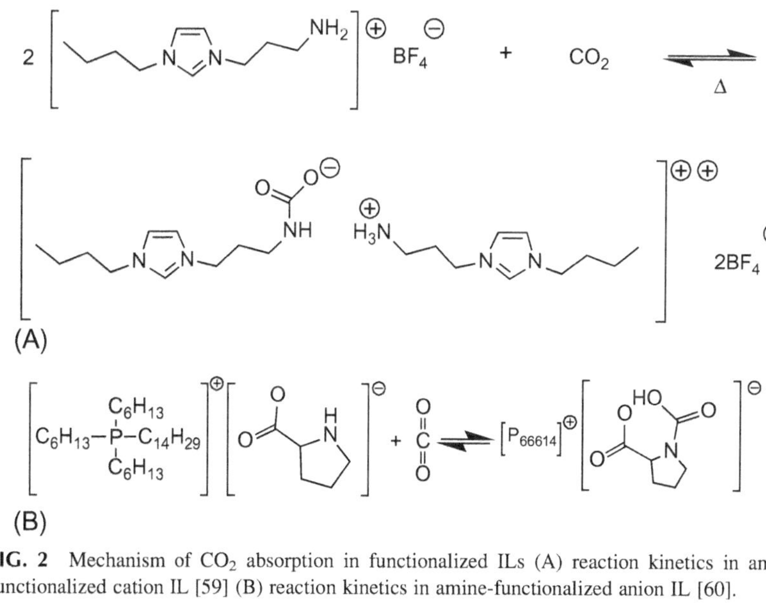

FIG. 2 Mechanism of CO_2 absorption in functionalized ILs (A) reaction kinetics in amine-functionalized cation IL [59] (B) reaction kinetics in amine-functionalized anion IL [60].

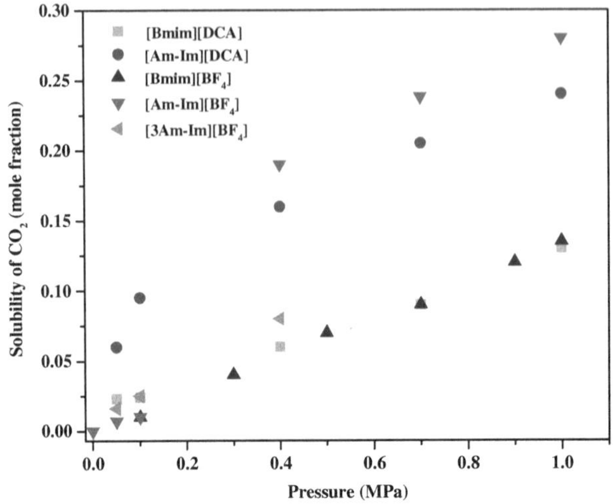

FIG. 3 CO_2 solubility in various functionalized and nonfunctionalized ILs in terms of mole fraction [61].

amine-functionalized [Am-Im][BF_4], tertiary amine-functionalized [3Am-Im][BF_4], hydroxyl-functionalized [OH-Im][BF_4], and primary amine-functionalized [Am-Im][DCA] [61]. CO_2 absorption in nonfunctionalized ILs was attributed mainly by physical mechanisms, while both chemical and physical mechanisms were observed in functionalized ILs. Less CO_2 solubility was noticed in [3Am-Im][BF_4] because of the lower reactivity of the tertiary amine group (Fig. 3). Moreover, the addition of the hydroxyl group in imidazolium tetrafluoroborate IL had no notable improvement in solubility. However, in the case of functionalized ILs, more time was required to reach the equilibrium condition and increased the viscosity because of the long-chain cation.

The presence of the amine group in the ILs increases the volumetric gas load of ionic liquid and overcomes the energy barrier because of its high activation energy. The incorporation of amine groups in the cation of ILs enhances their performance because of the reaction between the absorbed CO_2 and the primary amine of the IL [62]. However, the effects of surface tension are quite the opposite. Because of the high basicity of anion, ionic liquid/water binary solvent forms hydrogen-carbonate anion and carboxylic acid from the reaction between water and CO_2. The complex mixture of IL cation, water, hydrogen-carbonate anion, and carboxylic acid acts as a strong CO_2 absorber competitive with the commercial solvents [63]. Carboxylic functionality anion containing IL was studied for CO_2 separation from natural gas using the flash regeneration method and compared with physical solvents such as monoethanolamine diethanolamine [64]. The CO_2 solubility and selectivity in nitrile-functionalized ILs display a very small trade-off [65].

The CO_2 solubility using 1-(2-hydroxyethyl)-3-methyl-imidazolium dicyanamide-aqueous mixture of monoethanolamine absorbent and a 1-butyl-3-methylimidazolium-aqueous mixture of monoethanolamine absorbent were investigated under the CO_2 partial pressure of 10–800 kPa. The solubility in these functionalized ILs was considerably higher than that in nonfunctionalized ILs. The solubility in both ionic liquid-absorbents decreased slightly at a low concentration of ionic liquid (<0.1) however decreased significantly at higher concentration (<0.2) because of the salting-out effect [66]. The blend of ionic liquid and NHD absorbent showed promising results in CO_2 capturing from biogas compared to ionic liquid only. The solubility of CO_2 was investigated with [Bmim][BF_4]+NHD mixture, [Bmim][DCA]+NHD mixture, and [Bmim][NO_3]+NHD mixture. Among the binary solvents, [Bmim][NO_3]+NHD mixture exhibited the best result [20]. Therefore, the addition of cosolvents and functional groups in ILs enhances the solubility of CO_2, though in some cases increases the viscosity. However, the study related to the optimization of the IL-functional group ratio is nonexistent in the literature.

2.2.3 *Supported ionic liquid membranes treatment*

Recently, the use of room-temperature ionic liquids in replacement of traditional solvents for supported liquid membranes has gained attention toward various gas separations [67,68]. The loss of solvent due to evaporation is one of the major challenges of the supported liquid membrane, and thereby, ILs can be a potential alternative because of their unique properties. A number of microporous materials including polytetrafluoroethylene, polyethersulfone, polyvinylidene fluoride, and chitin are used as supports with IL for CO_2 separation from CH_4-rich gas mixture [69]. Carbon dioxide separation using the IL supported membrane was found economically competitive to amine scrubbing [50]. The CO_2 permeabilities vs partial pressure data indicated the following CO_2 solubility results in four non-hexafluorophosphate ([PF_6]$^-$), water-stable anion containing ILs: [Tf_2N]$^- \geq$[DCA]$^- \approx$[CF_3SO_3]$^- \geq$[Cl]$^-$ [70]. Surprisingly, the CO_2 permeability for [Emim][Tf_2N] membranes increases by 50% when the partial pressure of CO_2 decreases from 19 to 1.5 kPa. On the contrary, the presence of water breaks down the hexafluorophosphate [PF_6]$^-$ anion into [F]$^-$ which facilitates CO_2 transportation [71]. Deng et al. developed ether-functionalized pyridinium-based ionic liquid supported cellulose acetate composite membrane and investigated for CO_2 separation [72]. The IL supported membrane increased the CO_2 permeability by sevenfolds, when compared with that of cellulose acetate membrane only. The universal correlation model developed by Scovazzo suggested the increase of CO_2 selectivity with the decrease of the molar volume of IL for CO_2/CH_4 [73]. The CO_2 permeability through [Bmim][PF_6] and [Emim][Tf_2N]-based supported ionic liquid membranes was ranging between 168 and 9800 barrers based on the different supports [69]. Scovazzo also reported the CO_2 permeabilities in the range of [Emim][CF_3SO_3] and [Emim][Tf_2N] [73].

3 Treatment of liquid fuels

3.1 Treatment with traditional methods

A number of methods, including catalytic hydrodesulfurization, biodesulfurization, adsorption, and extraction have been explored for removing the sulfur compounds from liquid fuel [7,74,75]. The hydrodesulfurization is considered an established method in which the sulfur compounds react with H_2 in the presence of various catalysts such as $CoMo/\gamma\text{-}Al_2O_3$ and $NiMo/\gamma\text{-}Al_2O_3$ to form hydrocarbons and H_2S [76]. The main bottlenecks of this method include the requirement of high temperature and pressure, reduction of octane number, degradation of catalytic reactivity, and low efficiency in refractory sulfur compounds separation [5,77]. The use of various microorganisms for desulfurization of liquid fuel can reduce the capital and operating costs; however, the sensitivity of microorganisms and low process efficiency is considered the main drawbacks for the feasibility of the method [78,79]. The adsorption methods use various adsorbents, including zinc oxide, alumina, zeolites, and activated carbon for absorbing the sulfur compounds. This method requires low temperature and no H_2S gas is released from the method [80]. The adsorption methods are associated with several obstacles, for example, poor selectivity of sulfur compounds and low efficiency of the process [78]. Similar to the adsorption methods, the extraction process occurs at low temperature and pressure and requires no hydrogen and catalysts [80,81]. However, selective separation of sulfur compounds without extracting the desired hydrocarbons and poor efficiency of the process are the major challenges of the method [77].

The fatty acid methyl esters or fatty acid ethyl esters produced by transesterification from vegetable oils contain a relatively high amount of oxygen and need to blend with petroleum fuel to make it compatible with conventional diesel engines [82]. In addition, lignin-derived bio-oil contains guaiacol (39%), phenol, 3-methylphenol, 4-methylphenol, 2-ethylphenol, 4-propylphenol, cresol, anisole, 4-methylanisol, vanillin, eugenol, cyclohexanone, and other derivatives. However, the guaiacol is considered the model compound for hydrodeoxygenation because of its two types of C—O bonds (hydroxyl and methoxy) [83]. The low thermal stability of bio-oil limits the application of the distillation process on the separation of oxygenates; while the organic solvent is not feasible because of partly degradation in the bio-oil phase [84,85]. Therefore, the catalyst for deoxygenation of model bio-oil has been widely studied to produce oxygen-free or low oxygen hydrocarbon fuel [86]. Although the efficacy of acid and base catalysts were examined, the acid-base bifunctional catalysts were found favorable for the C—C bond-forming reaction, which strengthens the oxygen removal capacity [87–89]. The most commonly studied catalysts in the literature are CoMo/C [90], CoMo/Si [90], Rh/Al_2O_3 [91], Pd/SiO_2 [92], Ni/SiO_2 [93], Pd/C [94], and Pt/C [91]. The double bonds in bio-oil are saturated and converted to hydrocarbon by the deposition of oxygen in the form of CO, CO_2, and H_2O as shown in Fig. 4. The effects of various

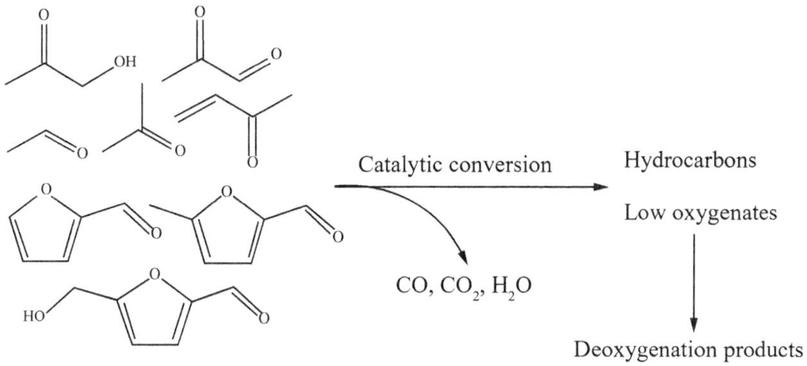

FIG. 4 Catalytic deoxygenation of model bio-oil for hydrocarbon fuel production [86].

catalysts such as silica gel, nanospring, and Al_2O_3-based Ni and Ru were examined for removing oxygen-based compounds from pyrolyzed bio-oil. The results depicted that the newly utilized nanospring-based catalysts show the promising result to upgrade the bio-oil into direct application transportation fuel [95].

However, the main issue related to hydrodeoxygenation is the high cost associated with the use of high-pressure H_2 [96]. The catalytic hydrodeoxygenation of bio-oil at low-pressure H_2 or no H_2 condition has been given attention [96,97]. The use of internally available H_2 in biomass-derived bio-oil can be a feasible approach for hydrodeoxygenation in case of a limited H_2 gas supply. Instead of traditional sulfide catalysts, the bimetallic catalysts containing oxophilic and active metal shows promising results [98]. The reaction mechanism of catalytic deoxygenation of triglycerides reported the breakdown of the triglycerides into fatty acids and propane known as β-elimination [96,97].

In a recent study, the catalytic effects of Ni-based mesoporous silica-supported catalysts (Ni/SBA-15 and Ni/Al-SBA-15) were evaluated at 300°C in 3 MPa H_2 pressure for 60 min in terms of hydrodeoxygenation reactivity with crude bio-oil. The result suggested that the degree of deoxygenation was about 54.9%. Fig. 5 outlines the degree of deoxygenation of some catalysts. The analysis of XRD patterns concluded that the mesoporous structure of SBA-15 and Al-SBA-15 lowers the dispersion compared to Ni/C [99]. The hydrodeoxygenation of real bio-oil extracted from the fast pyrolysis of cork oak was examined using platinum-loaded HY zeolites (Pt/HY) catalyst, which improved the properties of pyrolytic bio-oil [102]. The hydroprocessing of pine sawdust bio-oil with sulfided Ru/C catalyst at ~170°C produces upgraded oil of only 0.2–2.7 wt% oxygen and 0.82–0.92 g/mL density [103]. Su-Ping investigated the reaction kinetic model for deoxygenation and also the effects of reaction time, temperature, and hydrogen pressure on the hydrodeoxygenation bio-oil with cobalt molybdate catalyst [104]. The results indicated deoxygenation as

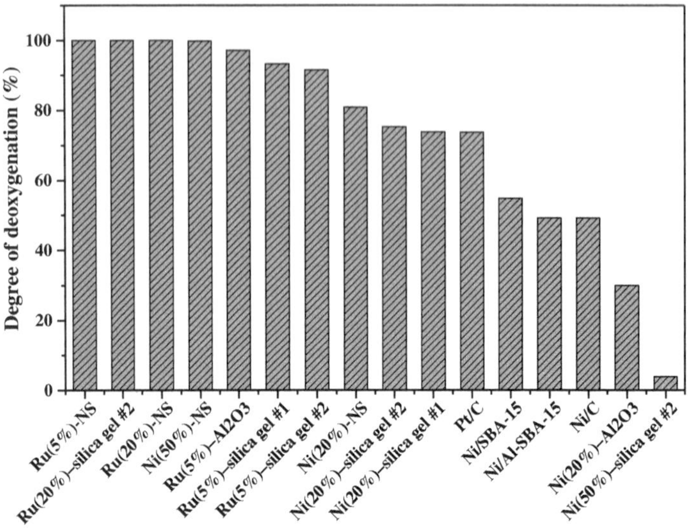

FIG. 5 Performance of different catalyst for deoxygenation of various vegetable oil [99–101].

a function of temperature and reaction time. Although catalytic deoxygenation significantly reduces the O_2 content in bio-oil or vegetable oil, they have a number of environmental, health, and safety issues, including sintering, dissolution, and poisoning. Apart from these, the low recovery of catalysts and deactivation of their catalytic property over the period of utilization increase overall operating costs.

3.2 Treatment with ionic liquids

3.2.1 Desulfurization

Desulfurization of petroleum-derived fuels such as gasoline and diesel using ILs has received remarkable attention from the scientific community and refining industry. Extractive desulfurization of diesel and gasoline using ILs exhibited promising removal efficiency of dibenzothiophene, which is very difficult to remove by hydrodesulfurization process and requires high temperature and energy [105]. Fig. 6 shows a simple schematic representation of liquid-liquid extraction of sulfur using ILs. Zhao et al. showed about 99% of dibenzothiophene extraction using ILs-based four steps extractive desulfurization methods [106]. The sulfur content was reduced from 500 ppm to <5 ppm. The room temperature treatment of gasoline using [Bmim][Cu_2Cl_3] removed 16.2%–37.4% of sulfur compounds (dibenzothiophene/thiophene) after 20 min, depending on the initial sulfur content in gasoline [107].

Zhu et al. studied catalytic oxidative desulfurization of dibenzothiophene using four ILs such as [Bmim][BF_4], [Omim][BF_4], [Bmim]PF_6], and

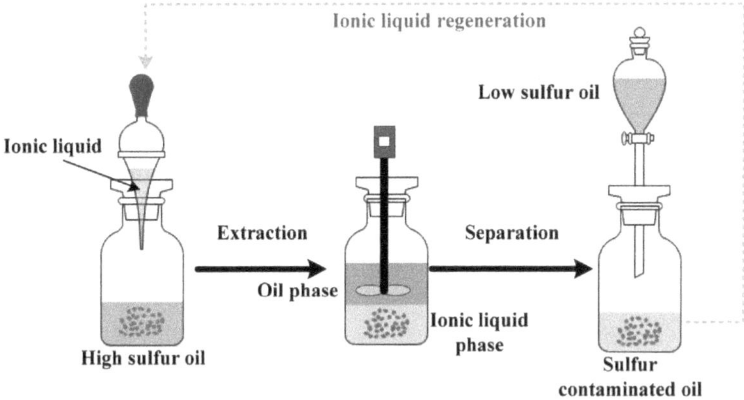

FIG. 6 Schematic representation of liquid-liquid extraction of sulfur using ILs.

[Omim][PF_6] [108]. A maximum sulfur removal of 99.6% was obtained using [Bmim][BF_4] at 30°C after 120min with vanadyl acetylacetonate as catalyst and H_2O_2 as oxidizing agent. Zhu et al. also reported about 99.0% sulfur (dibenzothiophene) removal using [Bmim][BF_4] as the solvent in the presence of molybdenum-based catalyst ($Na_2MoO_4.2H_2O$) and the oxidizing agent (H_2O_2) at 70°C after 180min [109]. The molybdic compound is first oxidized by H_2O_2 to form peroxomolybdic compound and dissolved in IL. Then the IL extracts dibenzothiophene and oxidizes to its corresponding sulfones ($DBTO_2$) by the peroxomolybdic compound and accumulated in IL as shown in Fig. 7. Rodríguez-Cabo et al. treated real diesel using [HmmPy][Tf_2N] and [Emim][Tf_2N] with acetic acid as catalyst and H_2O_2 as the oxidizing agent at 50°C and achieved almost 96.5% and 82.0% sulfur removal, respectively [110]. Lo et al. observed a higher extraction of dibenzothiophene using [Bmim][PF_6] compared to [Bmim][BF_4] due to the immiscible nature of [Bmim]

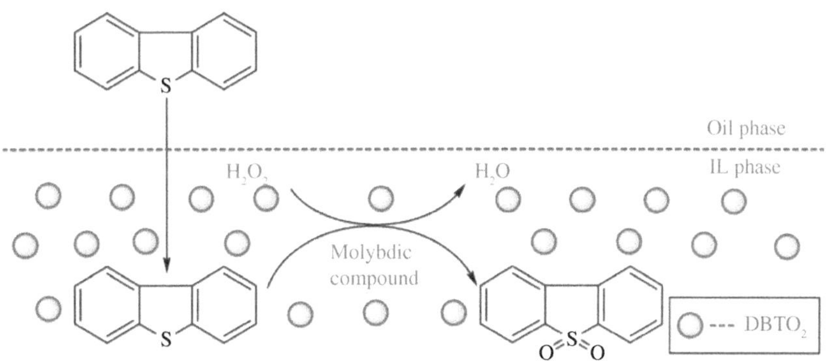

FIG. 7 Extractive oxidative desulfurization using catalyst [109].

$[PF_6]$ in water [111]. The acidic ILs also exhibited promising results up to 100% of sulfur removal under optimum process conditions [112].

So far, a number of interaction mechanisms between IL and sulfur compound, including π-π interaction, CH-π interaction, and hydrogen bonding have been proposed in the literature. Cassol et al. and Anantharaj and Banerjee identified the CH-π interaction between the cation and sulfur atom as the main driving force for the extraction of sulfur compounds [113,114]. Su et al. studied π-π interaction, CH-π interaction, and hydrogen bonding mechanism by ^1H NMR chemical shifts off thiophene [115]. The results of this investigation suggested that the π-π interaction was the dominant mechanism, which enhanced the extraction efficiency. On the other hand, hydrogen bonding between the aromatic structure of sulfur and IL was proposed by another research group as the key interaction mechanism [116]. Therefore, it can be concluded that the interaction mechanism between ILs and sulfur compounds is still contradictory and requires further in-depth investigation.

3.2.2 *Bio-oil esterification and removal of oxygenates*

The task-specific characteristic of ILs has stimulated the development and synthesis of these green solvents for the upgradation of bio-oil and vegetable oil. So far, the catalytic effect of various ILs in the esterification reaction for upgrading bio-oil to biodiesel has been explored well [117–119]. The role of acidic ionic liquid, $[C_6(mim)_2\text{-}HSO_4]$ as catalyst was investigated for upgrading the raw bio-oil produced from fast pyrolysis of rice husks. The esterification reaction was carried out at about 30, 60, and 110°C for 3 h and showed promising conversion efficiency by providing the desired Brønsted acidity [120]. Similarly, Ullah et al. utilized $[Bmim][HSO_4]$ for upgrading waste cooking oil to biodiesel [121]. However, the study related to the development of the feasible separation technique of oxygenates from bio-oil or other vegetable oil is deficient in the literature.

The oxygenates such as acetic acid, glycolaldehyde, and acetol in pyrolysis oil were separated using phosphonium and imidazolium-based ionic liquids from a model aqueous solution of pyrolysis oil [15]. However, it was quite difficult to separate completely in the first stage of extraction because of the difficulty of oxygenates evaporation. Two-stage extraction was suggested where acetic acid and acetol were extracted with $[Hmim][B(CN)_4]$ in the first stage. In the second stage, $[P_{666,14}][Phos]$ was used for separating glycolaldehyde from the bio-oil. Fig. 8 illustrates the conceptual process for removing oxygenates from bio-oil using IL solvent.

With the aim of exploring the possibility for the separation of oxygenates, for example, aldehydes/ketones/ethers from aliphatics/aromatics, Domańska et al. explained the liquid-liquid-phase equilibria of the binary system [122]. Kumełan et al. studied the O_2 solubility in $[Bmim][PF_6]$ experimentally and predicted by Monte Carlo simulations [123]. A recent patent indicated the

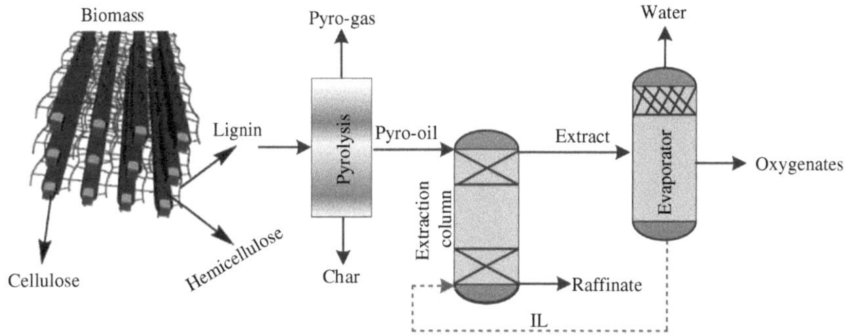

FIG. 8 Conceptual process flow for oxygenates separation using IL.

use of catalysts and ILs comprising imidazolium, phosphonium, and pyridinium cations for deoxygenation and removing nitrogen impurities from biomass-derived pyrolysis oil. The elemental analysis of deoxygenated bio-oil contained only 0.26% (wt%) of oxygen [124]. Table 3 shows the O_2 solubility in ILs available in the literature. Jaschik et al. experimentally determined the O_2 solubility in ILs and compared with that from COSMO-SAC model [127]. The authors found that the experimental value of solubility differed from the model predicted value and concluded that the model is not suitable for the prediction of O_2 solubility in ILs.

4 Influential factors for ionic liquid treatment

4.1 Factors affecting gas absorption

4.1.1 Henry's law constant

The pressure drop method was extensively studied to correlate Henry's law constant to the mole of CO_2 dissolved in ILs and to investigate the effect of temperature on Henry's constant [128–131]. The amount of CO_2 dissolved in ILs is calculated in terms of the mole using ideal gas law:

$$\Delta n_{CO_2} = \Delta pV/RT \tag{1}$$

where Δp is the difference between initial pressure (p_{in}) and equilibrium pressure (p_{eq}); R is the ideal gas constant; T is the operating temperature in K; and V is the cell volume in m^3.

Henry's law constant is determined by the ratio of the equilibrium pressure (p_{eq}) to the mole fraction of CO_2 (x) in the ionic liquid:

$$H_{CO_2} = p_{eq}/x \tag{2}$$

$$x = \Delta n_{CO_2}/(x_{IL} + \Delta n_{CO_2}) \tag{3}$$

TABLE 3 Oxygen solubility data in various ILs.

ILs name	Temperature (K)	Time (h)	H_2 pressure (MPa)	Analysis method	Remarks	Reference
[Bmim][PF_6]	293.2–373.1		9.1		Only about 0.16 mol of O_2 is dissolved in 1 kg of the ionic liquid.	[123]
[$P_{666,14}$][Cl]	25 < T < 60	18		Gravimetric microbalance	[$P_{666,14}$][Phos] has the highest glycolaldehyde distribution coefficient, whereas [Hmim][$B(CN)_4$] is suitable to extract acetic acid and acetol.	[15]
[$P_{666,14}$][$N(CN)_2$]	25 < T < 60	18				
[$P_{666,14}$][Phos]	25 < T < 60	18				
[Hmim][$B(CN)_4$]	25 < T < 60	18				
[Omim][$B(CN)_4$]	25 < T < 60	18				
[Bmim][PF_6]	50		13	Gravimetric microbalance	[Bmim][PF_6] is not suitable for dissolving O_2 at a low or moderate pressure.	[44]
[MOA][Tf_2N]	303.15–333.15		1.5	Gravimetric microbalance	[MOA][Tf_2N] is found the most efficient among the other ILs because of its high molar volume as well as its highly fluorinated anion and the larger cation chain.	[125]
[Bmim][Tf_2N]						
[Bmim][$MeSO_4$]						
[Bmim][BF_4]	303–343	8 h	0.015–0.1	Isochoric saturation	[Bmim][BF_4] shows greater solubility compared to [Bmim][PF_6]	[46]
[Bmim][PF_6]	298.15–323.15		0–13	Gravimetric microbalance	The IL containing $[Tf_2N]^-$ anion shows the most promising result for O_2 solubility. The positive enthalpy and entropy change indicate that the O_2 solubility is governed by the degree of disorder in the system.	[126]
[Bmim][Tf_2N]	283.15–323.15		0–13			
[Bm-Pyr][Tf_2N]	283.15–323.15		0–13			
[$N_{4,4,4,1}$][Tf_2N]	283.15–323.15		0–11.5			
[$P_{1,4,4,4}$][TOS]	323.15–323.15		0–13			

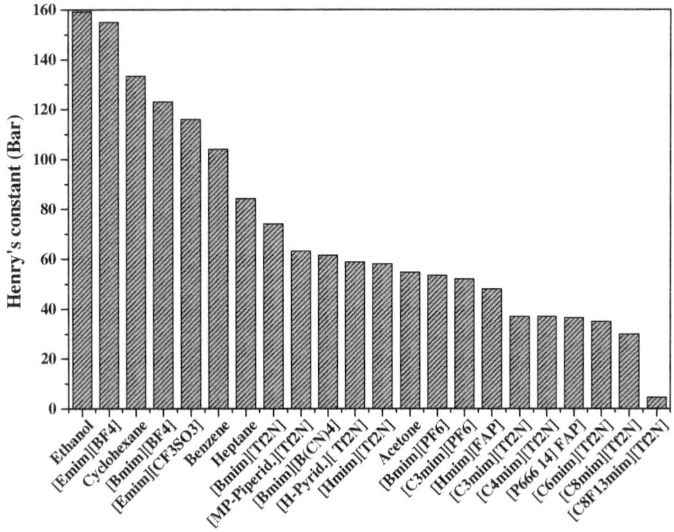

FIG. 9 Comparative study of Henry's constant value between various ILs and traditional solvents [132,133].

The effect of temperature on Henry's constant and CO_2 solubility in [Bmim][BF_4], [Emim][CF_3SO_3], [Emim][Tf_2N], [BM-Pyrrolid.][Tf_2N], and [Bmim][B$(CN)_4$] was studied in the pressure drop method over a temperature range of 30–260°C. Henry's constant value reduced with an increase in temperature hence decreased the CO_2 solubility [132]. Fig. 9 illustrates Henry's constant value of various ILs and traditional solvents. The cation has a very insignificant effect on Henry's constant; however, the effect is slightly increased for $[BF_4]^-$ compared to $[Tf_2N]^-$. On the other hand, Henry's constant is highly affected by the nature of anion. Despite the smaller size of $[BF_4]^-$ than $[FAP]^-$, $[BF_4]^-$ shows almost 2.5 times higher value of the Henry's constant compared to $[FAP]^-$ and the order as follows [132]:

$$[BF_4]^- > [CF_3SO_3]^- > \left[B(CN)_4\right]^- \sim [Tf_2N]^- > [FAP]^-$$

The solubility of oxygen in [Bmim][PF_6] is very low compared to other gasses such as CO_2 and ethane at low or moderate pressure. This is because of its low interaction with ionic liquid as well as high Henry's coefficient and low mass transfer [44]. The higher value of Henry's law coefficient indicates the lower solubility of oxygen in [Bmim][BF_4]. The value of Henry's law coefficient decreases with the increase of temperature hence enhances the solubility of oxygen in the ILs as presented in Fig. 10A. Although the behavior of [Bmim][BF_4] and [Bmim][PF_6] with the solute are almost the same, [Bmim][BF_4] shows a greater solubility because of its high O_2 gas affinity compared to [Bmim][PF_6] [46].

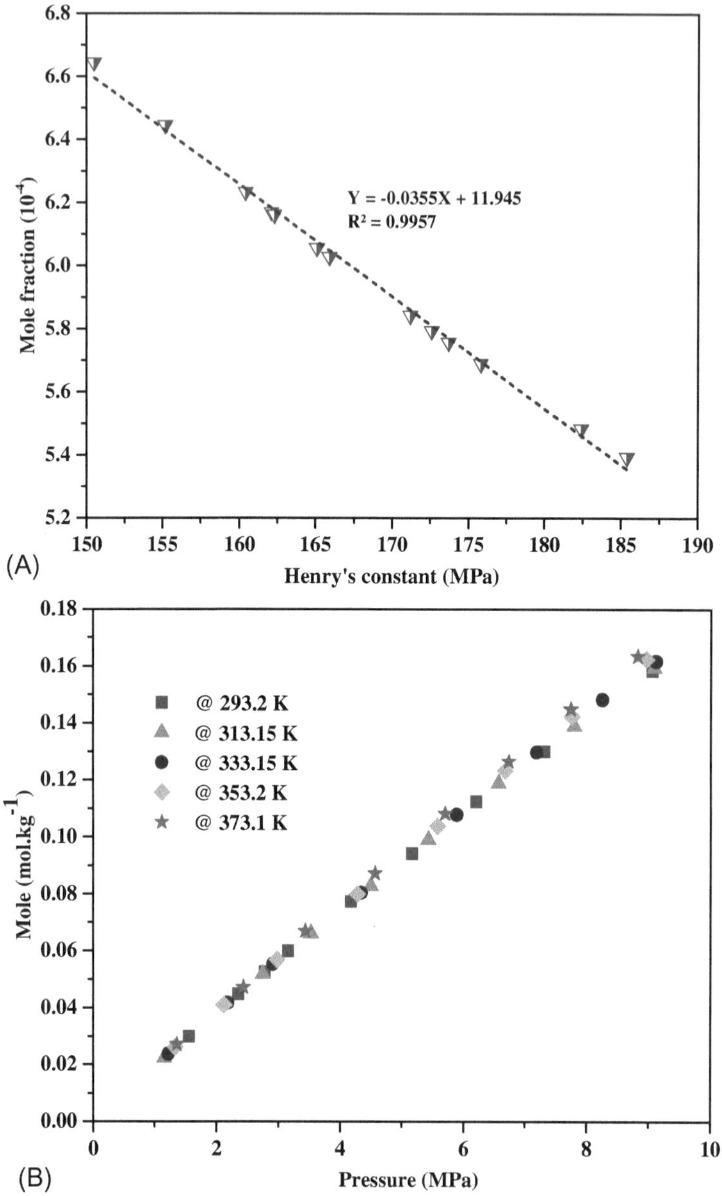

FIG. 10 Solubility of O_2 in ILs (A) effects of Henry's constant on solubility in [Bmim][BF_4]. (B) Effect of temperature and pressure on solubility in [Bmim][PF_6]. *(A: Prepared using data from Jacquemin J, Costa Gomes MF, Husson P, Majer V. Solubility of carbon dioxide, ethane, methane, oxygen, nitrogen, hydrogen, argon, and carbon monoxide in 1-butyl-3-methylimidazolium tetrafluoroborate between temperatures 283 K and 343 K and at pressures close to atmospheric. J Chem Thermodyn 2006;38:490–502. doi:10.1016/j.jct.2005.07.002. B: Prepared using data from Kumełan J, Kamps Á P-S, Urukova I, Tuma D, Maurer G. Solubility of oxygen in the ionic liquid [bmim][PF6]: experimental and molecular simulation results. J Chem Thermodyn 2005;37: 595–602. https://doi.org/10.1016/j.jct.2005.03.005.)*

4.1.2 Nature of cation and anion of ILs

The nature of the anion in ILs plays a key role on gas solubility. The interaction between the negative fluorine atoms of the anion and the positive charge on the carbon of carbon dioxide enhances the CO_2 solubility in ILs. Because of the highly fluorinated anion and larger cation, [MOA][Tf_2N] showed the highest CO_2 absorption capacity among [Bmim][$MeSO_4$] and [Bmim][Tf_2N] [125]. From the findings of the study, it can be concluded that ammonium-based cations achieve superior CO_2 absorption over the imidazolium and pyridinium-based cations. For the similar reason, the presence of fluoroalkyl groups in [Emim][Tf_2N] showed the better solubility of CO_2 than [Emim][PF_6] though they had exactly the same cation [39]. The CO_2 solubility in [Bmim][PF_6], [C_6mim][PF_6], [Emim][BF_4], [C_6mim][BF_4], [Emim][Tf_2N], and [C_6mim][Tf_2N] were investigated experimentally (Fig. 11) at 298.15 K under 1.0 MPa pressure. The ILs containing [Tf_2N] anion exhibited greater solubility compared to other anions. The developed GC-EOS model showed its suitability for CO_2 solubility prediction in ILs [40]. Therefore, the nature of anion of ILs compared to cation and the operational parameters such as temperature and pressure are the key factors that influence the CO_2 solubility in ILs.

The solubility of O_2 in ILs is determined by the interaction of anion of ILs with O_2 gas and the molar volume of the ILs. Higher molar volume increases the free space in ILs for dissolving the O_2 gas into that space. It was found that the molar volume of the ILs decreased as [MOA][Tf_2N] > [Bmim][Tf_2N] > [Bmim][$MeSO_4$] thereby reduced solubility accordingly. The highly fluorinated anion $[Tf_2N]^-$ enhanced the molecular interaction and the larger cation chain $[MOA]^+$ intensified the pores, which favor the O_2 absorption capacity [125]. Anthony et al. monitored the influence of different anions ($[Tf_2N]^-$, $[BF_4]^-$, $[PF_6]^-$) with the $[Bmim]^+$ cation on O_2 solubility. The presence of

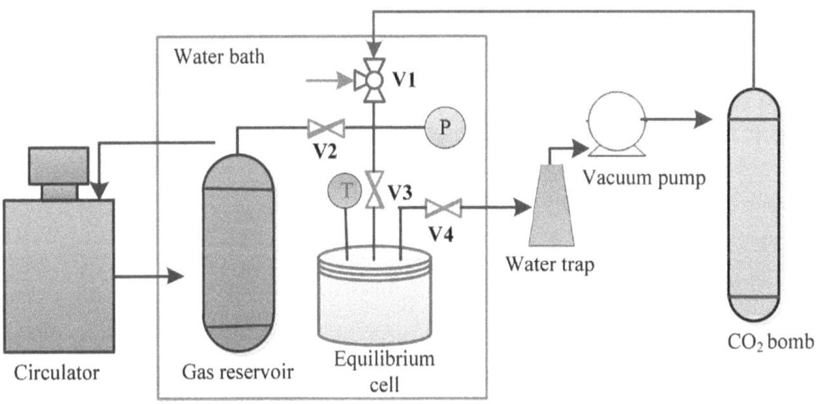

FIG. 11 Schematic illustration of the experimental apparatus for carbon dioxide solubility measurement in ionic liquids [40].

$[Tf_2N]^-$ anion significantly increased the O_2 solubility compared to the other two anions, whereas $[BF_4]^-$ had an insignificant effect on the solubility [126]. The substitution of the butyl group in cation of pyrrolidinium and piperidinium-based ILs with the 2-methoxyethyl group considerably decreases the oxygen solubility [134]. The C—F bonds and C—H bonds in ILs stimulate the formation of hydrogen bonds with dissolved O_2 in ILs, which importantly affects the solubility value [135].

4.1.3 Process parameters

The linear line graph between CO_2 solubility pressure vs CO_2 molality indicates that the gas molality in ILs increases with the rise of solubility pressure. Moreover, the solubility of CO_2 in [Hmim]$[Tf_2N]$ decreases with the increase in temperature [38]. The similar effects of temperature and pressure on the CO_2 solubility in ILs were observed in different studies [50,132,136,137]. On the other hand, the effect of temperature increase on the CO_2 solubility in [MOA]$[Tf_2N]$ is found to be insignificant [125]. The temperature and pressure have similar effects on the CO_2 solubility in both functionalized and nonfunctionalized ILs. The CO_2 solubility in functionalized ILs increases with the increase in pressure and decreases with the increase in temperature [20,62,66].

The oxygen solubility in the ILs slightly decreases with the increase in temperature [44]. On the contrary, another study concluded that the temperature has an insignificant effect on the solubility of O_2 in ILs [123]. It is also reported that the O_2 solubility in ILs increases with the increase in temperature [46]. The efficiency of the solvent depends on its affinity as well as on the distribution coefficient of solute. The solubility of glycolaldehyde and acetol increases with the increase of temperature because of their weak intermolecular interaction and instability at high temperatures. On the other hand, there is no significant effect on the solubility of acetic acid [15]. Bahadur et al. found that solubility of oxygen in ILs increased with a decrease in operating temperature and an increase in pressure [125]. Fig. 12B shows the temperature and pressure effects on O_2 solubility in [Bmim]$[PF_6]$. From the above literature, it can be concluded that both cation and anion play a significant role in the O_2 solubility in ILs. Surprisingly, the role of temperature on O_2 solubility in ILs is contradictory in the literature. Some studies reported an increase in solubility, while some observed decrease in solubility with increase in temperature.

4.2 Factors affecting the desulfurization of liquid fuel

Desulfurization of liquid fuel using IL is significantly influenced by IL types, catalyst, and oxidizing agent used in catalyzed oxidative desulfurization, and process parameters. Zhu et al. observed a slight increase in sulfur removal when H_2O_2 was added with IL [108]. However, a significant increase in sulfur removal was noticed after using a catalyst with H_2O_2, when compared with

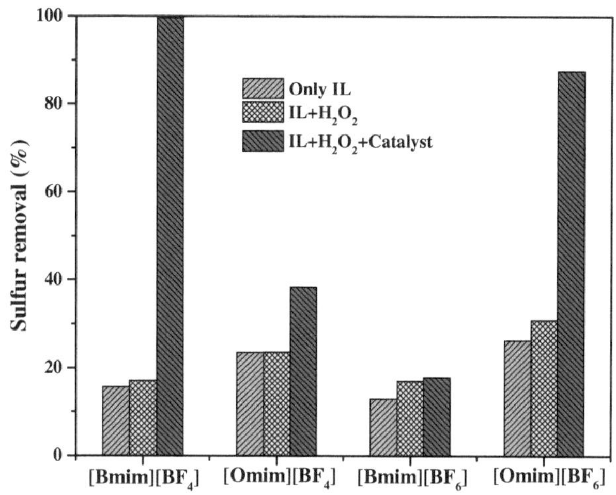

FIG. 12 Effect of catalyst on sulfur removal using different ILs [108].

that of IL only (Fig. 12). Among the four ILs investigated 1-butyl-3-methylimidazolium tetrafluoroborate with vanadyl acetylacetonate catalyst removed almost all sulfur as the IL dissolved in H_2O_2/catalyst system and formed a biphasic system. They also reported the increase in sulfur removal with the increase in reaction temperature and time. On the contrary, Asumana et al. observed the reduction in sulfur removal efficiency from diesel and gasoline with the increase in temperature using [Bmim][$N(CN)_2$] and keeping IL to oil ratio 1:1 [138]. When the IL to oil mass ratio was changed from 2:1 to 1:5, the extraction efficiency of dibenzothiophene from n-octane using [Bmth][N$(CN)_2$] reduced from 66.8% to 15%, while thiophene removal efficiency from n-hexane reduced from 60.4% to 18% using the same IL [139]. The amount catalyst significantly impacted the sulfur extraction efficiency of ILs-based solvent system. When the catalyst ($Na_2MoO_4{\cdot}2H_2O$) and dibenzothiophene molar ratio was changed from 1:100 to 1:20, the dibenzothiophene content in the oil reduced from 400 to 50 ppm [109]. The extraction experiment was conducted at 60°C for 180 min. Chen et al. achieved complete removal of sulfur compound from 1 L of oil sample after 45 min of treatment with [Bmim][BF_4] in the presence of tungsten-based catalyst ($Na_7H_2LaW_{10}O_{36}.32H_2O$) and H_2O_2 [140]. The use of catalyst ($(NH_4)_6Mo_7O_{24}.4H_2O$) along with [Bmim][Bu_2PO_4] and H_2O_2 extracted about 89.2% of sulfur at 50°C for 3 h, which was much higher when compared with IL only treatment (9.3%) of model oil [141].

The structure of ILs can also be important in the extraction of sulfur compounds from diesel and gasoline. Zhang et al. studied a series of imidazolium cation-based ILs with different anions for the removal of thiophene [142]. The results suggested that the increase in alkyl chain length imidazolium cation increases the thiophene extraction efficiency by ILs. The ILs with larger anions structure also showed better extraction efficacy compared to ILs with smaller anions. Accordingly, the ILs investigated on that study was ordered as follows: [Bmim][PF_6] > [Bmim][BF_4] > [Emim][BF_4]. Anantharaj and Banerjee screened the possibility of 168 ILs by COSMO-RS from the combinations of six cations and 28 anions [114]. The authors observed that the ILs consisting of smaller nonaromatic cations showed a better performance in the removal of sulfur compounds such as thiophene, benzothiophene, and dibenzothiophene. Kumar and Banerjee also suggested that ILs with smaller cations extracted more thiophene [143]. On the contrary, Holbrey et al. recommended that the higher aromatic structure of cation is favorable for the extraction of sulfur compounds [144]. They also noticed the dominating effects of cations structure on sulfur extraction compared to anions. Therefore, the role of cations structure in sulfur removal is contradictory in the literature and also varies due to the types of sulfur compounds structure.

5 Future prospects and challenges

The ILs have been utilized for the absorption of CO_2 from CO_2/CH_4 mixture and desulfurization of crude oil and the results are quite impressive; however, the investigations are only limited in laboratory scale. The research related to the separation of oxygenates from bio-oil and CO_2 from natural gas/biogas is very limited in the literature. The utilization of ILs in this sector is in the embryonic stage and more research is required for the process development. A feasible IL-based separation process development is a quite challenging task and needs to resolve a number of issues as highlighted in Fig. 13.

- The reaction mechanism during gas sweetening and liquid fuel treatment using ILs is still unclear. The gas absorption mechanism in ILs is mainly governed by mass transfer and thermodynamic characteristics [145,146]. However, there is still no significant advancement in the understanding of IL-gas interaction, mass transfer characteristics, molecular dynamic simulation, and other limitations associated with the solubility mechanism. To date, only a few study explained the CO_2 absorption mechanism and reaction kinetics in functionalized ILs. The literature also lacks the systematic evaluation and mechanistic understanding of the extraction of sulfur compounds in IL phase from oil phase. Hence, more research is the urgency to underpin the science of reaction kinetics during IL treatment.
- The solubility of gas and extraction of sulfur compounds in ILs is also enhanced by the nature of anions and cations as well as the various process parameters. Most of the literature investigated the effect of process parameters on the gas solubility in ILs. Till now, no process parameter optimization study has been carried out. Therefore, it is difficult to predict the operational parameters, energy requirement, and the possible problems associated with reactor design and full-scale industrial process development.

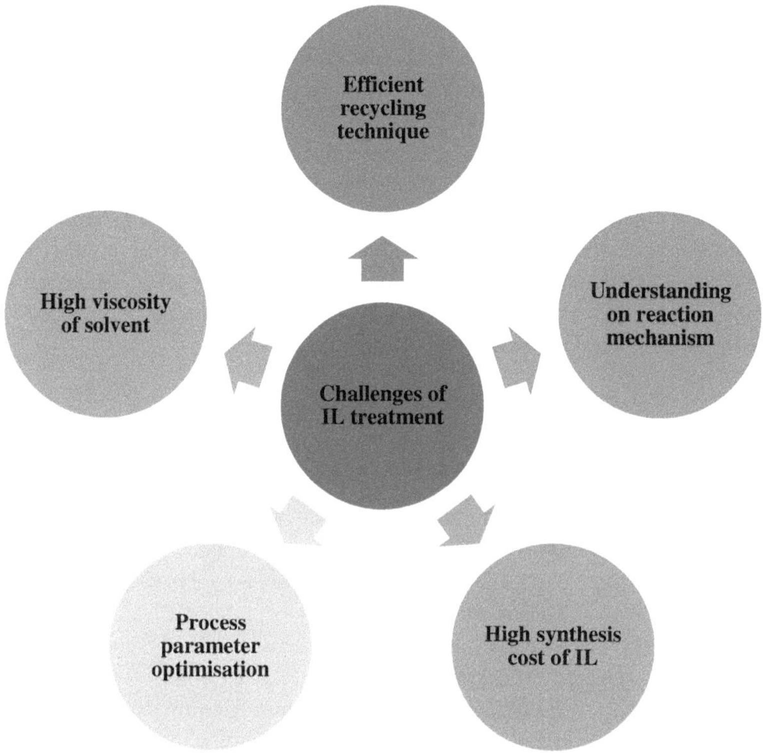

FIG. 13 Key challenges for the implementation of ILs gas sweetening and liquid fuel treatment.

- Moreover, the high viscosity of ILs is another critical challenge on the application as a solvent. With the aim of reducing viscosity and cost, Shannon et al. mixed commercially available solvents to ILs and desired properties in the blended ILs [147]. The addition of a certain amount of water in ILs can also reduce the viscosity as well as process cost. However, the role of water addition on the solubility of gas or extraction of sulfur compounds in ILs is yet to be explored in great detail. In this case of IL-based deep eutectic solvents can be researched in the near future.
- Although ILs are considered as the green solvent, the synthesis cost of ILs is the foremost barrier for its successful implementation in gas absorption. However, fortunately, ILs can be recycled and reused. The regeneration of ILs can reduce the cost of the overall process. So far, there is no literature suggesting efficient recycling technology and reuse for gas absorption. A recent study showed no significant reduction in sulfur removal using IL up to 5 times recycling [108]. However, the reuse of the hydrophilic ILs can degrade its solvation power. Therefore, the development of efficient recycling techniques or synthesis of low-cost ILs can make the technology viable for commercial scale-up applications.

6 Concluding remarks

To date, a wide range of ILs have been synthesized which can be used as the green solvent for the separation of CO_2 from natural gas/biogas and oxygenates from bio-oil. Nature of anions and cations and various process parameters such as temperature and pressure play a significant role in the separation efficiency. Greater attention is essential to explore the science behind the kinetics of the reaction during IL treatment. Furthermore, the high cost of the ILs synthesis is the major constraint for its successful utilization, and thereby future research should focus on the development of low-cost IL synthesis route. Therefore, the challenges associated with the use of ILs need to be addressed immediately to develop an industrially viable and sustainable process.

Abbreviations

[Am-Im][BF_4]	primary amine-functionalized 1-butyl-3-methylimidazolium tetrafluoroborate
[3Am-Im][BF_4]	secondary amine-functionalized 1-butyl-3-methylimidazolium tetrafluoroborate
[Am-Im][DCA]	primary amine-functionalized 1-butyl-3-methylimidazolium dicyanamide
[Bmim][Cu_2Cl_3]	1-butyl-3-methylimidazolium trichlorodicuprate
[Bmim][BF_4]	1-butyl-3-methylimidazolium tetrafluoroborate
[Bmim][Bu_2PO_4]	1-butyl-3-methylimidazolium dibutylphosphate
[Bmim][DCA]	1-butyl-3-methylimidazolium dicyanamide
[Bmim][HSO_4]	1-butyl-3-methylimidazolium hydrogensulfate
[Bmim][$MeSO_4$]	1-butyl-3-methyl imidazolium methyl sulfate
[Bmim][$N(CN)_2$]	1-butyl-3-methylimidazolium dicyanamide
[Bmim][NO_3]	1-butyl-3-methylimidazolium nitrate
[Bmth][$N(CN)_2$]	3-butyl-4-methylthiazolium dicyanamide
[Bm-Pyrrolid.] [Tf_2N]	1-butyl-1-methylpyrrolidinium bis(triflouromethylsulfonyl)imid
[Bmim][PF_6]	1-butyl-3-methylimidazolium hexafluorophosphate
[Bm-Pyr][Tf_2N]	1-butyl-1-methylpyrrolidinium bis(triflouromethylsulfonyl)imide
[Bmim][Tf_2N]	1-butyl-3-methylimidazolium bis(trifluoromethylsulfonyl)imide
[$C_6(mim)_2$-HSO_4]	1-hexyl-3-methylimidazolium double hydrogen sulfate
[C_1C_2Im][Tf_2N]	1-ethyl-3-methylimidazolium bis(trifluoromethylsulfonyl)imide
[C_1C_4pyrr][Tf_2N]	1-butyl-1-methylpyrrolidinium bis(trifluoromethylsulfonyl)imide
[$N_{1132\text{-}OH}$][Tf_2N]	propyl-cholinium bis (trifluoromethylsulfonyl)imide
[C_6mim][BF_4]	1-hexyl-3-methylimidazolium tetrafluoroborate

[C_6mim][PF_6]	1-hexyl-3-methylimidazolium hexafluorophosphate
[C_6mim][Tf_2N]	1-hexyl-3-methylimidazolium bis (trifluoromethylsulfonyl) imide
[C_8F_{13}mim] [Tf_2N]	1-methyl-3-(3,3,4,4,5,5,6,6,7,7,8,8,8-tridecafluorooctyl) imidazolium bis (trifluoromethylsulfonyl) amide
[$C_n$mim]	1-n-alkyl-3-methyl-imidazolium
[Emim][BF_4]	1-ethyl-3-methylimidazolium tetrafluoroborate
[Emim][CF_3SO_3]	1-ethyl-3-methylimidazolium trifluoromethanesulfonate
[Emim][PF_6]	1-ethyl-3-methylimidazolium hexafluorophosphate
[Emim][Tf_2N]	1-ethyl-3-methylimidazolium bis (trifluoromethylsulfonyl) imide
[FAP]$^-$	tris(perfluoroethyl) trifluorophosphate anion
[Hmim][$B(CN)_4$]	1-hexyl-3-methylimidazolium tetracyanoborate
[Hmim][Tf_2N]	1-hexyl-3-methylimidazolium bis (trifluoromethylsulfonyl) imide
[HmmPy][Tf_2N]	1-hexyl-3,5-dimethylpyridinium bis(trifluoromethyl sulfonyl)imide
[MOA][Tf_2N]	methyl trioctyl ammonium bis (trifluoromethylsulfonyl) imide
NHD	polyethylene glycol dimethyl ethers
[$N_{4,4,4,1}$][Tf_2N]	methyl(tributyl)ammonium bis(trifluoromethylsulfonyl) imide
[Omim][$B(CN)_4$]	1-octyl-3-methylimidazolium tetracyanoborate
[OH-Im][BF_4]	hydroxyl functionalized 1-butyl-3-methylimidazolium tetrafluoroborate
[Omim][BF_4]	1-octyl-3-methylimidazolium tetrafluoroborate
[Omim][PF_6]	1-octyl-3-methylimidazolium hexafluorophosphate
[$P_{666,14}$][Cl]	trihexyl(tetradecyl)phosphonium chloride
[$P_{666,14}$][$N(CN)_2$]	trihexyl(tetradecyl)phosphonium dicyanamide
[$P_{666,14}$][Phos]	trihexyl(tetradecyl)phosphonium bis-2,4,4-(trimethylpentyl)phosphinate
[$P_{1,4,4,4}$][TOS]	tributylmethylphosphonium tosylate

References

[1] British Petroleum (BP). BP statistical review of world energy. British: BP Plc; 2016. http://www.bp.com/. [Accessed 11 May 2017].

[2] Goli A, Shamiri A, Talaiekhozani A, Eshtiaghi N, Aghamohammadi N, Aroua MK. An overview of biological processes and their potential for CO_2 capture. J Environ Manag 2016;183:41–58. https://doi.org/10.1016/j.jenvman.2016.08.054.

[3] Ahmad F, Keong LL, Shariff A, Engineering C. Removal of CO_2 from natural gas using membrane process: design, fabrication and parametric study. In: EMS Summer School, GKSS Research center, 29 September–2 October 2010, Geeesthact, Germany; 2010. p. 1–3.

[4] Datta AK, Sen PK. Optimization of membrane unit for removing carbon dioxide from natural gas. J Membr Sci 2006;283:291–300. https://doi.org/10.1016/j.memsci.2006.06.043.

[5] Hayyan M, Ibrahim MH, Hayyan A, Alnashef IM, Alakrach AM, Hashim MA. Facile route for fuel Desulfurization using generated superoxide ion in ionic liquids. Ind Eng Chem Res 2015;54:12263–9. https://doi.org/10.1021/acs.iecr.5b03427.
[6] Ibrahim MH, Hayyan M, Hashim MA, Hayyan A. The role of ionic liquids in desulfurization of fuels: a review. Renew Sust Energ Rev 2017;76:1534–49. https://doi.org/10.1016/j.rser.2016.11.194.
[7] Boniek D, Aparecida M, Stoianoff DR. Biodesulfurization: a mini review about the immediate search for the future technology. Clean Techn Environ Policy 2015;17:29–37. https://doi.org/10.1007/s10098-014-0812-x.
[8] Sasser E, Hemby J, Adler K, et al. Report to congress on black carbon; 2012. https://doi.org/10.1016/j.icrp.2009.12.007.
[9] Renewable Energy Policy Network (REN21). Global status report 2016, http://www.ren21.net/. [Accessed 12 May 2017].
[10] Xiu S, Shahbazi A. Bio-oil production and upgrading research: a review. Renew Sust Energ Rev 2012;16:4406–14. https://doi.org/10.1016/j.rser.2012.04.028.
[11] Halder PK, Paul N, Joardder MUH, Khan MZH, Sarker M. Feasibility analysis of implementing anaerobic digestion as a potential energy source in Bangladesh. Renew Sust Energ Rev 2016;65. https://doi.org/10.1016/j.rser.2016.06.094.
[12] Shah MS, Halder PK, Shamsuzzaman ASM, Hossain MS, Pal SK, Sarker E. Perspectives of biogas conversion into bio-CNG for automobile fuel in Bangladesh. J Renew Energy 2017;2017:1–14. https://doi.org/10.1155/2017/4385295.
[13] Huber GW, Sara Iborra A, Corma A. Synthesis of transportation fuels from biomass: chemistry, catalysts, and engineering. Chem Rev 2006;2:4044–98. https://doi.org/10.1021/CR068360D.
[14] Balat M, Balat H. A critical review of bio-diesel as a vehicular fuel. Energy Convers Manag 2008;49:2727–41. https://doi.org/10.1016/j.enconman.2008.03.016.
[15] Li X, Kersten SRA, Schuur B. Extraction of acetic acid, glycolaldehyde and acetol from aqueous solutions mimicking pyrolysis oil cuts using ionic liquids. Sep Purif Technol 2017;175:498–505. https://doi.org/10.1016/j.seppur.2016.10.023.
[16] Khurshid M. Biogas development in rural areas of Pakistan: a sustainable option for domestic energy. Sci J COMSATS 2009;15(2):57–61.
[17] Lee K, Aitomäki Y, Berglund LA, Oksman K, Bismarck A. On the use of nanocellulose as reinforcement in polymer matrix composites. Compos Sci Technol 2014;105:15–27. https://doi.org/10.1016/j.compscitech.2014.08.032.
[18] Awe OW, Zhao Y, Nzihou A, Minh DP, Lyczko N. A review of biogas utilisation, purification and upgrading technologies. Waste Biomass Valoriz 2017;8:267–83. https://doi.org/10.1007/s12649-016-9826-4.
[19] Xu Y, Huang Y, Wu B, Zhang X, Zhang S. Biogas upgrading technologies: energetic analysis and environmental impact assessment. Chin J Chem Eng 2015;23:247–54. https://doi.org/10.1016/j.cjche.2014.09.048.
[20] Zhang X, Zhang S, Bao D, Huang Y, Zhang X. Absorption degree analysis on biogas separation with ionic liquid systems. Bioresour Technol 2015;175:135–41. https://doi.org/10.1016/j.biortech.2014.10.048.
[21] Chiaramonti D, Prussi M, Ferrero S, Oriani L, Ottonello P, Torre P, et al. Review of pretreatment processes for lignocellulosic ethanol production, and development of an innovative method. Biomass Bioenergy 2012;46:25–35. https://doi.org/10.1016/j.biombioe.2012.04.020.
[22] Halder P, Kundu S, Patel S, Parthasarathy R, Pramanik B, Paz-ferreiro J, et al. TGA-FTIR study on the slow pyrolysis of lignin and cellulose-rich fractions derived from imidazolium-

based ionic liquid pre-treatment of sugarcane straw. Energy Convers Manag 2019;200. https://doi.org/10.1016/j.enconman.2019.112067, 112067.

[23] Halder P, Kundu S, Patel S, Setiawan A, Atkin R, Parthasarthy R, et al. Progress on the pre-treatment of lignocellulosic biomass employing ionic liquids. Renew Sust Energ Rev 2019;105:268–92. https://doi.org/10.1016/j.rser.2019.01.052.

[24] Prado R, Erdocia X, Labidi J. Study of the influence of reutilization ionic liquid on lignin extraction. J Clean Prod 2016;111:125–32. https://doi.org/10.1016/j.jclepro.2015.04.003.

[25] Halder P, Kundu S, Patel S, Ramezani M, Parthasarathy R, Shah K. A comparison of ionic liquids and organic solvents on the separation of cellulose rich material from river red gum. Bioenergy Res 2019;12:275–91. https://doi.org/10.1007/s12155-019-09967-8.

[26] Rufford TE, Smart S, Watson GCY, Graham BF, Boxall J, Diniz da Costa JC, et al. The removal of CO_2 and N_2 from natural gas: a review of conventional and emerging process technologies. J Pet Sci Eng 2012;94–95:123–54. https://doi.org/10.1016/j.petrol.2012.06.016.

[27] Zhao Y, Wang J, Zhang H, Yan C, Zhang Y. Effects of various LED light wavelengths and intensities on microalgae-based simultaneous biogas upgrading and digestate nutrient reduction process. Bioresour Technol 2013;136:461–8. https://doi.org/10.1016/J.BIORTECH.2013.03.051.

[28] Niesner J, Jecha D, Stehlík P. Biogas upgrading technologies: state of art review in european region. Chem Eng Trans 2013;35:517–22. https://doi.org/10.3303/CET1335086.

[29] Burr B, Lyddon L. A comparison of physical solvents for acid gas removal. In: GPA Annu Conv Proc, 1; 2008. p. 100–13.

[30] Siriwardane RV, Ming-Shing Shen A, Fisher EP. Adsorption of CO_2, N_2, and O_2 on natural zeolites. Energy Fuel 2003;17:571–6. https://doi.org/10.1021/EF020135L.

[31] Alonso-Vicario A, Ochoa-Gómez JR, Gil-Río S, Gómez-Jiménez-Aberasturi O, Ramírez-López CA, Torrecilla-Soria J, et al. Purification and upgrading of biogas by pressure swing adsorption on synthetic and natural zeolites. Microporous Mesoporous Mater 2010;134:100–7. https://doi.org/10.1016/J.MICROMESO.2010.05.014.

[32] Yang L, Ge X, Wan C, Yu F, Li Y. Progress and perspectives in converting biogas to transportation fuels. Renew Sust Energ Rev 2014;40:1133–52. https://doi.org/10.1016/j.rser.2014.08.008.

[33] Hedayat M, Soltanieh M, Mousavi SA. Simultaneous separation of H_2S and CO_2 from natural gas by hollow fiber membrane contactor using mixture of alkanolamines. J Membr Sci 2011;377:191–7. https://doi.org/10.1016/j.memsci.2011.04.051.

[34] Faiz R, Al-Marzouqi M. Insights on natural gas purification: simultaneous absorption of CO_2 and H_2S using membrane contactors. Sep Purif Technol 2011;76:351–61. https://doi.org/10.1016/j.seppur.2010.11.005.

[35] Peters L, Hussain A, Follmann M, Melin T, Hägg M-B. CO_2 removal from natural gas by employing amine absorption and membrane technology—a technical and economical analysis. Chem Eng J 2011;172:952–60. https://doi.org/10.1016/j.cej.2011.07.007.

[36] Rossi F, Nicolini A, Palombo M, Castellani B, Morini E, Filipponi M. An innovative configuration for CO_2 capture by high temperature fuel cells. Sustainability 2014;6:6687–95. https://doi.org/10.3390/su6106687.

[37] Khedkar SV, Bhautik N. Studies in purification of biogas using Al gas in sewage waste water—a review paper. Int J Pure Appl Res Eng Technol 2015;3:95–108.

[38] Kumełan J, Pérez-Salado Kamps Á, Tuma D, Maurer G. Solubility of CO2 in the ionic liquid [hmim][Tf2N]. J Chem Thermodyn 2006;38:1396–401. https://doi.org/10.1016/j.jct.2006.01.013.

[39] Schilderman AM, Raeissi S, Peters CJ. Solubility of carbon dioxide in the ionic liquid 1-ethyl-3-methylimidazolium bis(trifluoromethylsulfonyl)imide. Fluid Phase Equilib 2007;260:19–22. https://doi.org/10.1016/j.fluid.2007.06.003.
[40] Kim YS, Choi WY, Jang JH, Yoo K-P, Lee CS. Solubility measurement and prediction of carbon dioxide in ionic liquids. Fluid Phase Equilib 2005;228-229:439–45. https://doi.org/10.1016/j.fluid.2004.09.006.
[41] Hong G, Jacquemin J, Deetlefs M, Hardacre C, Husson P, Gomes MFC. Solubility of carbon dioxide and ethane in three ionic liquids based on the bis{(trifluoromethyl)sulfonyl}imide anion. Fluid Phase Equilib 2007;257:27–34. https://doi.org/10.1016/j.fluid.2007.05.002.
[42] Shariati A, Peters CJ. High-pressure phase behavior of systems with ionic liquids: II. The binary system carbon dioxide+1-ethyl-3- methylimidazolium hexafluorophosphate. J Supercrit Fluids 2004;29:43–8. https://doi.org/10.1016/S0896-8446(03)00032-9.
[43] Kamps ÄP-S, Tuma D, Xia J, Maurer G. Solubility of CO_2 in the ionic liquid [bmim][PF6]. J Chem Eng Data 2003;48:746–9. https://doi.org/10.1021/je034023f.
[44] Anthony JL, Maginn EJ, Brennecke JF. Solubilities and thermodynamic properties of gases in the ionic liquid 1-n-butyl-3-methylimidazolium hexafluorophosphate. J Phys Chem B 2002;106:7315–20. https://doi.org/10.1021/JP020631A.
[45] Shariati A, Peters CJ. High-pressure phase behavior of systems with ionic liquids part III. The binary system carbon dioxide + 1-hexyl-3-methylimidazolium hexafluorophosphate. J Supercrit Fluids 2004;30:139–44. https://doi.org/10.1016/j.supflu.2003.09.001.
[46] Husson-Borg P, Vladimir Majer A, Gomes MFC. Solubilities of oxygen and carbon dioxide in butyl methyl imidazolium tetrafluoroborate as a function of temperature and at pressures close to atmospheric pressure. J Chem Eng Data 2003;48:480–5. https://doi.org/10.1021/JE0256277.
[47] Mortazavi-Manesh S, Satyro MA, Marriott RA. Screening ionic liquids as candidates for separation of acid gases: solubility of hydrogen sulfide, methane, and ethane. AIChE J 2013;59:2993–3005. https://doi.org/10.1002/aic.14081.
[48] Althuluth M, Kroon MC, Peters CJ. Solubility of methane in the ionic liquid 1-ethyl-3-methylimidazolium tris(pentafluoroethyl)trifluorophosphate. Ind Eng Chem Res 2012;51:16709–12. https://doi.org/10.1021/ie302472t.
[49] Shiflett MB, Niehaus AMS, Yokozeki A. Separation of CO_2 and H_2S using room-temperature ionic liquid [bmim][$MeSO_4$]. J Chem Eng Data 2010;55:4785–93. https://doi.org/10.1021/je1004005.
[50] Baltus RE, Counce RM, Culbertson BH, Luo H, DePaoli DW, Dai S, et al. Examination of the potential of ionic liquids for gas separations. Sep Sci Technol 2005;40:525–41. https://doi.org/10.1081/SS-200042513.
[51] Karadas F, Atilhan M, Aparicio S. Review on the use of ionic liquids (ILs) as alternative fluids for CO_2 capture and natural gas sweetening. Energy Fuel 2010;24:5817–28. https://doi.org/10.1021/ef1011337.
[52] Bara JE, Camper DE, Gin DL, Noble RD. Room-temperature ionic liquids and composite materials: platform technologies for CO_2 capture. Acc Chem Res 2010;43:152–9. https://doi.org/10.1021/ar9001747.
[53] Zhang X, Zhang X, Dong H, Zhao Z, Zhang S, Huang Y. Carbon capture with ionic liquids: overview and progress. Energy Environ Sci 2012;5:6668. https://doi.org/10.1039/c2ee21152a.
[54] Yokozeki A, Shiflett MB, Junk CP, Grieco LM, Foo T. Physical and chemical absorptions of carbon dioxide in room-temperature ionic liquids. J Phys Chem B 2008;112:16654–63. https://doi.org/10.1021/jp805784u.

[55] Mindrup EM, Schneider WF. Computational comparison of the reactions of substituted amines with CO_2. ChemSusChem 2010;3:931–8. https://doi.org/10.1002/cssc.201000060.
[56] Jing G, Zhou L, Zhou Z. Characterization and kinetics of carbon dioxide absorption into aqueous tetramethylammonium glycinate solution. Chem Eng J 2012;181:85–92. https://doi.org/10.1016/j.cej.2011.11.007.
[57] Gurkan BE, Gohndrone TR, McCready MJ, Brennecke JF, Nanjo H, Kato M, et al. Reaction kinetics of CO_2 absorption into phosphonium based anion-functionalized ionic liquids. Phys Chem Chem Phys 2013;15:7796–811. https://doi.org/10.1039/c3cp51289d.
[58] Galán Sánchez LM, Meindersma GW, de Haan AB. Kinetics of absorption of CO_2 in amino-functionalized ionic liquids. Chem Eng J 2011;166:1104–15. https://doi.org/10.1016/j.cej.2010.12.016.
[59] Bates ED, Mayton RD, Ntai I, Davis JH. CO_2 capture by a task-specific ionic liquid. J Am Chem Soc 2002;124:926–7. https://doi.org/10.1021/ja017593d.
[60] Goodrich BF, Gurkan BE, de la Fuente JC, Mindrup EM, Ficke LE, Price EA, et al. Equimolar CO_2 absorption by anion-functionalized ionic liquids. J Am Chem Soc 2010;132:2116–7. https://doi.org/10.1021/ja909305t.
[61] Galá Sánchez LM, Meindersma GW, De Haan AB. Solvent properties of functionalized ionic liquids for CO_2 absorption. Chem Eng Res Des 2007;85:31–9. https://doi.org/10.1205/cherd06124.
[62] Galán Sánchez LM, Meindersma GW, de Haan AB. Solvent properties of functionalized ionic liquids for CO_2 absorption. Chem Eng Res Des 2007;85:31–9. https://doi.org/10.1205/cherd06124.
[63] Anderson K, Atkins MP, Estager J, Kuah Y, Ng S, Oliferenko AA, et al. Carbon dioxide uptake from natural gas by binary ionic liquid–water mixtures. Green Chem 2015;17:4340–54. https://doi.org/10.1039/C5GC00720H.
[64] Chinn D, Vu D, Driver M, Boudreau L. CO_2 removal from gas using ionic liquid absorbents. US20050129598 A1; 2005.
[65] Carlisle TK, Bara JE, Gabriel CJ, Noble RD, Gin DL. Interpretation of CO_2 solubility and selectivity in nitrile-functionalized room-temperature ionic liquids using a group contribution approach. Ind Eng Chem Res 2008;47:7005–12. https://doi.org/10.1021/ie8001217.
[66] Xu F, Gao H, Dong H, Wang Z, Zhang X, Ren B, et al. Solubility of CO_2 in aqueous mixtures of monoethanolamine and dicyanamide-based ionic liquids. Fluid Phase Equilib 2014;365:80–7. https://doi.org/10.1016/j.fluid.2013.12.020.
[67] Neves LA, Crespo JG, Coelhoso IM. Gas permeation studies in supported ionic liquid membranes. J Membr Sci 2010;357:160–70. https://doi.org/10.1016/j.memsci.2010.04.016.
[68] Cserjési P, Nemestóthy N, Bélafi-Bakó K. Gas separation properties of supported liquid membranes prepared with unconventional ionic liquids. J Membr Sci 2010;349:6–11. https://doi.org/10.1016/j.memsci.2009.10.044.
[69] Akhmetshina AA, Davletbaeva IM, Grebenschikova ES, Sazanova TS, Petukhov AN, Atlaskin AA, et al. The effect of microporous polymeric support modification on surface and gas transport properties of supported ionic liquid membranes. Membranes 2015;6:1–18. https://doi.org/10.3390/membranes6010004.
[70] Scovazzo P, Kieft J, Finan D, Koval C, Dubois D, Noble R. Gas separations using non-hexafluorophosphate [PF6]− anion supported ionic liquid membranes. J Membr Sci 2004;238:57–63. https://doi.org/10.1016/j.memsci.2004.02.033.
[71] Quinn R, Appleby JB, Pez GP. New facilitated transport membranes for the separation of carbon dioxide from hydrogen and methane. J Membr Sci 1995;104:139–46. https://doi.org/10.1016/0376-7388(95)00021-4.

[72] Deng J, Bai L, Zeng S, Zhang X, Nie Y, Deng L, et al. Ether-functionalized ionic liquid based composite membranes for carbon dioxide separation. RSC Adv 2016;6:45184–92. https://doi.org/10.1039/C6RA04285F.

[73] Scovazzo P. Determination of the upper limits, benchmarks, and critical properties for gas separations using stabilized room temperature ionic liquid membranes (SILMs) for the purpose of guiding future research. J Membr Sci 2009;343:199–211. https://doi.org/10.1016/j.memsci.2009.07.028.

[74] Muhammad Y, Lu Y, Shen C, Li C. Dibenzothiophene hydrodesulfurization over Ru promoted alumina based catalysts using in situ generated hydrogen. Energy Convers Manag 2011;52:1364–70. https://doi.org/10.1016/j.enconman.2010.09.034.

[75] Hernández-Maldonado AJ, Yang RT. Desulfurization of diesel fuels by adsorption via π-complexation with vapor-phase exchanged Cu(I)-Y zeolites. J Am Chem Soc 2004; 126:992–3. https://doi.org/10.1021/ja039304m.

[76] Jantaraksa N, Prasassarakich P, Reubroycharoen P, Hinchiranan N. Cleaner alternative liquid fuels derived from the hydrodesulfurization of waste tire pyrolysis oil. Energy Convers Manag 2015;95:424–34. https://doi.org/10.1016/j.enconman.2015.02.003.

[77] Kulkarni PS, Afonso CAM. Deep desulfurization of diesel fuel using ionic liquids: current status and future challenges. Green Chem 2010;12:1139–49. https://doi.org/10.1039/c002113j.

[78] Srivastava VC. An evaluation of desulfurization technologies for sulfur removal from liquid fuels. RSC Adv 2012;2:759–83. https://doi.org/10.1039/c1ra00309g.

[79] Chen T, Shen Y, Lee W, Lin C, Wan M. An economic analysis of the continuous ultrasound-assisted oxidative desulfurization process applied to oil recovered from waste tires. J Clean Prod 2013;39:129–36. https://doi.org/10.1016/j.jclepro.2012.09.001.

[80] Mjalli FS, Ahmed OU, Al-Wahaibi T, Al-Wahaibi Y, AlNashef IM. Deep oxidative desulfurization of liquid fuels. Rev Chem Eng 2014;30:337–78. https://doi.org/10.1016/S1351-4180(14)70125-4.

[81] Abro R, Abdeltawab AA, Al-deyab SS, Yu G, Qazi AB, Gao S, et al. A review of extractive desulfurization of fuel oils using ionic liquids. RSC Adv 2014;4:35302–17. https://doi.org/10.1039/c4ra03478c.

[82] Lee S-P, Ramli A. Methyl oleate deoxygenation for production of diesel fuel aliphatic hydrocarbons over Pd/SBA-15 catalysts. Chem Cent J 2013;7:149. https://doi.org/10.1186/1752-153X-7-149.

[83] Saidi M, Samimi F, Karimipourfard D, Nimmanwudipong T, Gates BC, Rahimpour MR. Upgrading of lignin-derived bio-oils by catalytic hydrodeoxygenation. Energy Environ Sci 2014;7:103–29. https://doi.org/10.1039/C3EE43081B.

[84] Oasmaa A, Korhonen J, Kuoppala E. An approach for stability measurement of wood-based fast pyrolysis bio-oils. Energy Fuel 2011;25:3307–13. https://doi.org/10.1021/ef2006673.

[85] Mahfud FH, van Geel FP, Venderbosch RH, Heeres HJ. Acetic acid recovery from fast pyrolysis oil. An exploratory study on liquid-liquid reactive extraction using aliphatic tertiary amines. Sep Sci Technol 2008;43:3056–74. https://doi.org/10.1080/01496390802222509.

[86] Zhang J, Wang K, Nolte MW, Choi YS, Brown RC, Shanks BH. Catalytic deoxygenation of bio-oil model compounds over acid-base bifunctional catalysts. ACS Catal 2016;6:2608–21. https://doi.org/10.1021/acscatal.6b00245.

[87] Wang Z, Lu Q, Zhu X-F, Zhang Y. Catalytic fast pyrolysis of cellulose to prepare levoglucosenone using sulfated zirconia. ChemSusChem 2011;4:79–84. https://doi.org/10.1002/cssc.201000210.

[88] Torri C, Lesci IG, Fabbri D. Analytical study on the production of a hydroxylactone from catalytic pyrolysis of carbohydrates with nanopowder aluminium titanate. J Anal Appl Pyrolysis 2009;84:25–30. https://doi.org/10.1016/j.jaap.2008.10.002.

[89] Lin Y, Zhang C, Zhang M, Zhang J. Deoxygenation of bio-oil during pyrolysis of biomass in the presence of CaO in a fluidized-bed reactor. Energy Fuel 2010;24:5686–95. https://doi.org/10.1021/ef1009605.

[90] Centeno A, Laurent E, Delmon B. Influence of the support of CoMo sulfide catalysts and of the addition of potassium and platinum on the catalytic performances for the hydrodeoxygenation of carbonyl, carboxyl, and guaiacol-type molecules. J Catal 1995;154:288–98. https://doi.org/10.1006/jcat.1995.1170.

[91] Nunes P, Brodzki D, Bugli G, Djega-Mariadassou G. Hydrocracking of soybean oil under pressure: research procedure and general aspects of the reaction. Rev Inst Fr Pétrol 1986; 41:421–31.

[92] Maier WF, Roth W, Thies I, Schleyer PVR. Hydrogenolysis, IV. Gas phase decarboxylation of carboxylic acids. Chem Ber 1982;115:808–12. https://doi.org/10.1002/cber.19821150245.

[93] Gusmão J, Brodzki D, Djéga-Mariadassou G, Frety R. Utilization of vegetable oils as an alternative source for diesel-type fuel: hydrocracking on reduced Ni/SiO_2 and sulphided $Ni\text{-}Mo/\gamma\text{-}Al_2O_3$. Catal Today 1989;5:533–44.

[94] Kubičková I, Snåre M, Eränen K, Mäki-Arvela P, Murzin DY. Hydrocarbons for diesel fuel via decarboxylation of vegetable oils. Catal Today 2005;106:197–200. https://doi.org/10.1016/j.cattod.2005.07.188.

[95] Han Y, McIlroy DN, McDonald AG. Hydrodeoxygenation of pyrolysis oil for hydrocarbon production using nanospring based catalysts. J Anal Appl Pyrolysis 2016;117:94–105. https://doi.org/10.1016/j.jaap.2015.12.011.

[96] Santillan-Jimenez E, Crocker M. Catalytic deoxygenation of fatty acids and their derivatives to hydrocarbon fuels via decarboxylation/decarbonylation. J Chem Technol Biotechnol 2012;87:1041–50. https://doi.org/10.1002/jctb.3775.

[97] Gosselink RW, Hollak SAW, Chang S-W, van Haveren J, de Jong KP, Bitter JH, et al. Reaction pathways for the deoxygenation of vegetable oils and related model compounds. ChemSusChem 2013;6:1576–94. https://doi.org/10.1002/cssc.201300370.

[98] Rogers KA, Zheng Y. Selective deoxygenation of biomass-derived bio-oils within hydrogen-modest environments: a review and new insights. ChemSusChem 2016;9:1750–72. https://doi.org/10.1002/cssc.201600144.

[99] Oh S, Choi HS, Choi I-G, Choi JW. Evaluation of hydrodeoxygenation reactivity of pyrolysis bio-oil with various Ni-based catalysts for improvement of fuel properties. RSC Adv 2017;7:15116–26. https://doi.org/10.1039/C7RA01166K.

[100] Wang Y, De S, Yan N. Rational control of nano-scale metal-catalysts for biomass conversion. Chem Commun 2016;52:6210–24.

[101] Han Y, Mcilroy DN, Mcdonald AG. Hydrodeoxygenation of pyrolysis oil for hydrocarbon production using nanospring based catalysts. J Anal Appl Pyrolysis 2016;117:94–105. https://doi.org/10.1016/j.jaap.2015.12.011.

[102] Lee H, Kim H, Yu MJ, Ko CH, Jeon J-K, Jae J, et al. Catalytic hydrodeoxygenation of bio-oil model compounds over Pt/HY catalyst. Sci Rep 2016;6:28765. https://doi.org/10.1038/srep28765.

[103] Elliott DC, Hart TR, Neuenschwander GG, Rotness LJ, Olarte MV, Zacher AH, et al. Catalytic hydroprocessing of fast pyrolysis bio-oil from pine sawdust. Energy Fuel 2012;26:3891–6. https://doi.org/10.1021/ef3004587.

[104] Su-Ping Z. Study of hydrodeoxygenation of bio-oil from the fast pyrolysis of biomass. Energy Sources 2003;25:57–65. https://doi.org/10.1080/00908310290142118.

[105] Kowsari E. Recent advances in the science and technology of desulfurization of diesel fuel using ionic liquids. In: Kadokawa J, editor. Ionic liquids: new aspects for the future. InTech; 2013. https://doi.org/10.5772/51651.

[106] Zhao H, Baker GA, Wagle DV, Ravula S, Zhang Q. Tuning task-specific ionic liquids for the extractive desulfurization of liquid fuel. ACS Sustain Chem Eng 2016;4:4771–80. https://doi.org/10.1021/acssuschemeng.6b00972.

[107] Huang CP, Liu ZC, Xu CM, Chen BH, Liu YF. Effects of additives on the properties of chloroaluminate ionic liquids catalyst for alkylation of isobutane and butene. Appl Catal A Gen 2004;277:41–3. https://doi.org/10.1016/j.apcata.2004.08.019.

[108] Zhu W, Xu D, Li H, Ding Y, Zhang M, Liu H, et al. Oxidative desulfurization of dibenzothiophene catalyzed by VO(acac)2 in ionic liquids at room temperature. Pet Sci Technol 2013;3114. https://doi.org/10.1080/10916466.2010.545790.

[109] Zhu W, Li H, Jiang X, Yan Y, Lu J, He L, et al. Commercially available molybdic compound-catalyzed ultra-deep desulfurization of fuels in ionic liquids. Green Chem 2008;10:641–64. https://doi.org/10.1039/b801185k.

[110] Rodríguez-Cabo B, Rodríguez H, Rodil E, Arce A, Soto A. Extractive and oxidative-extractive desulfurization of fuels with ionic liquids. Fuel 2014;117:882–9. https://doi.org/10.1016/j.fuel.2013.10.012.

[111] Lo WH, Yang HY, Wei GT. One-pot desulfurization of light oils by chemical oxidation and solvent extraction with room temperature ionic liquids. Green Chem 2003;5:639–42. https://doi.org/10.1039/b305993f.

[112] Ge J, Zhou Y, Yang Y, Xue M. The deep oxidative desulfurization of fuels catalyzed by surfactant-type octamolybdate in acidic ionic liquids. Pet Sci Technol 2014;32:116–23. https://doi.org/10.1080/10916466.2011.580304.

[113] Cassol CC, Umpierre AP, Ebeling G, Ferrera B, Chiaro SSX, Dupont J. On the extraction of aromatic compounds from hydrocarbons by imidazolium ionic liquids. Int J Mol Sci 2007;8:593–605. https://doi.org/10.3390/i8070593.

[114] Anantharaj R, Banerjee T. COSMO-RS based predictions for the desulphurization of diesel oil using ionic liquids: effect of cation and anion combination. Fuel Process Technol 2011;92:39–52. https://doi.org/10.1016/j.fuproc.2010.08.018.

[115] Su B-M, Zhang S, Zhang ZC. Structural elucidation of thiophene interaction with ionic liquids by multinuclear NMR spectroscopy. J Phys Chem B 2004;108:19510–7. https://doi.org/10.1021/jp049027l.

[116] Nie Y, Li C, Meng H, Wang Z. N,N-dialkylimidazolium dialkylphosphate ionic liquids: their extractive performance for thiophene series compounds from fuel oils versus the length of alkyl group. Fuel Process Technol 2008;89:978–83. https://doi.org/10.1016/j.fuproc.2008.04.003.

[117] Liang X, Xiao H, Qi C. Efficient procedure for biodiesel synthesis from waste oils using novel solid acidic ionic liquid polymer as catalysts. Fuel Process Technol 2013;110:109–13. https://doi.org/10.1016/J.FUPROC.2012.12.002.

[118] Elsheikh YA, Man Z, Bustam MA, Yusup S, Wilfred CD. Brønsted imidazolium ionic liquids: synthesis and comparison of their catalytic activities as pre-catalyst for biodiesel production through two stage process. Energy Convers Manag 2011;52:804–9. https://doi.org/10.1016/J.ENCONMAN.2010.08.005.

[119] Muhammad N, Elsheikh YA, Mutalib MIA, Bazmi AA, Khan RA, Khan H, et al. An overview of the role of ionic liquids in biodiesel reactions. J Ind Eng Chem 2015;21:1–10. https://doi.org/10.1016/J.JIEC.2014.01.046.

[120] Xiong W-M, Zhu M-Z, Deng L, Fu Y, Guo Q-X. Esterification of organic acid in bio-oil using acidic ionic liquid catalysts. Energy Fuel 2009;23:2278–83. https://doi.org/10.1021/ef801021j.

[121] Ullah Z, Bustam MA, Man Z. Biodiesel production from waste cooking oil by acidic ionic liquid as a catalyst. Renew Energy 2015;77:521–6. https://doi.org/10.1016/j.renene.2014.12.040.

[122] Laskowska M, Marciniak A. Phase equilibria of (1-ethyl-3-methylimidazolium ethylsulfate + hydrocarbon, + ketone, and + ether) binary systems. J Chem Eng Data 2008;498–502.

[123] Kumełan J, Kamps ÁP-S, Urukova I, Tuma D, Maurer G. Solubility of oxygen in the ionic liquid [bmim][PF6]: experimental and molecular simulation results. J Chem Thermodyn 2005;37:595–602. https://doi.org/10.1016/j.jct.2005.03.005.

[124] Mezza BJ, Wang H, Bhattacharyya A, Nicholas CP. Decontamination of deoxygenated biomass-derived pyrolysis oil using ionic liquids. US 20140005451 A1; 2014.

[125] Bahadur I, Osman K, Coquelet C, Naidoo P, Ramjugernath D. Solubilities of carbon dioxide and oxygen in the ionic liquids methyl trioctyl ammonium bis (trifluoromethylsulfonyl) imide, 1-butyl-3-methyl imidazolium bis (trifluoromethylsulfonyl) imide, and 1-butyl-3-methyl imidazolium methyl sulfate. J Phys Chem B 2015;119:1503–14. https://doi.org/10.1021/jp5061057.

[126] Anthony JL, Anderson JL, Maginn EJ, Brennecke JF. Anion effects on gas solubility in ionic liquids. J Phys Chem B 2005;109:6366–74. https://doi.org/10.1021/jp046404l.

[127] Jaschik M, Piech D, Warmuzinski K, Jaschik J. Prediction of gas solubility in ionic liquids using the cosmo-Sac model. Chem Process Eng 2017;38:19–30. https://doi.org/10.1515/cpe-2017-0003.

[128] Bara JE, Carlisle TK, Gabriel CJ, Camper D, Finotello A, Gin DL, et al. Guide to CO_2 separations in imidazolium-based room-temperature ionic liquids. Ind Eng Chem Res 2009;48:2739–51. https://doi.org/10.1021/ie8016237.

[129] Camper D, Becker C, Koval C, Noble R. Diffusion and solubility measurements in room temperature ionic liquids. Ind Eng Chem Res 2006;45:445–50. https://doi.org/10.1021/IE0506668.

[130] Finotello A, Bara JE, Camper D, Noble RD. Room-temperature ionic liquids: temperature dependence of gas solubility selectivity. Ind Eng Chem Res 2008;47:3453–9. https://doi.org/10.1021/IE0704142.

[131] Scovazzo P, Camper D, Kieft J, Poshusta J, Koval C, Noble R. Regular solution theory and CO_2 gas solubility in room-temperature ionic liquids. Ind Eng Chem Res 2004;43:6855–60. https://doi.org/10.1021/IE049601F.

[132] Blath J, Christ M, Deubler N, Hirth T, Schiestel T. Gas solubilities in room temperature ionic liquids—correlation between RTiL-molar mass and Henry's law constant. Chem Eng J 2011;172:167–76. https://doi.org/10.1016/j.cej.2011.05.084.

[133] Hasib-ur-Rahman M, Siaj M, Larachi F. Ionic liquids for CO_2 capture-development and progress. Chem Eng Process Process Intensif 2010;49:313–22. https://doi.org/10.1016/j.cep.2010.03.008.

[134] Neale AR, Li P, Jacquemin J, Goodrich P, Ball SC, Compton RG, et al. Effect of cation structure on the oxygen solubility and diffusivity in a range of bis{(trifluoromethyl)sulfonyl}imide anion based ionic liquids for lithium–air battery electrolytes. Phys Chem Chem Phys 2016;18:11251–62. https://doi.org/10.1039/C5CP07160G.

[135] Chrobok A, Swadźba M, Baj S. Oxygen solubility in ionic liquids based on 1-Alkyl-3-methylimidazolium cations. Pol J Chem 2007;81:337–44.

[136] Davis JH. Task-specific ionic liquids for separations of petrochemical relevance: reactive capture of CO_2 using amine incorporating ions. In: Ionic liquids IIIB: fundamentals, progress, challenges, and opportunities; 2005. p. 49–56. https://doi.org/10.1021/bk-2005-0902.ch004.

[137] Camper D, Becker C, Koval C, Noble R. Low pressure hydrocarbon solubility in room temperature ionic liquids containing imidazolium rings interpreted using regular solution theory. Ind Eng Chem Res 2005;44:1928–33. https://doi.org/10.1021/IE049312R.

[138] Asumana C, Haque MR, Yu L, Wu X, Chen X, Yu G. Desulfurization of real fuel oils by extraction with ionic liquids. Sep Sci Technol 2013;48:2582–8. https://doi.org/10.1080/01496395.2013.804559.

[139] Chen X, Liu G, Yuan S, Asumana C, Wang W, Yu G. Extractive desulfurization of fuel oils with thiazolium-based ionic liquids. Sep Sci Technol 2012;47:819–26. https://doi.org/10.1080/01496395.2011.637281.

[140] Chen Y, Zhao S, Song YF. An efficient heterogeneous catalyst based on highly dispersed Na 7H2LaW10O36·32H2O nanoparticles on mesoporous silica for deep desulfurization. Appl Catal A Gen 2013;466:307–14. https://doi.org/10.1016/j.apcata.2013.06.030.

[141] Shao BB, Shi L, Meng X. Deep desulfurization of 4,6-dimethyldienzothiophene by an ionic liquids extraction coupled with catalytic oxidation with a molybdic compound. Ind Eng Chem Res 2014;53:6655–63. https://doi.org/10.1021/ie500236b.

[142] Zhang S, Zhang Q, Zhang ZC. Extractive desulfurization and denitrogenation of fuels using ionic liquids. Sep Purif Technol 2004;43:614–22. https://doi.org/10.1016/j.seppur.2019.115923.

[143] Kumar AAP, Banerjee T. Thiophene separation with ionic liquids for desulphurization: a quantum chemical approach. Fluid Phase Equilib 2009;278:1–8. https://doi.org/10.1016/j.fluid.2008.11.019.

[144] Holbrey JD, López-Martin I, Rothenberg G, Seddon KR, Silvero G, Zheng X. Desulfurisation of oils using ionic liquids: selection of cationic and anionic components to enhance extraction efficiency. Green Chem 2008;10:87–92. https://doi.org/10.1039/b710651c.

[145] Zhang X, Bao D, Huang Y, Dong H, Zhang X, Zhang S. Gas-liquid mass-transfer properties in CO_2 absorption system with ionic liquids. AIChE J 2014;60:2929–39. https://doi.org/10.1002/aic.14507.

[146] Lu B, Wang X, Xia Y, Liu N, Li S, Li W. Kinetics of carbon dioxide absorption into mixed aqueous solutions of MEA + [Bmim]BF4 using a double stirred cell. Energy Fuel 2013;27:6002–9. https://doi.org/10.1021/ef400976j.

[147] Shannon MS, Tedstone JM, Danielsen SPO, Bara JE. Evaluation of alkylimidazoles as physical solvents for CO_2/CH_4 separation. Ind Eng Chem Res 2012;51:515–22. https://doi.org/10.1021/ie202111k.

Section C

Hydrogen energy

Chapter 6

Hydrogen production via electrolysis: Mathematical modeling approach

Atta Sojoudi[a], Ali M. Sefidan[b], K.C.A. Alam[c], and Suvash C. Saha[c]
[a]*School of Mechanical Engineering College of Engineering, University of Tehran, Tehran, Iran,*
[b]*Department of Mechanical Engineering, University of Canterbury, Christchurch, New Zealand,*
[c]*School of Mechanical and Mechatronics Engineering, University of Technology Sydney, Sydney, NSW, Australia*

1 Hydrogen and methods to produce

Nowadays, fossil fuels supply the main part of the energy of the world, which led to some challenges. Limited resources, high-quality energy degradation into hazardous low-quality waste, and increasing energy demands are some issues related to this energy source. The world energy needs to be supplied by nonfossil or alternative clean energy resources to provide secure energy to meet the growing demands of developed and developing countries.

Among common alternative energy resources like solar, water, hydrogen, wind, waves, ocean, geothermal, etc., hydrogen has the desirable qualities to replace fossil fuel. Hydrogen as the intermediary or secondary form of energy is produced from a nonfossil energy system and used where fossil fuel is consumed today. Hydrogen is odorless, colorless, extremely flammable, and the smallest member of the chemical elements, which is the most abundant element of the universe. Hydrogen can be generated without greenhouse gases emission. It can be used to produce electricity directly by being fed into a fuel cell, through an electrochemical process without combustion. Due to the ability to be highly flammable, it could be used as a combustion fuel [1].

Hydrogen is the cleanest nonpoisonous and recyclable fuel, which produces only water. It can be stored and delivered in gas or in liquid phases, or combined with CO_2 and/or nitrogen and other gases to produce other fuels like ammonia or methane. It is also usually used as a chemical input in some industrial processes, helping to produce plastics, fertilizers, or pharmaceuticals. These merits make hydrogen an attractive fuel alternative for electricity

Bioenergy Resources and Technologies. https://doi.org/10.1016/B978-0-12-822525-7.00012-3
Copyright © 2021 Elsevier Inc. All rights reserved.

generation applications and transportation. It can be used in many more applications like in houses, for portable power, in cars, etc. Hydrogen is an energy that can be used to move, store, and deliver from other sources. It is considered to be a potential solution for current energy problems due to its higher energy conversion efficiency. It can contribute to less release of environmentally damaging materials and accordingly offer a good solution for global warming and air pollution issues [1–3].

For the same amount of energy production, hydrogen weighs one-third of the fossil fuels but it is bulkier and occupies 3.6 times the volume occupied by natural gas [3]. Wide flammability limits and high flame speed make hydrogen very desirable for internal combustion engines, jet engines, and gas turbines. Since hydrogen has high ignition temperature and low flame luminosity, it is also safer than other energy sources [2]. Some of the advantages of using hydrogen as an energy source are as follows [1–6]:

(i) it has a high energy conversion efficiency;
(ii) it is produced from water without emission;
(iii) it is found in earth abundantly;
(iv) it can be stored in different forms (e.g., gaseous, liquid, or together with metal hydrides);
(v) could be transported over long distances;
(vi) it is easily converted to other forms of energy;
(vii) it has a higher HHV (higher heating value) and LHV (lower heating value) than most of the conventional fossil fuels.

The most important part of the hydrogen energy system, which receives lots of attraction among researchers, is the production process. The most common methods employed for hydrogen generation are classified as [2]:

- steam reforming of fossil fuels,
- partial oxidation of hydrocarbons,
- coal gasification,
- electrolysis,
- photolysis,
- photovoltaic-electrolysis system,
- photochemical,
- photoelectrochemical,
- biophotolysis,
- thermal cracking of natural gas,
- photodecomposition of organic compounds,
- thermochemical/thermolysis,
- thermochemical process,
- hybrid system using photosynthetic and fermentative bacteria,
- fermentative hydrogen production from organic compounds,
- biomass.

Hydrogen is the lightest element in the Mendeleev table with the smallest size molecule which makes it easier to leak from a container or a vessel [6]. It has a very low critical temperature, and it is difficult to liquefy compared to methane and propane. Hydrogen has got lower energy density as well, which means that a vehicle with hydrogen fuel should have a bigger fuel tank rather than conventional petrol or diesel cars. Since the hydrogen element is commonly available as water in every place such as oceans, seas, lakes, and rivers, it is essential to make it more feasible and cheaper to generate hydrogen from water to meet future energy requirements [7].

H_2 can be properly reserved and supplied and conveniently accessible for distributing for a range of demands. The storage of hydrogen is one of the substantial necessities for the improvement of the related systems. In today's industrial world, three main techniques are available for hydrogen storage [6]: as a pressurized gas, as a cryogenic liquid, and as a metal hydride. According to the unit volume, hydrogen has one of the least energy storage densities. The density of hydrogen in gaseous form is much lower than that of gaseous methane. So it is necessary to reduce the volume of gas to a great extent. A key purpose is to pressurize the gas as much as possible to store a large volume of gas in a particular tank. On behalf of storing much hydrogen in a specific volume, compressing work must be done besides reducing the temperature below the critical temperature to liquefy it. Although the repulsive force of the molecules will decrease, it demands a reduction in temperature to around 22 K [6,7]. Covalent bonds of hydrogen with carbon, for example, and then behaving like metal to create alloys or distinctive compounds at ambient condition are a kind of chemical storage of hydrogen. However, the reversibility of the chemical mixture and the release of hydrogen from the host substance is a rather problematic issue. Oxidation or heating to above 800°C can break all covalent hydrogen-carbon [6–8].

Water splitting methods are recommended as economical industrial-scale methods for hydrogen generation. In these methods, a direct current passes between two electrodes of the cells generating oxygen (at the anode) and hydrogen (at the cathode) from water. The technology is efficient (70%) and works at high temperatures and pressures [8,9].

1.1 Water splitting methods

Water is the most abundant and inexhaustible raw material on the earth, which makes it a common source of hydrogen through water splitting. Water splitting is a chemical reaction in which water is broken into hydrogen and oxygen. There are three main methods to produce hydrogen through electrolysis [8,9]:

Electrolysis: Electrolyzing is a well-known method for water splitting, which includes a cathode and an anode immersed in an electrolyte. When a direct current is applied, water splits into hydrogen (cathode) and oxygen (anode side). Nowadays, an Alkaline Electrolyzer (AE), Proton Exchange Membrane Electrolyzer (PEME), and Solid Oxide Electrolyzer (SOE) are the most advanced and commonly used methods of Electrolyzing. Extremely pure hydrogen could be easily provided with the electrolyzing method.

Thermolysis: This is a process in which water is decomposed to hydrogen and oxygen by heating it to a high temperature. As the temperature needs to reach 2500°C, considerable energy should be supplied during this process. Heat as an energy source is converted to chemical energy in the form of hydrogen in the thermolysis process.

Photo-electrolysis: In this method, photocatalysts are used to absorb the energy of visible light to use in water decomposition into hydrogen and oxygen. The water-splitting process is the same as electrolysis, and sunlight energy is absorbed through semiconducting materials.

1.2 Electrolysis

Water electrolysis is the most frequent and popular procedure in pure hydrogen production in industries. Alkaline, polymer membrane, and solid oxide electrolyzers are the major electrochemical hydrogen manufacturing technologies in which efficiency and current density are the main characteristics. Catalysts (especially Platinum as a heterogeneous catalyst) are frequently employed to enhance the current density and rate of electrolysis reactions [9]. The purity of water is highly important in electrolyzers (especially PEM electrolyzers). Then desalination and demineralization have to be implemented before the electrolysis process [10]. For example, with an electrolyzer, it is highly possible to generate chlorine rather than oxygen if it is supplied by brine (or seawater) [9,10]. Side reactions during electrolysis like a chlorine evolving reaction could be stopped using several methods; one of them is ion-selective membranes utilization to water desalination.

1.3 Alkaline electrolyzers (AE)

Two water molecules are converted to one hydrogen molecule and two hydroxyl ions at the cathode. The ions transfer to the anode through the liquid electrolyte and are converted to half of an oxygen molecule and one water molecule. The hydrogen molecule recombines in a gaseous form as it escapes from the cathode surface. The oxygen also escapes from the electrode surface and recombines as a gas like a hydrogen. Pressureless electrolyzers are manufactured in the simplest construction principles, the monopolar arrangement; however, pressurized types are built in the bipolar filter press arrangement, which saves space and materials [11]. Inside the electrolyte ducts of bipolar electrolyzers, shunt current may occur, which is the minor disadvantage of them compared to monopolar. The new technologically advanced cells operate at temperatures up to 150°C with pressures ranging from 5 to 30 bars. These newly developed cells are three times more capable than the conventional low-pressure alkaline Electrolyzers allowing current densities up to 10 kA/m^2, which increases the cell efficiency to around 80% [11].

1.4 Proton exchange membrane electrolyzers (PEME)

In this method, the polymeric proton exchange membrane is used as the solid electrolyte. Polymer electrolyte technology has the following advantages over the alkaline method. (i) Since there is no caustic electrolyte circulated in the cell stack, it is much safer and more reliable. (ii) Materials used in this method can withstand more pressure without damage and then can easily prevent gas mixing [11]. The anode of the electrolyzer cell is made of porous titanium to which pure water is fed. Hydrated protons are conducted by the membrane from the anode to the cathode side. Low ohmic resistance due to a proper swelling technique enables high current density within the cell. The cathode of this electrolyzer is made up of a porous graphite current collector with a mixed oxide as the electrocatalyst [11]. Electroosmotic water transit takes place through the membrane from the anode side to the cathode as a consequence of the cell's operation. This type of electrolyzer has the main drawback of the restricted life of the cells. Nafion, as a Perfluorinated sulfonic acid membrane, showed high resistance against the oxidative power of oxygen or even ozone [11]. A Pt layer is also exerted to one side of the cathode as the cathode catalyst, but the anode catalyst for oxygen evolution is Ru/Ir mixed oxide [11]. Water flow through the anode maintains the temperature of the entire membrane/electrode at a constant level. The temperature of the system is recorded at the anode loop outlet, and online instrumentations are employed to measure the purity of the outlet gases.

1.5 Solid oxide electrolyzers (SOE)

High electricity consumption is one of the main problems of conventional electrolyzers. But the steam electrolysis method has much efficiency compared to the alkaline and proton exchange membrane electrolysis methods. Thermodynamics approves that it is more beneficial to break down water at higher temperatures (800–1000°C) [12]. The prepared energy is a mixed form of electricity and heat at higher temperatures, and most of it is added heat rather than electricity. Also, reaction kinetics is accelerated at high temperatures, which reduces energy waste due to electrode polarization and consequently increases total efficiency. Oxygen ion-conducting ceramics are employed as an electrolyte at high temperatures [11]. Steam is converted into hydrogen gas and also oxygen ions which are migrated to the anode through ceramic material. In the usual steam electrolyzers, a mixture of steam and hydrogen is prepared for the cathode, and the air is conveyed to the anode. At zero current, depending on the operating temperature and hydrogen steam ratio, the system has an open-circuit voltage of 0.8–0.9 V, which is higher for water electrolysis [11,12]. To water decomposition, an opposite voltage, which is higher than the open-circuit voltage, requires to be exerted to pump the oxygen from the steam side to the airside. Approximately 70% of the net electricity utilization is employed to impose the system to work against the high potential gradient of oxygen [11,12]. A newly developed method to reduce electricity consumption due to reducing the chemical potential gradient across the electrolyzer cells is employing natural gas, which is called the natural gas-assisted steam electrolyzer (NGASE). Natural gas in this method is replaced by the air supplied on the anode side to reduce the open-circuit voltage and consequently the electricity consumption [13].

1.6 Renewable hydrogen production

Although hydrocarbon is the main source of hydrogen production, renewable technologies need to be integrated into this process. There are many schemes to produce hydrogen from renewable resources, including biomass-based technologies and water splitting methods as the main methods. Biomass is a primary source of renewable energy-driven from plants or animal products. Hydrogen production using biomass could be divided into two modes, thermochemical and biological methods [10–12,14]. As sustainable development and waste minimization are important these days, the biological process has gained attention in recent years. This method, which could happen directly or indirectly, provides low rates of hydrogen but is environmentally friendly and less energy-intensive. In the direct process, hydrogen is directly produced using some bacteria or algae. The indirect method involves converting carbohydrate-containing materials into organic acids and then to hydrogen gas through bio-processing technologies. The thermochemical process is much faster and also economical and environmentally benign. Using this method, biomass is transferred into hydrogen or hydrogen-rich gases. Mainly CH_4 and CO are products of this method which can be transferred to hydrogen using a steam reforming process [10–12,14].

Water splitting methods can also be run through renewable energy systems. Solar hydrogen and wind to hydrogen are the main processes to supply electricity to decompose the water to hydrogen and oxygen. Since the energy efficiency of the water electrolyzing system is around 70%–75%, when they are combined as a solar-to-hydrogen system, the energy efficiency decreases to 8%–14% for commercial silicon PV cells [15]. Polymer electrolyte membrane and alkaline electrolyzer systems are two technologies used in this method. There are some usual ways to increase the efficiency of this system. Since the PV cells are efficient at lower temperatures, removing heat from their surfaces can help. Also, solar module skyward orienting improves solar capture by 40% [14,15]. New PV modules are made up of semiconductor materials and have an efficiency of 36%. The PV-electrolyzer system is used to produce hydrogen using solar energy. Solar hydrogen efficiency is calculated as the ratio of energy stored in the manufactured hydrogen to the energy delivered to the PV panel by solar radiation [14,15]. Using wind turbines is another way to extract electricity from the electrolyzer systems; however, it should be noticed that these systems should be completely analyzed through thermodynamic relations and economic status concerning other methods to ensure competitiveness in the market.

1.7 Cost analysis of electrolysis

Data shows that hydrogen production from renewable energies like wind and biomass is considerably costly compared to hydrogen generation using conventional sources like gasoline. Hydrogen production cost analysis in the United States shows that electrolytically generated hydrogen is considerably more expensive due to electricity consumption [10–14]. However, solar hydrogen production is significantly cheaper than using commercial electricity and even makes it competitive with the hydrogen produced by gasoline. Solar energy needs to cut down the module price by one-third to be competitive with the electricity produced by fossil fuel. Researchers are working to develop PV technologies to improve the module cost [10–14]. Efforts are on to a collection of organic and inorganic nanostructures into solar cells. Photochemical cells are also employed for the direct conversion of solar energy to hydrogen according to semiconductor materials. It is extremely important to match components when renewable energy systems are combined with the hydrogen generation system to have a smooth function. The package should be optimized in a way to maximize the system efficiency, reduce the capital cost, and lower the system complexity.

2 Background of electrolysis

Electrolysis is an electrochemical procedure in which a chemical reaction is commenced with electricity. This is often carried out with liquids and particularly with charged ions dissolved in water. Electrolysis is employed extensively nowadays in industry and is a key section of the manufacture of several commodities. It was first launched in the 1800s and has progressed into the understanding scientists have of it today. In the future, electrolysis may be more vital, and as scientific development proceeds, researchers will discover new and pivotal important applications for the process.

Electrolysis is operated by passing DC electricity through a liquid, often water. This makes the ions in the water obtain and release charges at the terminals. Both terminals are a cathode and an anode. The cathode is the terminal that the positively charged ions are moved toward and the anode is the terminal that the negatively charged ions are attracted to. When the required voltage is applied across the two terminals, the charged ions in the liquid will travel toward one of the terminals. The positive and negative ions will transmit to the cathode and the anode, respectively. When DC electricity passes through the circuit, the electrons will travel toward the cathode. This causes the cathode to be negative in charge of attracting the positive ions which will move toward the cathode. Reducing the number of cations at the cathode, they gain electrons producing atoms again and generate a mixed form of the elements they are.

After the discovery of the voltaic pile by Alessandro Volta in 1800, chemists employed a battery and merged the poles into water. They found out that when current is passed, hydrogen and oxygen elements are produced at the terminals. They performed the same experiment with various solutions of solids, and also they found out that electricity passed through particles of the solid attracted to the electrodes. This exciting invention motivated further efforts and experimental tests.

Two main theories were launched. One was introduced by Humphrey Davy. He believed that the electrochemical affinity of the particles contributing to the process is governed by a simple rule regarding the opposite charge attraction [12,14,15]. The other one was suggested by Jöns Jacob Berzelius, who believed that materials composed of collections of "electropositive" and "electronegative" elements categorized the units by the pole at which they accumulated during electrolysis [11,15]. Today, knowledge of electrolysis is beholden to those beliefs.

Later, Michael Faraday commenced carrying out several experiments on this science. He made intensive efforts to find out if DC electricity would pass through a solution even when one of the terminals of the circuit was taken out and electricity was subjected to the solution through a spark. He believed that the influences of electrolysis appeared from forces that are internally attributed to the matter under decomposition. He supposed that these effects are due to a change, by the electric current, of the chemical affinity of the particles through or by which the current is passing [8–10,15]. Faraday's empirical tests displayed that the electrolyte itself was a key section of the current in an electrochemical reaction, and it led him to the thoughts of oxidation and reduction mechanisms.

Electrolysis has several applications in the modern industrial world. Purifying aluminum is one of the pivotal important usages. Aluminum is often manufactured from the mineral bauxite. The initial level is to treat the bauxite so it becomes purer and aluminum oxide is produced. After melting the aluminum oxide, the compound dissolves into its corresponding ions and this is where the electrolysis technique is employed. The walls of the oven are operated as a cathode and blocks of carbon hovered from the above means as the anode. When there is electricity through the liquid aluminum oxide, the aluminum ions will travel to the cathode where they will obtain a negative charge and turn into aluminum metal. The negative oxygen ions will travel to the anode and release some of their electrons and produce oxygen and other mixtures. The electrolysis of aluminum oxide requires an abundant amount of energy, which is around 12–14 kWh per kg of aluminum [14].

Electroplating is another application of electrolysis where it is implied to locate a thin layer of a specific metal over another metal. This is certainly beneficial if you want to hinder corrosion in spectacular metals such as iron. Electroplating is operated by using the metal you want to have coated in a particular metal function as the cathode, and the cation of this solution would then be the metal that is desired as a layer for the cathode. To stop corrosion and improve the appearance of specific metals, Zinc is usually employed as the coating material. For instance, the Silver solution will coat a metal with a thin shell of silver so that the metal seems to be Silver. In the next decades, electrolysis will have several aspects. Our consumption of natural resources such as fossil fuels will finally finish and the world economy will become based on hydrogen.

The beauty of electrolysis can be revealed wherever there is an energy source in which scientists and industrial designers can harvest energies received from the sun and wind to generate hydrogen. However, they will not be dependable on particular geographic places and can generate hydrogen locally where they demand it.

3 Mathematical modeling of an alkaline electrolyzer (AE)

AE has been known as the cheapest and most robust electrolyzer rather than others, while PEM electrolyzers require attentive control and management. Although AE should be operated at a high level of liquid purity, it remains the most useful and conventional type [16]. In this type of electrolyzer, KOH with a concentration of 25%–35% in distilled water plays as the electrolyte of the H_2 producer in which OH^- is the electron transporter. There are two main types of AE: monopolar AE in which cells are arranged in a parallel configuration, each electrode is connected to the opposite terminal of the next cell's electrode, and bipolar AE that is composed of cells accumulation while two outer cells are connected to the power supply and the rest operate as anode and cathode at both sides [17].

The manufacturer's selection of AE type varies according to the energy consumption and place of use. The operating temperature of AE differs from 5°C to around 120°C while the pressure of the device can be regulated from atmospheric pressure up to the vicinity of 30 bar according to the design parameters, mainly the amount of H_2 produced per day. Larger units of H_2 production can store at least 68 kg of H_2 per hour [18]. Pressurized AE requires less compression energy to store the produced gas while they exhibit lower efficiency rather than low-pressure AE. Moreover, high-pressure AE should employ a very strong and highly chemically and mechanically stable separator ceasing the O_2 and H_2 mixing while exhibiting low ionic resistivity [19].

AE modeling has received great attention among researchers, and some of them have also validated their outcomes against experimental results including empirical and semiempirical modelings [20–26]. These empirical models mainly reported the current and voltage behavior of the AE along with illustrating thermal behavior, cell voltage, produced gas purities, and various efficiencies and types of the electrolyzer [20–26].

Multi-physics modeling of AE considering detected events while running the device was reported in several research projects [27,28] enhancing the clear understanding and diagnosis of any fault in the design. Although multidimensional modeling of the matter, including thermal and electrical management, is highly important, importing hybrid systems as energy supply will increase the complexity of the simulation. So, simple, understandable, and precise models have the main priority. A decomposing water molecule can be achieved by electricity according to the following chemical relation resulting in gaseous hydrogen and oxygen [29]:

$$H_2O_{(l)} + 2F \rightarrow H_{2(g)} + 1/2\,O_{2(g)}, \tag{1}$$

where F displays the Faraday constant representing the value of electric charge per mole of electrons (96,487.332 As/mol). Anode and cathode are the two main terminals in which electrons are transported from the anode to the cathode through an electrical wire while they are covered by an electrolyte (Fig. 1). Chemical reactions occurring at these portals are as follows:

$$\text{Anode}: 2OH^- \rightarrow 1/2\,O_2 + H_2O + 2e^-, \tag{2}$$

$$\text{Cathode}: 2H_2O \rightarrow 2H^+ + 2OH^- \text{ and } 2H^+ + 2e^- \rightarrow H_2 \tag{3}$$

According to the voltage difference between the two electrodes applied by a power supply and the leading electrical field, H^+ and OH^- travel to the cathode and anode, respectively.

Gas bubbles are created at the vicinity of the electrodes and may cover a certain area of electrode surfaces depending on the current density, intermediate fluid properties, and working pressure and temperature. The electrolyte inside the system should be circulated to keep its concentration at a constant level as well as distributing heat and pressure homogeneously. By considering an equal value of current density at both terminals and assuming uniformly distributed

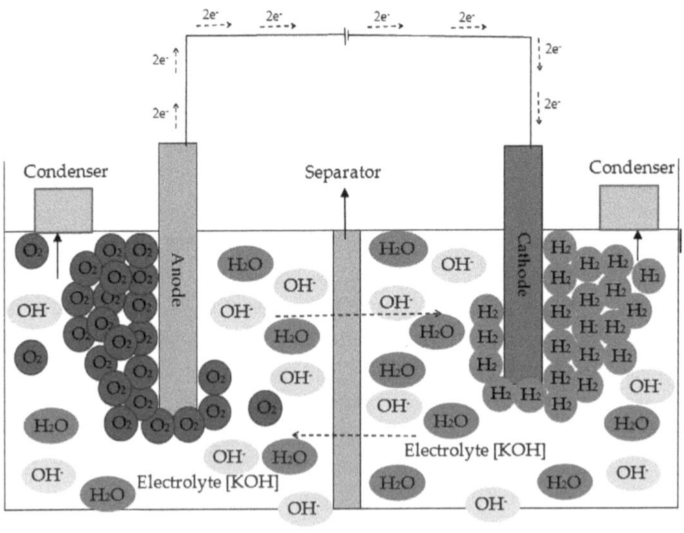

FIG. 1 Schematic of AE cell [16].

bubbles alongside the electrodes, we can write mole balance for the anode and cathode [16]:

$$\text{Anode}\quad \begin{aligned} &\frac{dN_{H_2O}}{dt}=\dot{N}^{in}_{H_2O}-\dot{N}^{out}_{H_2O}+\dot{N}^{gn}_{H_2O} \\ &\frac{dN_{O_2}}{dt}=\dot{N}^{in}_{O_2}-\dot{N}^{out}_{O_2}+\dot{N}^{gn}_{O_2} \\ &\dot{N}^{gn}_{H_2O}=\frac{I}{2F},\dot{N}^{gn}_{O_2}=\frac{I}{4F} \end{aligned} \tag{4}$$

$$\text{Cathode}\quad \begin{aligned} &\frac{dN_{H_2O}}{dt}=\dot{N}^{in}_{H_2O}-\dot{N}^{out}_{H_2O}+\dot{N}^{cons}_{H_2O} \\ &\frac{dN_{H_2}}{dt}=\dot{N}^{in}_{H_2}-\dot{N}^{out}_{H_2}+\dot{N}^{gn}_{H_2} \\ &\dot{N}^{cons}_{H_2O}=\frac{I}{2F},\dot{N}^{gn}_{H_2}=\frac{I}{2F} \end{aligned} \tag{5}$$

Generation of molar rates is presented using Faraday's law. After considering a homogeneous and one-dimensional distribution of current alongside the anode, we can state:

$$I=iA_e \tag{6}$$

Now, we can predict effective partial pressure using thermodynamics laws as we know the sum of the molar fractions of O_2 and H_2O at the anode is 1 [16]. Assuming uniform pressure distribution through the pressure channel and ideal gas thermodynamics laws:

$$\text{Anode}\qquad p_{O_2}=\frac{p_{H_2O}}{X_{H_{2O}}}\left(1-X_{H_{2O}}\right) \tag{7}$$

Utilizing the Stefan-Maxwell equation [30], the water molar fraction gradient at the anode terminal is:

$$\frac{dX_{H_2O}}{dx}=\frac{\varepsilon_{anode}}{\tau_{anode}}\frac{R^*\times T\times i}{2FP_{anode}D^{eff}_{anode}} \tag{8}$$

Then, taking the integral from the anode to the catalyst surface [16]:

$$X_{H_2O}=\frac{\varepsilon_{anode}}{\tau_{anode}}exp\left(\frac{l_{an-cay}R^*\times T\times i}{2FP_{anode}D^{eff}_{anode}}\right) \tag{9}$$

Researchers have demonstrated that effective water pressure is equivalent to the saturated vapor pressure of KOH solution; so the partial pressure of oxygen at the anode can be expressed as [16,31]

$$p_{O_2}=\left[1/\frac{\varepsilon_{anode}}{\tau_{anode}}exp\left(\frac{l_{an-cay}R^*Ti}{2FP_{anode}D^{eff}_{anode}}\right)-1\right]P^{saturation}_{H_2O,KOH}(T) \tag{10}$$

The partial pressure of hydrogen at the cathode can be expressed as

$$p_{H_2} = \left[1/\frac{\varepsilon_{cathode}}{\tau_{cathode}} exp\left(\frac{l_{ca-cay} R^* Ti}{2FP_{cathode} D^{eff}_{cathode}} \right) - 1 \right] P^{saturation}_{H_2O,KOH}(T) \tag{11}$$

To onset an electrochemical reaction, a specified amount of energy is required at the environmental temperature. The electrical current supplied this energy and fulfills the requirement of the process. Approximately 1.48 V at standard condition (ambient temperature and pressure of 25°C and 1 atm, respectively) is desired to maintain H_2 generation through the electrolyzer in an insulated condition [32]. Endothermic and Exothermic reactions can be expected if the amount of voltage is below or above that value, respectively. However, the mentioned amount of voltage is the ideal value within the standard condition, and there are several resistances through the process to start and keep producing hydrogen by electrochemical reactions [33,34]. So overpotential is essential to activate the reaction steps besides the ideal value. The following statement should be met when an electrochemical system is in thermodynamical equilibrium [35]:

$$\Delta G = z.F.V^{rev}_{P_0,T_0} \tag{12}$$

where ΔG represents a change in Gibbs energy which is 237.0 kJ/mol for the standard condition and reversible voltage ($z = 2$ for the water electrochemical splitting process). However, this electrolysis process is endothermic at this status and must absorb heat from the environment. Eventually, the entropic term should be involved to consider irreversibility and develop the thermodynamic relation for change of Gibbs energy [36]:

$$\Delta G = \Delta H - T.\Delta S, \tag{13}$$

where ΔH is the amount of energy needed to start a reversible water decomposition process [35]:

$$\Delta \mathrm{H} = z.F.V_{tn}, \tag{14}$$

where V_{tn} is the thermoneutral cell voltage.

The quantity of energy consumed for the electrochemical procedure directly relies on the values of pressure and temperature. With the increase of temperature, the amount of V_{rev} descends due to depletion of ΔG while it increases with the increment of pressure. However, V_{tn} remains almost constant because ΔG is compensated by a variation of $T.\Delta S$ [18].

According to resistances existing alongside the reaction kinetics and charge transport through the liquid and electrical conductors, almost all water spiriting processes are performed with a certain amount of voltage higher than the thermoneutral voltage. Predicting the precise value of the cell voltage is one of the important advantages of any mathematical modeling. Many scientists have researched and represented several empirical and semiempirical relations for this stage of mathematical modeling. To evaluate the working value of cell voltage, all phenomena taking place at the cell should be thoroughly realized and represented in mathematical form. Accordingly, the cell voltage can be stated as the sum of overpotentials [16]:

$$V = V_{OC} + V_{act} + V_{con} + V_{ohm} \tag{15}$$

where V_{OC} is the open-circuit voltage, which is the amount of reversible voltage needed to start the electrochemical process, and it changes according to the pressure and temperature. V_{act} is the activation overpotential resulting from activation phenomena related to the electrochemical kinetics of the anode and cathode half-reactions. In other words, the energy needed for the charges to move from electrodes to transporting species and vice versa is called activation overpotential, which is intensively associated with the electro-catalytic properties of the terminal's materials [22,33]. V_{con} represents the concentration overpotential and it is developed due to variation in the reactant concentration around the electrode surface [16]. The mass flow within the porous terminals is commonly recognized as a diffusion phenomenon, which can be expressed by Fick's law. This overpotential is ignored in some reports for the alkaline electrolyzer [22]. Finally, V_{ohm} is attributed to the ohmic resistances according to the process such as resistance to the electron flux of electrodes, bipolar plates, the current collectors and their inward connections, and the terminal's compartments opposing the ion flux. It is highly related to the current transmitting through the electrolytic cell [22]. Evaluating these overpotentials is an important challenge through the papers, and here we point out some precise models. The amount of Open-Circuit Voltage can be achieved according to the Nernst equation as follows [37,38]:

$$V_{OC} = V^{rev}_{P_0,T_0} + \left(T - T_{ref}\right) \times \frac{\Delta S^0}{nF} + \frac{R^*T}{2F}\left[ln\left(\frac{p_{H_2}\sqrt{p_{O_2}}}{\alpha_{H_2O,KOH}}\right)\right] \tag{16}$$

where the second term is added to evaluate the variation of the amount of reversible voltage because of temperature change and $\Delta S^0/nF$ is standard entropy change approximately equals -0.9×10^{-3} J/mol K. The water activity of the electrolyte aqueous is also shown by $\alpha_{H2O,KOH}$. However, there are some restrictions in employing this equation even for those dilute solutions or higher concentrations and even when there is a net current flow through the electrodes, because of the activity of ion changes at the surface of electrodes [38]. Another mostly employed relation is [27,28,31,32,39]:

$$V_{OC} = V^{rev}_{P_0,T} + \frac{R^*T}{2F} ln\left(\frac{\left(P^{saturation}_{H_2O,KOH}\right)^{3/2}}{\alpha_{H_2O,KOH}}\right) \tag{17}$$

where

$$V_{P_0,T}^{rev} = 1.5184 - 1.5421 \times 10^{-3} \times T + 9.526 \times 10^{-5} \times T \times lnT + 9.84 \times 10^{-8} \times T^2 \quad (18)$$

$$P_{H_2O,KOH}^{saturation} = exp\left(2.302a + blnP_{H_2O}^{saturation}\right)\,(in\,bar) \quad (19)$$

$$a = -0.0151m - 1.6788 \times 10^{-3}m^2 + 2.2588 \times 10^{-5}m^3$$

$$b = 1 - 1.2062 \times 10^{-3}m + 5.6024 \times 10^{-4}m^2 - 7.8228 \times 10^{-6}m^3$$

$$P_{H_2O}^{saturation} = exp\left(81.6179 - \frac{7699.68}{T} - 10.9\,\ln T + 9.5891 \times 10^{-3}T\right)(in bar) \quad (20)$$

And the water activity of the KOH solution can be calculated [28]:

$$\alpha_{H_2O,KOH} = exp\left(-0.05192m + 0.003302m^2 + \frac{(3.177m - 2.131m^2)}{T}\right) \quad (21)$$

These proposed empirical relations can be used for temperature, pressure, and the molar concentration range of 0–250°C, 1–200 bar, and 2–18 mol/kg, respectively.

To evaluate the activation overpotential, we utilize the Butler-Volmer equation according to the transition state theory [40,41]. So we can write:

$$V_{act}^{anode} = \frac{R^*T}{\alpha_{anode}F}\left[ln\left(\frac{i}{i_0^{anode}}\right)\right] \quad (22)$$

$$V_{act}^{cathode} = \frac{R^*T}{\alpha_{cathode}F}\left[ln\left(\frac{i}{i_0^{cathode}}\right)\right] \quad (23)$$

where i_0 is the effective exchange current density of the electrode and can be predicted at any temperature assuming an activated mechanism [40,41]:

$$i_0 = \gamma_M \times exp\left[-\frac{\Delta G_C}{R^*}\left(\frac{1}{T} - \frac{1}{T_{ref}}\right)\right] i_{0,ref} \quad (24)$$

To contribute the covered portion of the electrode with bubbles, activation overpotential is rewritten as follows:

$$V_{act}^{anode} = \frac{R^*T}{\alpha_{anode}F}\left[ln\left(\frac{i}{i_0^{anode}(1 - \Theta_{anode})}\right)\right] \quad (25)$$

$$V_{act}^{cathode} = \frac{R^*T}{\alpha_{cathode}F}\left[ln\left(\frac{i}{i_0^{cathode}((1 - \Theta_{cathode}))}\right)\right] \quad (26)$$

Evaluation bubble coverage factor (Θ) is extremely difficult and has many influencing parameters such as natural or forced circulation of aqueous, surface tension of the liquid, and so on. So an empirical equation is used and modified for pressure and temperature variation [4,16,27,42,43]:

$$\Theta = \left[-97.25 + 182\frac{T}{T_{ref}} - 84\left(\frac{T}{T_{ref}}\right)^2\right] \times \left(\frac{i}{i_{lim}}\right)^{0.3} \times \frac{P}{P - P^{sat}_{H_2O,KOH}(T)} \tag{27}$$

As already mentioned, concentration overpotential is neglected in some modelings. However, we present one model to evaluate this overpotential [16]. According to Fick's law at the electrolyte-electrode interface, ideal gas assumption of gaseous species, molecular and Knudsen diffusion of species in porous media, Kinetic theory of gases, Chapman-Enskog theory of the ideal gases, Lennard-Jones energy parameters for the binary species and solving relating parameters, we can write [44–50]:

$$V_{con} = \frac{R^*T}{4F} ln \frac{C^{anode}_{O_2,el}}{C^{anode}_{O_2,0}} + \frac{R^*T}{2F} ln \frac{C^{cathode}_{H_2,el}}{C^{cathode}_{H_2,0}} \tag{28}$$

where $C^{anode}_{O2,\, el}$ and $C^{cathode}_{H2,\, el}$ represent species concentration at the vicinity of the electrode and $C^{anode}_{O2,\, 0}$ and $C^{cathode}_{H2,\, 0}$ show species concentration in working condition at the electrode surface [44,51,52].

Ohmic overpotential is directly and linearly related to current and consists of three main ohmic resistances including electrolyte resistance, electrodes resistance, and separator resistance. Assuming insulated pores for electrode material and bubbles relative to the electrolyte itself, and according to the basic electrical resistance of a uniform conductor and solving Bruggeman's equation for the conductivity of a two-phase conductor [53], electrode resistance can be evaluated as follow:

$$\begin{aligned} R_{electrode} &= R_{anode} + R_{cathode} \\ &= \frac{\rho_0^{anode}}{(1-\varepsilon_{anode})^{3/2}} \frac{\delta_{anode}}{A_e} \left[1 + \kappa_{anode}\left(T - T_{ref}\right)\right] \\ &+ \frac{\rho_0^{cathode}}{(1-\varepsilon_{cathode})^{3/2}} \frac{\delta_{cathode}}{A_e} \left[1 + \kappa_{cathode}\left(T - T_{ref}\right)\right] \end{aligned} \tag{29}$$

To evaluate electrolyte resistance, it should be separated into two parts: a bubble zone and a bubble-free zone for each terminal, and according to the basic electrical resistance and Bruggeman's equation, we can write [16,53–55]

$$R_{electolyte} = R^{anode}_{electrolyte} + R^{cathode}_{electrolyte} = \frac{\rho_{electrolyte,ref}}{\left[1 + \kappa_{electrolyte}\left(T - T_{ref}\right)\right]} \times \left[\frac{l_{anode-s} - \beta_{anode}}{A_e} + \frac{1}{\left(1-\Theta_{anode}\right)^{3/2}}\frac{\beta_{anode}}{A_e} + \frac{l_{cathode-s}\beta_{cathode}}{A_e} + \frac{1}{\left(1-\Theta_{cathode}\right)^{3/2}}\frac{\beta_{cathode}}{A_e}\right] \tag{30}$$

Providing continuous ionic conduction and preventing gas mixture formation is the main duty of the separator. Considering uniform current distribution, we can write [16,56]:

$$R_{separator} = \rho_{electrolyte} \frac{\tau^2_{separator} \delta_{separator}}{\omega_{separator} \times \varepsilon_{separator} \times A_{separator}} \tag{31}$$

Researchers have reported that with the increase of temperature, the amount of open circuit, ohmic, and activation overpotentials decreases, and the total voltage demand will reduce [57]. Now, all unknown parameters have been correlated and can be evaluated using a computer program such as MATLAB.

Fig. 2 depicts the different voltage contributions in an ordinary alkaline electrolyzer [16] at a constant temperature and pressure, and it can be understood that the portion for concentration overpotential is negligible. Moreover, all kinds of voltages increase with the enhancement of current value except open-circuit overpotential, which is fixed for the mentioned pressure and temperature.

The energy efficiency of an electrolytic cell can be introduced according to Eq. (32) [58]:

$$\eta_C = \frac{\dot{H}_{H_2,c}}{P_{ele,c}} = \frac{\eta_F \times H_{S,H_2} \times \dot{m}_{H_2,c}}{V.I} \tag{32}$$

where η_F is the Faraday efficiency which is defined as the purity of the H_2 leaving the system. The amount of H_2 produced by the system can also be estimated according to the ideal gas law. Faraday efficiency and thermal efficiency at 25°C are defined as follows [17]:

$$\eta_F = \frac{\Delta G}{V} = \frac{1.23}{V} \tag{33}$$

$$\eta_{Thermal} = \frac{\Delta H}{V} = \frac{1.48}{V} \tag{34}$$

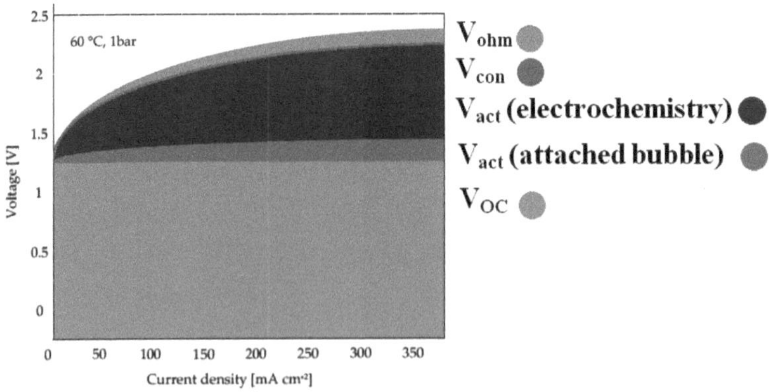

FIG. 2 AE *V-I* curve at constant temperature and pressure [16].

where V is the total operating voltage of the system. Some researchers have proposed an easier path to get the overall working voltage. They write the relation in terms of various coefficients in different mathematical forms and then perform some experiments to get those unknown variables which can be used for a wide range of working conditions. The following relation is an example [35]:

$$V = V_{P_0,T_0}^{rev} \times r_1 \frac{i}{A} + r_2 \times ln\left(r_3 \times \frac{i}{A} + 1\right) \tag{35}$$

where coefficients r_1, r_2, and r_3 depend on the operating pressure and temperature and should be experimentally determined. It should be noted that higher operating pressure consequents to a higher amount of H_2 loss depending on the manufacturing precision [59]. The energy efficiency of alkaline electrolyzers is typically varied from 70% to 80% and a higher electrical current reduces this efficiency most of the time [58,59]. The electrolyzer's characteristics showed that higher H_2 production leads to higher energy losses [58,59]. However, operating the cell at various conditions corresponds to the less purity of the production [58,59]. In another model, we can write [39]:

$$V = V_{P,T}^{rev} + sln\left(\frac{i}{t} + 1\right) + vln\left(\frac{i}{w} + 1\right) + \frac{k}{A} i \tag{36}$$

where

$$\begin{aligned} k &= k_1 + k_2 T + \frac{k_3}{T} + \frac{k_4}{T^2} \\ s &= s_1 + s_2 T + s_3 T^2 \\ t &= t_1 + t_2 T + t_3 T^2 \\ v &= v_1 + v_2 T + v_3 T^2 \\ w &= w_1 + w_2 T + w_3 T^2 \end{aligned} \tag{37}$$

All these unknown parameters should be determined after several experiments while employing the MATLAB optimization tool from the adjustment of the V-I curve. In another model, researchers proposed the following model [60,61]. This model is based on semiempirical equations which express the voltage cell, Faraday efficiency, and gas purity in terms of the current. The outcomes depict that enhancing the working temperature and decreasing the pressure can improve the overall efficiency of the system.

$$\begin{aligned} V = V_{P_0,T_0}^{rev} &+ [(r_1 + p_1 + q_1) + r_2 T + P_2 C + P_3 C^2 + q_2 d] i + s \\ &\times log\left[\left(t_1 + \frac{t_2}{T} + \frac{t_3}{T^2}\right) i + 1\right] \end{aligned}$$

And

$$\eta_F = \left(\frac{i^2}{f_{11} + f_{12} T + i^2}\right) . (f_{21} + f_{22} T) \tag{38}$$

where C is the electrolyte concentration and d is the electrode/diaphragm distance. Unknown constants for the Faraday efficiency were proposed as follows [35]:

$$f_{11} = 478645.74, f_{12} = -2953.15, f_{21} = 1.0396, f_{22} = -0.00104 \tag{39}$$

In this model, the voltage gradually reduces when the temperature varies from 50°C up to 80°C, and then less power supply is needed. Consequently, less H_2 is produced due to lower Faraday efficiency. However, changes may be different in other conditions.

The bubble phenomenon is another important resistance encountered around the electrodes. The overall amount is low; however, it can be eliminated through circulating the electrolyte, which enhances the bubble detachment. Hydrophilic surfaces are strongly recommended to reduce the bubble on the surface of the electrodes. Moreover, some appropriate additives can reduce the surface tension providing fewer bubbles generated around the terminals. It should be noted that these additives must not take part in the electrochemical process and have to be electrically, chemically, and mechanically neutral [17]. Another complete model considering the attached bubble size is as follows [62]:

$$\begin{aligned} V = V_{P,T}^{rev} &+ \frac{R^*T}{\alpha_{anode}F} log\left(\frac{i}{i_{0,anode}}\right) + \frac{R^*T}{\alpha_{cathode}F} log\left(\frac{i}{i_{0,cathode}}\right) + i\left(\frac{L}{K_{OH^-}^{eff}}\right) \\ &+ \frac{R^*T}{\alpha_{anode}F} log\left(\frac{1}{1-\Theta}\right) + \frac{R^*T}{\alpha_{cathode}F} log\left(\frac{1}{1-\Theta}\right) + 2\left(i \times \frac{D_{bubble}}{K_{OH^-}^{eff}(1-\Theta)^{1.5}}\right) \end{aligned} \tag{40}$$

The bubble size can be estimated by basic fluid mechanics rules according to the inside and outside pressure differences considering the surface tension. L is the cathode and anode distance, $K_{OH^-}{}^{eff}$ is the effective hydroxide ionic conductivity which should be determined through experimental tests. Θ is the bubble coverage factor which is identified in Eq. (27). The second and third terms are related to activation overpotentials. The fourth term is attributed to the ohmic resistance considering the electrodes distance and the last three-term emphasize the adverse effects of bubble attachment to the surfaces of the electrodes. They reported that the model overpredicts the overall voltage for higher current densities; however, it can be employed for the industrial application range [62].

One of the most applicable relations used in many research projects is as follows [63]:

$$V = V_{P_0,T_0}^{rev} + \frac{e_1 + e_2T}{A_e} I + \left(e_3 + e_4\text{T} + e_5T^2\right).log\left(\frac{e_6 + \frac{e_7}{T} + \frac{e_8}{T^2}}{A_e} I + 1\right) \tag{41}$$

$$\dot{N}_{H_2}^{gn} = \eta_F \frac{I}{2F} = 2\dot{N}_{O_2}^{gn} \tag{42}$$

$$\eta_F = g_1.exp\left(\frac{g_2 + g_3T + g_4T^2}{I/A_e} + \frac{g_5 + g_6T + g_7T^2}{(I/A_e)^2}\right) \tag{43}$$

The addition of the electrolyte flow rate to the abovementioned relation is also recommended [64], which are rarely used in the literature. These empirical models are used a lot due to their higher precision rather than completely theoretical and complicated expressions. The only difficulty of a semiempirical relation is running the system to obtain the constants of the equation, and then anyone can use it for the range of application.

Thermal modeling of the electrolyzer has also been discussed [27] to evaluate the bulk temperature variation of the cell using thermodynamic and heat transfer basic rules. The lump capacitance method is used to formulate the heat transfer problem. It means there is no temperature gradient within the system, and the entire system temperature is identical at any instant [65–68]. So we can write a thermal energy balance to the system [27]:

$$C_t\frac{dT}{dt} = P_{heat} - \dot{Q}_{loss} - \dot{Q}_{cooling} - \sum_i h_i\dot{m}_i \tag{44}$$

This equation states that temperature variation is related to the amount of electrical power input dissipated to the system or it can be called internal heat generation, heat loss to the ambient, heat removed by the cooling water, and enthalpy of the species leaving the system. C_t is a lumped thermal capacitance of the entire system. This quantity can be obtained through the sum of the thermal capacitance of each component multiplied by their volume divided by the total mass of the system (mass average). Internal heat generation can be achieved according to the difference in operating voltage and the thermoneutral cell voltage multiplied by the current. Empirical natural convection coefficients can certainly help determine the amount of heat loss to the environment. Exiting enthalpy can be calculated through two parts. The first one is sensible heat that can be evaluated through the mass flow rate of species multiplied by the temperature variation and thermal specific capacity. The second part is latent heat, which is the amount of heat absorbed to evaporate the water in the system. The mass balance of the materials cooperating in the process by the main thermal balance equation can reveal the temperature change [27]. Another easy path to get C_t is to omit any heat transfer to the environment, including enthalpy removal to the outside through species. Then the amount of heat generated can be divided by the rate of temperature change to achieve C_t.

There are two main ways for representing voltage variation of the electrolyzers [69]. The first one is the polarization curve which is depicted to show cell voltage change concerning the cell current density. This plot is widely employed to express the total loss in the cell; however, it is unable to give any information about each part of the loss. There are three main stages of this trend: activation polarization, ohmic polarization, and concentration polarization. Activation loss occurs at a low current density due to the low speed of the electrochemical kinetics (Butler-Volmer equation). Ohmic loss happens on behalf of the resistances through transporting the ions, and concentration loss which appears at higher current densities is because of the higher reaction rate concerning the transfer speed of the species and products. These concepts are mathematically discussed throughout the chapter. The second method to represent the electrolyzer performance is the current interrupt method [69]. In this method, after the steady operation of the electrolyzer at a specific time, the current is interrupted, and then there is a sudden drop of operating voltage due to the membrane resistance, and one can evaluate the amount of membrane resistance by dividing the value of voltage drop by the steady current. This fall occurs very quickly (within the order of a microsecond), so it should be operated at high frequencies [69]. When this voltage changes, it gradually reduces due to the slow electrochemical process and related overpotentials, which eventually hits the zero line of operating voltage.

4 Mathematical modeling of a proton exchange membrane electrolyzer (PEME)

The power supply is exerted to a PEM electrolyzer to onset the electrochemical procedure. Water is injected into the cell through the anode section, and it is broken down into oxygen gas, protons, and electrons. Then the protons (H^+) are transmitted through the highly conductive solid membrane to the cathode where they get to the traveled electrons arriving from the out circuit generating water and hydrogen gas. Fig. 3 depicts a schematic of a PEME.

So the following chemical reactions take place in the cell:

$$\begin{aligned} &Anode: H_2O \rightarrow 2H^+ + \frac{1}{2}O_2 + 2e^- \\ &Cathode: 2H^+ + 2e^- \rightarrow H_2 \end{aligned} \tag{45}$$

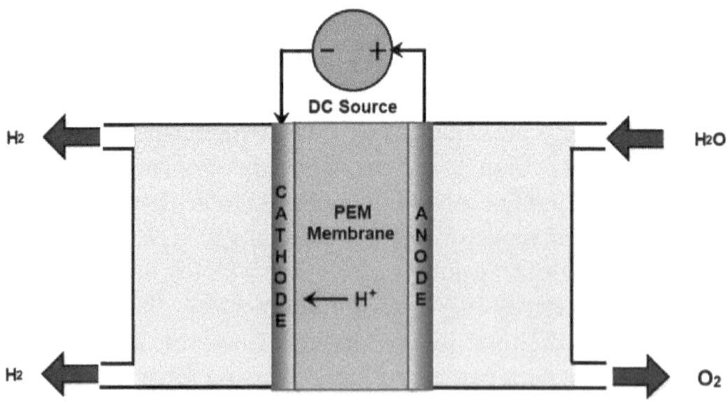

FIG. 3 Schematic of a PEM Electrolyzer [44].

Then the total reaction will be:

$$H_2O \rightarrow H_2 + \frac{1}{2}O_2 \tag{46}$$

Molar relations within the cell are described according to the conservation of mass:

$$Anode \quad \begin{array}{l} \dfrac{dN_{H_2O}}{dt} = \dot{N}_{H_2O}^{in} - \dot{N}_{H_2O}^{out} - \dot{N}_{H_2O}^{membrane} - \dot{N}_{H_2O}^{consumed} \\ \dfrac{dN_{O_2}}{dt} = \dot{N}_{O_2}^{in} - \dot{N}_{O_2}^{out} + \dot{N}_{O_2}^{gn} \\ \dot{N}_{H_2O}^{consumed} = \dfrac{I}{2F}, \dot{N}_{O_2}^{gn} = \dfrac{I}{4F} \end{array} \tag{47}$$

$$Cathode \quad \begin{array}{l} \dfrac{dN_{H_2O}}{dt} = \dot{N}_{H_2O}^{in} - \dot{N}_{H_2O}^{out} + \dot{N}_{H_2O}^{membrane} \\ \dfrac{dN_{H_2}}{dt} = \dot{N}_{H_2}^{in} - \dot{N}_{H_2}^{out} + \dot{N}_{H_2}^{gn} \\ \dot{N}_{H_2O}^{membrane} = \dot{N}_{H_2O}^{cathode}, \dot{N}_{H_2}^{gn} = \dfrac{I}{2F} \end{array} \tag{48}$$

And total pressure and partial pressure of species can be determined according to the ideal gas law and the molar fractions in the fundamental thermodynamics law.

PEM water electrolyzers (PEMWE) have many advantages compared to conventional alkaline electrolyzers. Solid electrolyte, the possibility of a more compact design, early start-up process, working at high pressure, simplicity of the technology, and high current density can be considered as the main positive aspects of PEMWE [70]. H_2 produced through AE needs to be purified due to the application of the electrolyte. One of the main assumptions in the mathematical modeling of the PEMWE is to consider the entire stack at an instant uniform temperature, and the membrane is to fill with saturated water. Consequently, the amount of conductivity is the only function of temperature. Likewise AE, there are overpotentials to define cell operating voltage besides the reversible potential (open-circuit voltage). The reversible voltage in PEME depends on water pressure entering the anode and resulting gas pressures and the reaction zone temperature, but its dependency on pressure can be neglected [71]. Its normal value at 25°C and 1 bar are 1.229 V. However, it can be evaluated for other temperatures. The reversible potential can be estimated according to the Nernst equation at constant atmospheric pressure or many other empirical relations [72,73]:

$$V_{P_0,T}^{rev} = 1.229 - 0.9 \times 10^{-3}(T - 298) \tag{49}$$

Activation overpotential can be achieved according to the Butler-Volmer expression [74]:

$$V_{act} = \frac{R^*T}{\alpha_{anode}F}\left[arcsinh\left(\frac{i}{2i_0^{anode}}\right)\right] + \frac{R^*T}{\alpha_{cathode}F}\left[arcsinh\left(\frac{i}{2i_0^{cathode}}\right)\right] \tag{50}$$

And to consider diffusion overpotential:

$$V_{act} = \frac{R^*T}{4F}\left[ln\left(\frac{C_{O_2}^{membrane}}{C_{O_2}^{ref}}\right)\right] + \frac{R^*T}{2F}\left[ln\left(\frac{C_{H_2}^{membrane}}{C_{H_2}^{ref}}\right)\right] \tag{51}$$

where the reference concentrations are the concentration of gas at the specific reference position and other concentrations can be evaluated according to [44,75,76].

Ohmic loss consists of ionic loss and out circuit ohmic loss. The out circuit loss can be evaluated according to the ohmic loss in AE and [44,76]. The polymeric membrane causes this loss for protons traveling to the cathode. Ionic loss is a function of membrane thickness and its conductivity [74]:

$$R_{ion} = \frac{t_m}{\sigma} \tag{52}$$

The membrane should be completely hydrated to ensure the proper displacement of the protons and dependence of the conductivity on the temperature. The amount of conductivity can be expressed according to the Arrhenius relation [74]:

$$\sigma(T) = \sigma_{ref} e^{\left[\frac{-E_{Pro}}{R^*}\left(\frac{1}{T_{ref}} - \frac{1}{T}\right)\right]} \tag{53}$$

It can also be calculated in terms of temperature and the humidification factor which is introduced as the fraction of molar number of H_2O to the molar number of SO_3^- changes between 14 and 25 (25 for an intensively hydrated membrane) [75,77]:

$$\sigma(T) = (0.005139\lambda - 0.00326)exp\left(1268\left(\frac{1}{303} - \frac{1}{T}\right)\right) \tag{54}$$

where E_{pro} is a temperature-independent parameter that expresses the activation energy for proton transition within the membrane. λ means molecules of water per sulphonic group. PEM conductivity at the reference temperature can be conveniently found from the guidance for ionomers represented by the manufacturing companies. So by adding ionic loss to the electrode loss, the overall ohmic loss can be achieved.

For each cell,

$$\dot{N}_{H_2}^{gn} = \eta_F \frac{I}{2F} = 2\dot{N}_{O_2}^{gn} = \frac{1}{1.25}\dot{N}_{H_2O}^{gn} \tag{55}$$

In most cases, the Faraday efficiency is assumed to be at least 99% [70,76]. Finally, there is a suggestion of the following relation to be fitted according to the polarization curve at atmospheric pressure [74]:

$$V = V_{P_0,T}^{rev} + \frac{R^*T}{F} ln\left(\frac{i}{P_1}\right) + P_2 i \tag{56}$$

where P_1 and P_2 are constant coefficients. The membrane material is very important to be a proper proton conductive. It is often made of a Perfluoro sulfonic acid (PFSA) polymer [78].

There are two key events within the membrane: diffusion and electroosmotic drag of the water. The electroosmotic drag of water is obtained according to [79,80]:

$$\dot{N}_{H_2O}^{electroosmotic} = \frac{n_d \times i \times M_{H_2O} \times A_{active}}{F} \quad (57)$$

And the electroosmotic drag coefficient is $n_d = 0.016T - 2.89556$. The water diffusion amount through the membrane can be achieved when the value of electroosmotic induced water is subtracted from the total water transporting through the membrane, or it can be estimated as follows [81–83]:

$$\begin{aligned}
\dot{N}_{H_2O}^{diffusion} &= D_w \times \frac{C_{wc} - C_{wa}}{t_m} \times M_{H_2O} \times A_{active}. \\
D_w &= D_\lambda exp\left[2416\left(\frac{1}{303} - \frac{1}{T}\right)\right] (T is\ in\ K) \\
D_\lambda &= 10^{-10}, \lambda_m < 2; \\
D_\lambda &= 10^{-10}[1 + 2(\lambda_m - 2)], 2 \le \lambda_m \le 3; \\
D_\lambda &= 10^{-10}[3 - 1.67(\lambda_m)], 3 \le \lambda_m \le 4.5; \\
D_\lambda &= 1.25 \times 10^{-10}, 4.5 \le \lambda_m; \\
\lambda_m &= (\lambda_{cathode} + \lambda_{anode})/2 \\
\lambda &= 0.043 + 17.81a - 39.85a^2 + 36a^3 \\
a &= \frac{P}{P_{sat}} \\
P_{sat}(Pa) &= -2846.4 + 411.24T - 10.554T^2 + 0.16636T^3 (T is\ in\ ^\circ C) \\
C_w &= \frac{\rho_{membrane}}{Weight_{membrane}} \lambda_m
\end{aligned} \quad (58)$$

5 Mathematical modeling of the solid oxide electrolyzer (SOE)

A solid oxide electrolyzer (SOE) operating at high temperatures has many advantages over the conventional AE and PEM. It can result in higher electrochemical reactions while using lower electrical energy [84]. However, the highly ionic conductive electrolyte and two porous electrodes matter and should be considered as important parameters for SOE design. In this cell, steam is fed into the porous cathode and it is decomposed into gaseous hydrogen and oxygen ions at the cathode-electrolyte interface. The resulting H_2 is collected through the cathode and oxygen ions travel to the anode side through an electrolyte creating an oxygen molecule (Fig. 4). The total energy required for the SOE cell, like the AE and PEM, consists of electrical energy demand (free Gibson energy

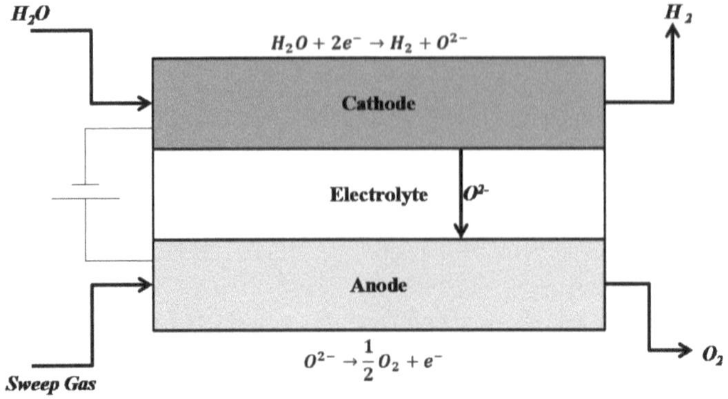

FIG. 4 The Planar configuration of SOE.

change) and thermal energy demand (temperature multiplied by the entropy change). With the increase in operating temperature, the thermal energy needed to complete the process reduces while the electrical energy demand reduces. Though this is favorable, high-temperature considerations should be taken into account [84].

As the SOE cell is working at a high temperature, so the following technical and economic statuses should be well taken into account [84,85]:

- High electronic conductivity, chemically, and physical stability of the electrolyte and electrodes.
- The gastight ability of the electrolyte to hinder H_2 and O_2 recombination.
- Reasonable pore size and porosity for electrodes to maintain proper transportation of gases and reaction sites.
- Similar thermal expansion coefficients of electrolyte and electrodes due to the higher thermal expansion and thermal stress.
- Low cost of manufacturing and initial materials.

Yttria-stabilized zirconia (YSZ) is one of the widely employed electrolytes for high-temperature SOE. Not only is it mechanically stable, but also it is an excellent oxygen ion conductor; however, it is expensive concerning other electrolytes such as LSGM (strontium/magnesium-doped lanthanum gallate). As these electrolytes are emerging within a high-temperature condition, the reaction risk with electrodes is very high. So we can say operating temperature, cost, and reactivity rate with other parts of the cell such as electrodes are the main characteristics to select an electrolyte for high-temperature SOE [84].

Unlike the conventional AE and PEM, there are two widely used configurations for the SOE cell: tubular and planar configurations. Fig. 5 depicts the schematic of the two models of SOE.

According to scientific reports, tubular SOE cells have better mechanical strength under high-temperature operation; however, planar cells have

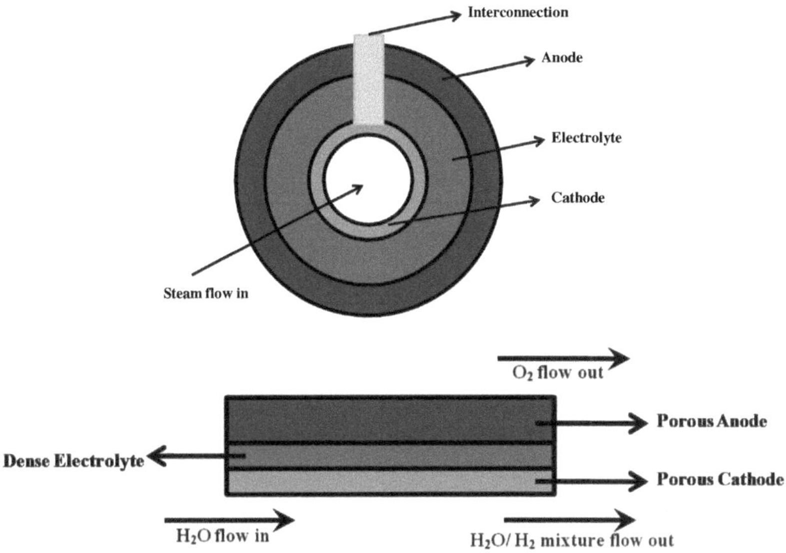

FIG. 5 SOE configurations (A) tubular (end view) (B) planar [84].

exhibited high manufacturability [84]. Moreover, according to the appropriate and uniform distribution of gaseous materials on the planar SOE and more convenient production of mass, there can be more advantages for planar cells rather than tubular ones [86]. Apart from various configurations suggested in the literature, researchers recommend specified shapes and configurations for each spectacular application [87–89]. The more active the area is involved within the electrochemical process, the more hydrogen is produced. Although increasing cell size has many limits, many cells can be connected in many models to compensate for the hydrogen production rate.

Since high-temperature steam is fed into SOE and the outcome products still have a high temperature, it seems to be reasonable to employ a high-efficiency heat exchanger to reuse the amount of leaving thermal energy to heat the entering steam and reduce the energy needed at the internal phase [90,91]. However, it needs further energy and exergy analysis to ensure the feasibility, applicability, and effectiveness.

Like the AE and PEM cells, besides the reversible potential, there are overpotentials to evaluate the amount of operating voltage. Reversible voltage can be achieved as follows [92,93]:

$$V_{P,T}^{rev} = V_{P_0,T}^{rev} + \frac{R^*T}{2F} ln\left(\frac{p_{H_2} p_{O_2}^{1/2}}{p_{H_2O}}\right) \quad where \quad V_{P_0,T}^{rev} = 1.253 - 2.4516 \times 10^{-4} T \tag{59}$$

Activation overpotential, which is responsible for energy loss due to the kinetic energy loss, can be evaluated through the Butler-Volmer equation [94,95]:

$$V_{act} = \frac{R^*T}{\alpha_{anode}F}\left[arcsinh\left(\frac{i}{2i_0^{anode}}\right)\right] + \frac{R^*T}{\alpha_{cathode}F}\left[arcsinh\left(\frac{i}{2i_0^{cathode}}\right)\right] \quad (60)$$

Ohmic loss can be mainly considered for the electrolyte layer. So ohmic overpotential is evaluated as follows [96]:

$$V_{ohm} = 2.99 \times 10^{-5} \times i \times La \times exp\left(\frac{10,300}{T}\right) \quad (61)$$

where *La* is the electrolyte layer. The amount of operating voltage should overcome the mass transfer resistance electrode and electrolyte, which is previously described as concentration overpotential and can be estimated as follows [97–99]:

$$V_{con} = \frac{R^*T}{4F} ln\left(\frac{\sqrt{(p_{H_2})^2 + (i \times R^* \times \mu \times d_a/2FB_g)}}{p_{O_2}}\right) + \frac{R^*T}{2F} ln\left(\frac{\left(p_{H_2} + i \times R^* \times T \times d_c/2FD_{H_2O}^{eff}\right)p_{H_2O}}{\left(p_{H_2O} - i \times R^* \times T \times d_c/2FD_{H_2O}^{eff}\right)p_{H_2}}\right) \quad (62)$$

where μ, B_g, and D_{H2O}^{eff} are oxygen dynamic viscosity, flow permeability, and effective diffusion coefficients of steam, respectively. These parameters can be achieved through analytical relations or experimental tests depending on the operational temperature, porosity, and tortuosity of the cathode [97–99]. The amount of overall operating voltage is the sum of all overpotentials mentioned and the reversible voltage. Results of mathematical modeling and experimental tests proved that thermal energy increases by decreasing the electrical energy supply. In addition, increasing the porosity and pore size of electrodes leads to higher voltage loss and temperature, and also steam molar fraction enhancement increases the electrical efficiency of the cell [98]. Some researchers have used air to control cell temperature [99,100]. Air, which mostly consists of oxygen and nitrogen molecules, enters the cell through the anode.

After neglecting the pressure drop through the channel and end effects of a stack and then assuming constant operating voltage across the channel and ideal gas behavior, high-temperature SOE can be mathematically modeled by writing electrochemical relations. Moreover, for the previously mentioned Nernst and Butler-Volmer equation for species x-dependent concentrations, valuable results have been achieved [100,101]. Various inlet air flows were successfully used to control the temperature; however, both endothermic and exothermic operations were reported with the variation of a steam molar fraction [100,101].

6 Mathematical modeling of electrolyzers powered by renewable resources

Renewable energy resources are a very cheap and sustainable way to supply voltage to the electrolyzers, especially solar energy, which is the most available and convenient one. However, there are some restrictions according to the night dark hours and cloudy weather. Regardless of these limits, many countries and research labs are trying to produce H_2 using solar energy. There are some mathematical models for supporting electricity generated through PV panels [102–104]. One of the widely used models is explained here. Photovoltaic cell operation is based on semiconductor performance and it can be shown as the following equivalent circuit (Fig. 6) [102]:

Where I_L, I_d, I_P, V_d, R_S, and R_P are photo-generated current, diode current, the current through the parallel resistor, diode voltage, series resistance, and parallel resistance, respectively. According to Kirchhoff's current law, we can write [102]:

$$I = I_L - I_d - I_P \tag{63}$$

and

$$I = I_L - I_0\left(e^{\frac{qV_d}{nKT_cN_s}} - 1\right) - \frac{V_d}{R_P} \tag{64}$$

where

$$I_L = \left(I_{SC} + \mu_{SC}\left(T_C - T_{ref}\right)\right)\frac{G_r}{G_{r,ref}} \tag{65}$$

$$I_0 = I_{0-ref}\left(e^{\left(\frac{1}{T_{ref}} - \frac{1}{T_C}\right)\frac{q\varepsilon_g}{nK}}\right)\left(\frac{T_C}{T_{ref}}\right)^3 \tag{66}$$

$$I_{0-ref} = I_{SC}\left(e^{\frac{qV_{OC}}{nN_SKT_C}} - 1\right) \tag{67}$$

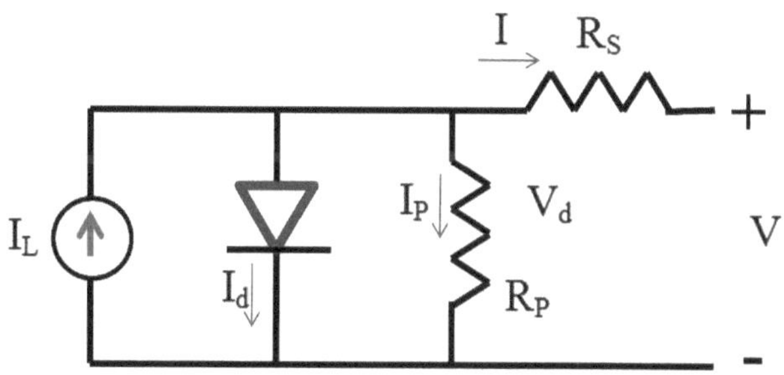

FIG. 6 PV module equivalent circuit.

and all these empirical parameters are reported through some experimental tests conducted on the PV modules [102].

Wind energy is another useful renewable energy that is applicable for H_2 production. It can supply the electricity needed for the electrochemical process through the electrolyzer. However, the zone where wind turbines are planted should essentially have a cost-effective energy storage process. Wind energy mainly depends on four parameters: the amount of air passing through the turbine blades, air speed, air density, and the amount of area for air flux, so the power supply is defined as follows:

$$P_m = \frac{1}{2} c_P \rho A v^3 \tag{68}$$

The power coefficient is the function of pitch angle and the tip speed ratio. Pitch angle depends on the shape of the blades, and tip speed ratio is the ratio of the tip rotational speed to the average air velocity. Various models have been suggested to point out the amount of power coefficient; however, the maximum amount of this constant is around 0.44. Wind turbine designers try to meet this value for their blade shapes [105,106].

Although there are many other renewable energy resources, these two are the most powerful and available among other sources and can transmit extra generated energy to the houses directly (Fig. 7). They are expensive when installed; however, over time, their great economic value will be amazing.

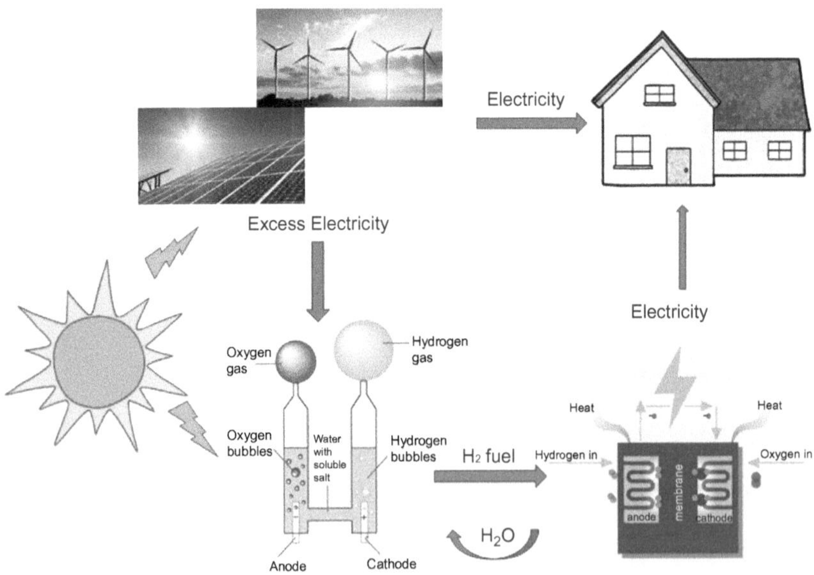

FIG. 7 PV module equivalent circuit.

7 Conclusion

In this chapter, the necessity and importance of H_2 production and its mathematical modeling through electrolyzers were discussed. After introducing three main electrolyzers, theoretical and semiempirical relations to predict the operating parameters of the device were presented. Operating voltage is one of the important pivotal characteristics of an electrolyzer which consists of several overpotentials including open-circuit overpotential, ohmic overpotential, concentration overpotential, and activation overpotential. Various models demonstrated for these overpotentials and affective phenomena such as attached bubbles to the electrodes or changing design parameters of the system (electrodes distance, for instance) were analyzed. There are some models containing several unknown parameters that should be diagnosed after a few experimental tests, and employing these models is highly recommended through several research projects. Renewable energy resources, mainly solar energy and wind energy, which are widely used to supply electricity, are abundant, free, and appear cost-effective in most cases.

Nomenclature

A	air flow area through wind turbines (m^2)
A_e	external electrode surface area (m^2)
$A_{seperator}$	uniform distribution cross section of separator (m^2)
A_{active}	active cross section of membrane (m^2)
B_g	flow preamibility (m^2)
C	electrolyte concentration (mol/m^3)
$C_{O2,\,el}^{anode}$ and $C_{H2,\,el}^{cathode}$	species concentration at the vicinity of the electrode (mol/m^3)
$C_{O2,\,0}^{anode}$ and $C_{H2,\,0}^{cathode}$	species concentration in working condition at the electrode surface (mol/m^3)
C_t	lumped thermal capacitance (J/kg K)
c_P	pressure coefficient of wind power (−)
D^{eff}	the effective binary diffusion coefficient (m/s)
d	electrode/diaphragm distance (m)
D_{bubble}	bubble diameter (m)
D_{H2O}^{eff}	effective diffusion coefficient of steam (m/s)
d_a	thickness of SOE anode (m)
d_c	thickness of SOE cathode (m)
E_{Pro}	Arrhenius temperature-independent parameter which demonstrates the activation energy for proton movement within the membrane (kJ)
F	Faraday constant, the amount of electric charge per mole of electrons (96,487 As/mol)
ΔG	change in Gibbs energy (kJ/mol)

ΔG_C	free energy of activation (kJ/mol)
G_r	irradiance input (W/m^2)
$G_{r,\ ref}$	radiance at STC = 1000 W/m^2
ΔH	enthalpy change (kJ)
$H_{S,\ H2}$	higher heating value of H_2 (kJ)
$\dot{H}_{H_2,c}$	amount of enthalpy leaving the system through H_2 production (kJ)
$\sum_i h_i \dot{m}_i$	enthalpy of the species leaving the system (kJ)
I	current through the cell (A)
i	current density (A/m^2)
i_0	effective exchange current density of the electrode (A/m^2)
$i_{0,ref}$	reference exchange current density at T_{ref} (A/m^2)
i_{lim}	limiting current density at 100% bubble coverage with typical value 300 kAm^{-2}
i_L^{anode}	current limiting term (A/m^2)
I_L	photo-generated current in PV model (A)
I_d	diode current in PV model (A)
I_P	current through parallel resistor in PV model (A)
I_L	photo-generated current (A)
I_0	the reverse saturation current of the diode (A)
I_{SC}	short-circuit current (A)
I_P	current through parallel resistor (A)
K	Boltzmann constant (J/K^{-1})
$K_{OH^-}^{eff}$	effective hydroxide ionic conductivity (s/m)
l_{an-cay}	anode channel to catalyst distance (m)
l_{ca-cay}	cathode channel to catalyst distance (m)
$l_{anode-s}$ $l_{cathode-s}$	bubble free zone width (m)
L	anode to cathode distance (m)
La	electrolyte thickness in SOE (m)
$\dot{m}_{H_2,c}$	theoretical mass flow of H_2 (kg/s)
M_{H2O}	molar mass of water (mol)
N	mole of substance (mol)
$\dot{N}$	molar rate (mol/s)
n_d	Electro-osmotic drag coefficient (–)
N_s	the number of cells connected in series (–)
p	partial pressure (Pa)
P	total pressure (Pa)
$P_{H2O,\ KOH}^{saturation}(T)$	saturation pressure of H_2O an KOH at temperature T (Pa)
$P_{H2O}^{saturation}$	saturation pressure of H_2O (Pa)
$P_{ele,\ c}$	amount of electricity supply to run the cell (W)
P_{heat}	electrical power input dissipation to the system (W)
P_m	power generated by wind energy (W)

$\dot{Q}_{loss}$	heat loss to the ambient (W)
$\dot{Q}_{cooling}$	heat removed by cooling water (W)
q	the electron charge (C)
R	electrical resistance (Ω)
R^*	world gas constant ($Jmol^{-1}\,K^{-1}$)
R_{ion}	ionic losses (Ω)
R_S	series resistance in PV model (Ω)
R_P	parallel resistance in PV model (Ω)
R_P	parallel resistance (Ω)
ΔS	change in entropy (kJ/mol K)
ΔS^0	standard entropy change (kJ/mol K)
T	temperature (K)
T_{ref}	reference temperature (K)
t_m	membrane thickness (m)
T_c	the cell temperature (K)
$V_{P0,\,T0}^{rev}$	reversible cell voltage at standard pressure and temperature, minimum voltage to drive water decomposition reaction (V)
V_{tn}	thermoneutral cell voltage (V)
V	cell working voltage (V)
V_{OC}	open-circuit voltage (V)
V_{act}	activation overpotential (V)
V_{con}	concentration overpotential (V)
V_{ohm}	ohmic overpotential (V)
V_d	diode voltage in PV model (V)
v	air velocity (m/s)
X	molar fraction (−)
x	distance (m)
z	number of moles of electrons transferred in the reaction (−)

Greek symbols

$\alpha_{H2O,KOH}$	water activity of the electrolyte solution (−)
$\alpha_{cathode}$	charge transfer coefficient (−)
α_{anode}	charge transfer coefficient (−)
$\hat{a}_{anode}\hat{a}_{cathode}$	bubble zone width (m)
$\ddot{a}_{anode}\ddot{a}_{cathode}$	thickness of the electrode (m)
$\ddot{a}_{sep}$	thickness of the separator (m)
ε_{anode}	anode porosity (−)
$\mathring{a}_{seperator}$	separator porosity(−)
ε_g	the material band gap energy, 1.12 eV for Si
η_C	energy efficiency of an electrolytic cell (−)

η_F	Faradic efficiency or current efficiency, ratio between the actual and theoretical maximum amount of H_2 produced (−)
$\eta_{Thermal}$	thermal efficiency of the cell (−)
Θ	fractional bubble coverage on the electrode surface (−)
$\hat{e}_{anode}\hat{e}_{cathode}\hat{e}_{electrolyte}$	temperature coefficient of resistivity (K^{-1})
μ	dynamic viscosity (Pa s)
μ_{SC}	coefficient temperature of short circuit current (A/K)
ρ	air density (kg/m^3)
$\tilde{n}_0^{anode}\tilde{n}_0^{cathode}$	resistivity of the 100% dense anode/cathode material at reference temperature (Ω. m)
$\tilde{n}_{electrolyte,\ ref}$	effective resistivity of electrolyte at reference temperature (Ω. m)
σ	electric conductivity of the membrane (s/m)
σ_{ref}	electric conductivity of the membrane at reference temperature (s/m)
$\hat{o}_{anode}$	anode tortuosity (−)
$\hat{o}_{seperator}$	tortuosity factor of the separator for electrolyte (−)
$\grave{u}_{seperator}$	wettability factor of the separator for electrolyte (−)
in	inlet molar flow
Out	outlet molar flow
gn	generation of substance
cons	consumed

References

[1] Wang J, Wan W. Factors influencing fermentative hydrogen production: a review. Int J Hydrogen Energy 2009;34(2):799–811.

[2] Kothari R, Buddhi D, Sawhney RL. Sources and technology for hydrogen production: a review. Int J Glob Energy Issues 2004;21(1–2):154–78.

[3] Wang M, Wang Z, Gong X, Guo Z. The intensification technologies to water electrolysis for hydrogen production – a review. Renew Sustain Energy Rev 2014;29:573–88.

[4] Turner J, et al. Renewable hydrogen production. Int J Energy Res 2008;32(5):379–407.

[5] Dincer I, Acar C. Review and evaluation of hydrogen production methods for better sustainability. Int J Hydrogen Energy 2015;40(34):11094–111.

[6] Turner JA. Sustainable hydrogen production. Science 2004;305(5686):972–4.

[7] Romm JJ. The hype about hydrogen: fact and fiction in the race to save the climate. Island Press; 2004.

[8] Wang M, Wang Z, Gong X, Guo Z. The intensification technologies to water electrolysis for hydrogen production – a review. Renew Sustain Energy Rev 2014;29:573–88.

[9] Safari F, Dincer I. A review and comparative evaluation of thermochemical water splitting cycles for hydrogen production. Energ Conver Manage 2020;205:112182.

[10] Bhandari R, Trudewind CA, Zapp P. Life cycle assessment of hydrogen production via electrolysis – a review. J Clean Prod 2014;85:151–63.
[11] Holladay JD, Hu J, King DL, Wang Y. An overview of hydrogen production technologies. Catal Today 2009;139(4):244–60.
[12] Dutta S. A review on production, storage of hydrogen and its sutilisationas an energy resource. J Ind Eng Chem 2014;20(4):1148–56.
[13] Martinez-Frias J, Pham A-Q, Aceves SM. A natural gas-assisted steam selectrolyserfor high-efficiency production of hydrogen. Int J Hydrogen Energy 2003;28(5):483–90.
[14] Abbasi T, Abbasi SA. 'Renewable' hydrogen: prospects and challenges. Renew Sustain Energy Rev 2011;15(6):3034–40.
[15] Chang H. Is water H_2O?: evidence, realism and pluralism. Vol. 293. Springer Science & Business Media; 2012.
[16] Abdin Z, Webb CJ, MacA Gray E. Modelling and simulation of an alkaline electrolyser cell. Energy 2017;138:316–31.
[17] Zeng K, Zhang D. Recent progress in alkaline water electrolysis for hydrogen production and applications. Prog Energy Combust Sci 2010;36(3):307–26.
[18] Ursua A, Gandia LM, Sanchis P. Hydrogen production from water electrolysis: current status and future trends. Proc IEEE 2011;100(2):410–26.
[19] Vermeiren PH, Leysen R, Beckers H, Moreels JP, Claes A. The influence of manufacturing parameters on the properties of macroporousZirfon® separators. J Porous Mater 2008;15(3): 259–64.
[20] Hug W, Bussmann H, Brinner A. Intermittent operation and operation modeling of an alkaline electrolyser. Int J Hydrogen Energy 1993;18(12):973–7.
[21] Griesshaber W, Sick F. Simulation of hydrogen-oxygen systems with PV for the self-sufficient solar house. Freiburg im Breisgau: FhG-ISE; 1991.
[22] Ulleberg Ø, Mørner SO. TRNSYS simulation models for solar-hydrogen systems. Solar Energy 1997;59(4–6):271–9.
[23] Meurer C, Barthels H, Brocke WA, Emonts B, Groehn HG. PHOEBUS—an autonomous supply system with renewable energy: six years of operational experience and advanced concepts. Solar Energy 1999;67(1–3):131–8.
[24] Gutiérrez-Martín F, García-De María JM, Baïri A, Laraqi N. Management strategies for surplus electricity loads using electrolytic hydrogen. Int J Hydrogen Energy 2009;34(20): 8468–75.
[25] Takahashi R, Kinoshita H, Murata T, Tamura J, Sugimasa M, Komura A, Futami M, Ichinose M, Ide K. Output power smoothing and hydrogen production by using variable speed wind generators. IEEE Trans Ind Electron 2009;57(2):485–93.
[26] Havre K, Borg P, Tommerberg K. Modeling and control of pressurisedelectrolyzer for operation in standalone power systems. In: Second Nordic symposium on hydrogen and fuel cells for energy storage, Helsinki; 1995.
[27] Hammoudi M, Henao C, Agbossou K, Dubé Y, Doumbia ML. New multi-physics approach for modelling and design of alkaline electrolyzers. Int J Hydrog Energy 2012;37(19): 13895–913.
[28] Henao C, Agbossou K, Hammoudi M, Dubé Y, Cardenas A. Simulation tool based on a physics model and an electrical analogy for an alkaline electrolyser. J Power Sources 2014;250:58–67.
[29] Bianchini C, Barbaro P, editors. Catalysis for sustainable energy production. Wiley-Vch; 2009.

[30] Krishna R, Wesselingh JA. The Maxwell-Stefan approach to mass transfer. Chem Eng Sci 1997;52(6):861–911.

[31] Balej J. Water vapour partial pressures and water activities in potassium and sodium hydroxide solutions over wide concentration and temperature ranges. Int J Hydrogen Energy 1985;10(4):233–43.

[32] LeRoy RL, Bowen CT, LeRoy DJ. The thermodynamics of aqueous water electrolysis. J Electrochem Soc 1980;127(9):1954–62.

[33] Bard AJ, Faulkner LR. Fundamentals and applications: electrochemical methods; 2001. p. 482.

[34] Rossmeisl J, Logadottir A, Nørskov JK. Electrolysis of water on (oxidized) metal surfaces. Chem Phys 2005;319(1–3):178–84.

[35] Ulleberg Ø. Modeling of advanced alkaline electrolyzers: a system simulation approach. Int J Hydrogen Energy 2003;28(1):21–33.

[36] Wendt H, Kreysa G. Electrochemical engineering: science and technology in chemical and other industries. Springer Science & Business Media; 1999.

[37] Amphlett JC, Baumert RM, Mann RF, Peppley BA, Roberge PR, Harris TJ. Performance modeling of the Ballard Mark IV solid polymer electrolyte fuel cell I. Mechanistic model development. J Electrochem Soc 1995;142:1–8.

[38] Larminie J, Dicks A, McDonald MS. Fuel cell systems explained. Vol. 2. Chichester, UK: J. Wiley; 2003.

[39] Ursua A, Sanchis P. Static–dynamic modelling of the electrical behaviour of a commercial advanced alkaline water electrolyser. Int J Hydrogen Energy 2012;37(24):18598–614.

[40] Thampan T, Malhotra S, Zhang J, Datta R. PEM fuel cell as a membrane reactor. Catal Today 2001;67(1–3):15–32.

[41] Lewis D, Glasstone S. Elements of physical chemistry. Macmillan; 1960.

[42] Piontelli R, Mazza B, Pedeferri P, Tognoni A. Ricerchesullosviluppoelettrodico di gas e suglieffettianomaliche lo accompagnano, Sviluppo da soluzioneacquose. Electrochim Metall 1967;2:257–87.

[43] Vogt H, Balzer RJ. The bubble coverage of gas-evolving electrodes in stagnant electrolytes. Electrochim Acta 2005;50(10):2073–9.

[44] Abdin Z, Webb CJ, MacA Gray E. Modelling and simulation of a proton exchange membrane (PEM) electrolyser cell. Int J Hydrogen Energy 2015;40(39):13243–57.

[45] Ni M, Leung MKH, Leung DYC. A modeling study on concentration overpotentials of a reversible solid oxide fuel cell. J Power Sources 2006;163(1):460–6.

[46] Hernández-Pacheco E, et al. A macro-level model for determining the performance characteristics of solid oxide fuel cells. J Power Sources 2004;138(1–2):174–86.

[47] Zhu H, Kee RJ. A general mathematical model for analyzing the performance of fuel-cell membrane-electrode assemblies. J Power Sources 2003;117(1–2):61–74.

[48] Reid RC, Prausnitz JM, Poling BE. The properties of gases and liquids; 1987.

[49] Chu LY, Park SH, Yamaguchi T, Nakao SI. Preparation of thermo-responsive core-shell microcapsules with a porous membrane and poly (N-isopropylacrylamide) gates. J Membr Sci 2001;192(1–2):27–39.

[50] Ni M, Leung MKH, Leung DYC. An electrochemical model of a solid oxide steam electrolyzer for hydrogen production. Chem Eng Technol 2006;29(5):636–42.

[51] Marangio F, Santarelli M, Cali M. Theoretical model and experimental analysis of a high pressure PEM water electrolyser for hydrogen production. Int J Hydrogen Energy 2009; 34(3):1143–58.

[52] Rahim AA, Tijani AS, Kamarudin SK, Hanapi S. An overview of polymer electrolyte membrane electrolyzer for hydrogen production: modeling and mass transport. J Power Sources 2016;309:56–65.
[53] Bruggeman DAG. Dielectric constant and conductivity of mixtures of isotropic materials. Ann Phys (Leipzig) 1935;24:636–79.
[54] Perron AL, Kiss LI, Poncsak S. Mathematical model to evaluate the ohmic resistance caused by the presence of a large number of bubbles in Hall-Héroult cells. J Appl Electrochem 2007;37(3):303–10.
[55] Hine F, Murakami K. Bubble effects on the solution IR drop in a vertical electrolyzer under free and forced convection. J Electrochem Soc 1980;127(2):292.
[56] Vermeiren PH, Leysen R, Beckers H, Moreels JP, Claes A. The influence of manufacturing parameters on the properties of macroporousZirfon® separators. J Porous Mater 2008; 15(3):259–64.
[57] Ursua A, Marroyo L, Gubía E, Gandía LM, Diéguez PM, Sanchis P. Influence of the power supply on the energy efficiency of an alkaline water electrolyser. Int J Hydrog Energy 2009;34(8):3221–33.
[58] Mori M, Mržljak T, Drobnič B, Sekavčnik M. Integral characteristics of hydrogen production in alkaline electrolysers. Strojniškivestnik J Mech Eng 2013;59(10):585–94.
[59] Górecki K, Górecki P, Zarębski J. Electrical model of the alkaline electrolyser dedicated for SPICE. Int J Circuit Theory Appl 2018;46(5):1044–54.
[60] Sánchez M, Amores E, Abad D, Rodríguez L, Clemente-Jul C. Aspen plus model of an alkaline electrolysis system for hydrogen production. Int J Hydrogen Energy 2020;45(7): 3916–29.
[61] Amores E, Rodríguez J, Carreras C. Influence of operation parameters in the modeling of alkaline water electrolyzers for hydrogen production. Int J Hydrogen Energy 2014;39(25): 13063–78.
[62] Chen Y, Mojica F, Li G, Chuang PYA. Experimental study and analytical modeling of an alkaline water electrolysis cell. Int J Energy Res 2017;41(14):2365–73.
[63] Ulleberg O. Stand-alone power systems for the future: optimal design, operation & control of solar-hydrogen energy systems. 225. Trondheim, Norvège: NTNU; 1998.
[64] Amores E, et al. Development of an operation strategy for hydrogen production using solar PV energy based on fluid dynamic aspects. Open Eng 2017;7(1):141–52.
[65] Wüthrich R, Comninellis C, Bleuler H. Bubble evolution on vertical electrodes under extreme current densities. Electrochim Acta 2005;50(25–26):5242–6.
[66] Fouad MG, Sedahmed GH. Mass transfer at horizontal gas-evolving electrodes. Electrochim Acta 1973;18(1):55–8.
[67] Elsner C, Coeuret F. Potential distribution along gas evolving electrodes. J Appl Electrochem 1985;15(4):567–74.
[68] Dau H, Limberg C, Reier T, Risch M, Roggan S, Strasser P. The mechanism of water oxidation: from electrolysis via homogeneous to biological catalysis. ChemCatChem 2010; 2(7):724–61.
[69] Martinson CA, Van Schoor G, Uren KR, Bessarabov D. Characterisation of a PEM electrolyser using the current interrupt method. Int J Hydrog Energy 2014;39(36): 20865–78.
[70] Barbir F. PEM electrolysis for production of hydrogen from renewable energy sources. Solar Energy 2005;78(5):661–9.

[71] Gößling S, Stypka S, Bahr M, Oberschachtsiek B, Heinzel A. Proton exchange membrane water electrolysis modeling for system simulation and degradation analysis. ChemieIngenieur Technik 2018;90(10):1437–42.
[72] Roy A, Watson S, Infield D. Comparison of electrical energy efficiency of atmospheric and high-pressure electrolysers. Int J Hydrogen Energy 2006;31(14):1964–79.
[73] Awasthi A, Scott K, Basu S. Dynamic modeling and simulation of a proton exchange membrane electrolyzer for hydrogen production. Int J Hydrogen Energy 2011;36(22): 14779–86.
[74] García-Valverde R, Espinosa N, Urbina A. Simple PEM water electrolyser model and experimental validation. Int J Hydrog Energy 2012;37(2):1927–38.
[75] Marangio F, Santarelli M, Cali M. Theoretical model and experimental analysis of a high pressure PEM water electrolyser for hydrogen production. Int J Hydrogen Energy 2009; 34(3):1143–58.
[76] Rahim AA, Tijani AS, Kamarudin SK, Hanapi S. An overview of polymer electrolyte membrane electrolyzer for hydrogen production: modeling and mass transport. J Power Sources 2016;309:56–65.
[77] Gottesfeld S, Zawodzinski T. Polymer electrolyte fuel cell model. J Electrochem Soc 1991;138:2334–42.
[78] Görgün H. Dynamic modelling of a proton exchange membrane (PEM) electrolyzer. Int J Hydrogen Energy 2006;31(1):29–38.
[79] Li X, Qu S, Yu H, Hou M, Shao Z, Yi B. Membrane water-flow rate in electrolyzer cells with a solid polymer electrolyte (SPE). J Power Sources 2009;190(2):534–7.
[80] Kopp M, Coleman D, Stiller C, Scheffer K, Aichinger J, ScheppatEnergiepark Mainz B. technical and economic analysis of the worldwide largest power-to-gas plant with PEM electrolysis. Int J Hydrog Energy 2017;42(19):13311–20.
[81] Yigit T, Selamet OF. Mathematical modeling and dynamic Simulink simulation of high-pressure PEM electrolyzer system. Int J Hydrog Energy 2016;41(32):13901–14.
[82] Mench MM. Fuel cell engines. John Wiley & Sons; 2008.
[83] Medina P, Santarelli M. Analysis of water transport in a high pressure PEM electrolyzer. Int J Hydrogen Energy 2010;35(11):5173–86.
[84] Ni M, Leung MKH, Leung DYC. Technological development of hydrogen production by solid oxide electrolyzer cell (SOEC). Int J Hydrogen Energy 2008;33(9):2337–54.
[85] Wendt H. Electrochemical hydrogen technologies: electrochemical production and combustion of hydrogen. Elsevier Science Ltd; 1990.
[86] Hino R, Haga K, Aita H, Sekita K. R&D on hydrogen production by high-temperature electrolysis of steam. Nucl Eng Des 2004;233(1–3):363–75.
[87] Riess I. On the single chamber solid oxide fuel cells. J Power Sources 2008;175(1):325–37.
[88] Hao Y, Shao Z, Mederos J, Lai W, Goodwin DG, Haile SM. Recent advances in single-chamber fuel-cells: experiment and modeling. Solid State Ion 2006;177(19–25):2013–21.
[89] Liu Y, Hashimoto SI, Nishino H, Takei K, Mori M, Suzuki T, Funahashi Y. Fabrication and characterization of micro-tubular cathode-supported SOFC for intermediate temperature operation. J Power Sources 2007;174(1):95–102.
[90] Ni M, Leung MKH, Leung DYC. Energy and exergy analysis of hydrogen production by solid oxide steam electrolyzer plant. Int J Hydrogen Energy 2007;32(18):4648–60.
[91] Doenitz W, Schmidberger R, Steinheil E, Streicher R. Hydrogen production by high temperature electrolysis of water vapour. Int J Hydrogen Energy 1980;5(1):55–63.
[92] Salzano FJ, Skaperdas G, Mezzina A. Water vapor electrolysis at high temperature: systems considerations and benefits. Int J Hydrogen Energy 1985;10(12):801–9.

[93] Appleby AJ. Fuel cell handbook; 1988.
[94] Chan SH, Xia ZT. Polarization effects in electrolyte/electrode-supported solid oxide fuel cells. J Appl Electrochem 2002;32(3):339–47.
[95] Choi P, Bessarabov DG, Datta R. A simple model for solid polymer electrolyte (SPE) water electrolysis. Solid State Ion 2004;175(1–4):535–9.
[96] Ferguson JR, Fiard JM, Herbin R. Three-dimensional numerical simulation for various geometries of solid oxide fuel cells. J Power Sources 1996;58(2):109–22.
[97] Im-orb K, Visitdumrongkul N, Saebea D, Patcharavorachot Y, Arpornwichanop A. Flowsheet-based model and exergy analysis of solid oxide electrolysis cells for clean hydrogen production. J Clean Prod 2018;170:1–13.
[98] Ni M, Leung MKH, Leung DYC. Parametric study of solid oxide steam electrolyzer for hydrogen production. Int J Hydrogen Energy 2007;32(13):2305–13.
[99] Laurencin J, Kane D, Delette G, Deseure J, Lefebvre-Joud F. Modelling of solid oxide steam electrolyser: impact of the operating conditions on hydrogen production. J Power Sources 2011;196(4):2080–93.
[100] Cai Q, Luna-Ortiz E, Adjiman CS, Brandon NP. The effects of operating conditions on the performance of a solid oxide steam electrolyser: a model-based study. Fuel Cells 2010;10(6):1114–28.
[101] Cai Q, Adjiman CS, Brandon NP. Optimal control strategies for hydrogen production when coupling solid oxide electrolysers with intermittent renewable energies. J Power Sources 2014;268:212–24.
[102] Martinez D, Zamora R. MATLAB simscape model of an alkaline electrolyser and its simulation with a directly coupled PV module for Auckland solar irradiance profile. Int J Renew Energy Res 2018;8(1):552–60.
[103] Burhan M, Chua KJE, Ng KC. Sunlight to hydrogen conversion: design optimization and energy management of concentrated photovoltaic (CPV-hydrogen) system using micro genetic algorithm. Energy 2016;99:115–28.
[104] Đukić A, Firak M. Hydrogen production using alkaline electrolyzer and photovoltaic (PV) module. Int J Hydrog Energy 2011;36(13):7799–806.
[105] Kordkandy MN, Arash A, Kordkandy MN. Hydrogen gas production in a stand-alone wind farm. Eng Technol Appl Sci Res 2017;7(2):1444–9.
[106] Minh HQ, Frederic N, Essounbouli N, Abdelaziz H. A new MPPT method for stand-alone wind energy conversion system. In: 2012 2nd international symposium on environment friendly energies and applications. IEEE; 2012.

Chapter 7

Techno-economic evaluation methodology for hydrogen energy systems

Venkatakrishnan Balasubramanian[a,c], Nawshad Haque[c], Suresh Bhargava[b], Srinivasan Madapusi[a,b], and Rajarathinam Parthasarathy[a,b]
[a]*Chemical Engineering, School of Engineering, City Campus, RMIT University, Melbourne, VIC, Australia,* [b]*Centre for Advanced Materials and Industrial Chemistry, RMIT University, Melbourne, VIC, Australia,* [c]*CSIRO Energy, Clayton, VIC, Australia*

1 Introduction

Techno-economic evaluation (TEE) is effective throughout the entire life cycle of technology growth. A techno-economic model generally includes a process model, capital cost model, operating cost model, and cash flow analysis. The process model is built using mass and energy balances which results in a stream table. The data from the stream table is the foundation for equipment sizing calculations which are then used for estimating capital and operating costs [1]. A well-developed TEE model would benefit scientists, industries, and investors for developing an optimized technology and assist in identifying key technical and economic inputs and changes the course of direction at the early stage of the research and development. Scientists and innovators can use TEE to assess the economic viability and efficiency during the evaluation of novel concepts. At the laboratory scale, TEE can be used to identify the most efficient process parameters and potential scale-up opportunities for pilot scale. Engineers can use TEE during the process design to compare the financial effects of various factors and optimize the plant setup via equipment sizing. TEE includes information from both these phases of development and provides a foundation for rational and strategic decision-making for the investors. An effective TEE model is represented in Fig. 1.

2 Process model of a hydrogen energy system

Building a TEE for the hydrogen energy system involves numerous challenges. Hydrogen has significant potential in addressing the quest for alternatives to

Bioenergy Resources and Technologies. https://doi.org/10.1016/B978-0-12-822525-7.00016-0
Copyright © 2021 Elsevier Inc. All rights reserved.

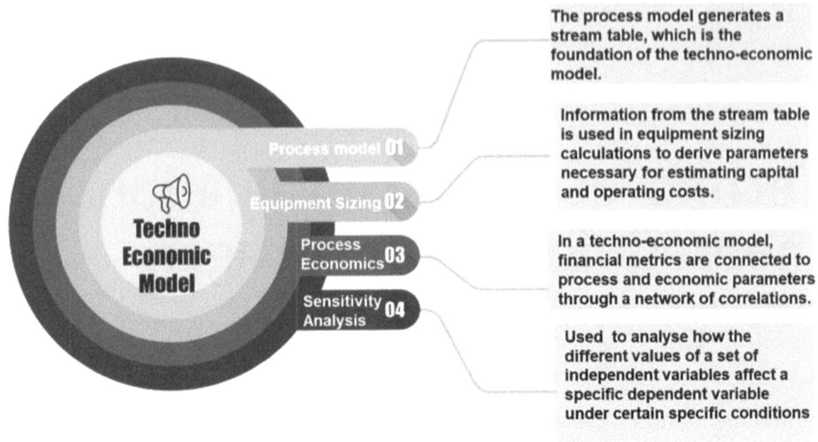

FIG. 1 Techno-economic evaluation (TEE) model.

carbon-intensive fossil fuels as it effectively acts as a carrier of energy for several applications including transportation fuel and energy storage. While hydrogen can be produced from various sources, the production of 100% renewable hydrogen faces a lot of economic, production, and transportation challenges. In addition, a comparison between renewable and nonrenewable hydrogen based on economics would give a clear picture in the interest of global energy sustainability. The entire process can be assessed in the following steps namely hydrogen source, production method, carrier route, storage, transportation, and energy utilization for renewable hydrogen production. One such scenario is shown in Fig. 2 and this system is used for the TEE.

A process model involves mass and energy balances of the system that enables us to estimate the detailed material and energy flows in the process [2]. It comprises a process flow diagram (PFD), process conditions, and calculations. The proposed hydrogen energy system involves several different processes namely desalination, electrolysis, NH_3 formation, NH_3 cracking, air separation unit, hydrogen and ammonia storage, transportation, and the usage of solar and wind energy. The combined process model leads to a process stream table and stream data from the stream table are used for equipment sizing calculations which in turn is linked to the economic analysis of the entire system [2]. The process model can be developed using Microsoft Excel in which the "iterative calculation" tools such as "Goal Seek," and "Solver" can be used [3]. Goal Seek is used for mass and energy balances especially when the recycle stream is involved and the Solver is used for sensitivity analysis. Moreover, in the assumed scenario, each process is considered as a unit operation for a detailed analysis. The PFD for each process is built using the details of each equipment and the chemicals used. A specialized process simulation software such as Aspen Plus can also be used to develop a PFD and carry out mass and

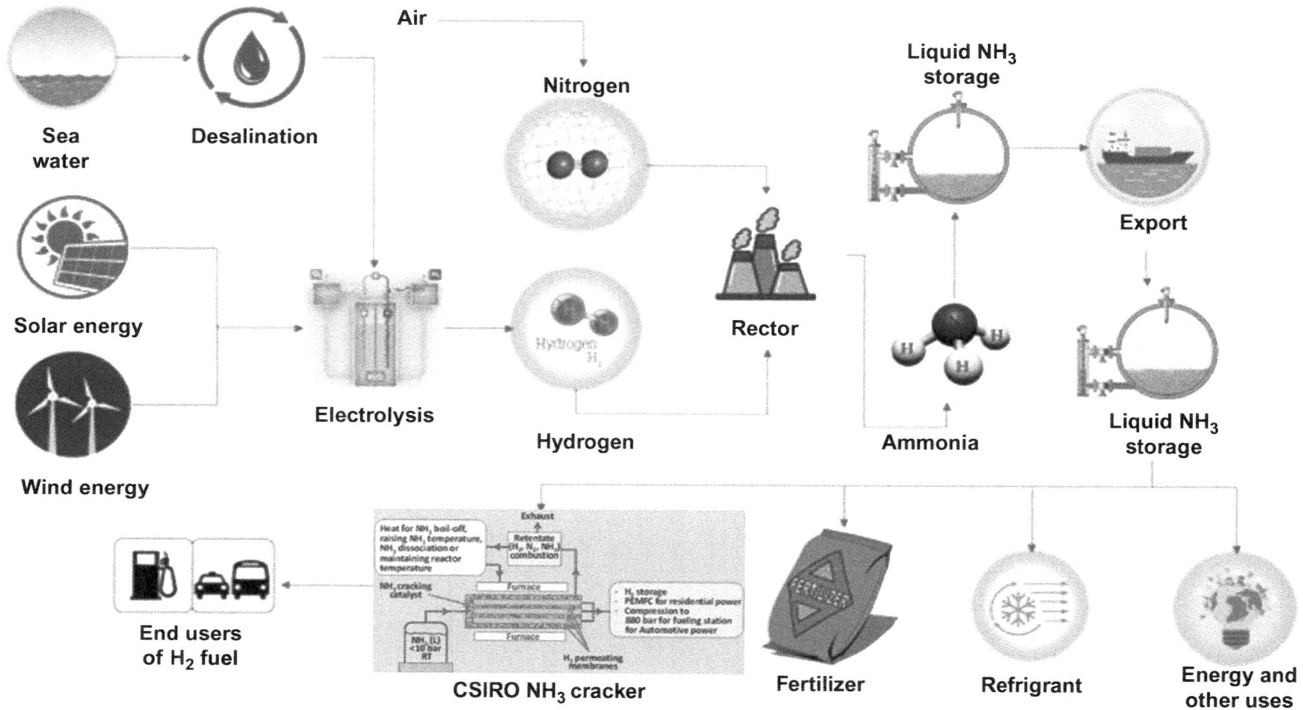

FIG. 2 Renewable hydrogen supply chain via NH_3 scenario.

energy balances of the processes resulting in a stream table and data. ASPEN Economic Analyzer module can be used to generate techno-economic indicators of the proposed process.

To know the boundary conditions for the overall process model, the block diagram for the entire system is built and the mass balance calculations for the entire process are done to determine the current flow rate of the feed and product for various unit operations. Then, the capital cost (CAPEX) is calculated using the exponential method and the operating cost (OPEX) is calculated from the unit price of the inputs for each process. The exponential method is a preliminary analysis done based on the location factor and CAPEX of an existing plant. The cash flow analysis is then carried out using the CAPEX and OPEX values. CAPEX and OPEX estimations are briefly discussed in the following sections. A typical process block diagram used for the mass and energy balance calculations and economic analysis is shown in Fig. 3.

Once the boundary conditions are known, a detailed analysis is done by listing all the process equipment, and carrying out the detailed mass and energy balances, equipment sizing, CAPEX and OPEX calculations, and the cash flow analysis. The detailed mass and energy balance and equipment sizing calculations contribute to both CAPEX and OPEX calculations. While doing the energy balance calculations, "pinch analysis" [3] can be done for energy optimization. Pinch analysis is a well-established technique used for optimizing energy usage in a process by considering operating conditions, energy supply methods, and recycling energy between streams (e.g., energy recovered from a hot stream can be supplied to a cold stream). The proposed process model is compared with the existing process (conventional), which is also simulated using both exponential and detailed methods. The conventional model will help the user to identify the overall cost and the energy-rich and deficit zones thus leading to a better-optimized process. A simple methodology involved in process modeling and economic analysis is shown in Fig. 4.

3 The capital cost model

Capital cost is an upfront expense which is calculated at the beginning of a project. The range varies from thousands to billions of dollars depending on the plant size. The capital costs are the investment on which return is produced from the sales of the product.

In techno-economic analysis, capital costs are usually best estimated using a list of equipment included in the entire process and applying historical-data-derived multiplying factors to purchased equipment costs. Major equipment purchase costs can be estimated using correlations from literature or vendor quotes or literature values for equipment, using equipment size and other factors. The accuracy of these methods is typically in the range of −30% to +50% [4,5]. Capital cost estimation gives a better understanding of the capital requirement of a new project when techno-economic modeling is undertaken. It also

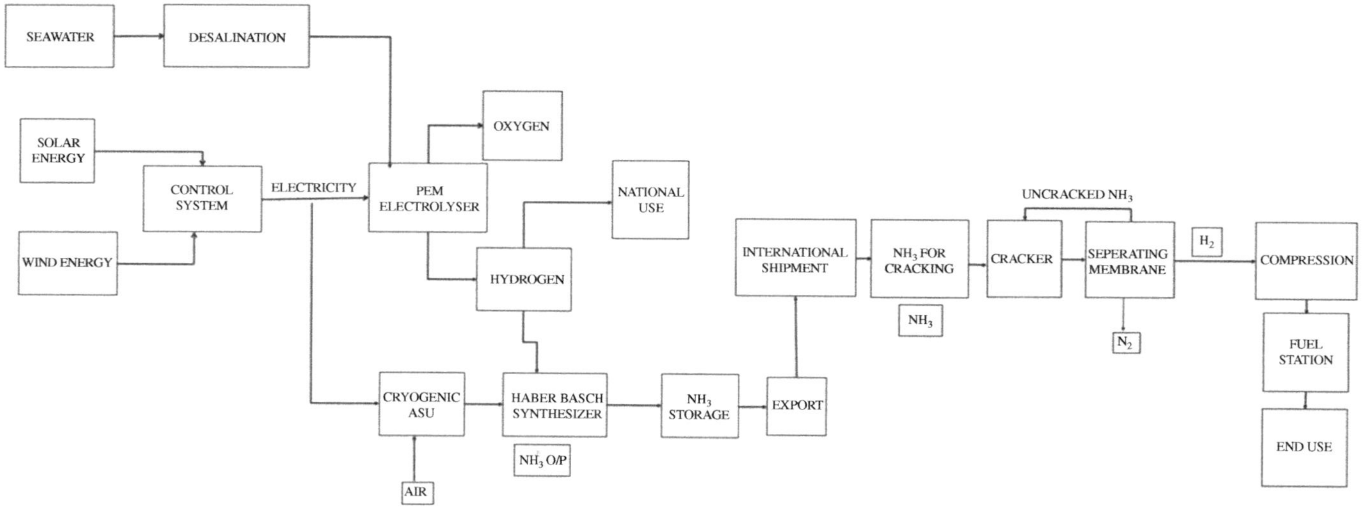

FIG. 3 Process block diagram—H_2 via renewable NH_3 carrier.

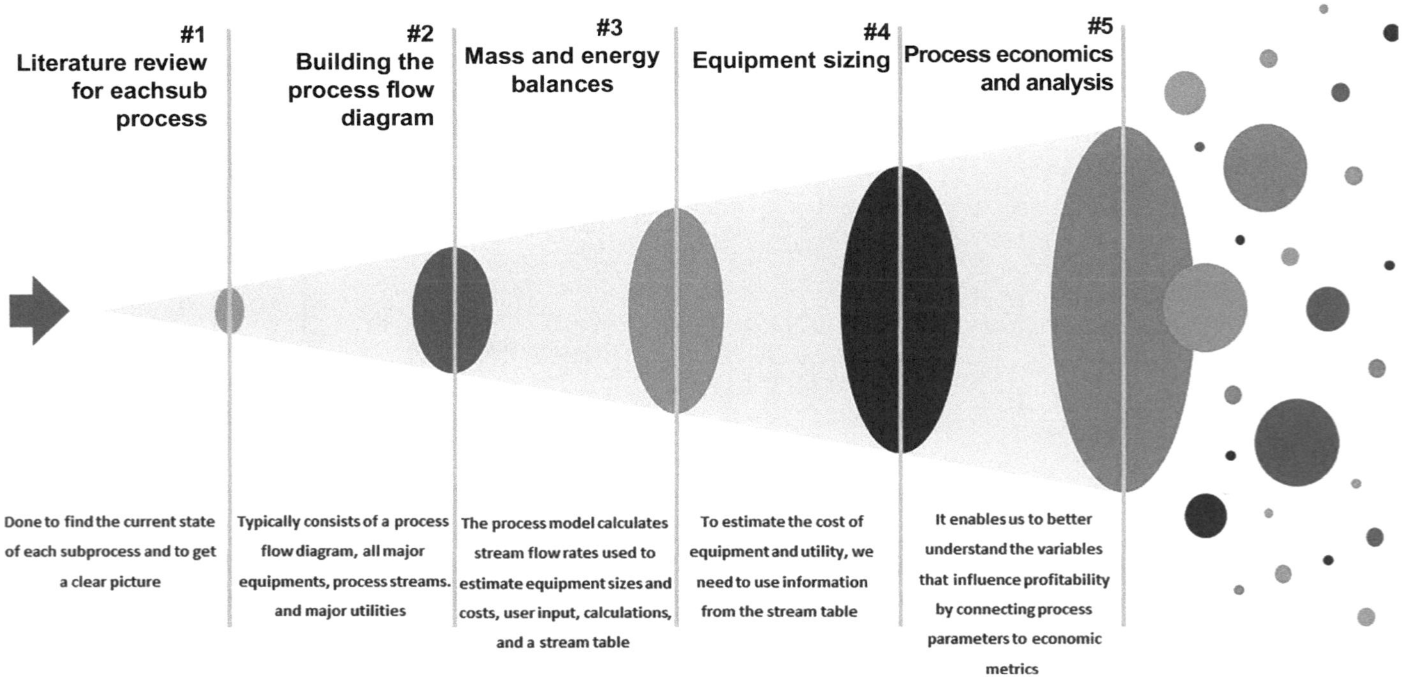

FIG. 4 Methodology for building TEE model.

provides an estimation of the capital that is to be spent during the operating life of an existing plant due to the modification and retrofitting required to accommodate changes in demand, feedstock, product quality, operating cost. A reliable estimate of the capital cost is also mandatory to decide between the modification of a brownfield project or to set up a greenfield project, planning funds allocation to an approved project, and manage their subsequent expenditure. Capital cost includes land, fixed capital for plant, start-up, and working capital.

$$\text{Total capital expenditure} = \text{land capital} + \text{fixed capital} + \text{start-up capital} + \text{working capital} \quad (1)$$

Capital cost can be estimated using five different methods namely the exponential method, study method, preliminary method, and definitive and detailed method [5]. The accuracy and the cost of capital cost estimation are mentioned in Table 1. This chapter focuses on the exponential and study methods.

3.1 Exponential method

This method is used in the preliminary stage of a project. The data for the plant capacity is acquired from the overall or generic mass balance calculation and the data for the reference plant is taken from the literature. The main objective is to find the fixed capital for the new project. It is an approximate method and is applicable over a limited range. It involves extrapolation from an existing plant cost data and has a potential accuracy of ±30%–50% [4,5].

The general cost ratio used in the exponential method is shown in the following equation:

$$\frac{C_P}{C_R} = \left(\frac{K_P}{K_R}\right)^e \left(\frac{CI_P}{CI_R}\right) L \quad (2)$$

TABLE 1 Capital cost estimate classification.

Classification of capital cost estimation methods	Accuracy range	% of project expenditure
Detailed method	±2 to 5%	5 to 10%
Definitive method	±5 to 15%	1 to 3%
Preliminary method	±10 to 25%	0.4 to 0.8%
Study method	±20 to 30%	0.1 to 0.2%
Exponential method	±30 to 50%	0 to 0.1%

where C_P and C_R are investments, K_P and K_R are plant capacities, CI_P and CI_R are chemical engineering plant cost index values for the years of construction of the proposed (P) and reference plants (R), respectively, and e is the cost exponent ranging 0.8 to 0.9 for parallel stream processes and 0.5 to 0.6 for a single stream process.

For plants with mixed parallel/single stream, the e value may be calculated as follows [4,5]:

$$e = \frac{\sum_{i=1}^{j} c_i e_i}{\sum_{i=1}^{j} c_i} \tag{3}$$

where j is the number of equipment items (excluding multiples of a given item) and c is the purchased cost of main equipment items of base capacity.

For plants with mixed parallel/single streams, the e value may also be calculated according to capacity distribution:

$$e = Q_{\text{mix}}\, e_{\text{mix}} + Q_{\text{single}}\, e_{\text{single}}$$

where Q_{mix} and Q_{single} are the capacities of mixed-parallel and single-stream facilities, respectively, and e_{mix} and e_{single} are the exponents for mixed-parallel and single-stream facilities, respectively. L in Eq. (2) is the location factor [4].

$$L = \frac{\text{Cost of a plant built in the proposed location}}{\text{Cost of an identical plant built in the reference plant location}} \tag{4}$$

3.2 Study method

For the detailed analysis [5], each subprocess is assessed based on unit operations involved. The bare module cost of each equipment is calculated, and it is the sum of direct and indirect costs related to equipment purchase and installation. In this approach, depending on the market trend, the purchased cost C_p for each piece of equipment is found from graphs based on the following conditions:

(i) **Base condition**—If the equipment is fabricated using carbon steel and operated at ambient pressure condition.
(ii) **Non-Base condition**—If the equipment is fabricated using different materials and operated at different conditions.

The bare module equipment cost , also known as the installed cost, is calculated using the following equation [5]:

$$C^{o}_{\text{BMC}} = C_P F^{o}_{\text{BM}} \tag{5}$$

where C^{o}_{BMC} is the bare module equipment cost, C_P is the purchased equipment cost, and F^{o}_{BM} is the bare module cost factor.

The bare module cost factor F^{o}_{BM} is determined using two other factors, namely, the construction material factor F_M and the pressure factor F_P. These factors are determined for the construction material and the operating pressure for the equipment using appropriate graphs and tables [5] for non-base conditions while for the base conditions they are taken as unity.

The installed plant cost is the sum of the bare module costs of all the equipment in the process plant. Contingency and contractor fee costs are also added to the total bare module cost. Contingency covers unforeseen circumstances and it is usually assumed as 15% of the total bare module cost. Contractors' fees are taken as about 3% of the total bare module cost. Thus, the fixed capital cost C_{TMC} [5] (also known as total module cost) is evaluated as

$$C_{TMC} = 1.18 \sum_{i=1}^{n} C^{o}_{BM,i} \tag{6}$$

where n is the total number of process equipment.

For a greenfield project, the fixed capital cost should include costs for land, area development, auxiliary buildings, off-sites, and utilities. Under base case conditions, the auxiliary cost is assumed to be 50% of the bare module costs. The sum of the total module cost and the 50% of bare module cost is known as the grass roots cost C_{GRC} [5] which is evaluated as

$$C_{GRC} = C_{TMC} + 0.50 \sum_{i=1}^{n} C^{o}_{BMC,i} \tag{7}$$

3.3 Estimation of working capital

Working capital [4,5] is the start-up capital for a plant before the plant starts to generate revenues from production. It includes raw materials inventory, finished products inventory, materials in process, balance between debtors and creditors, and cash. It is recovered without loss at the end of the plant's operating life and thus, it cannot be depreciated. The working capital is directly proportional to production and is expressed as a percentage of the annual sales revenue. Working capital is assumed to be in the range from 10% to 20% of the fixed capital investment or 30% of the annual sales. These assumptions are however approximate.

Working capital (I_{wc}) for the detailed analysis is estimated using the following relationship [4]:

$$I_{wc} = S_{RM} + S_{FP} + S_{MIP} + A_R - A_P + \text{Cash} \tag{8}$$

where S_{RM} is raw material stocks, S_{FP} is finished product stocks, S_{MIP} is materials in progress inventory, A_R is accounts receivable from product sales (debtors), A_P is accounts payable due to the purchase of raw materials and utilities (creditors). Generally, a raw materials stockpile of 1 month of the plant production capacity is recommended although this can vary depending on the industry and goods produced.

4 Operating costs

Operating costs are per diem expenditures encompassing the maintenance and administration of the plant. It is categorized as variable, fixed, and general costs. It varies every year depending on the operating routines, production rate as per market demand, maintenance, and plant modifications. Variable operating costs depend on the production rate which is associated with inventory, and fixed operating costs are associated with the maintenance and administration related issues. General operating costs are not directly related to operation, but still need to be considered for a complete economic analysis.

The four major variable operating costs are for raw materials, operating labor, utilities, and waste treatment. These are estimated based on the process stream table, equipment sizing, and market prices. Fixed and general operating costs are then calculated by applying multiplication factors to the variable operating costs and capital costs [4,5].

The economic feasibility of a process industry project depends strongly on the manufacturing cost of the proposed product. The nonmanufacturing costs include distribution and selling, R&D and corporate administration as shown in Fig 5. The overall costs, which are known as "total operating costs" or "total production cost," is the sum of direct manufacturing costs and nonmanufacturing costs. An understanding of various components that contribute to these costs is essential for process engineers involved in designing and operating a process plant. Estimation of operating costs is an important step before the economic profitability of the proposed project is evaluated using cash flow analysis. In general, operating costs are estimated on annual basis. The various components contributing to the operating cost are shown in Fig. 5.

4.1 Simplified model for operating cost estimation

Manufacturing/production costs are estimated by considering four major components namely raw materials, utilities, personnel and personnel-related costs, and capital-related costs. A simplified model [4] based on these four components is given by

$$C = \sum \mathrm{Rr} + \sum \mathrm{Eg} + \frac{(\mathrm{Mm})}{\mathrm{QU}} + \frac{(\mathrm{kI})}{\mathrm{QU}} \tag{9}$$

where C is the manufacturing/production cost, R is the consumption of raw material consumption, E is the consumption of utilities, M/QU is the number

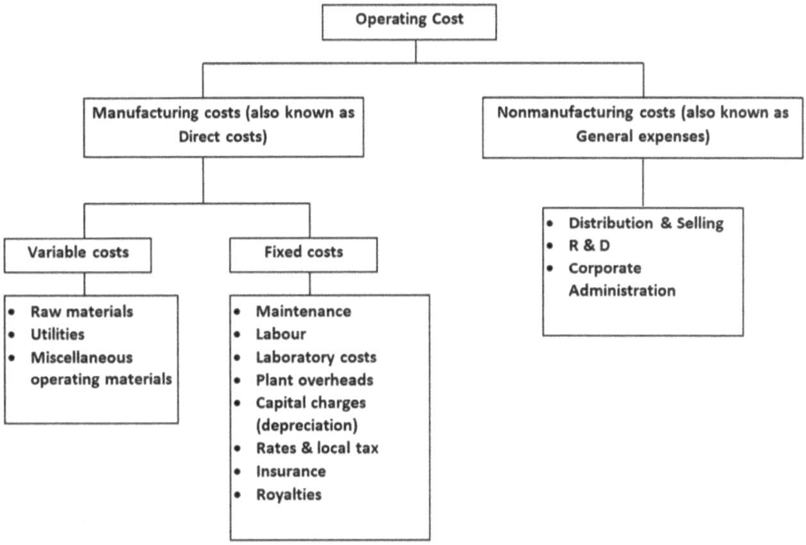

FIG. 5 Operating cost flow diagram.

of employees per tonne of product, I/QU is the fixed capital per tonne of product, r is the unit cost of raw material, g is the unit cost of utility, m is the average salary per employee, k is the factor dependent on fixed capital, Q is the production capacity, U is the capacity utilization, and QU is the annual production.

4.2 Variable manufacturing cost estimation

(a) Raw material costs:

Raw material consumption is identified from the PFD. Annual requirements of the raw materials are obtained by multiplying the consumption rate with the operating hours per year. In raw material consumption calculation, losses due to operating inefficiencies must also be accounted for. Data on raw materials prices are available in journals, reference books, and cost databases. Transportation cost is also included to raw material cost.

(b) Utilities costs:

Steam, fuel, electricity, cooling and process water, compressed air, nitrogen, coolant, and effluent treatment costs are included in the utilities cost for the current scenario. Consumption rate of utilities must be known before their costs are estimated. Consumption rate may be read from PFD or determined from the mass and energy balance calculations.

(c) Estimation of utilities consumption:

Utilities like steam and cooling water consumption rate can be estimated from process flowsheets. Allowances for losses and contingencies must be made in utilities consumption. Energy (heat) loss could be up to 20% of theoretical calculations. Cooling water loss can range up to 30% of theoretical calculations. Electricity loss can range up to 20% of estimated value.

Units costs are obtained from public utility company or from adjacent plant or from company-owned facility. Unit cost is a function of plant location.

4.3 Fixed manufacturing cost estimation

(a) Personnel costs (Labor costs):

It includes the costs for the following process labor (or) operating labor, maintenance labor, supervision labor, and contract labor. Among these, estimation of "process labor" cost is important because other personnel and related costs are estimated based on it.

(b) Process Labor:

It is the human resource required for plant operation. It is an estimate from the number of shift and day personnel required. Usually, three shifts and four shift crews are required for a 40-h working week. The process labor cost includes wages and allowances. The process labor cost is usually about 15% of the total operating cost. The process labor sometimes, is estimated from unit operations and the number of large equipment used within a unit operation.

(c) Plant (or) works overheads:

They include the costs of staff such as engineers, scientists, accountants, senior managers, clerks, secretarial personnel, and specialist personnel. They also include the costs of vehicles, cafeteria, and workshop facilities. They are estimated as a percentage of the process labor or a percentage of (process + maintenance) labor. They range from 50% to 150% of the process labor or from 20% to 100% of the process and maintenance labor. They are occasionally estimated as a percentage of the fixed capital.

(d) Depreciation:

Depreciation is regarded as an operating cost and included in the fixed manufacturing costs. Straight-line method of depreciation is calculated as follows [4]:

$$\text{Annual depreciation allowance} = \frac{\text{Fixed capital investment} - \text{Salvage value}}{\text{Economic life in years}} \tag{10}$$

However, companies may specify or use alternate methods of depreciation that is permissible in the taxation laws of the country of plant operation to minimize tax liability in a given year.

(e) Distribution cost

Distribution cost is directly related to the end user but it is included in the production cost. It includes transport cost, product handling cost and storage cost. It is a function of the product type, scale of transportation, hazards in transport, and transportation distance.

(f) Selling expenses

Selling expenses include the costs of distribution and merchandising of the products and it depends on the product quality and market demand. Selling expenses range from 5% to 25% of the manufacturing costs or from 2% to 20% of the sales revenue.

(g) Research and development costs

These costs depend on the plant and its capacity, ownership structure and policy, product nature and demand, innovation, and technology used. The R&D costs are higher for industries that are undergoing transient technological change involving products with short life cycles or processes such as NH_3 cracking using higher risk new technologies such as "membrane technology" and change of catalysts. The R&D costs include lab-scale & pilot-scale costs and allowance for skilled personnel. The R&D costs range from 0% to 5% of manufacturing costs or 0%–6% of the sales revenue. Regardless of whether a company has R&D or not, royalty for intellectual property fees or patent or license purchase fees need to be included within the operating cost.

(h) Corporate administration

It includes costs due to management, public relations, finance, and corporate planning. It usually ranges from 2% to 6% of the manufacturing costs. Alternatively, it may be calculated as varying from 1% to 4% of the fixed capital.

5 Cash flow analysis

Cash flow is similar to a material balance or an energy balance [1]. A material balance gives an account of what happens to the raw material entering a process whereas an energy balance gives an account of energy supplied to and generated by a process. Similarly, cash flow represents the inflow and outflow of all the money spent and generated within the plant. For a process plant, this ranges from the construction of the plant until it has been shut down, decommissioned, and sold. The cash flow analysis should show the exchange of funds on how it is spent and received every year in the project life since the value of money is time dependent.

Cash flow analysis of a chemical plant is usually done after the completion of the preliminary plant design. The initial expenditure taken into account is the land and the fixed capital, followed by the working capital and start-up expenses. The costs for manufacturing the product and maintenance are added in the subsequent stages. Income is generated after the commissioning and continues until the plant is shut down or sold. When the plant service life is over, the working capital is recovered, the plant is sold for its salvage value, and the land is sold. Time value of money is not considered at this stage. It will be incorporated in the analysis after various profitability measures and indices have been discussed [1,6].

The relationships and equations presented in the following section [1] are used to calculate the annual cash flow.

Annual total expenses (operating costs) A_T is given by the equation

$$A_T = A_M + A_N \tag{11}$$

where A_M and A_N are annual manufacturing expenses and annual general (non-manufacturing) expenses, respectively.

Annual manufacturing expenses A_M is related to its two components, annual fixed (A_{FE}) and variable costs (A_{VE}), as

$$A_M = A_{FE} + A_{VE} \tag{12}$$

Substituting for A_M from Eq. (12) in Eq. (11)

$$A_T = A_N + A_{FE} + A_{VE} \tag{13}$$

Let A_S is the total annual sales revenue and A_{CI} is the annual cash income (equivalent to annual profit).

A_S and A_{CI} are related by the following equation:

$$A_{CI} = A_S - A_T \tag{14}$$

Annual net cash income A_{NCI} is now calculated as.

$$A_{NCI} = A_{CI} - A_{IT} \tag{15}$$

where A_{IT} is the annual income tax amount which is calculated as follows:

$$A_{IT} = (A_{CI} - A_D - A_A)\,t \tag{16}$$

where A_D is the annual depreciation, A_A is the other allowances, t is the fractional tax rate.

Net annual cash flow (after tax) A_{CF} can be calculated as

$$A_{CF} = A_{NCI} - A_{TC} \tag{17}$$

where A_{TC} is annual capital expenditure (sum of fixed and working capital).

Schematically annual cash flow can be represented as shown in Fig. 6.

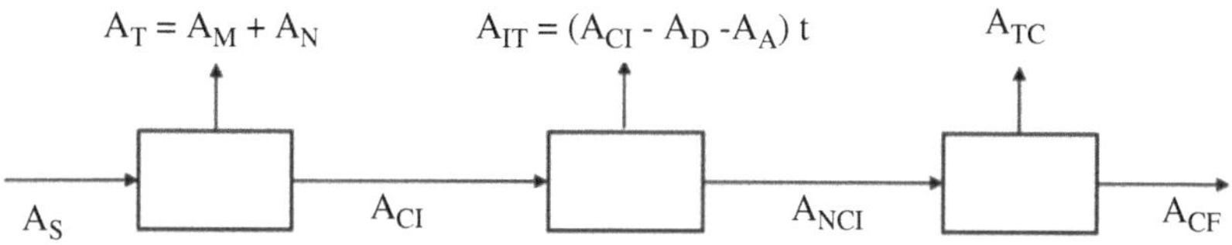

FIG. 6 Schematic representation of annual cash flow.

5.1 Profitability evaluation

Profitability is a measure of the financial gain that is generated from a given investment. A company desires to generate the highest profit it can from the money it invests. It may want to compare the prospective earnings it may get from different ventures using profitability measures. Two of the fundamental measures are the rate of return on investment (RORI) and payout period.

5.2 Rate of return on investment

$$\text{RORI}\% = \frac{\text{Average yearly profit}}{\text{Total investment}} \times 100 \tag{18}$$

Eq. (18) is used to calculate the percentage return that investors may expect to earn on their money [4]. Yearly profit could be before or after tax. In calculating the yearly profit, depreciation is considered an expense. The average annual profit may be calculated by dividing the cumulative profit over the life of the plant by the operating life. In calculating the yearly profit, the plant is assumed to be operating at its full capacity [1,4].

5.3 Payout period

It is the time taken take to recover all expenses involved in a project, except for land and working capital. Land and working capital are not included because these items could always be sold or recovered for their original value. Payout period in cash flow diagram would be the time which elapses from the start of the project (0 year) to the breakeven point where the escalating part of the cumulative cash flow curve passes the zero cash position line. Payout period measures the time needed for the cumulative project investment and the expenditure to be exactly balanced by the cumulative income. Prior to the breakeven point, the project is in debt; beyond it, it shows profit. For an attractive project, the payout period has to be shorter. Factors that influence the payout period are the fixed capital, operating expenses, and the selling price. The project manager can control the fixed and operating expenses to a certain extent but the selling price entirely depends on the market forecast. Thus, the reduction of the payout period depends mainly on the demand and nature of the product.

The profitability indices considered so far, payout period and the RORI, take no account of the cash flow pattern during a project, and the time value of money. The profitability indices that take these factors into account are net present value (NPV) and internal rate of return (IRR). To evaluate the profitability of alternative projects or investments, the projects are compared at their equivalent values at the same instant of time. The time of evaluation is usually the present time (zero time).

5.4 Net present value (NPV)

NPV is a measure of the net cash gain generated at the required rate of investment by a project. It is the sum of the present values of the cash flow for each year in the plant life. The time of evaluation is usually the start of the project. The NPV concept uses an approach to bring the values of future cash flows to present value by taking the interest-earning capability of the money [1,4].

The present value P at year zero of a cash flow C_t in year t at an annual discount rate of i is expressed [4,5] as

$$P = \frac{C_t}{(1+i)^t} = C_t\left[\frac{1}{(1+i)^t}\right] \tag{19}$$

The group $\frac{1}{(1+i)^t}$ is known as the discount factor f_D because it is used to discount a future value to its equivalent present value at a discount rate of i. For a complete project, the net present value (NPV) is the sum of the individual present values of all the yearly cash flows [5]:

$$\text{NPV} = \sum P = \sum_{t=0}^{t=n} \frac{C_t}{(1+i)^t} \tag{20}$$

where n is the total number of years over which the cash flow occurs.

NPV can also be described as

$$\text{NPV} = \text{total income} - \text{total expenditure} \tag{21}$$

where all income and expenditure are expressed as present values.

5.5 Internal rate of return (IRR)

It is defined as the discount rate i at which NPV is equal to zero. The following equation represents IRR [1,4].

$$\text{If NPV} = \sum P = \sum_{t=0}^{t=n} \frac{C_t}{(1+i)^t} = 0, \text{then } i = \text{IRR}. \tag{22}$$

IRR is to be determined iteratively. The higher the value of IRR for a project, the more attractive it is. For a profitable project, IRR earned by a project should be greater than the interest rate for the current investment and vice versa. Therefore, the IRR's minimum acceptable value is the cost of capital or market interest rate.

IRR for a project is calculated by trial-and-error calculations. A value of i is selected and the corresponding net present value is calculated. If NPV is positive, the value of i is too small. If NPV is negative, the value of i is too large. Graphical interpolation between two or three values gives i for NPV = 0 with sufficient accuracy. IRR can also be determined using the Goal-Seek tool in MS Excel spreadsheet. Mathematically it is possible to obtain more than one value of IRR for the same cash flow values. However, usually only one root will be real and has practical significance.

5.6 Comparison of NPV and IRR in economic evaluation

NPV and IRR serve different purposes while evaluating the profitability. NPV (Fig. 7) is a measure of the cash profit (in terms of the present value) generated by the project after recovering the initial investment. Thus, it is an absolute determinant of the entire project's profit. The IRR for a project represents the RORI. It is a measure of the efficiency with which the capital is employed and indicates the earning power of the project investment. The NPV requires an estimate of the cost of capital whereas IRR does not require the cost of capital. For a single project considered in isolation, NPV and IRR both lead to the same conclusion if a project has a positive NPV. It is bound to have an IRR above the discount rate used in NPV calculation. If a company has potentially more acceptable projects and limited capital to finance them, projects are usually selected in a way to maximizes the total NPV. The ranking of the projects is done in the decreasing order of IRR. If a choice is to be made between alternative proposals of the same project, then the alternative with the greatest NPV

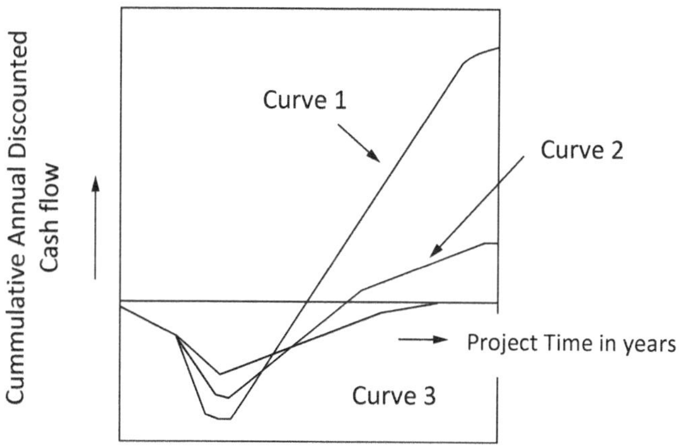

FIG. 7 Genaral net present value (NPV) diagram.

should be selected. Note that although IRR is widely used in economic evaluation of projects, it is more limited than NPV in its applications. NPVs are additive when dealing with multiple project selection whereas IRRs are not additive.

5.7 Economic evaluation of projects: Effect of inflation

Inflation can be defined as the erosion of the purchasing power of money with time [4]. For buying the same equipment, more money is required as time passes (or) a constant number of money units purchase less as time passes.

To deal with the inflation, cash flows occurring at different times must be compared using money having the same purchasing power. Cash flows expressed in terms of constant purchasing power are called "real cash flows" or "real money." Cash flows expressed in terms of changing purchasing power are called "nominal cash flows" or "nominal money."

Let C_{nt} is the cash flow at the end of year t in nominal money of year t. C_{rt} is the cash flow at the end of year t in real money.

C_{nt} and C_{rt} can be related as follows:

$$C_{\mathrm{nt}} = C_{\mathrm{rt}}(1+f)^t \tag{23}$$

where f is the yearly inflation rate which is assumed constant in this case and t is the year in which nominal cash flow occurs.

If yearly inflation f is changing,

$$C_{\mathrm{rt}} = \frac{C_{\mathrm{nt}}}{(1+f_1)(1+f_2)\dots(1+f_t)} \tag{24}$$

where $f_1, f_2, \dots$ are the inflation rates for years 1, 2...

5.8 Discounting the C_{rt} to get the present value

The present value of money units with constant purchasing power (real money) can be calculated as

$$\mathrm{PV} = \frac{C_{rt}}{(1+i)^t} \tag{25}$$

where PV is the present value, C_{rt} is the cash flow in year t in real money, and i is the discount rate.

6 Sensitivity analysis

Sensitivity analysis is a detailed economic evaluation of a project incorporating the uncertainty and consequent economic risk, testing the robustness, and identifying optimal parameter settings in the model quantitatively. Sensitivity analysis examines the relationships between input and output parameters by varying the key variables. It gives a better understanding of the process dependencies, opportunities, uncertainties, and threats which affect the model. Two types are presented below.

6.1 Univariate optimization

Univariate optimization is a simple method to analyze the impact of a single variable on the model [2]. A single input parameter is varied over a range and the robustness is tested based on the output parameter which is plotted. This is quick and easy and is suitable for dependent variables. It gives a clear picture of how each variable affects the entire system and makes a pathway for optimization. An example is shown in Fig. 8.

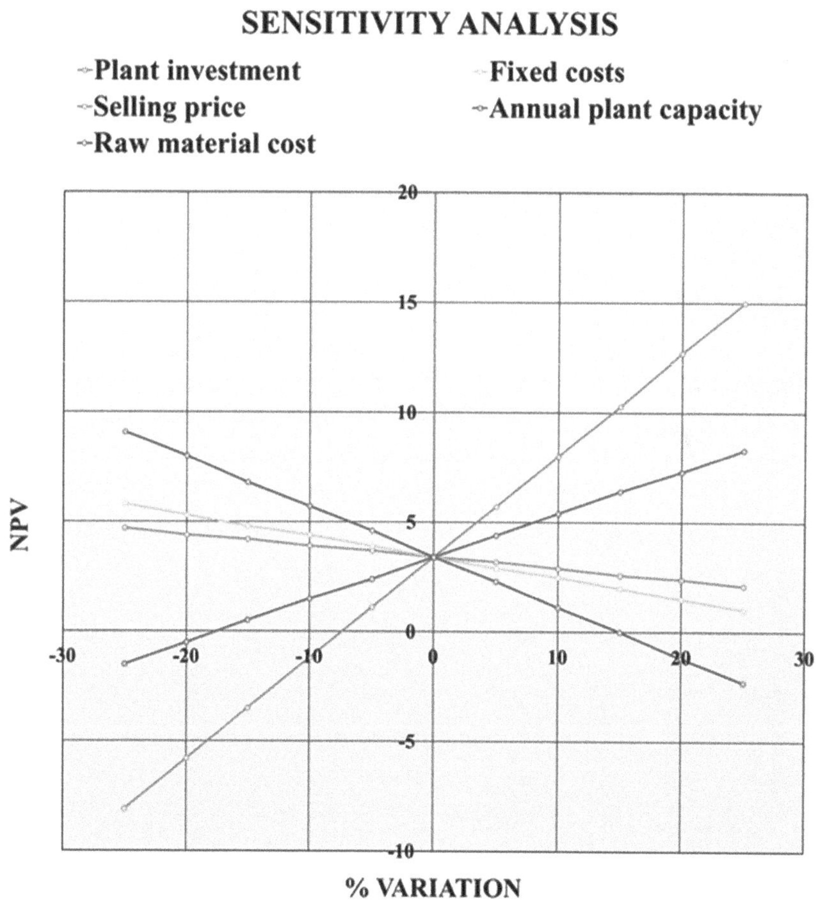

FIG. 8 Sensitivity analysis.

6.2 Monte Carlo analysis

Monte Carlo analysis is a probabilistic method which quantifies the magnitude and nature of uncertainty in our estimates. It is a detailed iterative method that uses probability distributions as both input and output thus, examining the impact of them on the model. This information is important for both project management and project evaluation. It varies all the parameters in the model simultaneously and calculates the expected output by analysis of variance of the input parameters [2,7]. An example is shown in Fig. 9.

7 Results and discussion

There are number of technologies for hydrogen energy production which needs evaluation in the same way to compare and understand the key contributing inputs within a technology. Thus, a generic tool and template sets are required to evaluate hydrogen energy systems. The H_2 export market for Australia will be complex, involving several geographic locations in Australia (e.g., in Western Australia, South Australia, Queensland). Several primary energy resources (renewable vs fossil + carbon capture and storage) are possible. Several methods for desalination of seawater and several electrolysis technologies can be considered. Several proposed hydrogen vectors (liquid H_2 vs liquid NH_3) at a range of scales can be considered. There are other ways of transporting H_2 such as in the form of methylcyclohexane, hydrocarbons, or metal hydride. For liquid NH_3 route, H_2 needs reconversion by cracking and decomposition of NH_3.

Among the various types of hydrogen separating inorganic membranes, hydrogen-permeable metal membranes made of palladium (Pd) and its alloys are the most widely studied due to their high hydrogen permeability but they

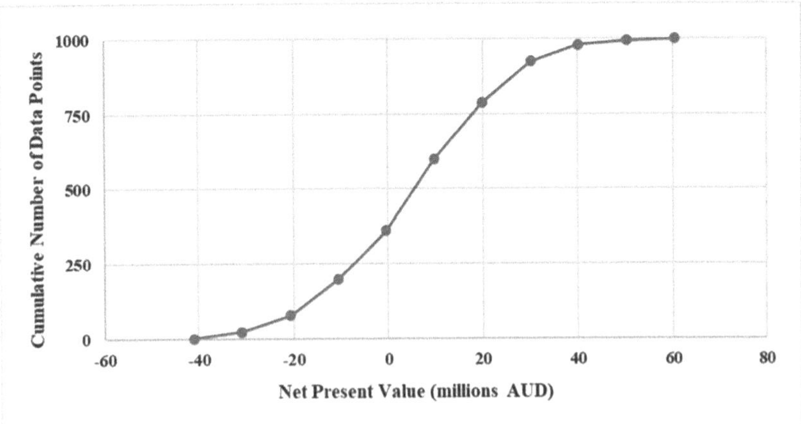

FIG. 9 Monte Carlo analysis example.

are still at the developing stage. For efficient technologies to be utilized for commercial production and the modification of existing technologies, both exponential and detailed methods of the capital cost estimations are highly useful. The former gives the boundary in which optimization is achieved in terms of both process and economics, and the latter gives a detailed estimation for the same. The challenge in the overall techno-economic analysis is to interlink the process model, equipment sizing, cost models, and the sensitivity analysis so that system is generic, and consolidates the input and results on a single spreadsheet, thus enabling an efficient evaluation of the process economics.

The following case study is an example of the polymer electrolyte membrane (PEM) electrolysis process (from Fig. 3) for hydrogen production from desalinated water. The economic analysis is carried out for 57 t/day of hydrogen production using the exponential method for capital cost estimation, literature data for the operating cost, and the cash flow analysis to find the payout period and the net present value (NPV). Assumptions considered and the output of the techno-economic evaluation for this case study are shown in Table 2.

An example output table for the process in Fig. 3 is shown in Table 3. This table is needed to compare various process options and choose one that will have economic feasibility while meeting key technical requirements. The inputs and outputs shown in the table were obtained using the assumptions and calculation procedure similar to those shown in Table 2 for PEM electrolysis for hydrogen production.

8 Summary

The global energy situation is going through a profound transition due to the requirements for the reduction of CO_2 emissions. Renewable hydrogen production is a promising way to solve the energy crisis amidst this transition. Renewable electricity can be used to produce hydrogen, which in turn can be used as a low emission source. Australia has a unique advantage in utilizing its renewable energy sources for renewable hydrogen production. Australia represents one of the world's best solar energy opportunities owing to its maximum solar radiation as 58 million PJ. Moreover, its actual use of solar energy is minimal at present amounting to 0.1% of its use [9]. In order to maintain renewable energy usage throughout the year, electricity generated from wind energy provides an alternative solution. Due to the increased usage of wind energy-based electricity, Australia has demonstrated a successful transition to a low carbon economy [10]. These resources can be effectively utilized for the generation of renewable electricity via hydrogen pathway.

The hydrogen sector is maturing due to the years of experience from R&D innovations and it is moving into the large-scale and real-time applications. In this new phase, there are a lots of challenges to overcome for a large-scale production and transportation of renewable hydrogen. TEE assesses and optimizes the process by considering the market forecast and profitability, and provides a

TABLE 2 Output table for hydrogen energy system shown in Fig. 3.

Parameter	Calculation method	Value	Unit
Fixed capital	Eq. (2)	160	Million
Working capital	15% of the fixed capital	24	Million
Operating life	Assumption	30	years
Production capacity (H_2)	Calculated using overall mass balance	19,950	t.p.a.
Operating costs	2% of fixed capital	4000	(\$/t)
% of fixed capital spent in the 1st year	Assumption	60	
% of fixed capital spent in the second year	Assumption	40	
First year sales volume (% of capacity)	Assumption	90	
Second year sales volume (% of capacity)	Assumption	95	
Third and other years sales volume (% of capacity)	Assumption	100	
Discount rate	Assumption	10%	
Selling price of the product	Obtained from literature [8]	7.5	(\$/kg)
Depreciation rate factor	Straight line method	3%	
Tax rate	Assumption	30%	
Net present value (NPV)	Obtained using Eq. (20)	29.1	\$ Million
Payout period for best case	Obtained from the cash flow diagram	7	years

reliable pathway by accounting for all the risks involved in a large-scale production. Hence, TEE is done from the starting stage by forming a boundary and then working towards the output via detailed estimation. The TEE discussed in this work provides a generic methodological tool involving different template sets. It can be used to evaluate different designs of hydrogen energy systems.

TABLE 3 Output table for hydrogen energy system shown in Fig. 3.

Process	Production capacity (tonnes per annum)	CAPEX ($millions)	Annual OPEX ($millions/year)	NPV ($million)	Payout period (years)	
					Best case	*Worst case*
Desalination	310,000	23	0.9	0.7	10	13
PEM electrolysis—H_2 production	19,950	160	80	29.1	7	12
Cryogenic ASU	93,146	59	23.3	24.4	5	11
NH_3 production	113,183	520	39	21.4	10	19
NH_3 cracking—H_2 production	19,950	633	16.8	30.1	12	18
Solar and wind energy	210MW	N/A	12	N/A	N/A	N/A
Total		1395	172	105.7		

Note: These values are shown as an example only. Operating life of each process is assumed to be 30years. The production capacity for each process is used in the mass balance of the each process, CAPEX is calculated using the exponential method, OPEX is calculated using the unit price of the product, NPV is from the cash flow analysis, and pay out period for best and worst case is done for different cash inflows.

References

[1] Allen DH. Economic evaluation of projects, a guide. Rugby: Institution of Chemical Engineers; 1991.

[2] Burk C. Techno-economic modeling for new technology development [Internet]. Aiche.org; 2020. Available from: https://www.aiche.org/resources/publications/cep/2018/january/techno-economic-modeling-new-technology-development. [Accessed 17 April 2020].

[3] Moran S. An applied guide to process and plant design. San Diego: Elsevier; 2019.

[4] Brennan D. Process industry economics. Institution of Chemical Engineers: Rugby, England; 1998.

[5] Turton R, Shaeiwitz J, Bhattacharyya D, Whiting W. Analysis, synthesis, and design of chemical processes. Boston: Prentice Hall; 2018.

[6] Peters M, Timmerhaus K, West R. Plant design and economics for chemical engineers. Boston: McGraw-Hill; 2006.

[7] Dienemann P. Estimating cost uncertainty using Monte Carlo techniques. Santa Monica, CA: Rand Corp.; 1966.

[8] Ball M, Weeda M. The hydrogen economy—vision or reality? In: Compendium of hydrogen energy, vol. 4. Woodhead Publishing Series in Energy; 2016. p. 237–66.

[9] Aera.ga.gov.au. Australian energy resources assessment (AERA). [online]. Available at: https://aera.ga.gov.au/#!/solar; 2020. [Accessed 14 August 2020].

[10] Aera.ga.gov.au. Australian energy resources assessment (AERA). [online]. Available at: https://aera.ga.gov.au/#!/wind; 2020. [Accessed 14 August 2020].

Chapter 8

Hydrogen production from municipal solid waste (MSW) for cleaner environment

Apinya Chanthakett[a], M.T. Arif[a], Mohammad Masud Kamal Khan[b], and Aman M.T. Oo[a]

[a]*School of Engineering, Deakin University, Waurn Ponds, VIC, Australia,* [b]*School of Engineering and Technology, CQUniversity, Melbourne, VIC, Australia*

1 Introduction

Hydrogen (H_2) is one of the most abundant elements and always found with other substances such as water, hydrocarbon, or alcohol. It is also found in biomass sourced from plants or animals. It is a high-quality energy carrier that can be produced from various sources such as natural gas, coal, biomass, or water. The required processing energy can be sourced from renewable energy such as wind or solar. Currently, H_2 is receiving significant attention due to the growing concern of climate change and emission from conventional energy sources. Therefore, hydrogen has a huge potential to decarbonize the energy market.

The application of hydrogen is predominantly in the area of ammonia production, which has major application of fertilizer production. It is also heavily used in oil refineries to clean crude oil. Considering the advantages of hydrogen, and to meet the clean energy target, the application of hydrogen is increasing in other areas such as; electricity generation, energy storage, and as fuel in electric vehicles. Hydrogen can help tackle various critical energy challenges. It can work as a buffer for energy from renewable energy sources such as solar and wind and can reduce all intermittent challenges that renewable energy develops and can better support the power network. Therefore, the demand for hydrogen is growing and it will be a critical factor for the clean energy market. Fig. 1 shows the global demand for hydrogen which has grown more than 3 times since 1975 and it is increasing [1].

Demand for hydrogen is increasing with the increased population in the world; also, hydrogen has the potential to reduce greenhouse gas (GHG) emission from the energy sector. However, at this stage almost the entire generation

Bioenergy Resources and Technologies. https://doi.org/10.1016/B978-0-12-822525-7.00017-2
Copyright © 2021 Elsevier Inc. All rights reserved.

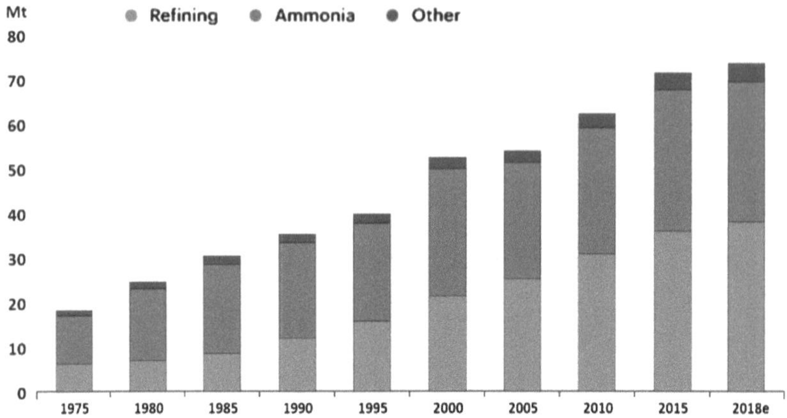

FIG. 1 Global demand of pure hydrogen from 1975 to 2018 [1].

of H_2 is based on fossil fuels (76% from natural gas and 23% from coal) by using steam methane reforming [2] and releasing CO_2 as by-product. Therefore, the production of H_2 contributes to CO_2 emission of approximately 830 million tons per year, which is equivalent to CO_2 emission of the United Kingdom and Indonesia combined [1]. On the other hand, with the population growth the waste disposal, particularly municipal solid waste (MSW) which is a complex mix of hydrocarbon-based biodegradable components with high calorific values is also increasing. MSW generated by households constitutes 11% of all global methane (CH_4) emissions which is considered the third largest source of CH_4 emission [3]. A number of waste management strategies are in practice, which also includes waste to energy conversion application to generate syngas that is a mix of carbon monoxide (CO) and hydrogen (H_2). By generating energy from MSW, it helps to reduce direct emission from waste while producing clean hydrogen. This chapter explains the gasification process for the efficient conversion of hydrogen from waste.

The chapter illustrates the energy potential of hydrogen as explained in Section 2. Applications of hydrogen are listed in Section 3, various sources of hydrogen are explained in Section 4. Gasification process of hydrogen production from MSW is explained in Section 5, benefits of hydrogen production from MSW are explained in Section 6. The future prospects of waste to hydrogen production are detailed in Section 7 and finally concluding remarks are presented in Section 8.

2 Energy potential of hydrogen

Hydrogen is the lightest chemical substance, which is highly flammable with no odor and color. Hydrogen is one of the most accessible resources on the

Earth but it is typically not found in its pure form, and it usually exists as a combination of other elements. Therefore, hydrogen can be extracted from different hydrocarbon compounds such as methane (CH_4), propane (C_3H_8), ethane (C_2H_6), butane (C_4H_{10}), heptane (C_7H_{16}), etc. [4]. Hydrogen is also found in fossil fuel including gasoline, kerosene, diesel, natural gas, and coal. Furthermore, hydrogen fuel can be generated from alternative resources, for instance; municipal solid waste (MSW), water, agriculture waste, and organic materials.

The global liquid fuel energy market is 4 times larger than the global electricity market. The major source of these forms of energy is based on hydrocarbon conversion, which is the major cause of emission. However, H_2 is carbon free and can be available in liquid or gas form. Therefore, H_2 has great potential to be an alternative energy source that can reduce the emission. Hydrogen has high energy value compared with several other fuels. Its higher heating value is approximately 142 MJ/kg and a lower heating value (LHV) of 120 MJ/kg or 33 kWh/kg [5]. Whereas natural gas and gasoline has a heating value of around 45–51 and 44–47 MJ/kg, respectively. Moreover, emission released from these fuels are measured by carbon intensity or emission intensity (CO_2-eq per MJ of LHV) where hydrogen contributes zero carbon emission to the environment. Other sources release carbon emissions in the range of 61.0–93.6 gCO_2-eq per MJ at LHV. The heating potential and emission rate of different fuels are summarized in Table 1 [6].

TABLE 1 Heating values of different fuels [6].

Fuels	Higher heating value (HHV): MJ/kg	Lower heating value (LHV): MJ/kg	Carbon intensity:kg CO_2-eq/GJ	Ref.
Hydrogen	141.9	120.1	0	[7]
Gasoline/ Petrol	47.00	44.50	67.4	[9]
Diesel	44.80	42.50	69.9	[9]
Kerosene	46.20	43.10	68.9	[9]
Natural gas	50.84	45.86	61.0	[9]
Liquefied petroleum gas (LPG)	50.15	46.61	60.2	[9]
Coal	23.97	22.73	93.6	[9]
Crude Oil	45.54	42.67	69.6	[9]

3 Application of hydrogen

Hydrogen is used in various applications including electricity generation, industrial process, manufacturing ammonia, metalworking, flat glass production, electronics industry, and transportation industry. Hydrogen is also used in the refinery to refine crude oil. Around 55% of the total H_2 produced around the world is used for ammonia production, 25% in refineries, 10% for methanol production, and remaining 10% for various other promising applications [14]. The main advantage of using hydrogen as a fuel is its emission-free property.

Hydrogen helps in energy transition, suitable for versatile applications and most importantly, it can contribute to decarbonize existing economies. Some of the promising applications are in electricity generation, energy transportation, energy storage, and heating application as shown in Fig. 2.

- **Ammonia production**

  Around 55% of worldwide hydrogen production is used for the synthesis of ammonia (NH_3). Almost 90% of ammonia is used goes into fertilizer production. Ammonia is also used in refrigeration plants.

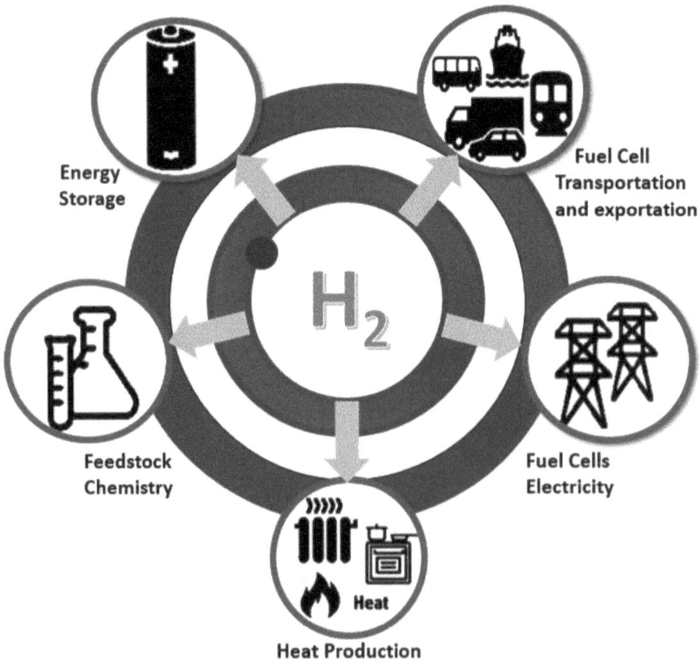

FIG. 2 The applications of hydrogen.

- **Fuel production**
  Hydrogen is used in oil refineries to refine crude oil into gasoline and diesel and also to remove contaminants from these fuels. Hydrogen is also used to produce methanol (CH_3OH) which can be used as fuel.
- **Electricity generation**
  A fuel cell is an electrochemical device that combines H_2 and O_2 (oxygen) to produce electricity which can be used as distributed power, backup power in stand-alone, or grid-connected application [15]. Hydrogen fuel cell is emission free and generates water and heat as by-product. In 2018, Japan tested hydrogen turbine to produce 1100 kW of electricity and 2800 kW of heat using hydrogen [16].
- **Transportation**
  Hydrogen fuel cell electric vehicles (FCEVs) are a good candidate to decarbonize the road transport sector. FCEV technology uses PEM (proton exchange membrane) fuel cell stack and hydrogen storage. Currently, many car manufacturers are integrated in fuel cell systems, for example, B-Class 300 Mercedes, Hyundai Santa Fe FCEV, Honda FCX clarity, Toyota Mirai, and so on. In addition, hydrogen is currently in use for maritime systems, aircraft, trucks, trains, and buses. The key challenge of FCEVs is the limited availability of hydrogen refueling station [17].
- **Heating application**
  Hydrogen can be used in space heating applications instead of natural gas. In this process, hydrogen is burnt with oxygen to produce heat and water as by-product. Therefore, no carbon emission occurs in this process and hydrogen can be used for decarbonizing the gas grid [18]. However, the invisible flame of hydrogen is a risk issue, hydrogen, and oxygen or air (with nitrogen) combustion could contribute to nitrogen oxide (NOx) emissions [19].

- **Chemical feedstock**
  Hydrogen compound can combine with other components to produce new products and there are a number of chemical industries utilizing hydrogen, for methanol generation, ammonia product, petrochemical product, metallurgical process, lipid of solid, and oil hydrogenation [20].
- **Energy storage and transport**
  Hydrogen is a good quality energy carrier and can be stored and used in gaseous or liquid form. Moreover, hydrogen can be combined with nitrogen to form ammonia (NH_3) to store energy from renewable energy sources and can be transported to different locations and extracted back to hydrogen for different applications.

All the above applications and many more new applications of hydrogen help to decarbonize various greenhouse gas emitting sectors [14]. Moreover, research is growing to improve the efficiency of hydrogen production from various energy sources, and hydrogen transportation for use in various forms including existing gas network.

4 Different sources and processes of hydrogen production

Hydrogen has a growing demand and different countries have adopted policies to support investment in developing H_2 technology from various sources and applications. There are a number of target applications for H_2, however, the most prominent applications are in electric vehicle (H_2 fuel cell), electricity generation, building heating application, and chemical industries. The principal sources of hydrogen fuel can be found in both fossil fuel and nonfossil fuel. Natural gas (CH_4) is currently the primary source of H_2 production accounting for around 75% of annual global H_2 production of around 70 million tons [1]. The cost of H_2 production is highly influenced by the fuel (natural gas) cost and that is around 45%–75% of production cost. The countries that are rich with natural gas such as the Middle East, Russia, and North America produce the lowest cost H_2 but gas importers such as Japan, Korea, China, and India have very high cost of H_2 production as shown in Fig. 3 [1]. Fig. 3 shows the H_2 production cost (capital and operational cost and natural gas cost) with and without CCUS (carbon capture utilization and storage) in some countries in 2018.

Coal is the second common source for H_2 production. Due to the high cost, water electrolysis contributes only around 0.1% of global dedicated H_2 production today. With the declining cost of renewable electricity, particularly from solar PV and wind, the growing demand for electrolytic H_2 is increasing, however, it is still expensive compared to natural gas and coal. Considering the location-specific renewable energy potential, the installation of water electrolyzer close to the renewable energy source could reduce the global H_2 production cost.

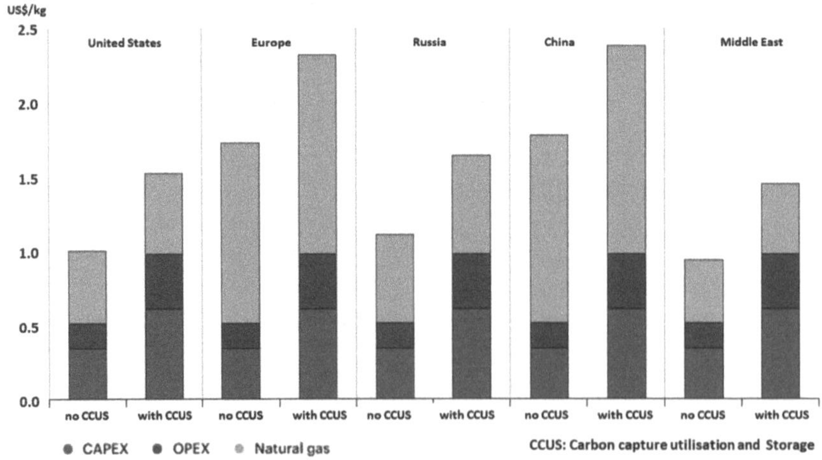

FIG. 3 Hydrogen production cost using natural gas, in 2018 [21].

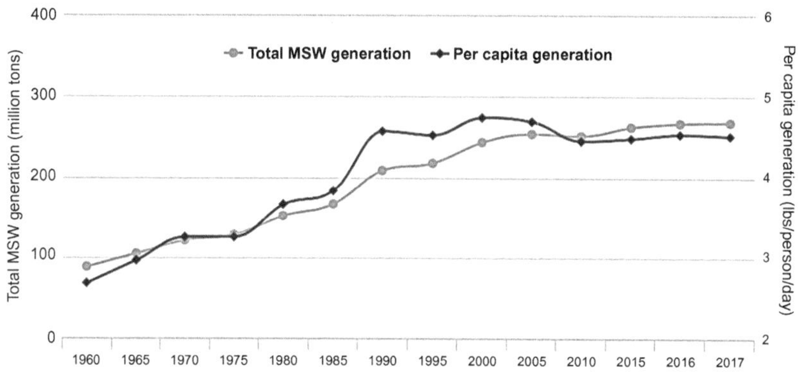

FIG. 4 MSW generation rate in USA from 1960 to 2017 [21].

Considering the limitations of the common sources, the waste, particularly municipal solid waste (MSW) can be a potential source for H_2 production. As the world population grows, the volume of waste is also increasing. Fig. 4 shows the MSW generation trend in the United States. As seen, in 2017 a total of 267.8 million tons of waste was generated and per-capita waste generation was 4.51 pounds/person/day [21]. A similar trend of waste generation persists in all other countries.

Different processing technologies are used for H_2 production from various sources. A short overview of different methods for H_2 production are listed below considering fossil fuel and nonfossil fuel sources.

4.1 Hydrogen production from fossil fuels

- **Steam Reforming method**

  This is the most commonly used method due to high efficiency as well as low operation cost. The common sources used in this method are coal, food, natural gas/methane, methanol, oxygenated, and light hydrocarbons. The steam-methane reforming operates at high temperature (700–1000°C) where methane reacts with steam to produce hydrogen (H_2), carbon monoxide (CO), and small amount of carbon dioxide (CO_2) [22].
- **Autothermal Reforming method**

  This method is an integration of steam methane reforming with partial oxidation to produce synthesis gas or syngas. The advantage of this method is that no or little additional heat is required due to its endothermic and exothermic reactions, therefore, this system has an economic benefit. It is widely used to convert higher hydrocarbon sources especially gasoline [23].
- **Partial Oxidation method**

  In this method, natural gas/methane or heavy hydrocarbon fuels are mixed with limited oxygen in an exothermic process. In this process, feed is partially combusted with oxygen and heat is released. This method consumes less external heat and as heat is released during oxidation the operating temperature reaches 1300–1500°C without catalyst and produces hydrogen, carbon monoxide (CO), and relatively low amounts of other compounds. In this method, methane, propane, gasoline, coal, and natural gas can be used as feed [24].

- **Plasma Reforming method**

  Plasma reforming has been developed from the conventional reforming method; however, it has many benefits including high conversion efficiency, accepts a broad range of feeds/fuels, fast response, or short residence time but its energy consumption is high [25]. Its operating condition includes very high temperature (more than 2000°C) and pressure. In this method, diesel, biomass, gasoline, oil, jet fuel, and natural gas can be used as fuel.

- **Water-Gas Shift method**

  Water-gas shift (WGS) reactor is used in many industries to increase hydrogen production. In this method, a reaction occurs between carbon monoxide (CO) and water (H_2O) to form carbon dioxide (CO_2) and hydrogen (H_2). It is a catalytic process at high and low temperatures and the common catalysts are molybdenum, copper, zinc, chromium, iron, and nickel. The process has high performance to purify hydrogen yield when operating with the Fischer-Tropsch process [26]. This method uses a mixture of product gas from coal, waste, biomass, etc. as feed or fuel [27].

4.2 Hydrogen production from renewable sources

- **Aqueous-Phase Reforming method**

  This method produces hydrogen from carbohydrates especially biomass derived from glycerol, sugar, and sugar alcohols. The process of aqueous phase is the reaction of hydrocarbons and water that runs at lower temperature between 220°C and 270°C and high pressure to decompose the material. Platinum (Pt) and nickel (Ni)-based catalysts are used to improve the performance of hydrogen production. This method has accomplished hydrogen conversion efficiency of more than 55%. However, the challenge of this process is the activation of catalysts in short time that requires a large-scale reactor [2].

- **Photoelectrolysis method**

  Photoelectrolysis is an electrolysis process by using sunlight as an energy source and splitting water into hydrogen (H_2) and oxygen (O_2). A photoelectrode gets energy from solar radiation and splits water molecules. Unlike the fossil fuel process, photoelectrolysis is not a combustion process so there is no emission effect. The efficiency of photoelectrolysis system is highly dependent on the electrode material and the most commonly used electrode materials are TiO_3, Fe_2O_3, and WO_3 [28].

- **Electrolysis method**
  Electrolysis is an advanced technology of splitting water into hydrogen and oxygen by passing electric current into cathode and anode in water. Furthermore, sulfuric acid is widely used as electrolyte and platinum base as electrodes. The process is environmentally friendly due to no emission release from the method. Water electrolysis approach has been developed into three main mature technologies: PEM (proton exchange membrane), SOE (solid oxide electrolyzer), and electrolyte alkaline [2].
- **Thermochemical Water Splitting**
  Thermochemical water splitting or thermolysis is involved with water and heat used in a single reaction to dissociate hydrogen (H_2) and oxygen (O_2). Furthermore, this process uses high-temperature heat (500–2000°C) to drive a series of chemical reactions to produce H_2. In this thermochemical cycle, a variety of chemical materials such as sulfur-iodine, zinc-zinc oxide, nickel-manganese, and copper-chlorine are used. The efficiency of the process is around 50% but it depends on the used technique as well [29].
- **Dark and Photo fermentation**
  Hydrogen fuel can be generated in biological methods from organic substances. Dark fermentation is an anaerobic process where in the absence of oxygen, organic substances can be decomposed by a number of microorganisms into hydrogen (H_2), alcohol, and volatile fatty acid. This process produces relatively low hydrogen. In the case of photo fermentative, it exploits sunlight and two main types of bacteria (green sulfur (GS) and purple non-sulfur (PNS) bacteria) to decay organic substances to produce hydrogen. However, the conversion efficiency of photo fermentation is very low around 1%–5%. In order to increase hydrogen productivity, a combination of dark and photo fermentation is recommended [30].
- **Pyrolysis method**
  Pyrolysis is a process of chemically decomposing organic material into liquid and synthesis gas. The pyrolysis process operates under endothermic reaction with a lack of oxygen and at a temperature between 500 and 550°C [31]. Pyrolysis transforms MSW into gas that consists of synthesis gas (carbon monoxide: CO and hydrogen: H_2), CH_4, N_2, H_2O, and other hydrocarbons [32]. Furthermore, hydrogen output is impacted by the influence of diverse condition parameters such as temperature, reactor, waste category, and so on.
- **Gasification method**
  Gasification is a thermochemical process that converts any carbon-based raw material or feed into fuel gas or synthesis gas also known as syngas. A primary feed/feedstock of gasification is biomass, natural gas, coal, municipal solid waste (MSW), and a variety of solid fuel. The process runs in a confined and controlled chamber known as a gasifier where endothermic and exothermic reactions convert the feed into syngas and residue or

slag. The syngas is mainly composed of carbon monoxide (CO) and hydrogen (H_2). The syngas can be further converted to H_2 and carbon dioxide (CO_2) by the steam reaction. The CO_2 can be efficiently captured from syngas to minimize emissions. The process in gasifier operates at different temperature between 500°C and 1000°C at different conditions (both in the presence and the absence of oxygen) [33]. As the gasification runs endothermic and exothermic reactions within the closed chamber, therefore it has less energy demand during its operation. Gasification is an advanced technology that produces high energy efficiency and low greenhouse gas emission [34].

Table 2 summarizes various hydrogen generation applications and their conversion efficiency with various feedstock. Many hydrogen technologies are used commercially in numerous industries including steam reforming, partial oxidation, gasification, and electrolysis. Hydrogen from fossil-based systems would contribute to higher emission leading to adverse impacts on climate change. Currently, the majority of hydrogen production is from natural gas and coal that uses steam reforming method and contributes to greenhouse gas emissions [2]. On the other hand, renewable source-based systems are becoming popular due to low emission during hydrogen production. Also, new applications can be

TABLE 2 The summary of hydrogen production technologies.

Technology	Feedstock	Technological efficiency	Reference
Fossil fuel			
Steam reforming	Coal, Food, natural gas, methane, methanol, oxygenated and light hydrocarbons	70%–85%	[2]
Auto thermal reforming	Hydrocarbons	60%–65%	[2]
Partial oxidation	Methane, propane, gasoline, coal, natural gas and other hydrocarbons	60%–65%	[35]
Plasma reforming	Diesel, biomass, gasoline, oil, jet fuel, natural gas and hydrocarbons	9%–85%	[35]
Water-gas shift	A mixture of product gas from coal, waste, biomass	NA	[27]

TABLE 2 The summary of hydrogen production technologies—cont'd

Technology	Feedstock	Technological efficiency	Reference
Renewable source			
Aqueous-phase reforming	Carbohydrates, biomass	35%–55%	[35]
Photo-electrolysis	Water	12.4%	[35]
Electrolysis	Water	50%–70%	[35]
Thermochemical water splitting	Water	NA	[35]
Dark fermentation	Organic feedstock	60%–80%	[35]
Photo fermentation	Organic feedstock	0%–5%	[30]
Pyrolysis	MSW, organic feedstock	35%–50%	[36]
Gasification	MSW, organic feedstock, coal, natural gas	69%	[37]

developed by integrating with one another to improve hydrogen yield and emission reduction during the process.

Some other H_2 production method includes dark and photofermentation, which is based on a biological process and organic feed is preferable. Pyrolysis process mainly converts municipal solid waste (MSW) into liquid products which result in a lower amount of hydrogen production. The gasification has higher conversion performance due to endothermic and exothermic processes. The process of gasification consists of steam reforming and water gas shift, which are the main reactions to produce hydrogen. Due to higher conversion efficiency, gasification is one of the recommended H_2 production processes from coal and natural gas [38]. Considering MSW is hydrocarbon-based feed the waste to H_2 gasification process is explained in the next section.

5 Gasification process for waste to hydrogen production

Gasification is a partial oxidation process, which means it requires less oxygen (only 25%–40% compared to normal combustion) to generate enough heat to gasify the feed and produce syngas (CO and H_2). The heat released during partial oxidation provides most of the energy to break up the chemical bonds in the feedstock. This process drives the other endothermic gasification reactions, which increase the temperature of the final gasification products. The output of the gasification, syngas, and slag can be used in different applications. Fig. 5 shows the gasification process and the output linked with respective applications. The syngas can be primarily used for power generation through gas turbines or steam turbines. It can be used in chemical production and as fuel. Hydrogen, however, can be separated and purified to use in many applications; such as electricity generation by using fuel cells and can be used as fuel. While the residue of the gasification process can be used as a construction material.

Gasification is a complex process that is accomplished through a series of physical transformation and chemical reactions within the gasifier. In gasification, feeds pass through different temperature stages and different air or oxygen (gasifying agent) levels where the reactions occur to convert the feed into synthesis gas and residue (slag and/or ash) [40]. MSW is a feed with high moisture, therefore, the gasification starts with drying the feeds. After drying, the feeds go through pyrolysis, oxidation, and gasification stages as explained below.

5.1 Chemical reactions and conversion process

The conversion process runs within the gasifier; therefore, this section explains all the conversion steps and respective chemical reactions that occur in the gasifier. In the gasifier, MSW feed is supplied from the top and the oxidizing agent is supplied from the bottom of the gasifiers while the syngas product is taken out from the top and the residue or slag is taken out from the bottom of the gasifier as shown in Fig. 6. Chemical reactions are the principal function of the gasification process that transform feed into syngas products. The complex thermodynamic reactions occur in the gasifier that includes heat absorption by MSW (endothermic reaction) and heat released by MSW (exothermic reaction). The conversion process of gasification can be classified into four processes as; drying, pyrolysis, oxidation, and gasification process [41]. To explain these processes, the gasifier is divided into respective four zones as presented in Fig. 6.

- **Drying Zone**: The free water or moisture in MSW evaporates in this stage leaving the dry material. This process dehydrates the moisture of raw material by applying temperature in the range of 100–135°C and brings the dry material for the next process to start the chemical reaction. This simple process reaction is shown in the following equation:

$$\text{MSW (Raw material and moisture)} + \text{Heat} \rightarrow \text{Dry material} + H_2O \quad (1)$$

- **Pyrolysis Zone:** The dry feed is exposed to the elevated temperature at this stage. This process operates at a temperature of around 500–550°C in the absence of oxygen. This endothermic process devolatizes and breaks the weaker chemical bonds of the feed material and releases volatile gases

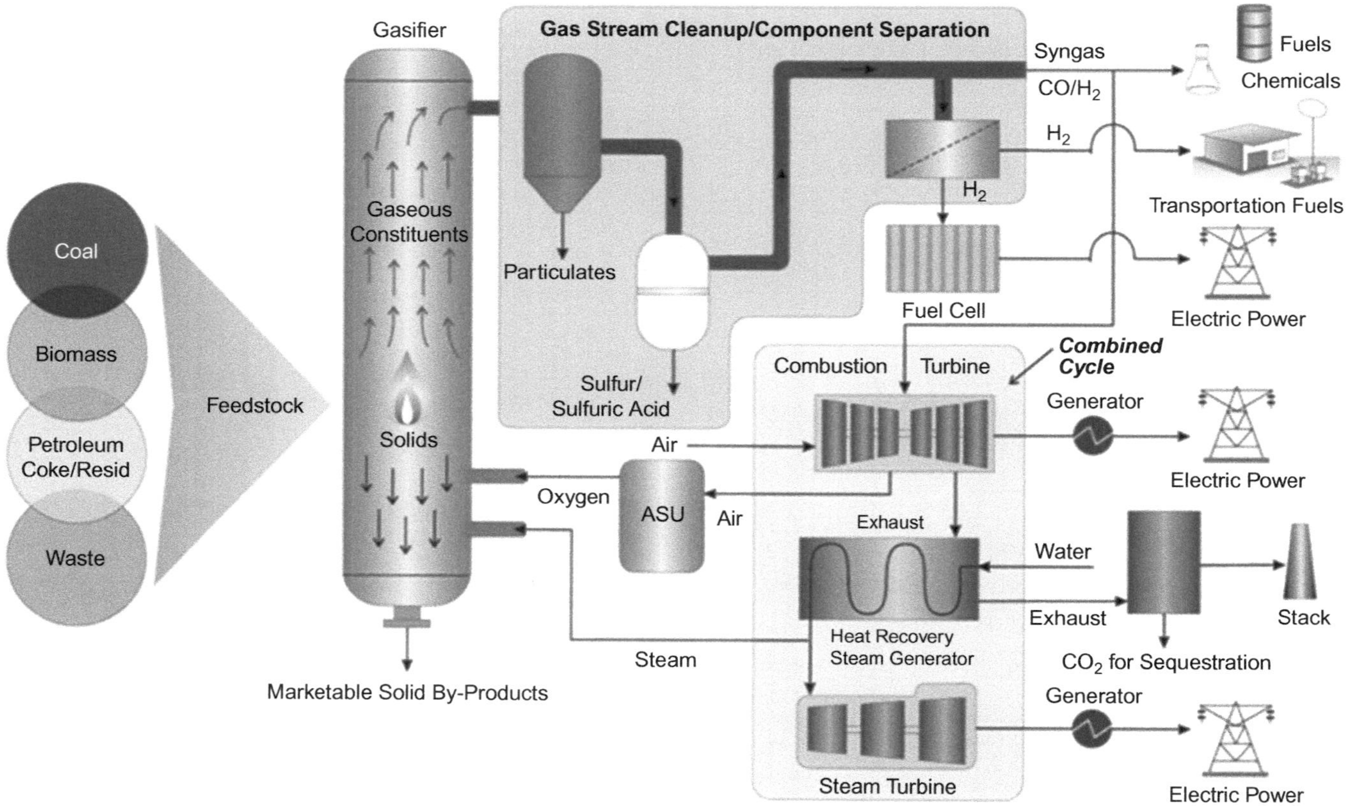

FIG. 5 Gasification process and applications [39].

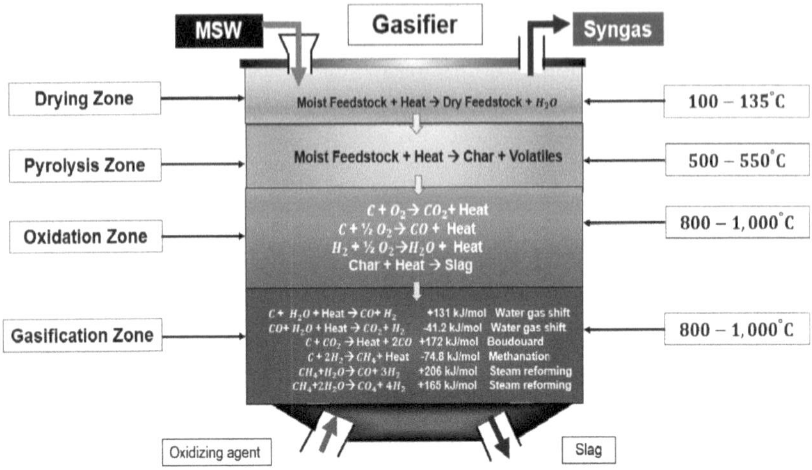

FIG. 6 Gasification conversion process and chemical reactions in a gasifier.

(tar vapor, CH_4, H_2) along with high-molecular-weight char. This process is explained in the following equation:

$$\text{Dry Material} + \text{Heat} \rightarrow \text{Char} + \text{Volatile gases} \tag{2}$$

- **Oxidation Zone:** The volatile product along with the char reacts with limited oxygen in this zone. This combustion process results in producing various gases (CO, CO_2), slag as residue, and releases heat. This heat accumulates in the confined chamber and increases the temperature needed for the gasification zone. The temperature is maintained between 800°C and 1000°C to accelerate the reactions. At this stage, slag is moved to the bottom of the gasifier. The oxidation reactions are listed in the following equations:

$$C + O_2 \rightarrow CO_2 + \text{Heat} \tag{3}$$

$$C + \frac{1}{2}O_2 \rightarrow CO + \text{Heat} \tag{4}$$

$$H_2 + \frac{1}{2}O_2 \rightarrow H_2O + \text{Heat} \tag{5}$$

$$\text{Char} + \text{Heat} \rightarrow \text{Slag} \tag{6}$$

- **Gasification Zone:** At this stage, in the gasification process, the volatiles and remaining char react with CO_2 and steam at high temperature (800–1000°C) and produce syngas (CO and H_2) and minor amount of other gases via various reactions. A number of separate reversible gas-phase reactions take place simultaneously based on gasifier conditions. The reactions with steam known as water-gas shift (WGS) reaction and steam reforming produce H_2, CO, and CO_2. The Boudouard reaction produces CO and hydrogasification of methanation methane (CH_4) [42]. The chemical reactions in gasification are presented in the following equations:

$$C + H_2O \rightarrow CO + H_2 \ (\Delta H) + 131\text{kJ/mol Water} - \text{gas shift reaction} \quad (7)$$

$$C + CO_2 \rightarrow 2CO \ (\Delta H) + 172\text{kJ/mol Boudouard reaction} \quad (8)$$

$$C + 2H_2 \rightarrow CH_4 \ (\Delta H) - 74.8\text{kJ/mol Hydrogasification} \quad (9)$$

$$CO + H_2O \rightarrow H_2 + CO_2 \ (\Delta H) - 41.2\text{kJ/mol Water} - \text{gas shift reaction} \quad (10)$$

$$CH_4 + H_2O \rightarrow CO + 3H_2 \ (\Delta H) + 206\text{kJ/mol Steam Reforming} \quad (11)$$

$$CH_4 + 2H_2O \rightarrow CO_2 + 4H_2 \ (\Delta H) + 165\text{kJ/mol Steam Reforming} \quad (12)$$

Gasification stage consists of multiple reactions in order to transform waste into hydrogen; however, the principal chemical reactions of hydrogen production are water-gas shift (WGS) reaction and steam reforming. In addition, the WGS reaction and steam reforming are considered economical and more mature reaction for hydrogen generation from MSW [43].

As mentioned, during the process, gasifying agents are introduced into the gasification chamber in order to control the process and temperature, particularly for the oxidation and gasification stage. The common injected agent includes steam, CO_2, pure oxygen, and air [42]. During the low oxygen environment in the gasifier, the impurities of the feed such as nitrogen converts to gaseous nitrogen (N_2), ammonia (NH_3), small amount of hydrogen cyanide (HCN), sulfur converts to hydrogen sulfide (H_2S), carbonyl sulfide (COS), and chlorine converts to hydrogen chloride (HCL). In general, the overall quantity of these undesired components is sufficiently small that they have a negligible effect on the main syngas components of H_2 and CO.

Finally, the output gases from gasification are transferred to the cleaning system, which consists of an acid catalyst, selected non-catalytic reduction (SNCR), air filtering system, acid gas scrubbing, and activated carbon to get improved syngas. The purified syngas can be used in various ways such as power generation, hydrogen generation, and chemical product. The syngas is then cooled to remove condensable hydrocarbons. The cooled syngas is heated again to pass through the membrane to separate pure H_2 and CO [44]. The efficiency of the membrane system is very high and can separate 85%–95% (by volume) of pure H_2 from syngas.

6 Benefits of hydrogen generation from MSW

Municipal solid waste (MSW) is primarily waste that is collected from households and councils. MSW consists of everyday items such as paper, bottle, food scrap, clothing, furniture, and a wide range of other materials. A large volume of this waste goes directly to landfill and a waste management system operates to

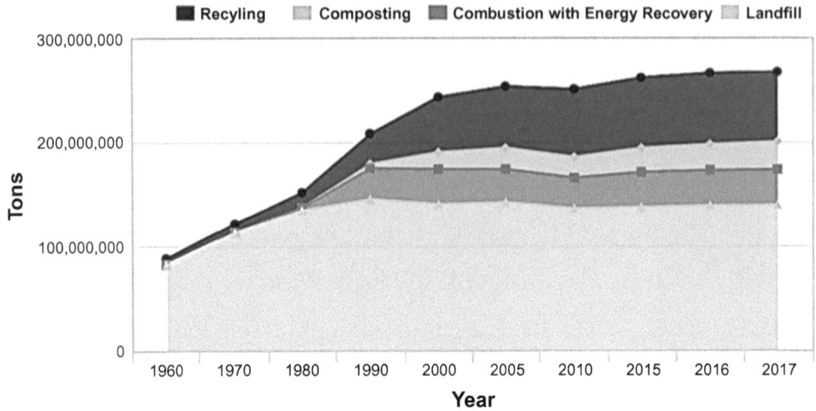

FIG. 7 MSW management in the United States from 1960 to 2017 [21].

reduce the landfill volume. In most of the countries in the world, waste goes to landfills; however, many countries cannot afford waste management. Fig. 7 shows that waste management in the United States from 1960 to 2017 [21] and this clearly shows that a large volume of waste still goes to landfills.

Landfills can impact air, water, and land quality. In the landfill, organic waste is decomposed and it produces landfill gas mainly methane (CH_4) that contributes to greenhouse gas emission. Water moving from or through the landfill waste forms leachate that can contaminate nearby land and ground water. Therefore, the key benefit of H_2 generation from MSW is an improved waste management system, reduction of emission, and production of new sources of clean energy. Brief details of the benefits are explained below:

- **Investment cost**

  Generally, landfill has the lowest cost for terminating waste while waste to hydrogen generation cost is much higher. However, it was found that the investment cost of landfill is around \$208 per ton of waste whereas waste to hydrogen-based gasification cost is around \$306–\$4012 per ton of waste [45,46]. Therefore, MSW to H_2 generation can be competitive in the future by adopting favorable environmental policy considering the environmental benefit and the opportunity to get a new clean energy source. Moreover, efficiency improvement of gasification and more use of MSW to H_2 applications will reduce the gap in current investment cost.

- **Emission reduction**

  As mentioned, landfill produces greenhouse gas during the decomposition of waste. The released gas from landfills includes CH_4 up to 55.63% and CO_2 around 37.14% [47]. Landfill also contaminates the surrounding soil and water. Whereas, MSW gasification for H_2 production can reduce emission and contamination. It is found that H_2 production from MSW gasification generated less amount of greenhouse gas emissions such as 0.1% of CH_4 and 15% of CO_2 [48] and released less toxic compounds than landfill. Therefore, the MSW gasification for H_2 generation has positive impacts on the environment.

- **Waste material reduction**

  The significant advantage of H_2 production from MSW is the reduction of substantial amounts of waste to landfill. With the growing population and living standards, the waste management is becoming a critical issue [49]. Landfill takes a longer time to decompose and with the increased population, there will be difficulty in finding more landfill sites. On the other hand, waste to H_2 can reduce landfill volume and requirement and generated H_2 can be a sustainable source of energy for the future.

- **Clean energy production**

  The gasification process to treat MSW provides syngas where H_2 is the clean form of energy, as when it is used in energy application it has zero direct emission. Syngas can be directly used in power generation or it can be integrated into the national gas grid to reduce emission from the gas use. Moreover, H_2 with fuel cell can produce electricity and it also can be used in fuel cell-based electrical vehicle. Hydrogen as a clean source contains high energy value and can be used in many applications, for example, using 1 kg of hydrogen an electric car can travel up to 100 km whereas a conventional car takes 9.3 L of petrol to travel 100 km. Moreover, 1 kg of hydrogen can operate a 1400 W air conditioner for around 14.5 h [5]. Therefore, generating H_2 from MSW not only can reduce the waste volume and emission but can also be a new source of clean energy.

- **Alternative application**

  Recent years have witnessed a growing commercial interest in the waste industry that has shaped more options of waste disposal schemes. Due to environmental issues, many countries are not able to export waste to other countries for recycling/reusing, therefore, most of the countries are facing challenges for a better solution of waste management and policies are now driving the waste strategy into a circular economy. The concept of waste into hydrogen generation is one of such waste management strategies that contributes to the country's economy and this will possibly shift the energy market in the coming years [50]. Moreover, the residue of the H_2 generation from waste is also useful for construction applications. Therefore, waste to H_2 generation can contribute to sustainable future.

- **Financial benefit**

  There is always a significant demand for H_2 in various industrial applications and due to its environmental benefit H_2 demand will grow in the future. Currently, H_2 produced from natural gas and coal has a price range of \$1–3/kg; however, this process contributes to GHG emission. With the use of renewable sources such as solar or wind in water electrolysis application to produce H_2 is still costly and roughly, it is \$11–16/kg.

One California-based company Ways2H is currently producing H_2 from MSW at a cost of \$5/kg and expecting to reduce the cost to \$3/kg in the next 5 years [51]. This indicates the scope of cheaper H_2 production as well as better waste management.

- **Power System application**

  As H_2 is a good energy carrier, it can be used to store energy from fluctuating renewable energy sources. This stored H_2 can be used to generate electricity and can contribute to power grid load balancing, improving voltage conditions, and be a stable source of energy. Therefore, H_2 from MSW can be integrated with renewable energy-based power grid for reliable operation of the power network.

6.1 Comparison of MSW gasification with landfill disposal

As mentioned above, the landfill has a significant impact on the environment; therefore, the cost of waste disposal in the landfill does not reflect the actual cost without considering the impacts of landfills. The cost of landfills includes capital cost and operational and maintenance cost of the site. The impact cost of landfills is somewhat complex and is difficult to assess and this needs to be considered when comparing with other waste management techniques. The waste to energy technology, particularly MSW to energy using gasification, has a relatively higher cost compared to landfill but it has a direct benefit in emission reduction as well as extracting usable energy from the waste. Table 3 presents and indicates a comparative advantage of MSW gasification compared to landfill.

From the comparison shown in Table 3, it is found that landfill is an economical pathway for MSW disposal (\$208–\$297 per ton). The gases that are produced in landfills are mainly methane (CH_4) and carbon dioxide (CO_2) and it is easy to collect them from the central storage without requiring any advanced technology, although the gas generation process is very slow. Currently, landfill gas (CH_4) is used as fuel in power generation in replacing or offsetting coal, natural gas, and oil [53]. Although the landfill gas product has been exploited in multiple applications, it has negative impacts on the environment. Landfill process is designed for decomposing biodegradable materials, which is a long-time process. The non-biodegradation waste remains in the landfill site and toxic waste would be able to contaminate the surrounding environment. Furthermore, due to the long-time process of landfill, the greenhouse gas emission is a major concern of MSW landfill as the system releases a significant amount of CH_4 and CO_2. Although CH_4 can be captured and used, the gas collection is not 100% effective, so landfill constantly produces CH_4 emission [54]. In fact, the effect of CH_4 is 28–36 times of CO_2 on global warming potential. The United States is going to reduce 75% MSW deposits in landfill by 2025, as it contributes to emissions [52].

TABLE 3 Comparison between landfill and gasification for MSW.

Technology	Product	Cost	Environmental impact	Benefit	Drawback	Reference
Landfill disposal	Mainly CH_4 and CO_2 Liquid and solid remains	\$208–\$297 per ton (includes capital, operational and maintenance cost)	High greenhouse gas emission (based CH_4 and CO_2) CH_4 55.63% CO_2 37.14% Heavy metal, toxic gases, nonbiodegradation, and leachate	Economical way to eliminate MSW at large scale	Long-time process Adverse impact on environment	[45,47]
Gasification (MSW/ Biomass)	Mainly Syngas (H_2 and CO) Residue or slag	\$306–\$4012 per ton	Less Greenhouse gas emission (based on CH_4 and CO_2) CH_4 0.1% CO_2 15% Small level of Particulate matters, H_2S, HCN, some toxic gases	Avoid 58–76 million tons CO_2 /year with cost between \$14 to \$92/ tons Benefit in economies in large-scale industry Short-time process	Cost of process	[48,52]

The MSW gasification converts waste into energy source in a short residence time with low-emission contribution. The gasification system reduces emissions (nearly 0.1% CH4, 15% CO2) as shown in Table 3 and also reduces fossil fuel utilization. Moreover, carbon capture and storage (CCS) can be applied with the MSW gasification to further reduce CO_2 emission and produce pure H_2. However, as landfill is an open dump site, the CCS method has a significant challenge to implement in reality. The gasification is able to reduce a significant amount of CO_2 from MSW which is around 58–76 million tons of CO_2/year at a low cost between \$14 and \$92/ton of waste [52]. Although the gasification would release particulate matters and diverse toxic gases, it is significantly less than that released in the landfill and other thermal waste conversion [55]. The key weakness of gasification technology is its high investment and maintenance cost. However, the minimum cost of waste gasification to H_2 is \$308 per ton, which is nearly the landfill cost of \$208–\$297 per ton, and it seems competitive to the energy market. In addition, gasification is considered economically viable for both small- and large-scale installation [52,56]. Therefore, the MSW gasification would be a significant alternative application for MSW reduction.

6.2 Comparison of H_2 generation from different sources

H_2 can be generated from fossil fuel as well as from renewable energy sources using various technologies. As mentioned in Section 1, almost entire H_2 production is currently based on the use of fossil fuel (natural gas and coal); however, emerging new technologies are being used to produce it from other sources while also reducing emission. The selection of future H_2 production will be based on the criteria of emission reduction as well as competitive production costs and other benefits. Table 4 summarizes various H_2 production sources with key comparative criteria.

- **Cost comparison**

  Fossil fuel-based H_2 production is prevalent due to the low production cost. As mentioned in Section 4, the steam methane reforming (SMR) and the gasification method of natural gas and coal are widely used in many industries with efficiency between 70% and 85% and gasification at 69%, respectively. Both SMR and coal gasification consume lower amount of input sources to generate 1 kg of H_2 product. SMR requires 4.5 m^3 of natural gas to produce 1 kg of H_2 and gasification uses 7.6 kg of coal to produce 1 kg of H_2 [58]. The cost of H_2 from SMR with natural gas and the coal gasification is around \$2.08–\$2.27/kg and \$1–3/kg [58], respectively. On the other hand, renewable sources for H_2 production demand more input feedstock and require more energy. For instance, to produce 1 kg H_2, gasification requires 13.5 kg biomass or MSW, photo fermentation requires 23 kg organic material, and electrolyzer uses 9 L of water and power up to 48

TABLE 4 Hydrogen production from various sources with cost and emission performance.

Source/ Feed	Technology	Process	Feed per kg H_2	Cost (USD)	Emission	Challenge	Reference
Fossil fuel							
Natural gas	Steam methane reforming	Steam methane reforming	4.5 normal m^3	\$2.08–2.27/kg	8.5 kg CO_2/kg H_2	Purification requirement	[57,58]
Coal	Coal gasification	Water gas shift and reforming reaction	7.6 kg	\$1–3/kg	12.7–16.8 kg CO_2/kg H_2 Particulate matter, toxic chemical and heavy metals	Environment issue Purification requirement	[58,59]
Renewable source							
MSW, biomass	Gasification	Water gas shift and reforming reaction	13.5 kg	\$3–5/kg	Less CO_2 emission, Particulate matters, H_2S, HCN, some toxic gases	Operational and maintenance cost	[60,61]
MSW, biomass	Pyrolysis	Bio-oil reforming to hydrogen	NA	\$2.20/kg	Heavy metals, toxic gases	Suitable catalyst Tar formation Less hydrogen yield	[52,57]
Organic material	Photo-Fermentation	Fermentation	23 kg	\$2.83/kg	NA	Low H_2 yield	[57,60]

Continued

TABLE 4 Hydrogen production from various sources with cost and emission performance—cont'd

Source/ Feed	Technology	Process	Feed per kg H_2	Cost (USD)	Emission	Challenge	Reference
Water	Electrolysis	Electrolysis	9 L of water and 39–48 kWh of electricity	$11–16/kg	40. 5 kg CO_2/kg H_2 (from grid electricity)	High cost	[57,62]
Sunlight and water	Solar PV-electrolysis	Electrolysis	9 L of water and 39–48 kWh of electricity	$5.78–23.27/kg	0 (100% renewable energy)	High cost	[57,62]
Solar and water	Photo-electrolysis	Electrolysis	9 L of water and 39–48 kWh of electricity	$10.36/kg	0 (100% renewable energy)	High cost	[57,62]

kWh [57,62]; therefore, the price of H_2 production goes higher. The electrolysis application has the highest cost of H_2 production, which is water electrolysis (grid electricity) between \$11 and \$16/kg, solar PV electrolysis around \$5.78–\$23.27/kg, and photoelectrolysis at \$10.36/kg [57,62]. H_2 production cost from the MSW/biomass gasification (\$3–\$5/kg), the MSW/biomass pyrolysis (\$2.20/kg), and fermentation (\$2.83/kg) is found to be competitive [60].

- **Emission comparison**

  The coal gasification during H_2 production releases significant emissions due to the combustion of coal and produces carbon dioxide (12.7–16.8 kg CO_2 /kg of H_2 production), tar, organic compounds, toxic chemical, heavy metals, particulate matter, sulfur dioxide (SO_2), and nitrogen oxide (NOx) [63]. SMR uses natural gas and generates 8.5 kg CO_2/kg of H_2 production [58]. Renewable sources, such as MSW/biomass gasification, has less pollution in terms of heavy metal, toxic gases, and slag compared to coal gasification. H_2 from the MSW/biomass gasification can be purified easily compared with that from coal gasification but results in less emission [52]. Pyrolysis can decompose MSW/biomass and effectively produce H_2 that emits less amount of toxic residue and heavy metal [55]. Water electrolysis though produces zero emission but if the grid electricity is used, then the equivalent grid electricity releases 40.5 kg CO_2 to produce 1 kg H_2. Various renewable sources (solar or wind) can be used in water electrolysis for emission-free H_2 production; however, the production cost becomes higher.

From the above discussion as well as from Table 4, it is found that the cheaper H_2 production from coal and natural gas has a higher emission and an additional purification cost is required. The solar PV and photoelectrolysis systems have challenges of gas, liquid and heat control, and flexible loading for H_2 production [64]. The disadvantage of pyrolysis and photo fermentation is low H_2 production [65]. Pyrolysis application also needs to use a catalyst for improved performance and deal with tar formation.

For abundant MSW material, H_2 from waste seems to be valuable input source for energy security. Gasification has high performance in H_2 production and MSW gasification can produce H_2 from moderate amount of feed and release relatively less CO_2. The main challenge of the MSW gasification and electrolysis systems is relatively high cost of operation and maintenance [61]. Considering all pros and cons and the environmental benefits, the MSW gasification has a great future prospect for H_2 production.

7 Future prospects of waste to hydrogen production

Global energy demand is anticipated to increase over 50% by 2050 due to the growth of population and global economy [66]. Fossil fuel including coal, natural gas, and petrol is still the primary source of energy consumption [43].

Among these fossil fuel sources, the world's energy-related carbon dioxide (CO_2) emission increased from 20.5 Gt CO_2 to 33.1 Gt CO_2 between 1994 and 2018 [67]. Moreover, these sources are not sustainable and secure as they are projected to be depleted by 2060 [68]. Alternative energy sources including renewable energy as well as hydrogen will play a crucial role in future energy economy due to emission reduction and its renewable properties.

During the last decade, the development in the field of energy source has renewed its interest in hydrogen production. Hydrogen is a key energy carrier with zero emission; therefore, it has become a potential energy source for many applications. Considering the benefits of hydrogen, many countries have revised their energy policy targeting the development of hydrogen from heterogeneous sources. Table 5 shows the list of target applications based on hydrogen use. Moreover, the demand for the hydrogen-based product is expected to grow up to 200 MtH_2/year by 2050 [69]. Therefore, hydrogen production is seen as a key factor to meet this target for many countries.

As mentioned in Sections 1 and 4, hydrogen production is currently dependent mostly on natural gas and then coal, and the major cost of hydrogen production also depends on these sources. It was also found in Sections 4 and 6 that MSW generation is increasing and it emits GHG if directly goes to landfills. It was recorded that carbon dioxide emissions from MSW in 2016 were estimated at 1.6 billion tons CO_2 and it will increase to 2.6 billion tons of CO_2 by 2050 [70]. To resolve this problem from MSW, various waste management methods are utilized in different countries.

The gasification process can convert MSW into synthesis gas as explained in Section 5. The synthesis gas consists of CO and H_2 and pure H_2 can be extracted by using separator devices. In the United States, the MSW gasification for hydrogen production is estimated to be supplying hydrogen by 2021 [71]. Therefore, the MSW gasification for hydrogen production has a good prospect. This also reduces the risk of landfill emission due to MSW.

8 Conclusions

Hydrogen is likely to be a promising source in the energy market in the coming years. In this chapter, the potential of hydrogen production from various sources and its applications along with a detailed explanation about the gasification process has been presented. The benefits of hydrogen which can reduce the emission from the MSW sector have also been explained. The key findings are listed below:

- There is a constant demand for hydrogen and this demand is increasing
- Hydrogen is a good carrier of energy and it has zero emission
- Currently hydrogen is generated mostly from fossil fuels, therefore emission still exists

TABLE 5 Hydrogen targets set by different countries [69].

Country	Renewable energy targets	Hydrogen specific targets			
		FCEV targets	Hydrogen refueling station (HRS) targets	Hydrogen flow (MtH_2/year)	Other
United States-Federal	20% (2020)	40,000 (2023)	100 (2023)	–	–
United States-California	100% (2045)	1 million (2030)	1000 (2030)	–	33% green hydrogen
Germany	18% (2020)	500	100 (2019)	–	–
France	32% (2020)	50,000 (2023)	100 (2023)	–	–
Netherlands	14.5% (2020)	50,000 (2023)	100 (2023)	–	–
Norway	67.5% (2020)	50,000	200	–	–
Denmark	30% (2020)	75	10	–	–
China	770 GW (2020)	1 million (2030)	500 (2030)	0.2 (2030)	–
South Korea	11% (2030)	630,000 (2030)	520 (2030)	–	–
Japan	22%–24% (2030)	800,000 (2030) and 1200 buses (2030)	320 (2030)	0.3 (2030)	–
India	175 GW (2022)	1 million (2020)	–	–	–
Australia	33,000 GWh (2020)	–	–	0.5 (2030)	–
New Zealand	100% (2035)	–	–	0.7 (Taranaki only, 2030)	Taranaki proposes exporting around 0.5–1 GW, or 40% of production

- Research and demand are increasing to generate hydrogen from other sources
- Considering the properties of hydrogen, a range of new applications are developing based on hydrogen as fuel
- Application of hydrogen in electricity generation and transportation sector can reduce significant emission from energy market
- Waste and MSW is also a source of GHG emission, but hydrogen generation from MSW can reduce significant emission from the waste sector
- Gasification can generate hydrogen from MSW
- Hydrogen production from MSW is promising in terms of cost and emission reduction
- A number of countries put target of hydrogen generation for various future applications

Quite clearly, hydrogen generation from MSW has the prospect to become an alternative option for both waste management and energy production.

References

[1] Birol F. The future of hydrogen. Available from: https://www.iea.org/reports/the-future-of-hydrogen; 2019.

[2] Kalamaras CM, Efstathiou AM. Hydrogen production technologies: current state and future developments. In: Conference papers in energy. Hindawi; 2013.

[3] Singh CK, Kumar A, Roy SS. Quantitative analysis of the methane gas emissions from municipal solid waste in India. Sci Rep 2018;8(1):2913.

[4] Energy, U.S.D.o. Hydrogen fuel cell engines and related technologies. In: Module 1: Hydrogen properties. Office of Energy Efficiency and Renewable Energy; 2001.

[5] Department of Industry, S., Energy and Resources. Australia's National Hydrogen Strategy; 2019.

[6] Staffell I. The energy and fuel data sheet. UK: University of Birmingham; 2011.

[7] Suh MP, et al. Hydrogen storage in metal–organic frameworks. Chem Rev 2012;112(2):782–835.

[8] National Research Council. The hydrogen economy: opportunities, costs, barriers, and R&D needs. National Academies Press; 2004.

[9] Department of the Environment and Energy. National Greenhouse Accounts Factors. In: Australian National Greenhouse Accounts; 2017.

[10] Nikolaidis P, Poullikkas A. A comparative overview of hydrogen production processes. Renew Sust Energy Rev 2017;67:597–611.

[11] Annamalai K, Puri IK. Combustion science and engineering. CRC Press; 2006.

[12] Armstrong GT, Jobe Jr TL. Heating values of natural gas and its components. National Standard Reference Data System; 1982.

[13] Boundy, B., et al., Biomass Energy Data Book, US Dep. of Energy, 2011. ORNL/TM-2011/446, http://cta.ornl.gov/bedb/download.shtml.

[14] Europe H. Hydrogen applications. Available from: https://hydrogeneurope.eu/hydrogen-applications.

[15] Genevieve S, et al. Stationary fuel cell evaluation. The National Renewable Energy Laboratory; 2016.

[16] NEDO, N.E.a.I.T.D.O. World's first heat and electricity supplied in an urban area using 100% hydrogen. In: Towards establishing optimal energy control technology in local communities. New Energy and Industrial Technology Development Organization (JCN); 2018.

[17] Arbizzani C, De Giorgio F, Mastragostino M. Battery parameters for hybrid electric vehicles. In: Advances in battery technologies for electric vehicles. Elsevier; 2015. p. 55–72.

[18] Pollet BG, et al. 22 – Fuel-cell (hydrogen) electric hybrid vehicles. In: Folkson R, editor. Alternative fuels and advanced vehicle technologies for improved environmental performance. Woodhead Publishing; 2014. p. 685–735.

[19] E4tech. H_2 emission potential literature review; 2018.

[20] Konieczny A, et al. Catalyst development for thermocatalytic decomposition of methane to hydrogen. Int J Hydrogen Energy 2008;33(1):264–72.

[21] EPA, U.S.E.P.A. National overview: facts and figures on materials, wastes and recycling. Available from: https://www.epa.gov/facts-and-figures-about-materials-waste-and-recycling/national-overview-facts-and-figures-materials#main-content.

[22] Onozaki M, et al. Hydrogen production by the partial oxidation and steam reforming of tar from hot coke oven gas. Fuel 2006;85(2):143–9.

[23] Joensen F, Rostrup-Nielsen JR. Conversion of hydrocarbons and alcohols for fuel cells. J Power Sources 2002;105(2):195–201.

[24] Energy, U.S.D.O. Hydrogen & our energy future. Available from: https://www.energy.gov/eere/fuelcells/downloads/hydrogen-our-energy-future.

[25] Bromberg L, Cohn DR, Rabinovich A. Plasma reformer-fuel cell system for decentralized power applications. Int J Hydrogen Energy 1997;22(1):83–94.

[26] Iulianelli A, et al. Water gas shift membrane reactors. In: Membrane reactors for energy applications and basic chemical production. Elsevier; 2015. p. 3–29.

[27] Kraussler M, Hofbauer H. Development and experimental validation of a water gas shift kinetic model for Fe-/Cr-based catalysts processing product gas from biomass steam gasification. Biomass Convers Biorefin 2017;7(2):153–65.

[28] Mor GK, et al. A review on highly ordered, vertically oriented TiO_2 nanotube arrays: fabrication, material properties, and solar energy applications. Solar Energy Mater Solar Cells 2006;90(14):2011–75.

[29] Funk JE. Thermochemical hydrogen production: past and present. Int J Hydrogen Energy 2001;26(3):185–90.

[30] Akroum-Amrouche D, et al. Biohydrogen production by dark and photo-fermentation processes. In: 2013 International renewable and sustainable energy conference (IRSEC). IEEE; 2013.

[31] Jahirul MI, et al. Biofuels production through biomass pyrolysis—a technological review. Energies 2012;5(12):4952–5001.

[32] Kumar A, Samadder SR. A review on technological options of waste to energy for effective management of municipal solid waste. Waste Manag 2017;69:407–22.

[33] Miller BG, Tillman D. Combustion engineering issues for solid fuel systems. Academic Press; 2008.

[34] Saidi M, Gohari MH, Ramezani AT. Hydrogen production from waste gasification followed by membrane filtration: a review. Environ Chem Lett 2020;1–28.

[35] El-Shafie M, Kambara S, Hayakawa Y. Hydrogen production technologies overview. J Power Energy Eng 2019;7(01):107.

[36] Bakhtyari A, Makarem MA, Rahimpour MR. Hydrogen production through pyrolysis. In: Meyers RA, editor. Encyclopedia of sustainability science and technology. New York, NY: Springer New York; 2017. p. 1–28.

[37] Matthias B, et al. Hydrogen from biomass gasification; 2018.
[38] Balat M, Balat M. Political, economic and environmental impacts of biomass-based hydrogen. Int J Hydrogen Energy 2009;34(9):3589–603.
[39] NETL NETL. Gasification introduction. Available from: https://www.netl.doe.gov/research/Coal/energy-systems/gasification/gasifipedia/intro-to-gasification.
[40] Dejtrakulwong C, Patumsawad S. Four zones modeling of the downdraft biomass gasification process: effects of moisture content and air to fuel ratio. Energy Procedia 2014;52:142–9.
[41] Seo Y-C, Alam MT, Yang W-S. Gasification of municipal solid waste. In: Gasification for low-grade feedstock. IntechOpen; 2018.
[42] Deng N, et al. Simulation analysis of municipal solid waste pyrolysis and gasification based on Aspen plus. Front Energy 2019;13(1):64–70.
[43] Balat H, Kırtay E. Hydrogen from biomass–present scenario and future prospects. Int J Hydrogen Energy 2010;35(14):7416–26.
[44] MTR, M.T.R. Hydrogen separations in syngas processes. Available from: https://www.mtrinc.com/our-business/refinery-and-syngas/hydrogen-separations-in-syngas-processes/.
[45] Assamoi B, Lawryshyn Y. The environmental comparison of landfilling vs. incineration of MSW accounting for waste diversion. Waste Manag 2012;32(5):1019–30.
[46] Majidi SS, Kamalan H. Economic and environmental evaluation of waste to energy through gasification; case study: Tehran. Environ Energy Econ Res 2017;1(1):113–24.
[47] Waldheim L, Consulting W. Gasification of waste for energy carriers: A review; 2018.
[48] Arena U. Process and technological aspects of municipal solid waste gasification. A review. Waste Manag 2012;32(4):625–39.
[49] Zheng X, et al. Hydrogen and syngas production from municipal solid waste (MSW) gasification via reusing CO_2. Appl Therm Eng 2018;144:242–7.
[50] Till B. The future of waste-to-hydrogen: Waste-to-hydrogen and alternative technologie. Available from: https://boegelsack.energy/2019/03/31/the-future-of-waste-to-energy-waste-to-energy-and-alternative-technologies/; 2019.
[51] Collins L. It's much cheaper to produce green hydrogen from waste than renewables. In: Rechargenews. Rechargenews; 2020.
[52] Baker S, et al. Getting to neutral: options for negative carbon emissions in California. Livermore, CA, United States: Lawrence Livermore National Lab (LLNL); 2019.
[53] EPA, U.S.E.P.A. Basic information about landfill gas. Available from: https://www.epa.gov/lmop/basic-information-about-landfill-gas.
[54] Gies E. Landfill have a huge greenhouse gas problem. Here's what we can do about it. Available from: https://ensia.com/features/methane-landfills/; 2016.
[55] Dong J, et al. Life cycle assessment of pyrolysis, gasification and incineration waste-to-energy technologies: theoretical analysis and case study of commercial plants. Sci Total Environ 2018;626:744–53.
[56] Behrend P, Krishnamoorthy B. Considerations for waste gasification as an alternative to landfilling in Washington state using decision analysis and optimization. Sustain Product Consumpt 2017;12:170–9.
[57] Kayfeci M, Keçebaş A, Bayat M. Chapter 3: Hydrogen production. In: Calise F, et al., editors. Solar hydrogen production. Academic Press; 2019. p. 45–83.
[58] Milbrandt A, Mann M. Hydrogen resource assessment: hydrogen potential from coal, natural gas, nuclear, and hydro power. Golden, CO, United States: National Renewable Energy Lab (NREL); 2009.
[59] Burmistrz P, et al. Carbon footprint of the hydrogen production process utilizing subbituminous coal and lignite gasification. J Clean Prod 2016;139:858–65.

[60] Mehmeti A, et al. Life cycle assessment and water footprint of hydrogen production methods: from conventional to emerging technologies. Environments 2018;5(2):24.

[61] Campbell RJ. C. Library of, and S. congressional research. In: Hydrogen in electricity's future; 2020.

[62] Gardner D. Hydrogen production from renewables. Available from: http://www.renewableenergyfocus.com/view/3157/hydrogen-production-from-renewables/; 2009.

[63] America G. Coal: why is it dirty? Available from: https://www.greenamerica.org/fight-dirty-energy/amazon-build-cleaner-cloud/coal-why-it-dirty.

[64] Grigoriev S, et al. Current status, research trends, and challenges in water electrolysis science and technology. Int J Hydrogen Energy 2020;45:26036–58.

[65] Mishra P, et al. Outlook of fermentative hydrogen production techniques: an overview of dark, photo and integrated dark-photo fermentative approach to biomass. Energ Strat Rev 2019;24:27–37.

[66] Bioenergy I. Bioenergy – a sustainable and reliable energy source. Paris, France: International Energy Agency Bioenergy; 2009.

[67] IEA. Global energy & CO_2 status report 2019. Paris: IEA; 2019.

[68] Howarth J. When will fossil fuels run out? Available from: https://octopus.energy/blog/when-will-fossil-fuels-run-out/; 2019.

[69] Deloitte DTT. Australian and global hydrogen demand growth scenario analysis; 2019.

[70] Kaza S, et al. What a waste 2.0: a global snapshot of solid waste management to 2050. The World Bank; 2018.

[71] Oberhaus D. Will the hydrogen revolution start in a garbage dump? Available from: https://www.wired.com/story/will-the-hydrogen-revolution-start-in-a-garbage-dump/; 2020.

Section D

Biofuels energy

Chapter 9

A comprehensive investigation on the effects of ceramic layering and cetane improver with an avocado seed oil biodiesel fueled diesel engine

Karthickeyan Viswanathan[a], Muhammad Ikhsan Taipabu[a], Wei Wu[a], and Shuang Wang[b]
[a]*Department of Chemical Engineering, National Cheng Kung University, Tainan, Taiwan,*
[b]*School of Energy and Power Engineering, Jiangsu University, Zhenjiang, Jiangsu, China*

1 Introduction

Due to continuous improvement in the automobile sector, production is estimated to reach 2800 million automobiles by 2036 [1]. This global increment leads to high fuel consumption to cater to the need for 112 million barrels of oil requirement per day [2–4]. It is noteworthy to mention that the compression-ignition engine operating with diesel as an operational fuel leads to 38% of the consumption rate for different development sectors [5–7]. This high rate of consumption paves the way for acute environmental concerns, namely global warming and climatic changes [8]. In addition to the transportation sector, the fossil fuels are used in power generation, agricultural equipment, mine locomotives, and so forth [9–11]. Due to the cumulative usage of fossil fuels, the price of the fuel is increased with the demand [12–14].

Fig. 1A depicts the global crude oil production statistics [15]. The predicted usage of oil was illustrated in terms of million barrels per day (Mb/d) as it was a reliable entity for examining the distinct kinds of energy reserves like oil, gas, coal, and so forth in conformity with the energy content proportional to a barrel of oil. To fuel around 4% of the light-duty means of transportation on the world's roads, one million oil-equivalent barrels per day was sufficient [16]. The expected demand for crude oil in forthcoming years was depicted from 2000 to 2050 [17]. In 2017, the need for fuel was presented as 93 million barrels

Bioenergy Resources and Technologies. https://doi.org/10.1016/B978-0-12-822525-7.00008-1
Copyright © 2021 Elsevier Inc. All rights reserved.

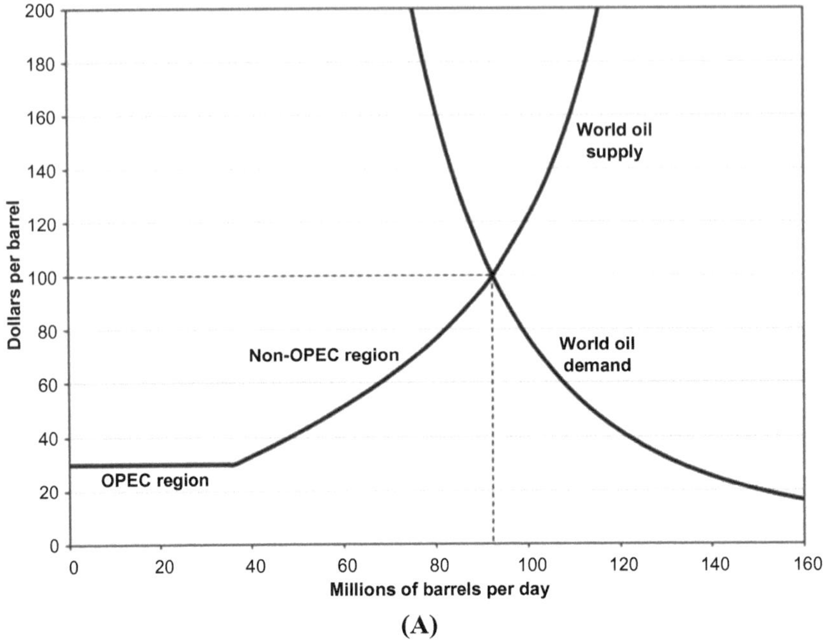

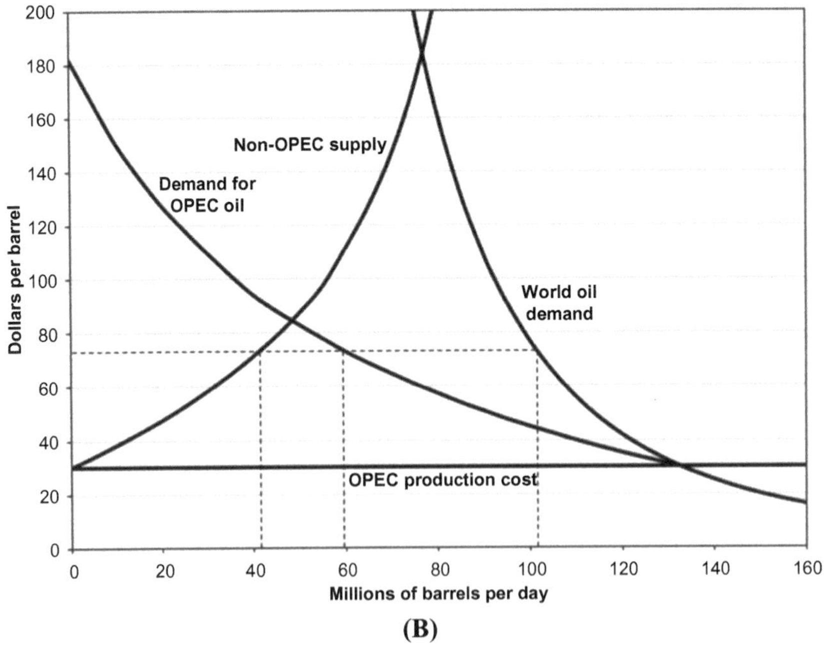

FIG. 1 (A) Global crude oil production statistics and (B) production rate of crude oil from Organization of Petroleum Exporting Countries (OPEC) and non-OPEC [15].

per day. From 2030 to 2050, the need for crude oil was predicted as 104, 112, and 118 million barrels per day, respectively. This increment was owing to extensive accessibility, profitable benefits, and elevated concentration of energy [18]. This successive increment was reported as 10 million barrels per day due to the requirement of energy. OPEC represents the Organization of Petroleum Exporting Countries. Fig. 1B depicts the production of crude oil from OPEC and non-OPEC. Under the non-OPEC category, the United States of America exhibited a higher level of oil production than Russia and Canada since 2010. This high level of oil production was due to stable investment in investigation and expansion [19]. By 2030, it was predicted that the United States would incur crude oil production of about 5.5 million barrels per day. This production accounted for about half the production rate with the global energy requirement. On comparing with Canada, Russia showed high crude oil production. In the present time, OPEC countries like Saudi Arabia, Iraq, and Iran exhibited an equal amount of crude oil production rate. It was predicted that the production rate for Saudi Arabia, Iraq, and Iran might increase due to the need to cater to the requirements. It was foreseen that the production rate of crude oil for Saudi Arabia, Iraq, and Iran was over 15.1, 7.5, and 6.7 million barrels per day, respectively.

Fig. 2 depicts the overview of biomass and biofuel production and utilization statistics [20]. For any nation, the transportation sector accounts for a major part since it incurs consumption of fossil fuel like petrol, diesel, and the like. Thus, biofuels have emerged as a preferred option, especially for the transport sector in India. In any nation, half of the total energy was consumed by industry and transport sectors. The industry and transport sectors were the largest end

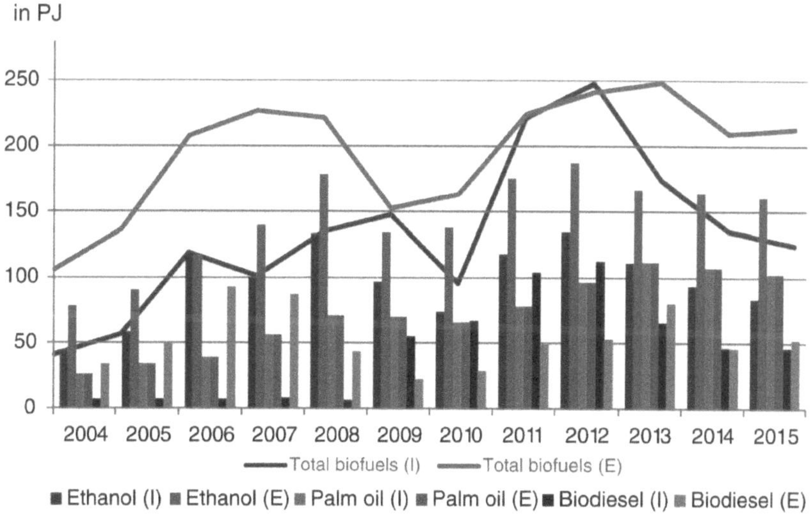

FIG. 2 Overview of biomass and biofuel production and utilization statistics [20].

users of energy. The key fuels contributing to this growth in demand were petroleum (transport), coal (industry and power supply), and electricity (in buildings, industry, and agriculture) [21,22]. In the transport division, the application of petroleum and diesel continues to develop its growth predominantly in road transport. The present development in transportation and subsequent increase in petroleum and diesel utilization increase the environmental concern too. Development in economic activities and personal income shows the way to increasing trade of goods and services, which in turn leads to advanced energy demand in the commercial transportation sectors. To attain enhanced fuel security, alleviate climate change by achieving decreased greenhouse gas (GHG) emission, environmentally sustainable development, and steps to achieve financial development, the application of biofuels as a renewable fuel for transport was said to be the important strategy in several nations.

Traditionally, biomass has been used widely due to their economic potential conditions. On the other hand, energy sustainability was considered as a key solution to eradicate the energy-related problems. At present, a large part of the research has been focused on environmentally friendly and sustainable energy from biomass to supplement conventional fossil fuels. The practice of using biomass for the provision of energy was the most ancient and most recognized strategy, which was intricately associated with the improvement of the human race. The requisite to obtain decreased greenhouse emission, the initiation of novel and effective biomass transformation technologies, practically continued high fossil fuel price, and unpredictability of high fuel price leads to further progression with the usage of biomass. As carbon dioxide was sequestered yet again, the production of electricity from biomass-based fuels was said to be renewable and sustainable. Energy shifts reflect the increasing living standards and rising urbanization through 2050. It was foreseen that the demand for electricity between 2017 and 2050 increases up to 70%, accounting for about 90% of the rise in demand, and achieving a portion of 40%.

1.1 Biodiesel

Today, diesel-driven engines are used for power generation, farming, and transportation, building construction, production, and manufacturing. Progressive growth with the application of the diesel engine in the aforementioned sectors was due to their upgraded fuel economy, elevated efficiency, persistent consistency, and minimum emissions. One of the pillars of viable and ecological development was mobility, which incurred very substantial energy demand. For the air, water, and land transport engine, fuels were required. The practice of using conventional, crude oil-based blended oil leads to the exhaustion of oil reserves in the long term. Thus, deficit natural resources, reliance on fossil fuel reserves, and fluctuating oil price leads to the identification of an alternative fuel. The continuous usage of fossil fuel causes the emission of greenhouse gases like CO, HC, NOx, etc., which leads to global warming. The conventional diesel used in a diesel engine consists of a high concentration of aromatics and sulfur, which leads to environmental pollution. High amounts of Particulate Matter (PM), HC, NO_x, CO, and SO_x were produced from a diesel engine. Moreover, NO_x and CO were the greenhouse gases, and SO_x causes acid rain. In order to obtain reduced effects of greenhouse gases and avert global warming, an effective substitute for fossil fuel was essential.

At present, biodiesel is considered as one of the alternatives to diesel fuel in the world. The production of biodiesel and its demand were rapidly increasing and will continue to do so. Among the distinct alternatives, biodiesel was CO_2 neutral. Biodiesel liberates the same quantity of CO_2 as the floras absorb in growth, and thus there is no greenhouse effect and global warming. Biofuels/biodiesel was considered as a way to afford a privileged degree of national energy security in an eco-friendly, cost-efficient, and sustainable manner. Nowadays, many countries trust biofuels for the process of decreasing dependence on imported fossil fuels, providing an enhancement to traditional energy resources. It also satisfied the energy requirements of the nation's immense population by the use of nonedible feedstocks. As biodiesel was nonhazardous, sulfur-free, renewable, and eco-friendly, it was considered as the most prominent and sustainable alternative to fossil fuel. The presence of 12% oxygen in biodiesel causes complete combustion of fuel, which in turn leads to less emission. Biodiesel can be used as an additive agent to petroleum fuels to improve the lubrication properties. Biodiesel extends the engine component life and can be used in existing diesel engines with only minor modifications.

Biodiesel is an eco-friendly substitute liquid fuel which can be used in any diesel engine with and without any modification. Biodiesel is blended with diesel or can be used directly in the diesel engine. Biodiesel, an alternate fuel consisting of fatty acid-methyl/ethyl esters of long fatty acids, is acquired by transesterification process. Usually, the diesel engine is used in power generation, automobiles, and numerous industrial applications. Accordingly, the diesel engine was deliberated as the basic unit of the present-day human civilization. The application of biodiesel in the diesel engine has gained considerable attention due to its ecological advantages and as a persistent alternative for diesel fuel.

1.2 Overview about avocado

Avocado (*Persea americana*), a tree, belongs to the family Lauraceae. It was extensively grown in south-central Mexico. All over the world, avocadoes were commercially valued and grown in tropical and Mediterranean climates with a nominal height growth of 30 m. Based on the different varieties, the weight of the avocado seed ranges with an average of 80–2500 g. Fig. 3 depicts the global status of avocado production on a country-wise basis [23]. The United States holds a high production rate of 70.3%; the second highest production rate was reported in Africa with 15.2%, and the third highest production rate was

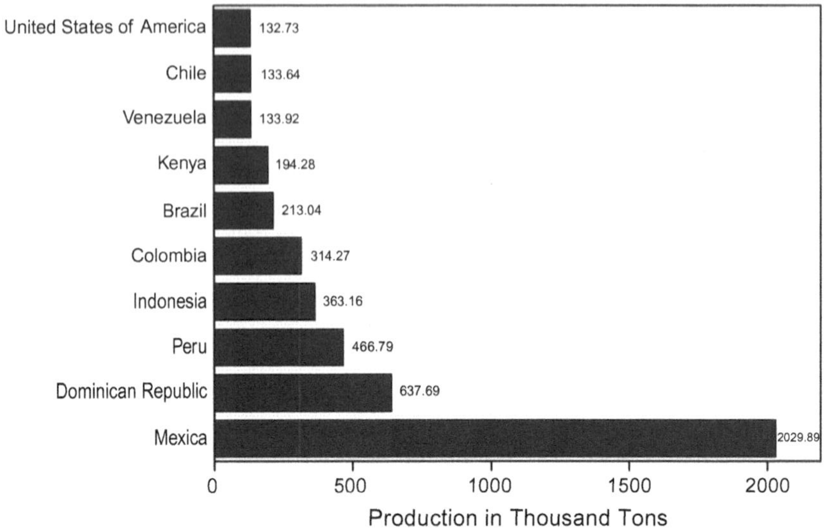

FIG. 3 Global status of avocado production on a country-wise basis [23].

reported in Asia with 10.9% [24]. The top avocado producing areas were identified as Mexico with 2029.89 thousand tons, the Dominican Republic with 637.69 thousand tons, and Peru with 466.69 thousand tons. Globally, the production of avocado fruit was increased from 2.71 million metric tons in 2000 to a high record of 5.92 million metric tons in 2017. Fig. 4 depicts the Avocado production worldwide from 2000 to 2017 [25]. The useful parts of fruits were converted into required materials as food essentials, and the seeds were considered as waste products. Usually, avocado fruit was edible and comprised of high-fat content. After the consumption of fresh fruit, the seed has remained as a residue. In the context of waste to energy conservation and management, the waste avocado seeds gained attention for the production of sustainable and renewable alternative fuel. Conversion of the avocado seed oil into eco-friendly, sustainable alternative fuel was said to be a promising strategy from the viewpoint of green and ecological protection. Thus, in the present investigation, nonedible oil like avocado seed oil was selected as the principal source for the production of biodiesel. The major consideration for selecting the aforesaid nonedible oil for biodiesel production includes consistent availability of the feedstock. Avocado seed oil was considered worldwide as a cost-effective and easily accessible source for the production of biodiesel. Avocado seed oil was one of the most available renewable sources of energy. A well-grown fruit of avocado contains nearly 20%–30% oil with seed. Avocado seed oil consists of triglycerides, carotenoids, and significant amounts of phenolic compounds. Avocado seed oil has a higher viscosity, density, and flash point than diesel fuel. Thus, the aforesaid characteristics of avocado seed oil make it a feasible source for biodiesel production.

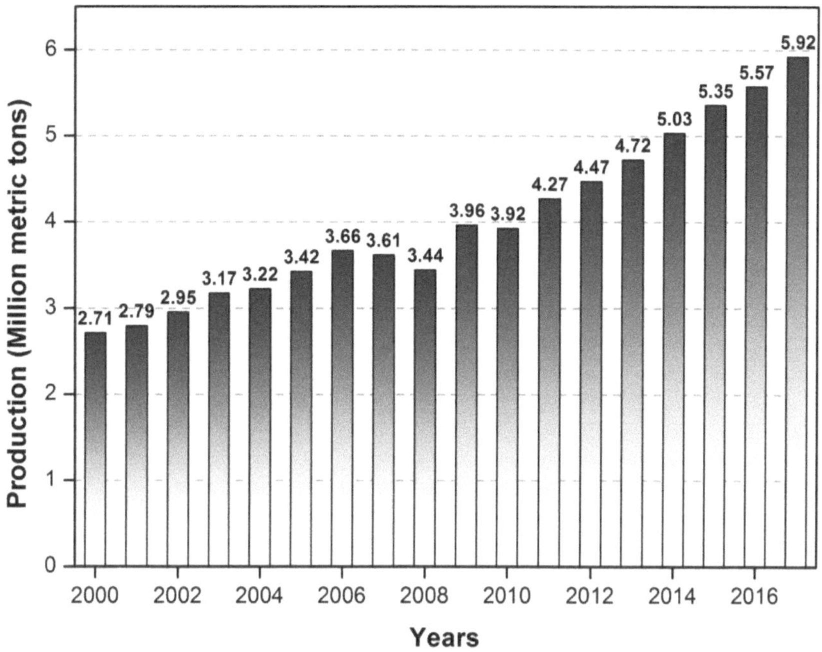

FIG. 4 Avocado production worldwide from 2000 to 2017 [25].

1.3 Overview of cetane improver and ceramic coating

Specifically, with diesel composition, aromatic components have gained extensive consideration among researchers. Any variation in the aromatic content of the fuel might have a significant impact on exhaust emissions in an intricate manner. The reason for this was that aromatics have a high impact on the physical and chemical properties of the fuel, which in turn incurred a substantial influence on the combustion process. Such fuel property includes its chemical structure, cetane number, density, viscosity, and the like. Thus, to obtain improved fuel properties, the cetane number was progressively increased with the enrichment of cetane enhancer or improver. Fig. 5 depicts the role of cetane enhancer with liquid-based alternative fuels.

In general, distinct types of enhancer or additives, namely metal-based, oxygenated additive like dimethyl ether (DME), diethyl ether (DEE), and alcohol like ethanol and methanol, antioxidants like di-tert-butyl peroxide (DTBP) and BHA (Butylated Hydroxy Anisole), cetane enhancer and the like were used to obtain reduced NOx emission during the combustion of biodiesel. The cetane improver was used to acquire decreased NOx emissions with biodiesel samples. This declined NOx emission was due to the reduced engine ignition delay and premixed combustion phase, which indicated the burning of less fuel in the premixed combustion phase. The high combustion chamber temperature was

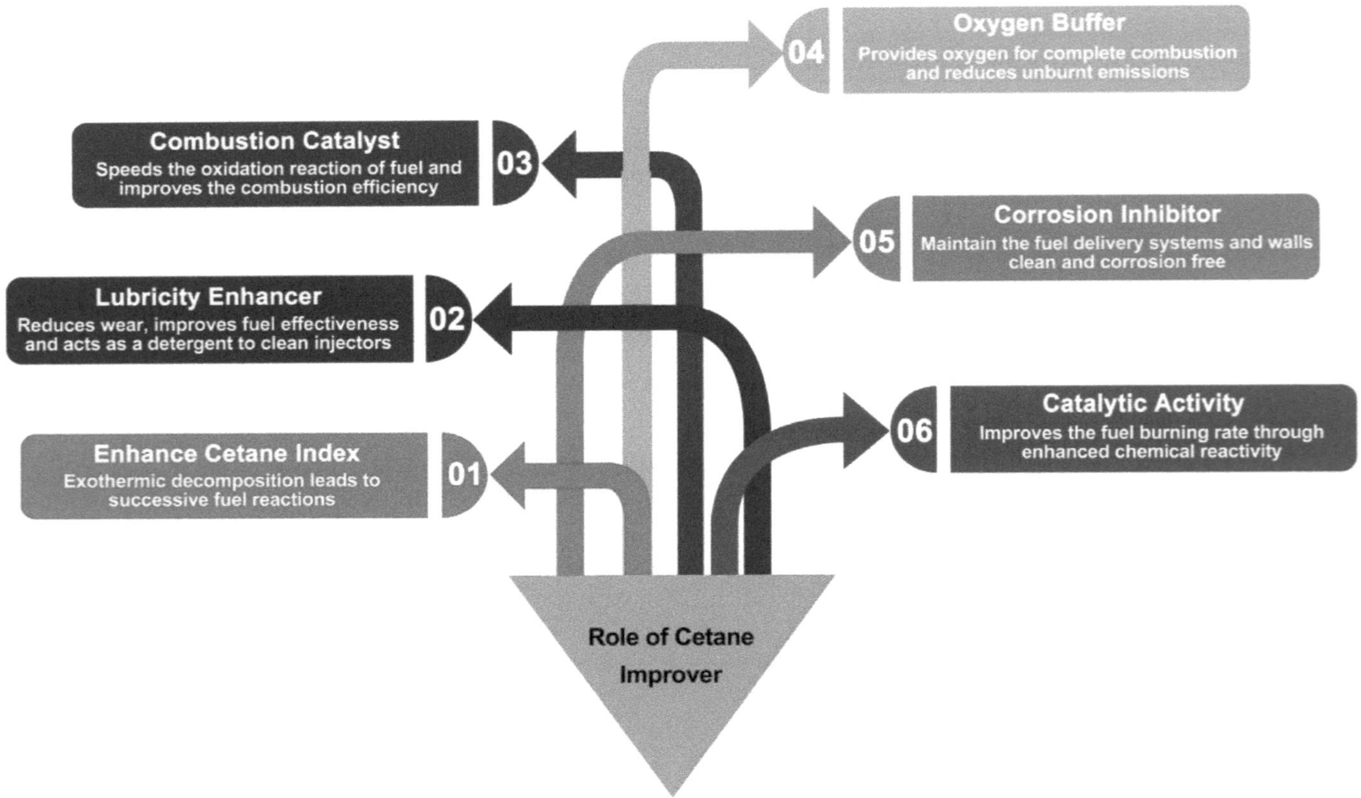

FIG. 5 Role of cetane enhancer with liquid-based alternative fuels.

decreased, which hinders the NO formation. Some of the efficient cetane improvers, such as 2-ethyl hexyl nitrate (EHN) and di-tert-butyl peroxide (DTBP), were frequently used with biodiesel to obtain reduced NOx emissions. Moreover, the existence of innate oxygen in a cetane improver overpowers the formation of soot which was considered to be a favorite for application in the diesel engine. In the diesel engine, the practice of using biodiesels has gained much glare of publicity due to their environmental advantages and as a renewable alternative for diesel fuel. In this view, a cetane improver along with fatty acid alkyl ester appears to be one of the evolving forthcoming fuels for diesel engines.

The investigation of biodiesel as an operating fuel leads to decreased engine characteristics owing to its inferior physiochemical properties. As a result, the fuel properties were examined with maximum attention in support of the persistent usability of biodiesel in a compression ignition engine. So the fuel properties modification technique such as the addition of a cetane improver gained more significance in realizing the better application of fuel to obtain the rated power and declined emission characteristics. The addition of a fuel additive with biodiesel leads to increased cetane number, which in turn leads to enhanced combustion and declining emission characteristics. In general, the cetane number of fuel defines its ignition characteristics. In addition, fuel additives unclog blocked injectors and even reduce or eliminate knocking, which leads to the smooth running of the engine.

Energy conservation and management have been the vision for researches associated with Internal Combustion Engines (ICE). In a compression ignition engine, from the total heat supplied, just one-third of supplied energy was transferred to positive work at the output end; one-third has vanished as exhaust gases and the remaining as cooling water. Thus, the heat transfer from the working fuel to cooling system accounts for up to 30% of the fuel energy. Thus, thermally insulating the combustion chamber components like piston head, cylinder head, and inlet and exhaust valves with ceramic coating material led to improved performance and emission characteristics of the engine and reduced heat loss and exhaust gas emission. The combustion chamber was thermally insulated and altered into a semi-adiabatic engine. The conventional diesel engine with its insulated combustion chamber by heat resistant coating was called a low heat rejection engine. The process of converting a conventional engine into an LHR diesel engine offered effective combustion and converted the accumulated heat energy into positive work. On comparison with a conventional diesel engine, the semi-adiabatic engine experienced a significant reduction of about 40% in specific weight and volume. The compact size and high power/weight ratio of the semi-adiabatic engine considerably shrank the size and weight of the vehicle.

Different materials such as fly ash, silicon carbide, silicon nitride, aluminum, magnesium silicate, and other ceramic materials like Zirconia oxide, NiCrB, MgO-ZrO_2, Yttria stabilized zirconia, and Partially stabilized zirconia were used as a thermal insulating material for an LHR engine. The merits of thermal barrier coating were fuel economy, increased thermal efficiency, multi-fuel capability using low-cetane biodiesel fuels, and reduced engine exhaust emissions. Thermal barrier coating was used to attain uniformity and firmness of hot section metal components and enhanced engine performance and effectiveness in diesel engines. The combustion chamber temperatures of ceramic-coated engines are higher than those of uncoated engines.

The ceramic-coating material was selected based on two aspects, namely low thermal conductivity and high coefficient of thermal expansion. Fig. 6A depicts the Evolution of Thermal barrier coating technology [26]. From 1940 to 2010, there existed different approaches like wrought materials, conventionally cast materials, a directionally solidified and single-crystal approach to obtain a low heat rejection phenomenon. The aforesaid approaches can withstand up to 1000°C. From early 1970, thermal barrier coating started to evolve and persist since it can withstand up to 1200°C. Fig. 6B depicts the cross-sectional view of the thermal barrier coating technique [27]. Important aspects like combustion gas, ceramic coating, bond coat, component, and cooling airflow were plotted against change in temperature. The sticking property of ceramic layering mainly depends on the bond coat initially applied to the component. From the figure, it was significant to mention that the heat transfer was high in the combustion gas area and low in the ceramic-coated area and the component. This decreased heat transfer was due to the low thermal conductivity nature of ceramic material. The heat transfer from the combustion area to the component was based on the thickness of the coating. The direct application of the coating on engine combustion chamber components was not suitable due to varying operating conditions. Thus, a certain amount of material removal was necessary before coating, and also the layering of ceramic coating was restricted with the bond coat. To achieve improved working, the components were coated with ceramic coating materials just above the bond coat. The major advantages of using bond coat were improved adhesion properties, reduced interface stresses, and corrosion behavior of a component. The important features of ceramic layering were reduced thermal conductivity, high co-efficiency of thermal expansion, and high phase stability during high-temperature operation. In the plasma coating process, the plasma gun comprised powdered ceramic material at high temperature for about 6000–15,000 °C and the material was allowed to coat on the component at a velocity up to 1500 m/s. These operating parameters vary based on the selected component and working operation considered.

1.4 Objective and novelty of the present work

An alternative fuel, namely renewable biodiesel, could be the substitute for fossil fuels and offset the emission of greenhouse gases. In this experimentation, raw avocado seed oil was converted into novel avocado seed oil biodiesel using

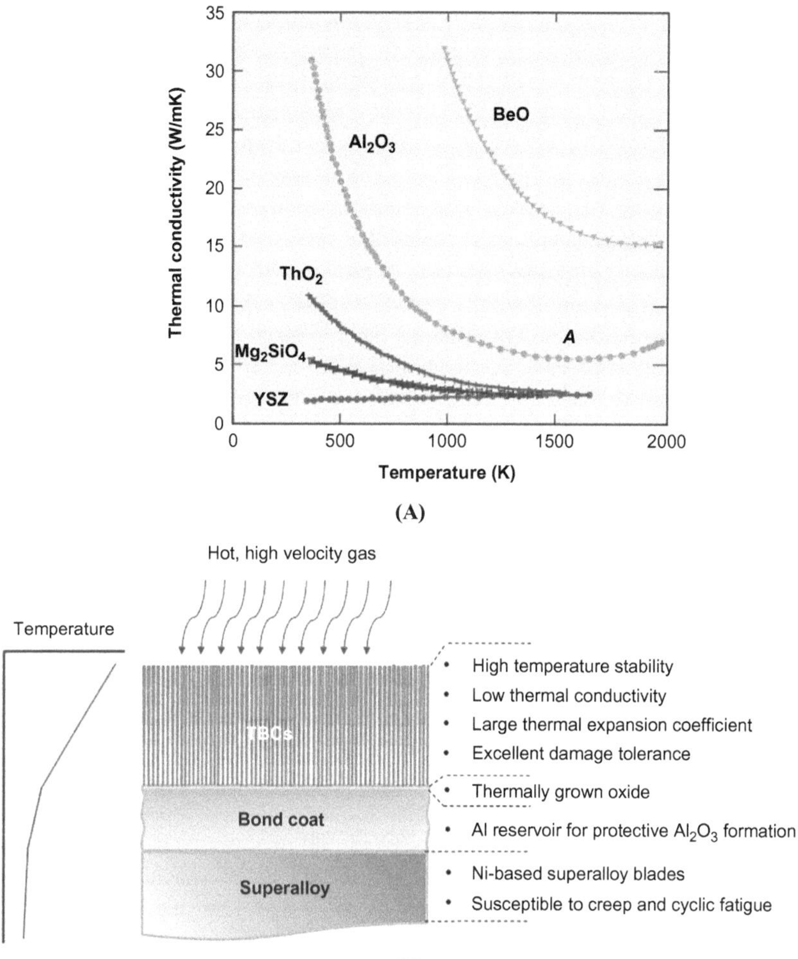

FIG. 6 (A) Evolution of thermal barrier coating technology [26] and (B) cross-sectional view of thermal barrier coating technique [27].

the transesterification process. In an unmodified internal combustion engine, the performance was profoundly affected by the application of renewable biodiesel as a working fuel. Many researchers experimented with distinct engine modification techniques in order to obtain enhanced engine performance characteristics in a biodiesel operated diesel engine. Thermal barrier coating was one of the techniques used to attain improved engine performance and emission characteristics by decreasing the heat loss incurred from the combustion chamber. The investigation of avocado seed oil methyl ester as an operating fuel leads to decreased engine characteristics due to its inferior physical and chemical properties. Consequently, a certain fuel properties modification technique such as the addition of a cetane improver gained more importance in attaining the better usage of fuel to acquire rated power and diminished emission characteristics. The addition of a cetane improver, namely Pyrogallol (PG), has highly influenced the cetane number of fuel. The cetane number of A100 greatly affects the engine combustion characteristics and emissions like CO, HC, and smoke. A detailed analysis revealed that few research investigations were observed in the combination of cetane improver with the engine modification technique. Many researchers have examined biodiesel production from distinguishing feedstocks, but no investigators have selected raw avocado seed oil for the production of biodiesel. A broad search did not reveal any research work about the effect of A100 + cetane improver+ YSZ coating on engine performance and emission characteristics. The uniqueness of this research work was to examine the engine performance and emission characteristics with special importance on emission by way of using A100 + PG in a YSZ-coated engine. The aim of this research work was to enhance the characteristics of a diesel engine using a thermal barrier-coated piston operated with avocado biodiesel and cetane improver (Pyrogallol).

1.5 Methodology

Fig. 7 depicts the methodology performed in the present study. The practice of choosing feedstock for the production of biodiesel relies on two factors, namely accessibility and price. The seeds of a perennial avocado were processed to obtain a high yield of avocado oil. The acquired oil was chosen as the main source for the production of biodiesel. With the operating conditions like the selection of a catalyst, the behavior of alcohol and reaction condition (temperature and operating speed), the production process was carried out. In a diesel engine, the prospect of selecting appropriate operating conditions was to acquire reduced emission and improved performance parameters. In addition, the calculation of acid value plays an important role in determining the type of esterification (either base esterification or acid esterification) process for biodiesel production. The physiochemical properties of the A100 were measured and evaluated based on the ASTM biodiesel standard under laboratory evaluation conditions. The acquired properties were compared with the readily available nonrenewable source of energy called diesel. Later, the obtained biodiesel samples have been experimented with in the diesel engine. At distinct loads, the engine performance characteristics were observed as brake thermal efficiency and brake specific fuel consumption, and emission characteristics were observed as CO, HC, NOx, and smoke that were examined to investigate the behavior under various operations. The present work was progressed with three different approaches like normal engine operation, ceramic-coated engine operation, and application of cetane improver. The produced biodiesel blends were

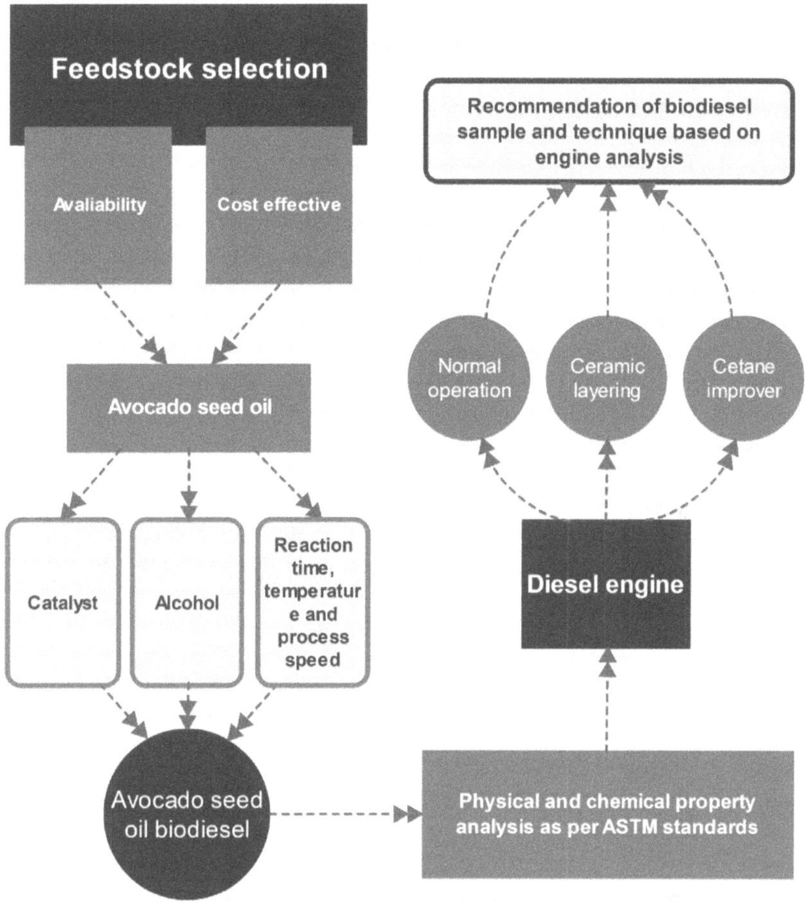

FIG. 7 Methodology used in the present study.

evaluated with the aforementioned strategies, and the obtained observations were compared with diesel in normal engine condition. This experimental methodology revealed the complete stage of processes involved in the present work and offered the recommendation for a biodiesel sample with an advanced technique based on the engine investigation.

2 Materials and methods

The experimentation comprising of conversion of feedstock into alternative fuel, analysis of physiochemical properties of produced biodiesel in line with ASTM biodiesel standards, exploring the engine characteristics with thermally coated combustion chamber components, investigating the enhanced fuel properties of the obtained biodiesel with cetane improver in the direct injection compression engine was discussed briefly in this section.

2.1 Production of biodiesel

Fig. 8 depicts the biodiesel production from avocado seed oil. Initially, the acid value of obtained raw avocado seed oil was measured using the titration method and observed within the standard ASTM prescribed bound. Since the acid value of the avocado seed oil was small, base esterification was carried out in the present work. The first stage comprises of preprocessing, oil filtration, and preheating process. The process of obtaining avocado oil from avocado seed includes drying the seeds for more than 3 weeks under direct sunlight to obtain reduced moisture content and the oil was extracted using an oil expeller. This machine was operated under normal speed and a twofold technique was used to obtain improved oil yield from the raw seeds. The extracted seed oil comprises solid particles due to crushing of dried seeds in between the roller crushers. Therefore, the oil was filtered and semisolid impurities were removed from the oil, and the remaining crushed oil was considered for the next stage. Later, the oil was preheated at 60°C for about 10 min to remove the moisture content and water particles. A single-stage base esterification process was carried out for the production of biodiesel samples. The transesterification process was said to be a reliable process as it was a cost-effective approach and incurred less time for oil conversion. During the course of the transesterification process, the condenser was used to lessen the escape of alcohol. The homogeneous catalyst was prepared by blending the catalyst with alcohol and its pilot-scale production was considered. The homogeneous catalyst was allowed to mix with preheated oil. The processed oil after 1 h of reaction time was transferred to a phase separation funnel. This settling process was allowed to perform for more than one complete day. The processed solution attained phase change based on density difference. The alternative fuel was separated and the quality improvement of the solution was performed using de-neutralized water washing. This process was performed more than four times to obtain high-quality fuel and further heating of processed oil was performed to eradicate the presence of moisture. The obtained solution was termed as avocado seed oil biodiesel or methyl ester of avocado seed oil biodiesel (A100).

2.2 Property analysis

The determination of physical and chemical properties of the obtained biodiesel plays an important role in measuring all the engine output variables. In the present work, the fuel properties were measured based on ASTM biodiesel standards. The density of both avocado seed oil biodiesel and diesel was performed based on the ASTM D1298 procedure. On comparing with diesel, the avocado seed oil biodiesel experienced high density due to conversion from triglycerides

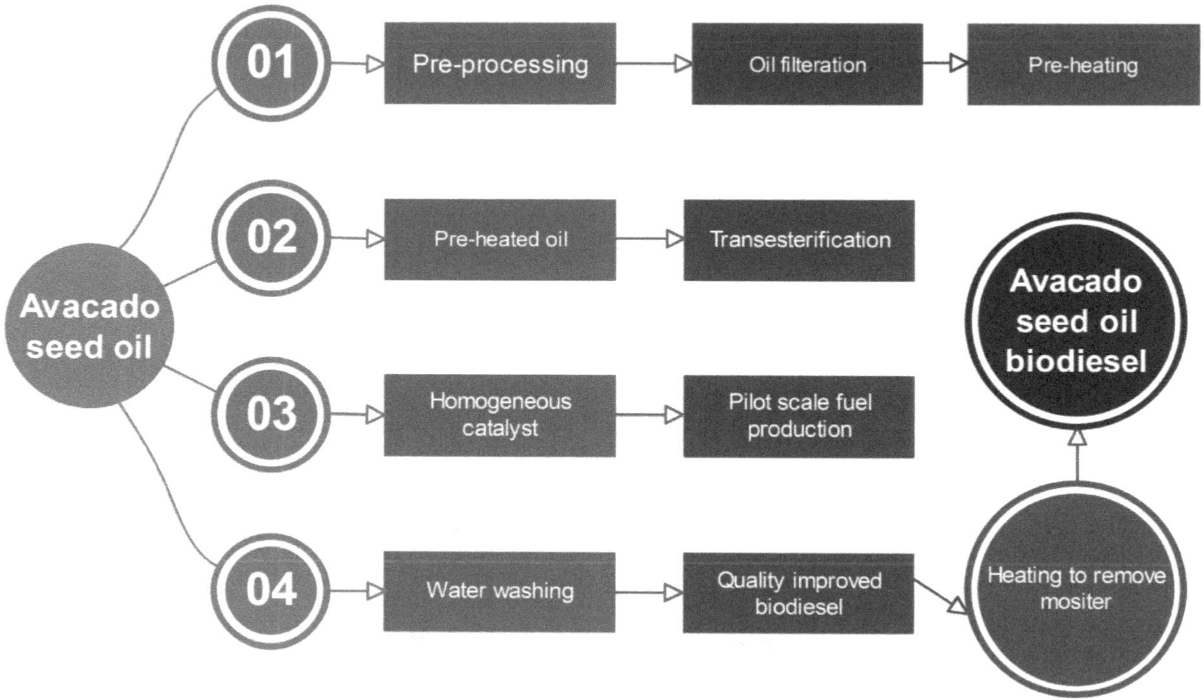

FIG. 8 Biodiesel production from avocado seed oil.

to monoglyceride. Further, the kinematic viscosity was measured in accordance with the ASTM D445 standard testing procedure. This testing was performed using a viscometer and the diesel exhibited lower viscosity than avocado seed oil biodiesel. In CI engine, both density and viscosity were considered as vital parameters which determine the fuel flow condition. In addition, the aforementioned properties highly influence the emissions acquired from the engine.

Followed by, flash point of fuel was considered as an important parameter owing to the occurrence of ignition delay period based on start of injection and start of combustion. This testing process was performed in an open cup apparatus in conformity with the ASTM D93 testing condition. At low temperature, diesel showed low flash and fire point due to the promising chemical compositions. On the other hand, the avocado seed oil biodiesel incurred flash followed by fire at increasing temperature. This high-temperature requirement to perform flash and fire was due to the calorific value of the fuel. Among all fuel properties, calorific value or heating value was considered to be the characteristic property which was measured by a Bomb calorimeter in line with the ASTM D240 testing procedure. The calorific value of both diesel and avocado seed oil biodiesel was measured and found within the bounds of ASTM standards. The performance characteristics of avocado seed oil biodiesel highly rely on calorific value. Thereby, low heating value leads to high fuel consumption and low thermal efficiency in the diesel engine. This variation was due to the necessity of producing the rated engine power output. In a diesel engine, cetane number plays an important role in examining the combustion characteristics of fuel. Due to the absence of oxygen, diesel exhibited a low cetane number. Conversely, avocado seed oil biodiesel perceived a high cetane number due to the inherent presence of oxygen content in the fuel. The chemical analysis was performed based on measuring elemental compositions like carbon, hydrogen, and oxygen using Perkin Elmer 2400 Series II. The composition of both diesel and biodiesel was compared, and it is worth mentioning that the oxygen in biodiesel was identified with 11.45 mass%. On the other hand, diesel showed higher carbon and hydrogen content than avocado seed oil biodiesel. On comparing with diesel, avocado seed oil biodiesel experienced a high ratio of carbon to hydrogen. Table 1 depicts the physical and chemical properties for conventional fuel and avocado seed oil biodiesel.

2.3 Ceramic layering

Thermal barrier coating was used to transform the conventional engine into a semi-adiabatic engine and supports to improve the shaft work by reducing the heat rejection from the combustion chamber. Ceramic layering was generally performed by using an atmospheric plasma spray coating technique which allows ceramic layering on the substrate. The ceramic layering on the combustion chamber components leads to reduced heat rejection and utilizes the accumulated heat to achieve improved combustion characteristics under a wide range of operating conditions. The bond coat between the substrate and ceramic

TABLE 1 Physical and chemical properties for conventional fuel and avocado seed oil biodiesel.

Property	ASTM standards	D100	A100
Density (kg/m^3)[a]	D1298	842	868
Kinematic viscosity at 40°C (cSt)[a]	D445	2.51	3.97
Flash point (°C)[a]	D93	53	148
Fire point (°C)[a]	D93	59	160
Gross calorific value (MJ/kg)[a]	D240	43.16	39.58
Calculated cetane number	D613	48	53
Proximity analysis (2400 Series II Perkin Elmer)			
C (mass %)	–	85	76.31
H (mass %)	–	15	12.22
O (mass %)	–	0	11.45
C/H	–	5.6	6.26

[a]*All properties were determined based on ASTM test conditions under laboratory condition.*

coating layer leads to enhanced adhesive property between the materials. The ceramic layer was introduced at high temperature ambience and durability plays a vital role to analyze the lifetime capability of the layer. The atmospheric plasma spray coating machine contains two major parts called the anode and cathode. The introduction of plasma gas and current was considered as the cathode, whereas the circulation of water to maintain the temperature of the plasma gun was termed as the water-cooled anode. At high temperature and under a pressurized plasma jet, the powdered ceramic material was introduced and converted into a molten state which was deposited on the surface of the substrate. This technology approach was considered as an advancement in the ceramic layering process and reliable with a cost-effective technology to achieve improved performance characteristics in internal combustion engines.

Direct layering of ceramic material on combustion chamber components was not recommended as the operation in a closed environment affects the specifications provided by the internal combustion engine manufacturer. Importantly, the compression ratio of the direct injection engine highly influenced the operation of the admitted fuel during the closed chamber combustion process. Therefore, any variation in compression ratio directly influenced all the engine output parameters directly or indirectly. Thus, before proceeding to ceramic layering, a certain amount of material was removed to achieve the same engine operating condition. The removal of material was possible with two mechanisms, namely sandblasting and micromachining. In the present study, the material removal was performed by using micromachining based on the coating layer thickness. Also, it is important to mention that the surface roughness of materials plays an important role in adhesion properties of ceramic material with substrate. Generally, the surface roughness of the material was considered as 4. Similarly, the compression ignition engine piston substrate material removal was performed by grit blasting to attain the surface roughness of 4 before ceramic coating [28]. Also, partially stabilized zirconia coating in compression ignition engine components, before coating substrate materials were removed and ceramic material was layered with a surface roughness value of 4 [29]. Based on the principle of material removal and surface roughness condition, the surface substrate was cleaned with anhydrous ethanol in the present work. Later, drying and cleaning were achieved under atmospheric conditions. The process of ceramic coating was planned initially with a bond coat of 100 μm using NiCrAl material. The coating was performed at a constant speed and material applying rate to reduce irregularity in the surface of the coating. After coating of bond coat, the surface of the material was allowed to cool and deposit on the surface under atmospheric conditions for one complete day. In the present work, yttria stabilized zirconia (YSZ) was designated as the ceramic material for the coating of the combustion chamber components. The reasons for selecting YSZ as a ceramic material were low thermal conductivity and high coefficient of thermal expansion. The coating of ceramic layer was created above the bond coat and a steady flow of material and travel speed for the entire coating process was measured to attain a reliable coating layer. Later, the coating components were allowed to settle in atmospheric conditions. Fig. 9 depicts the (a) YSZ-coated piston crown (b) YSZ-coated cylinder head, intake, and exhaust valve. It is important to mention that removal of material by micromachining was performed only in the piston head region and the other combustion unit parts like intake and exhaust valve and cylinder head were used as per the manufacturing standards.

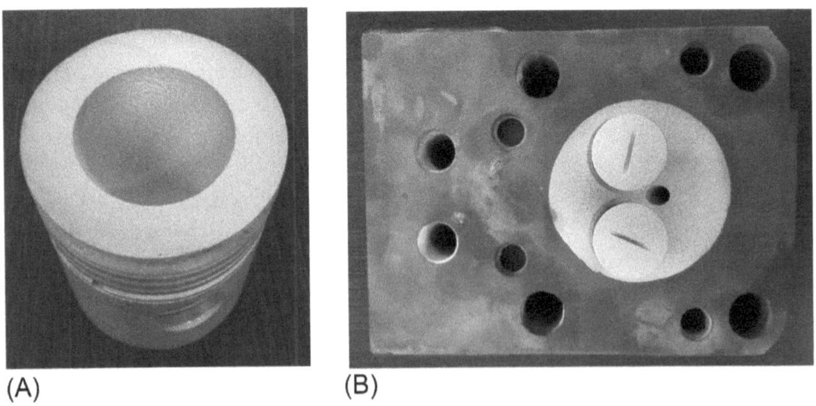

(A) (B)

FIG. 9 (A) YSZ-coated piston crown (B) YSZ-coated cylinder head, intake, and exhaust valve.

2.4 Cetane improver

With the inferior physical and chemical properties, the experimentation with avocado seed oil biodiesel might lead to a decline in engine characteristics. The fuel properties were observed with utmost care in good deed of the persistent usability of alternative fuel in engines. The fuel modification technique, like the addition of cetane enhancer shows the better utilization of fuel was achieved with high rated power and reduced emission characteristics. To attain the enhanced quality of fuel, the addition of cetane improver to the biodiesel was said to be a prominent technique. Reduced ignition delay and premixed combustion were incurred with fuel possessing a high cetane number. This high cetane number leads to enhanced performance and declining emission characteristics. Generally, the cetane number of the fuel depicts the ignition characteristics. The cetane number of the fuel highly affects the engine performance and emission characteristics. Avocado seed oil biodiesel possesses a higher cetane number than diesel. This increased cetane number with A100 leads to better combustion efficiency than conventional diesel. The importance of pyrogallol was depicted in Table 2. In the present work, Pyrogallol was used as a cetane enhancer. Pyrogallol (PG) is a white, water-soluble solid. Normally, the samples were brownish in color because of its sensitivity toward oxygen. The cetane enhancer leads to an improved fuel flow, cetane number, and pour point of biodiesel. Pyrogallol ($C_6H_6O_3$) is one such cetane improver

TABLE 2 Properties of PY.

Property	Pyrogallol
Molecular structure	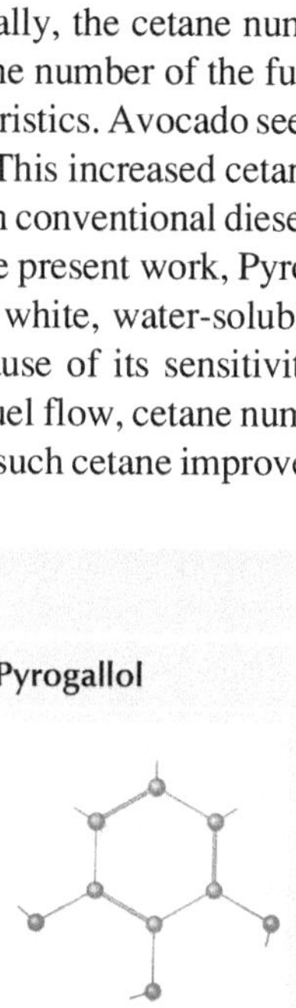
Chemical formula	$C_6H_6O_3$
CAS number	87-66-1
Molecular weight (g/mol)	126.11
Density (kg/m^3)	1.45
Flash point (°C)	230
Boiling point (°C)	309
Hydrogen bond donor count	3
Hydrogen bond acceptor count	3

TABLE 3 Properties of A100 and A100 + PY.

Properties	A100	A100 + PY
Density (kg/m^3)[a]	868	865.4
Kinematic viscosity at 40°C (cSt)[a]	3.97	3.58
Flash point (°C)[a]	148	140
Fire point (°C)[a]	160	152
Calorific value (MJ/kg)[a]	39.58	39.92

[a]*All properties were determined based on ASTM test conditions under laboratory condition.*

which helps to achieve upgraded properties of fuel. Table 3 depicts the properties of avocado seed oil biodiesel and PG enriched avocado seed oil biodiesel.

2.5 Fuel blends

In the present experimentation work, the following fuel blends were investigated:

- D100-Mineral diesel
- Neat Avocado seed oil biodiesel-A100
- Neat Avocado seed oil biodiesel with cetane improver Pyrogallol (PG)-A100 + PG
- Neat Avocado seed oil biodiesel in Yttria stabilized zirconia (YSZ)-coated engine-A100 + CE
- Pyrogallol enriched Neat Avocado seed oil biodiesel in a YSZ-coated engine-A100 + PG + CE.

2.6 Engine selection

Fig. 10 depicts the schematic diagram of the experimental setup. In the present work, the experimentation was carried out in a single-cylinder, water-cooled direct injection diesel engine. Table 4 shows the important specification of the experimental setup. As per the manufacturer's catalog, hemispherical bowl geometry, fuel injection pressure and injection timing, and engine compression ratio were used. Different sensors like airflow sensor, fuel flow sensor, temperature sensors were used for the measurement of data during engine experimentation. By means of the eddy current dynamometer, the engine brake power was produced. In order to reduce the heat, water was circulated persistently around the dynamometer for cooling purposes. Further, the circulation of water for engine and dynamometer cooling was carried out by using pump arrangement

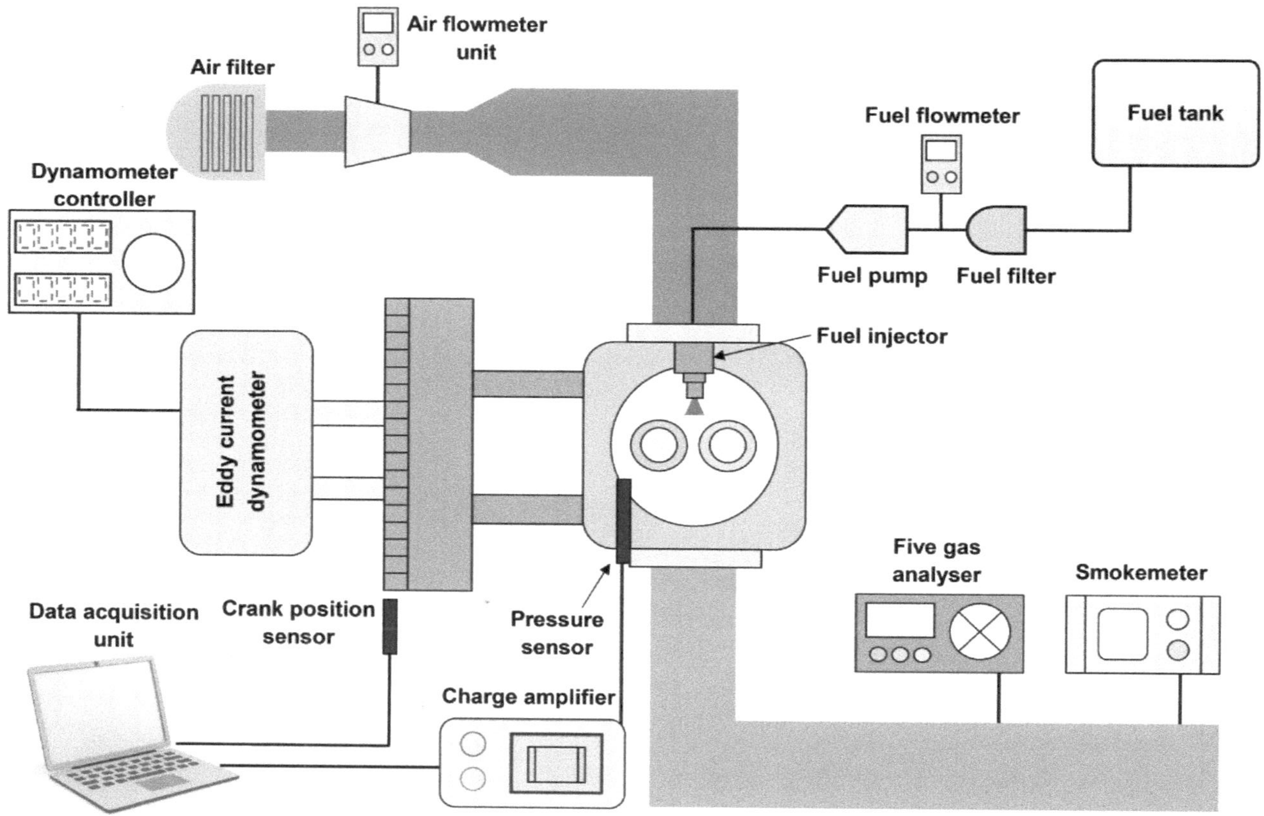

FIG. 10 Schematic diagram of experimental setup.

TABLE 4 Specification of experimental setup.

Details of engine	
Make and model	Kirloskar/TV1
Type	Direct injection, water cooled, vertical, diesel, naturally aspired engine
No. of cylinder/stroke	01/04
Rated power	5.2 kW
Bore (mm)/stroke (mm)	87.5/110
Ignition type	Compression Ignition (CI)
Compression ratio	17.5:1
Injection pressure	210 bar
Injection timing	23° before Top Dead Centre (bTDC)
Speed	1500 Rev./min
Diameter of fuel injector opening	0.3 mm
No. of nozzle hole in fuel injector	3
Details of dynamometer	
Make and model	Technomech Pvt. Ltd./TMEC10
Type	Eddy current dynamometer
Cooling	Water cooled
Speed	1500–6000 Rev./min
Details of combustion analyzer	
Cylinder pressure sensor	Kistler (M111A22), Piezo electric sensor
Crank angle encoder	Kistler (2614C11)
Details of data acquisition system	
Make and model	National instruments/USB-6210
Type	16AI; 4DI; 4DO USB-multifunction I/O device
Details of other instruments	
Air flow measurement	Make: Wika/Model: SL1
Fuel flow measurement	Make: Broiltech/Model: FCM and Differential pressure transmitter
Load cell	Make: Sensortronics/Model: 60001

and rotometers. The fuel was initiated from the storage tank to the fuel filter, then allowed to the fuel pump, and then permitted into the fuel injector. The engine setup was computerized to acquire a reduced effect of uncertainty during engine investigation. The measurements of various sensors were a junction to the NI USB 6210 model data acquisition unit. The various data were allowed to record online using MatLab based software for evaluation of engine characteristics. The regulated emissions like carbon monoxide (CO), oxides of nitrogen (NOx), and hydrocarbon (HC) were observed using the AVL 444 di-gas analyzer and the smoke emission was recorded using the AVL 437C smoke meter setup. Table 5 shows the specification of exhaust analyzers.

To achieve steady-state condition, the engine was allowed to run with mineral diesel for about 20 min at minimum load. In the present work, the engine was permitted to work with battery ignition technology at 1500 rpm constant speed and varying load. Once the steady state was attained, the load was varied, and corresponding engine observations were measured. Also, the engine was operated three times in each operating condition to obtain decreased influence of error during experimentation. The engine operation recording with diesel as working fuel was benchmarked as a baseline observation. After completing the observation with diesel, the fuel in the tank was discharged and avocado seed oil methyl ester (A100) was refueled in the tank. Then the same procedure was performed to obtain the performance and emission characteristics. Before the start of the observation, the engine was allowed to operate for about 10 min at minimum load to eradicate the presence of the former fuel in the fuel line of the engine. After completing the observation with A100 sample, the fuel was drained out and cetane improver (Pyrogallol-PG) was mixed with A100 to acquire improved fuel properties. PG enriched A100 refueled in the fuel storage tank and the same procedure were performed to record the performance and emission characteristics. After data recording with A100 + PG, the remaining samples were drained out of the fuel tank and the engine was allowed to cool for one complete day. The engine was reinstalled with thermally coated combustion chamber components to acquire a low heat rejection engine. Then neat biodiesel, namely A100, was fueled in the storage tank and the engine was operated to investigate the engine characteristics. Later, the storage tank was once again refueled with a modified alternative fuel sample A100 + PG to obtain various engine performance and emission characteristics. The data obtained from experimentation was presented for engine brake thermal efficiency and brake specific fuel consumption parameters using three-dimensional representations, and the same approach was employed for emission characteristics representation.

2.7 Uncertainty analysis

For the selected system, uncertainty or error analysis was performed by using the J.P. Holman proposed backward propagation technique. The different

TABLE 5 Specification of exhaust analyzers.

AVL 444 DIGAS analyzer						
Measured parameter	*Measurement principle*	*Measuring range*	*Resolution*	*Accuracy*		*% uncertainty*
CO	NDIR	0%–10% vol.	0.01% vol.	<0.6% vol:	±0.03% vol	±0.2
				≥0.6% vol:	±5% of value	±0.3
HC	NDIR	0–20,000 ppm vol.	≤2000: 1 ppm vol.	<200 ppm vol:	±10 ppm	±0.1
				≥200 ppm vol:	±5% of value	±0.2
NOx	Electro chemical sensor	0–5000 ppm vol.	1 ppm vol.	<500 ppm vol:	±50 ppm vol	±0.2
				≥500 ppm vol:	±10% of value	±0.9
Warm up time: Less than 10 min						
Response time: Less than 15 s						
Relative humidity: Less than 95% (non-condensing)						
AVL 437C free accelerator smokemeter						
Type					Hartridge smokemeter	
Measuring range					0–100 Hartridge Smoke Unit (HSU)	
Accuracy					±2% relative	
Resolution					0.1%	
Ambient temperature range					−5 to +45°C	
Warm up time					10 min (self-controlled) at 20°C	
Response time					15 s for more than 90% of response sample	
Sampling					Direct from engine exhaust tail pipe	

TABLE 6 Experimental uncertainty of various measuring units.

Parameters	Measuring technique	Accuracy	Errors (±)
Load	Strain gauge-type load cell	±10 N	±0.2
Speed	Magnetic pickup principle	±10 rpm	±0.1
Fuel flow measurement	Volumetric measurement	±0.1 cc	± 1
Temperature	Thermocouple	±1°C	±0.15
Crank angle encoder	Magnetic pickup principle	±1 deg	±0.2
Pressure	Magnetic pickup principle	±0.1 kg	±0.1
Time	Stop watch (Manual)	±0.1 s	±0.2
Manometer deflection	Balancing of column of liquid	±1 mm	±1
CO	NDIR technique	±0.02 vol%	±0.2
HC	NDIR technique	±10 ppm	±0.1
NOx	NDIR technique	±12 ppm	±0.2
Smoke	Opacimeter	±1 HSU	±1

parameters associated with the error were surrounding condition, data recording, and sensitivity of the measuring device, equipment selection, and so forth. The detailed error analysis was already mentioned in the previous publication [12–14] and the uncertainty value for the selected system was obtained as ±2.27%. Table 6 shows the experimental uncertainty for various measuring units.

3 Results and discussions

At varying loads, the engine performance characteristics like brake thermal efficiency and brake specific fuel consumption and emission characteristics like carbon monoxide (CO), hydrocarbon (HC), oxides of nitrogen (NOx), and smoke were reported for diesel (baseline), A100, A100 + PG, A100 + CE, and A100 + PG + CE mode of operations.

Fig. 11 depicts the engine brake thermal efficiency against load and working conditions. Varying engine load exhibited a gradual change in thermal efficiency. At increasing load, high thermal efficiency was observed for all fuel samples. At all loads, diesel (D100) experienced an increasing trend of thermal efficiency. This increment was due to its promising fuel properties. High thermal efficiency with diesel indicated the conversion of constituents in diesel into the mechanical shaft work of the system. Similarly, diesel in a Kirloskar TV1

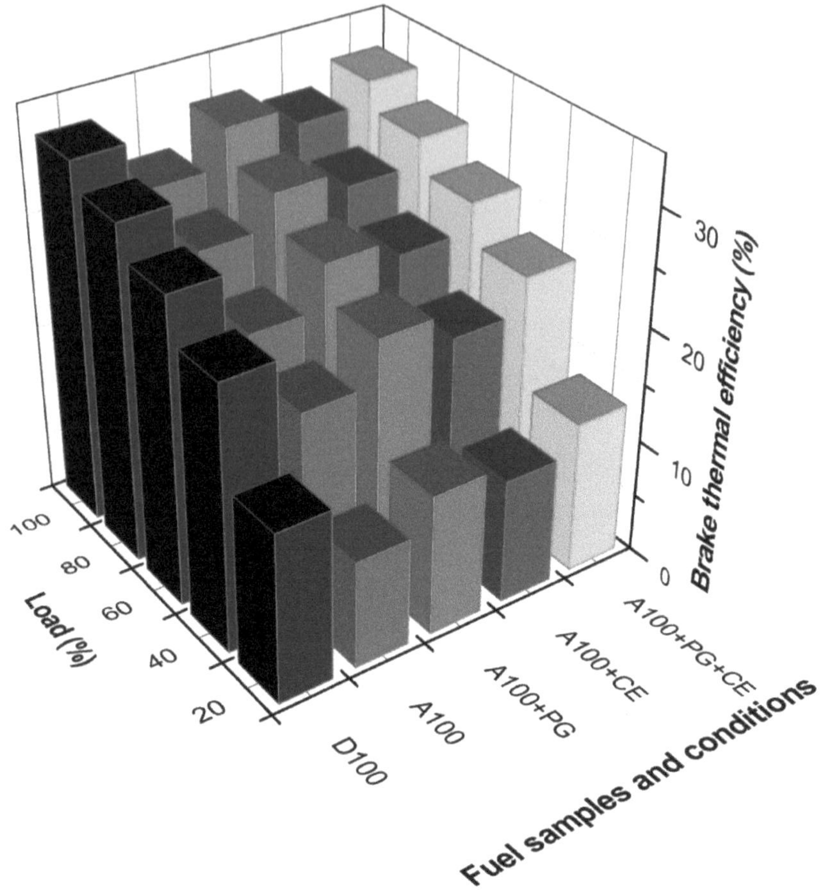

FIG. 11 Engine brake thermal efficiency against load and working conditions.

model diesel engine experienced high thermal efficiency. This high BTE was due to its potential fuel properties [30]. Also, experimentation of diesel in a six-cylinder diesel engine in comparison with alternative fuel exhibited high thermal efficiency [31]. On comparing with diesel, A100 showed low thermal efficiency at all loads. This decreased thermal efficiency was due to its inferior physiochemical properties. Similarly, rapeseed oil biodiesel under ternary fuel conditions in a four-cylinder turbocharged engine exhibited lower thermal efficiency than diesel [32]. At all loads, PG enriched A100 experienced a higher trend of thermal efficiency than neat biodiesel (A100). This efficiency improvement was due to enhanced fuel properties of A100. In the YSZ-coated engine, A100 showed high thermal efficiency at all loads. This increment with a layered ceramic engine was mainly due to the reduced heat transfer rate of the engine with cooling water. Similarly, a thermal barrier-coated engine investigation

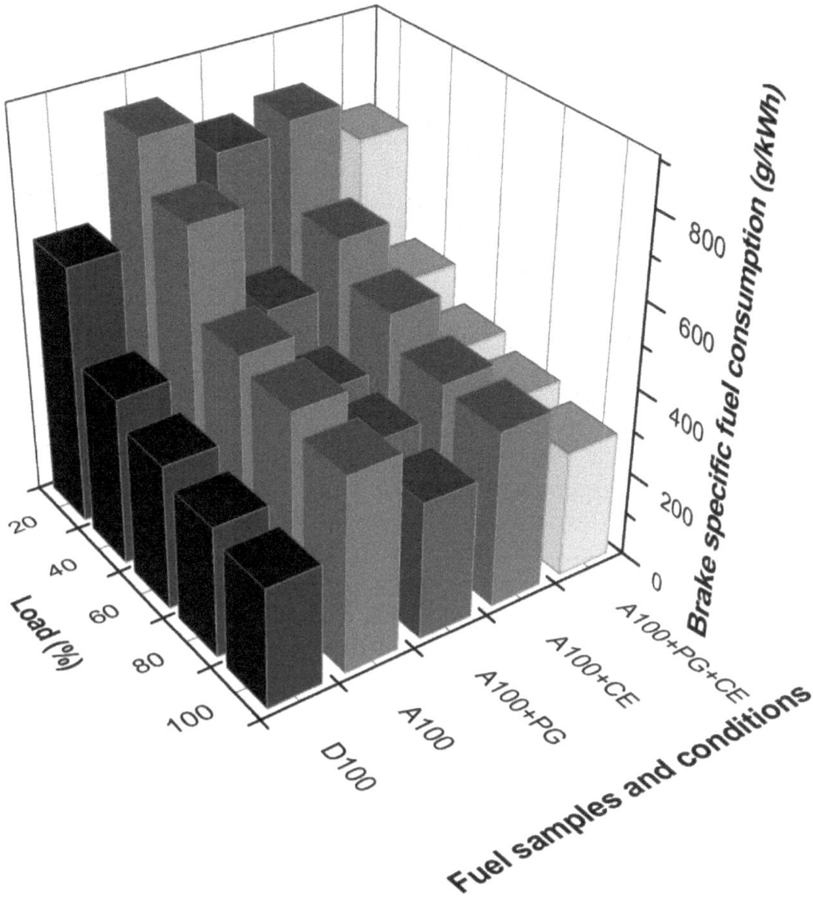

FIG. 12 Engine brake specific fuel consumption against load and working conditions.

with neem oil biodiesel showed higher efficiency than an uncoated engine due to improved heat utilization conditions [33]. In the coated engine, PG enriched A100 showed higher thermal efficiency than other fuel blends on all loads. This notable enhancement was due to improved fuel properties and utilization of accumulated heat in useful piston work. A similar trend of increased efficiency was reported with papaya biodiesel examination with an ignition promoter in a semi-heat rejection engine [34].

Fig. 12 depicts the engine brake specific fuel consumption against load and working conditions. During the course of engine operation, brake specific fuel consumption directly represents the ratio of the amount of fuel consumed to the brake power development. On comparing with A100 biodiesel, diesel exhibited a lower brake specific fuel consumption. This decreased BSFC was owing to its promising fuel properties. At an increasing load, low brake specific fuel consumption was observed with all fuel samples. Similarly, reduced fuel consumption was perceived by many researchers with diesel as a working fuel due to its supporting fuel properties [35–37]. On comparing with diesel, A100 showed higher fuel consumption at all loads. As the heating value of A100 was lower than diesel, A100 experienced high fuel consumption to acquire the rated engine power. Similar behavior of increased fuel consumption was reported with biodiesel examination in Lombardini 15 LD 350 compared to diesel [38]. On all loads, PG enriched A100 experienced lower fuel consumption than A100 due to its enhanced physical and chemical properties. With A100 biodiesel, the coated engine experienced lower brake specific fuel consumption than the uncoated engine. This decreased fuel consumption was due to enhanced combustion efficiency with the coated engine. A similar observation of fuel consumption was reported with peanut oil biodiesel in an Fe_2B-coated compression ignition engine [39]. In the coated engine, A100 + PG showed lower fuel consumption than other fuel blends and operating condition. In an LHR engine, PG enriched A100 experienced improved combustion efficiency and physiochemical properties. Similarly, ignition promoter enriched lemon oil biofuel experienced low brake specific fuel consumption [40].

Fig. 13 depicts the engine carbon monoxide emission against load and working conditions. Carbon monoxide (CO) was produced from a CI engine due to insufficient oxygen content in fuel during the oxidation process. This gas is colorless, dangerous, and tasteless. Due to deficiency in oxygen, carbon dioxide conversion is limited and CO emission produced. In the diesel engine, factors like engine load, speed, fuel-air ratio, geometry of the piston and combustion chamber, physiochemical properties of fuel and kind of injection, injection timing and pressure, compression ratio, cetane number, and the like influence the gases in the engine exhaust emission. CO emission highly relies on the presence of innate oxygen in the fuel-air mixture. Deficit oxygen during the course of combustion leads to high CO emission. Like HC emission, CO depicts the effectiveness of the combustion process, i.e., transformation of available chemical energy into useful piston work (mechanical energy). At an increasing load, CO emission was gradually increased for all fuel samples. On all loads, diesel exhibited higher CO emission than A100. This high emission was owing to lack of oxygen in diesel. In the uncoated engine, A100 showed lower CO emission than conventional diesel fuel. This decline in emission was due to the occurrence of inherent oxygen in A100 biodiesel. Similarly, low CO emission was perceived with alkyl esters of brassica, cardoon, and coffee in an AVL single-cylinder 5402 model compression ignition engine due to the carbon-hydrogen ratio and high cetane number conditions [41]. At all loads, PG enriched A100 showed lower CO emission than diesel and A100. This emission was due to enhanced fuel properties of A100. Also, increased cetane number leads to improved combustion efficiency of the system. At all loads, A100 in the YSZ-coated engine showed low CO emission due to the high combustion chamber temperature and presence of innate oxygen in A100. The trapping of heat by a

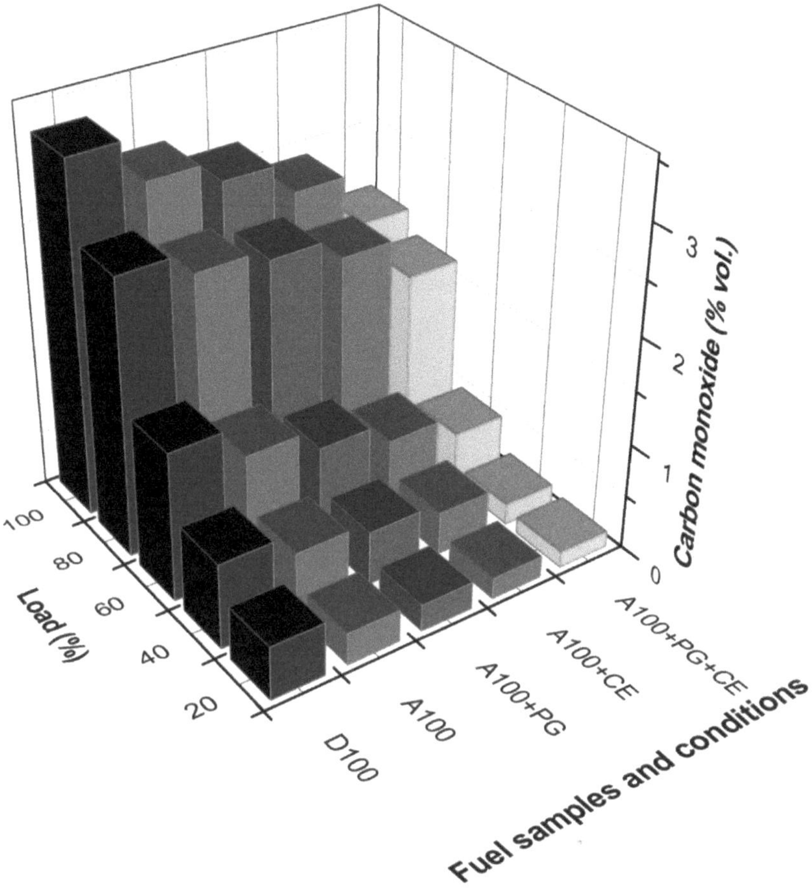

FIG. 13 Engine carbon monoxide against load and working conditions.

ceramic coating inside the combustion chamber leads to the utilization of the available heat to achieve complete combustion. A similar observation was noticed with esterified jatropha biodiesel in a mullite material-coated compression ignition engine [30]. In the YSZ-coated engine, PG enriched A100 showed lower CO emission than other fuel blends. This decline in CO emission was owing to the availability of high heat inside the combustion unit and improved fuel properties, which in turn lead to complete combustion of fuel.

Fig. 14 depicts the engine hydrocarbon emission against load and different working conditions. In a diesel engine, hydrocarbon (HC) emission is due to the partial combustion of fuel. This is due to the rich combustion of fuel or low combustion chamber temperature at the start of combustion. Both unburned hydrocarbon and incompletely oxidized hydrocarbon play a major role in exhaust emission. The factors affecting HC emission were physiochemical properties,

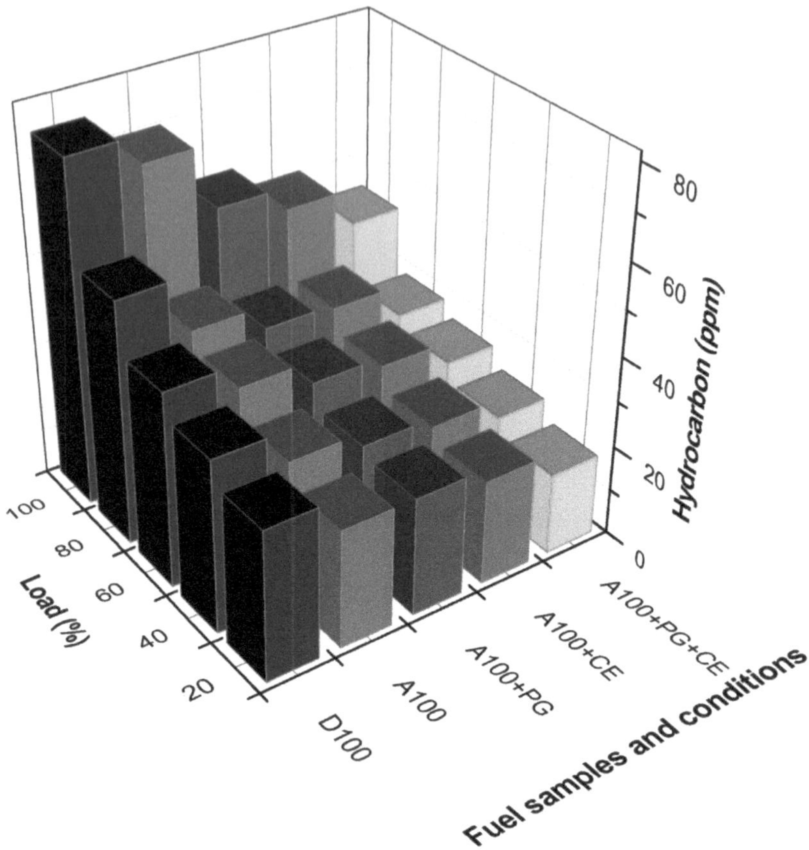

FIG. 14 Engine hydrocarbon against load and working conditions.

engine operating conditions, and fuel spray characteristics. Also, the important chemical components like aromatic hydrocarbon and alkenes in the fuel blend characterize the HC emission [42]. These chemical components were termed as volatile organic components. HC was nonhazardous by itself until it reacted with other pollutants in the air. In a CI engine, HC emission was due to the low combustion chamber temperature. At increasing loads, gradually rising HC emission was observed. At all loads, diesel showed a higher HC emission than A100 due to the absence of oxygen content in diesel fuel. Similarly, increasing hydrocarbon emission was reported with diesel in a single-cylinder compression ignition engine owing to the presence of oxygen and temperature parameters [35]. On comparing with diesel, A100 showed low HC emission due to the existence of inbuilt oxygen in A100. The existence of inherent oxygen in A100 biodiesel offered some beneficial conditions like post-flame oxidation, higher flame speed and improved mixing of fuel-air supports in the fuel-rich regions for improved the oxidation of unburned HC and thus leads to decreased HC emission significantly. Similarly, waste cooking oil biodiesel examination in compression ignition was found to decrease the development of hydrocarbon emission in all load conditions compared to neat diesel operation due to the influence of oxygen molecules [43]. At all loads, PG enriched A100 showed lower HC emission than A100 and diesel due to enhanced physiochemical properties and the presence of innate oxygen in A100. Similarly, diminished hydrocarbon emission was reported with *Pistacia khinjuk* biodiesel in a compression ignition engine with cetane improver. The application of cetane enhancer leads to increased combustion efficiency [44]. High combustion temperature was considered as the factor to obtain decreased HC emission. A thermally coated engine exhibited a lower HC emission than a conventional engine. In a TBC engine, A100 exhibited lower HC emission than diesel due to the high combustion chamber temperature and presence of inbuilt oxygen in A100. A similar trend of results was observed with pine oil biofuel in a low heat rejection engine at all loads owing to the improved combustion chamber temperature [45]. In a compression ignition engine, the combined influence of A100 with PG in the YSZ-coated engine showed lower HC emission than other fuel blends at different engine operating loads. This decrement in HC emission was due to enhanced fuel properties with cetane improver and high in-cylinder temperature with a ceramic-coated engine. A similar trend of observation noticed with *Moringa oleifera* was examined with a cetane improver in a ceramic-coated engine [46].

Fig. 15 depicts the engine oxides of nitrogen against different loads and working conditions. An oxide of nitrogen (NOx) was deliberated as one of the major pollutants from a diesel engine. They affect human health and cause a serious effect on the atmosphere, leading to water and soil pollution. Production of NOx emission from a diesel engine was mainly due to the high combustion chamber temperature and presence of more oxygen content in the combustion chamber. In the diesel engine, oxides of nitrogen emission were said to be the important issue as it was harmful to living beings and the environment. In general, NO (nitric oxide) and NO_2 (nitrogen dioxide) were collected under NOx emission. The formation of NOx emission was represented as thermal NOx. During the course of combustion, oxidation of atmospheric nitrogen was said to be the prime source of NOx emission in an IC engine. At high combustion temperature, the strong triple bond of nitrogen molecules was disrupted and the disassociated atomic state of nitrogen takes part in a series of reactions with oxygen, which results in thermal NOx; this mechanism is also known as the Zeldovich mechanism. NOx formation usually relies on the presence of oxygen, in-cylinder temperature, air surplus coefficient, and residence time. At all loads, diesel revealed a lower NOx emission than other fuel samples due to the absence of oxygen in diesel. Similarly, many research works experienced low levels of oxides of nitrogen emission with diesel in comparison with alternative fuels like biodiesel [47,48]. On all loads, A100 showed higher concentration of NOx

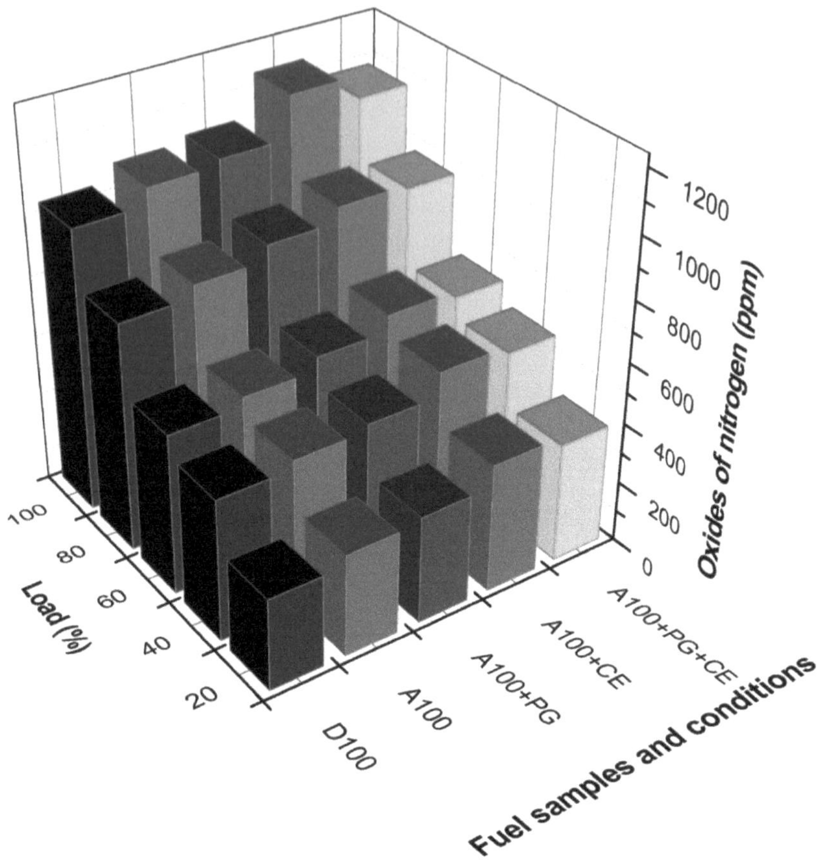

FIG. 15 Engine oxides of nitrogen against load and working conditions.

emission than diesel. The elemental analysis of A100 revealed that the presence of inherent oxygen as an important constituent in a biodiesel sample leads to high NOx emission. A similar trend of high NOx emission was observed with biodiesel from waste frying oil in a TecQuipment TD212 diesel engine [49]. Further, A100 was investigated with a cetane improver (PG) at all working loads. A100 + PG experienced higher NOx than other fuel blends. This increased NOx emission was due to the high cetane number and low heat of vaporization. The high cetane number of A100 provided a short ignition period and minimum time for fuel-air mixing, which in turn leads to high in-cylinder temperature. Thus, A100 + PG showed higher NOx emission than other fuel samples. A similar trend of observations was noticed with turpentine oil with cetane improver in a compression ignition engine [50]. Later, A100 was investigated in a ceramic-coated engine at different loads. In a coated engine, A100 showed a gradual increase in NOx emission compared to diesel. This increased NOx emission was due to the high in-cylinder temperature and presence of oxygen in A100. A similar trend of high NOx was perceived with an Fe_2B-coated engine with peanut oil biodiesel due to the high combustion temperature by a heat trapping technique [39]. In the YSZ-coated engine, PG enriched A100 exhibited low NOx emission. This reduced NOx emission was due to the increased cetane number, high combustion efficiency acquired by A100, and decreased free radical formation in the coated engine. Similarly, pomegranate biodiesel with cetane improver in a ceramic-coated engine showed lower oxides of nitrogen emission in all loads than other fuel samples owing to improved fuel properties and high combustion temperature [51].

Fig. 16 depicts the engine smoke against different loads and working conditions with diesel and alternative fuel samples. Smoke emission from a diesel

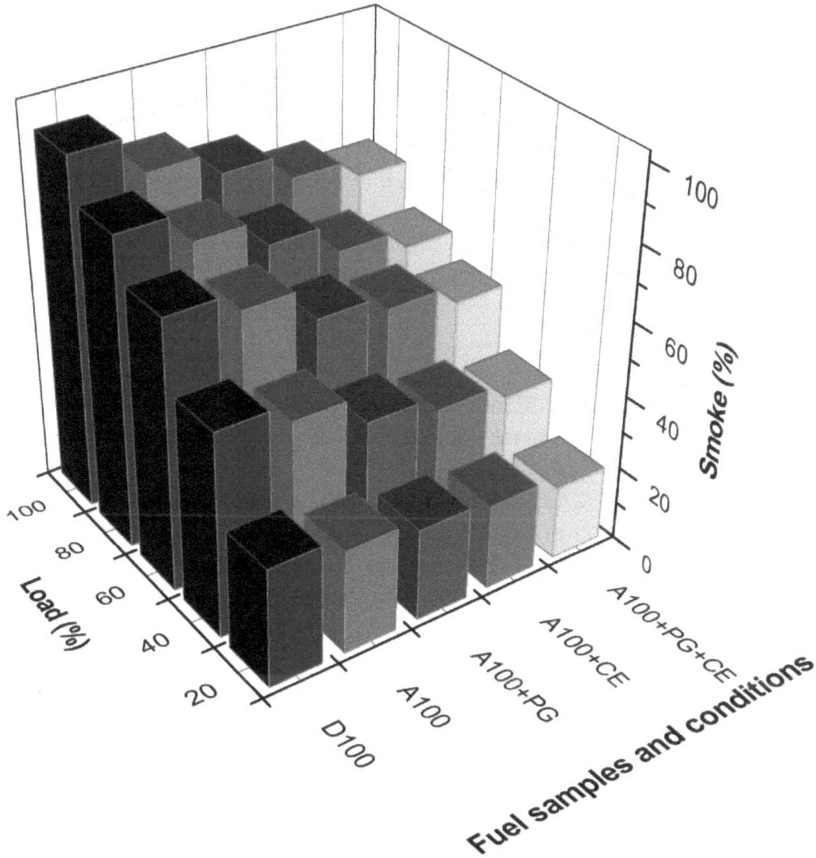

FIG. 16 Engine smoke against load and working conditions.

engine occurs due to soot formation and HC combination from working fuel and lubrication oil. This emission has a direct effect on humans by affecting the respiratory organs and causes breathing problems. The deposition of a large amount of admitted fuel for combustion causes incomplete combustion of fuel, which in turn leads to smoke emission [52]. The deposition of a rich air-fuel mixture inside the combustion unit was considered as one of the important reasons for development of smoke emission. In addition, the factors influencing smoke emission were unavailability of oxygen molecules, injection timing, atomization quality, air-fuel mixing, and so forth. The measurement of smoke from different operating conditions was performed using smokemeter equipment based on the principle of smoke obstructs light. In a conventional engine, diesel perceived higher smoke emission than A100 due to the unavailability of oxygen. From low to high load, a gradual increase in smoke emission was observed. In a conventional engine, A100 experienced lesser smoke emission than diesel. The trend of emission from minimum to maximum load was found to increase the level of smoke, and high load was found to produce peak smoke emission. This decreased trend of smoke emission with avocado seed oil biodiesel was achieved due to the presence of inbuilt oxygen and its elemental composition. Similarly, hybrid fuel composition with pine oil and soapnut oil biodiesel showed lower smoke emission than mineral diesel in a compression ignition engine owing to the presence of oxygen in the fuel [53]. In a conventional engine, PG enriched A100 exhibited lower smoke emission than A100 at all loads. This decreased smoke emission was due to the enhanced fuel properties and presence of oxygen in A100. Also, the formation of a solid soot particle with biodiesel in a conventional engine showed lower smoke emission owing to the high temperature of combustion unit and elemental compositions. Similarly, palm oil enhanced with di-tert-butyl peroxide showed lower smoke emission than other fuel blends owing to improved combustion efficiency and fuel properties [54]. At all loads, A100 exhibited lesser smoke emission than diesel in the YSZ-coated engine due to the high in-cylinder temperature procured by a ceramic-coated engine. Further, the presence of innate oxygen in A100 causes complete combustion of fuel, which in turn leads to declining smoke emission. Similarly, investigation of neem oil methyl ester in a ceramic-coated engine exhibited lower smoke emission due to the high in-cylinder temperature [33]. Besides, PG supplemented A100 in a ceramic-coated engine showed lower smoke emission than A100 and mineral diesel fuel. The enhanced physiochemical properties and elevated in-cylinder temperature incurred by PG enriched A100 in the YSZ-coated engine causes complete combustion of fuel, which in turn leads to minimum concentration of smoke emission. At increasing loads, a gradual decrement in smoke emission was observed with PG enriched A100 in a ceramic-coated engine owing to the improved combustion behavior of the admitted fuel sample. Similarly, reduced smoke emission was reported with a pomegranate biodiesel investigation in the diesel engine with the application of additive and ceramic coating [51].

4 Conclusion

In the present research work, raw avocado seed oil was preferred as the principal feedstock for the production of avocado seed oil biodiesel (A100) by means of the transesterification process. The fuel properties of A100 were analyzed in accordance with the ASTM biodiesel standard and equated with conventional diesel fuel. Initially, the engine investigation was performed with baseline diesel in a single-cylinder direct injection diesel engine. Further, the work was progressed with the application of Pyrogallol (PG) as a cetane improver and a thermal barrier coating technique. In the engine, the tests were performed by using the abovementioned fuels at varying loads and a constant engine speed. Engine exhaust emissions like CO, HC, NOx, and smoke were calculated for each test. Brake thermal efficiency and brake fuel consumption were measured to assess the engine performance. From the experimentation, the inferences associated with engine performance and emission characteristics were drawn as follows:

- A100 + PG exhibited superior performance characteristics than other fuel samples due to improved physiochemical properties and high in-cylinder temperature in the YSZ-coated engine.
- The thermally coated engine showed lower CO, HC, and smoke emissions than the uncoated engine. In both the coated and uncoated engines, A100 + PG perceived lower CO, HC, and smoke emissions than diesel as PG enriched the cetane number of A100. On comparing with diesel, A100 + PG experienced minimum CO and HC emissions due to high in-cylinder temperature, short ignition delay period, and better fuel-air mixing.
- Cetane improver enriched A100 has a significant impact on declined NOx emission. On comparing with diesel and A100, PG enriched A100 showed the minimum concentration of NOx emission due to decreased free radicals formation.
- In the YSZ-coated engine, declined smoke emission was perceived with A100 + PG due to the high in-cylinder temperature, enhanced fuel properties, and intrinsic availability of oxygen in A100.

Thus, the present research work was evolved to endorse a suitable method to attain enhanced engine performance and emission characteristics with the better energy usage of alternative fuel. Hence, the practice of using avocado seed oil biodiesel (A100) with a cetane improver (Pyrogallol-PG) in an Yttria Stabilized Zirconia ceramic-coated engine may be deliberated as the beneficial approach to attain upgraded engine performance and emission characteristics.

4.1 Future enhancement

The present study involves the production of biodiesel from raw avocado seed oil using the transesterification process in the presence of a catalyst. The inferences obtained from the study revealed that the avocado seed oil biodiesel might be considered as a renewable substitute to the conventional diesel with possible engine and fuel property modifications. The proposed work can be developed further with the following intentions:

- The study can be performed in a four-cylinder diesel engine to investigate the chances of commercialization of the avocado seed oil biodiesel.
- Optimization on engine characteristics for the effective operation of avocado seed oil biodiesel in a DI diesel engine may be performed.
- Study on the long-term resilience of the thermally coated engine with distinct biodiesel blends. Determination of the impact of different types of coating materials and coating thickness on the engine characteristics.
- Investigation on combustion chamber geometry in the thermally coated engine. Detailed analysis on a post-capture combustion system with a thermal barrier-coated engine.
- An inclusive analysis on modification of different engine operating parameters like injection pressure, injection timing, and compression ratio in a thermally coated engine.
- The research on distinct advanced combustion technologies like homogeneous compression ignition (HCCI), premixed controlled compression ignition (PCCI), reactivity controlled compression ignition (RCCI), and low-temperature combustion (LTC) may be analyzed to acquire low NOx emission. Also, investigation relevant to exhaust after-treatment strategy for a biodiesel driven engine may be performed for explicit commercial application.

Acknowledgments

Authors wish to thank the National Cheng Kung University Grants—NCKU 90 and Beyond for Research Fellow with grant number HUB109-22T-3-173.

References

[1] Krishnamoorthi M, Sreedhara S, Prakash Duvvuri P. Experimental, numerical and exergy analyses of a dual fuel combustion engine fuelled with syngas and biodiesel/diesel blends. Appl Energy 2020;263. https://doi.org/10.1016/j.apenergy.2020.114643, 114643.

[2] Xu Q, Zhang L, Sun K, Shao Y, Tian H, Zhang S, et al. Cross-polymerisation between the model furans and carbohydrates in bio-oil with acid or alkaline catalysts. J Energy Inst 2020. https://doi.org/10.1016/j.joei.2020.02.005.

[3] Xu S, Cao B, Uzoejinwa BB, Odey EA, Wang S, Shang H, et al. Synergistic effects of catalytic co-pyrolysis of macroalgae with waste plastics. Process Saf Environ Prot 2020;137:34–48. https://doi.org/10.1016/j.psep.2020.02.001.

[4] Feng Y, Zhang W, Niaz H, He Z, Wang S, Wang X. Parametric analysis and thermo-economical optimization of a supercritical-subcritical organic Rankine cycle for waste heat utilization. Energ Conver Manage 2020;212. https://doi.org/10.1016/j.enconman.2020.112773, 112773.

[5] Viswanathan K, Wang S, Esakkimuthu S. Impact of yttria stabilized zirconia coating on diesel engine performance and emission characteristics fuelled by lemon grass oil biofuel. J Therm Anal Calorim 2020. https://doi.org/10.1007/s10973-020-10364-z.
[6] Viswanathan K, Wang S. Experimental investigation on the application of preheated fish oil ethyl ester as a fuel in diesel engine. Fuel 2021;285:119244. https://doi.org/10.1016/j.fuel.2020.119244.
[7] Viswanathan K, Paulraj A. A comprehensive study on the performance and emission characteristics of a diesel engine with the blends of diesel, jojoba oil biodiesel, and butylated hydroxyl anisole as an alternative fuel. Energy Sources, Part A Recover Util Environ Eff 2020;00:1–15. https://doi.org/10.1080/15567036.2020.1849451.
[8] İlhak Mİ, Doğan R, Akansu SO, Kahraman N. Experimental study on an SI engine fueled by gasoline, ethanol and acetylene at partial loads. Fuel 2020;261. https://doi.org/10.1016/j.fuel.2019.116148.
[9] Balasubramanian D, Kamaraj S, Krishnamoorthy R. Synthesis of biodiesel from waste cooking oil by alkali doped calcinated waste egg shell powder catalyst and optimization of process parameters to improve biodiesel conversion. SAE Tech Pap 2020; 2020. https://doi.org/10.4271/2020-01-0341.
[10] EL-Seesy AI, He Z, Hassan H, Balasubramanian D. Improvement of combustion and emission characteristics of a diesel engine working with diesel/jojoba oil blends and butanol additive. Fuel 2020;279. https://doi.org/10.1016/j.fuel.2020.118433, 118433.
[11] Ramalingam K, Balasubramanian D, Chellakumar PJTJS, Padmanaban J, Murugesan P, Xuan T. An assessment on production and engine characterization of a novel environment-friendly fuel. Fuel 2020;279. https://doi.org/10.1016/j.fuel.2020.118558, 118558.
[12] Wang S, Shang H, Abomohra AEF, Wang Q. One-step conversion of microalgae to alcohols and esters through co-pyrolysis with biodiesel-derived glycerol. Energ Conver Manage 2019;198. https://doi.org/10.1016/j.enconman.2019.111792, 111792.
[13] Esakkimuthu S, Krishnamurthy V, Wang S, El-Fatah Abomohra A, Shanmugam S, Ramakrishnan SG, et al. Simultaneous induction of biomass and lipid production in tetradesmus obliquus BPL16 through polysorbate supplementation. Renew Energy 2019;140:807–15. https://doi.org/10.1016/j.renene.2019.03.104.
[14] Xu S, Elsayed M, Ismail GA, Li C, Wang S, Abomohra AEF. Evaluation of bioethanol and biodiesel production from scenedesmus obliquus grown in biodiesel waste glycerol: a sequential integrated route for enhanced energy recovery. Energ Conver Manage 2019;197. https://doi.org/10.1016/j.enconman.2019.111907, 111907.
[15] Brown SPA, Huntington HG. OPEC and world oil security. Energy Policy 2017;108:512–23. https://doi.org/10.1016/j.enpol.2017.06.034.
[16] Dhinesh B, Annamalai M. A study on performance, combustion and emission behaviour of diesel engine powered by novel nano *Nerium oleander* biofuel. J Clean Prod 2018;196: 74–83. https://doi.org/10.1016/j.jclepro.2018.06.002.
[17] Viswanathan K, Ashok B, Pugazhendhi A. Comprehensive study of engine characteristics of novel biodiesel from curry leaf (*Murraya koenigii*) oil in ceramic layered diesel engine. Fuel 2020;280. https://doi.org/10.1016/j.fuel.2020.118586, 118586.
[18] Vigneswaran R, Annamalai K, Dhinesh B, Krishnamoorthy R. Experimental investigation of unmodified diesel engine performance, combustion and emission with multipurpose additive along with water-in-diesel emulsion fuel. Energ Conver Manage 2018;172:370–80. https://doi.org/10.1016/j.enconman.2018.07.039.
[19] Dhinesh B, Maria Ambrose Raj Y, Kalaiselvan C, KrishnaMoorthy R. A numerical and experimental assessment of a coated diesel engine powered by high-performance nano biofuel. Energ Conver Manage 2018;171:815–24. https://doi.org/10.1016/j.enconman.2018.06.039.

[20] Proskurina S, Junginger M, Heinimö J, Vakkilainen E. Global biomass trade for energy—Part 1: Statistical and methodological considerations. Biofuels Bioprod Biorefining 2019;13:358–70. https://doi.org/10.1002/bbb.1841.

[21] Yuan C, Wang S, Cao B, Hu Y, Abomohra AEF, Wang Q, et al. Optimization of hydrothermal co-liquefaction of seaweeds with lignocellulosic biomass: merging 2nd and 3rd generation feedstocks for enhanced bio-oil production. Energy 2019;173:413–22. https://doi.org/10.1016/j.energy.2019.02.091.

[22] Cao B, Sun Y, Guo J, Wang S, Yuan J, Esakkimuthu S, et al. Synergistic effects of co-pyrolysis of macroalgae and polyvinyl chloride on bio-oil/bio-char properties and transferring regularity of chlorine. Fuel 2019;246:319–29. https://doi.org/10.1016/j.fuel.2019.02.037.

[23] Anon. Global avocado production in 2017, by country (in 1,000 tons), https://www.statista.com/statistics/593211/global-avocado-production-by-country/; 2017.

[24] Perea-Moreno AJ, Aguilera-Ureña MJ, Manzano-Agugliaro F. Fuel properties of avocado stone. Fuel 2016;186:358–64. https://doi.org/10.1016/j.fuel.2016.08.101.

[25] Anon. Avocado production worldwide from 2000 to 2017 (in million metric tons), https://www.statista.com/statistics/577455/world-avocado-production/; 2017.

[26] Clarke DR, Phillpot SR. Thermal barrier coating materials. Mater Today 2005;8:22–9. https://doi.org/10.1016/S1369-7021(05)70934-2.

[27] Liu B, Liu Y, Zhu C, Xiang H, Chen H, Sun L, et al. Advances on strategies for searching for next generation thermal barrier coating materials. J Mater Sci Technol 2019;35:833–51. https://doi.org/10.1016/j.jmst.2018.11.016.

[28] Vedharaj S, Vallinayagam R, Yang WM, Chou SK, Chua KJE, Lee PS. Experimental and finite element analysis of a coated diesel engine fueled by cashew nut shell liquid biodiesel. Exp Therm Fluid Sci 2014;53:259–68. https://doi.org/10.1016/j.expthermflusci.2013.12.018.

[29] Vedharaj S, Vallinayagam R, Yang WM, Saravanan CG, Chou SK, Chua KJE, et al. Reduction of harmful emissions from a diesel engine fueled by kapok methyl ester using combined coating and SNCR technology. Energ Convers Manage 2014;79:581–9. https://doi.org/10.1016/j.enconman.2013.12.056.

[30] Jagtap SP, Pawar AN, Lahane S. Improving the usability of biodiesel blend in low heat rejection diesel engine through combustion, performance and emission analysis. Renew Energy 2020. https://doi.org/10.1016/j.renene.2020.03.115.

[31] Wang P, Li Y, Duan X, Liu J, Wang S, Zou P, et al. Experimental investigation of the effects of CR, hydrogen addition strategies on performance, energy and exergy characteristics of a heavy-duty NGSI engine fueled with 99% methane content. Fuel 2020;259. https://doi.org/10.1016/j.fuel.2019.116212, 116212.

[32] Mikulski M, Ambrosewicz-Walacik M, Duda K, Hunicz J. Performance and emission characterization of a common-rail compression-ignition engine fuelled with ternary mixtures of rapeseed oil, pyrolytic oil and diesel. Renew Energy 2019. https://doi.org/10.1016/j.renene.2019.10.161.

[33] Karthickeyan V. Experimental analysis on thermally coated diesel engine with neem oil methyl ester and its blends. Heat Mass Transf 2018;54:1961–74. https://doi.org/10.1007/s00231-018-2286-6.

[34] Karthickeyan V. Experimental investigation on combined effect of ignition promoters and ceramic coating fuelled with papaya seed oil methyl ester in DI diesel engine. Renew Energy 2020. https://doi.org/10.1016/j.renene.2019.10.163.

[35] Reang NM, Dey S, Debbarma B, Deb M, Debbarma J. Experimental investigation on combustion, performance and emission analysis of 4-stroke single cylinder diesel engine fuelled with neem methyl ester-rice wine alcohol-diesel blend. Fuel 2020;271. https://doi.org/10.1016/j.fuel.2020.117602, 117602.

[36] Maawa WN, Mamat R, Najafi G, De Goey LPH. Performance, combustion, and emission characteristics of a CI engine fueled with emulsified diesel-biodiesel blends at different water contents. Fuel 2020;267. https://doi.org/10.1016/j.fuel.2020.117265.

[37] Wang S, Karthickeyan V, Sivakumar E, Lakshmikandan M. Experimental investigation on pumpkin seed oil methyl ester blend in diesel engine with various injection pressure, injection timing and compression ratio. Fuel 2020;264. https://doi.org/10.1016/j.fuel.2019.116868, 116868.

[38] Ağbulut Ü, Sarıdemir S, Albayrak S. Experimental investigation of combustion, performance and emission characteristics of a diesel engine fuelled with diesel–biodiesel–alcohol blends. J Braz Soc Mech Sci Eng 2019;41. https://doi.org/10.1007/s40430-019-1891-8.

[39] Öztürk U, Hazar H, Yılmaz F. Comparative performance and emission characteristics of peanut seed oil methyl ester (PSME) on a thermal isolated diesel engine. Energy 2019;167:260–8. https://doi.org/10.1016/j.energy.2018.10.198.

[40] Karthickeyan V, Thiyagarajan S, Geo VE, Ashok B, Nanthagopal K, Chyuan OH, et al. Simultaneous reduction of NOx and smoke emissions with low viscous biofuel in low heat rejection engine using selective catalytic reduction technique. Fuel 2019;255. https://doi.org/10.1016/j.fuel.2019.115854, 115854.

[41] Jaliliantabar F, Ghobadian B, Carlucci AP, Najafi G, Ficarella A, Strafella L, et al. Comparative evaluation of physical and chemical properties, emission and combustion characteristics of brassica, cardoon and coffee based biodiesels as fuel in a compression-ignition engine. Fuel 2018;222:156–74. https://doi.org/10.1016/j.fuel.2018.02.145.

[42] Kozina A, Radica G, Nižetić S. Analysis of methods towards reduction of harmful pollutants from diesel engines. J Clean Prod 2020;262. https://doi.org/10.1016/j.jclepro.2020.121105.

[43] Raju P, Masimalai SK, Ganesan N, Karthic SV. Engine's behavior on hydrogen addition of waste cooking oil fueled light duty diesel engine – a dual fuel approach. Energy 2020;194. https://doi.org/10.1016/j.energy.2019.116844, 116844.

[44] Karthickeyan V, Ashok B, Thiyagarajan S, Nanthagopal K, Geo VE, Dhinesh B. Comparative analysis on the influence of antioxidants role with Pistacia khinjuk oil biodiesel to reduce emission in diesel engine. Heat Mass Transf 2019.

[45] Viswanathan K, Balasubramanian D, Subramanian T, Varuvel EG. Investigating the combined effect of thermal barrier coating and antioxidants on pine oil in DI diesel engine. Environ Sci Pollut Res 2019. https://doi.org/10.1007/s11356-019-04649-6.

[46] Karthickeyan V. Effect of cetane enhancer on *Moringa oleifera* biodiesel in a thermal coated direct injection diesel engine. Fuel 2019;235:538–50. https://doi.org/10.1016/j.fuel.2018.08.030.

[47] Thiyagarajan S, Sonthalia A, Edwin Geo V, Prakash T, Karthickeyan V, Ashok B, et al. Effect of manifold injection of methanol/n-pentanol in safflower biodiesel fuelled CI engine. Fuel 2020;261. https://doi.org/10.1016/j.fuel.2019.116378.

[48] Sharma V, Duraisamy G, Arumugum K. Impact of bio-mix fuel on performance, emission and combustion characteristics in a single cylinder DICI VCR engine. Renew Energy 2020;146:111–24. https://doi.org/10.1016/j.renene.2019.06.142.

[49] Elwardany AE, Marei MN, Eldrainy Y, Ali RM, Ismail M, El-kassaby MM. Improving performance and emissions characteristics of compression ignition engine: effect of ferrocene nanoparticles to diesel-biodiesel blend. Fuel 2020;270. https://doi.org/10.1016/j.fuel.2020.117574, 117574.

[50] Jeevanantham AK, Madhusudan Reddy D, Goyal N, Bansal D, Kumar G, Kumar A, et al. Experimental study on the effect of cetane improver with turpentine oil on CI engine characteristics. Fuel 2019. https://doi.org/10.1016/j.fuel.2019.116551, 116551.

[51] Karthickeyan V, Thiyagarajan S, Ashok B, Edwin Geo V, Azad AK. Experimental investigation of pomegranate oil methyl ester in ceramic coated engine at different operating condition in direct injection diesel engine with energy and exergy analysis. Elsevier; 2020. https://doi.org/10.1016/j.enconman.2019.112334.
[52] Kadir M, Aydin M, Yilbasi Z, Arslan M. Investigation on the structural effects of the addition of alcohols having various chain lengths into the vegetable oil-biodiesel-diesel fuel blends: an attempt for improving the performance, combustion, and exhaust emission characteristics of a compre. Fuel 2020;269. https://doi.org/10.1016/j.fuel.2020.117455, 117455.
[53] Venkatesan V, Nallusamy N. Pine oil-soapnut oil methyl ester blends: a hybrid biofuel approach to completely eliminate the use of diesel in a twin cylinder off-road tractor diesel engine. Fuel 2019. https://doi.org/10.1016/j.fuel.2019.116500, 116500.
[54] Musthafa MM, Kumar TA, Mohanraj T, Chandramouli R. A comparative study on performance, combustion and emission characteristics of diesel engine fuelled by biodiesel blends with and without an additive. Fuel 2018;225:343–8. https://doi.org/10.1016/j.fuel.2018.03.147.

Chapter 10

Effect of low carbon biofuel on carbon emissions in biodiesel fueled CI engine

S. Thiyagarajan[a], Ankit Sonthalia[b], V. Edwin Geo[a], Karthikeyan Viswanathan[c], and Dhinesh Balasubramaniyam[d]
[a]Department of Automobile Engineering, SRM Institute of Science and Technology, Kattankulathur, Tamil Nadu, India, [b]Department of Automobile Engineering, SRM Institute of Science and Technology, NCR Campus, Modinagar, Uttar Pradesh, India, [c]Department of Mechanical Engineering, Sri Krishna College of Engineering and Technology, Coimbatore, Tamil Nadu, India, [d]Department of Mechanical Engineering, Mepco Schlenk Engineering College, Virudhunagar, Tamil Nadu, India

1 Introduction

The ever-increasing population of the world and its modern lifestyle have led to an increase in energy consumption. The human population is expected to surpass 9 billion by 2050, and there will be an 84% increase in energy demand [1,2]. The transportation sector needs to grow at a rapid rate for the growing population. Today, diesel engines are the prime mover of the sector due to their high fuel economy, better reliability, and handling facilities. However, the diesel engines use fossil fuel for producing power and release large amounts of gaseous pollutants into the atmosphere [3]. Among these gases, carbon dioxide is the most potent pollutant that instigates global warming. Since the industrial revolution, the concentration of CO_2 in the atmosphere has increased by nearly 43%, and if the same trend continues, it is expected to increase to 60% by 2100 [4]. An intergovernmental panel of climate control has projected that due to the surge in CO_2 in the atmosphere, the planet's temperature may rise by 2–5°C by 2070 [5]. Due to the increase in CO_2 level in oceans, their acidity has also increased by 30%. There are three possible strategies for mitigating CO_2: (1) proper utilization of fossil fuel, which can be achieved by upgrading the existing engine technology, (2) sequestration of CO_2, and (3) use of alternates to fossil fuels [6].

Among the three methods, many researchers have focused on the use of alternative fuels in the diesel engine. Biofuel is one such energy enriched

Bioenergy Resources and Technologies. https://doi.org/10.1016/B978-0-12-822525-7.00006-8
Copyright © 2021 Elsevier Inc. All rights reserved.

alternative which is produced from biomass abundantly available in nature. The biomass uses photosynthesis for converting the carbon dioxide present in the atmosphere into sugars by using water and solar energy, which makes it a renewable source of energy [7]. The developing economies are trying to increase the use of local renewable sources so that they can reduce their dependence on the fossil fuels which they are importing from crude oil-exporting countries (OPEC). The biofuel production of the world was 27.85 million tons of oil equivalent (MToe) in 2006, which increased to 95.37 MToe in 2018, an increase of 242.4%. In India, as compared to 2006, the biofuel production in 2018 increased by 600%, whereas in the same period, the biofuel production in China increased by 235% [8].

The biofuels can be classified as first-generation [9], second-generation [10,11], third-generation [12], and fourth-generation [13] fuels. The source, feedstock, and end products of a different generation of biofuels are shown in Table 1. The first-generation biofuel feedstocks are rapeseed, sugarcane, palm, barley, wheat, corn, and other oil-containing crops [21]. These crops

TABLE 1 Feedstock and processes of a different generation of biofuel.

Generation	Feedstock	Processing technology	Example
First [14,15]	Animal fats, food crops, and edible oil seeds	Thermochemical process, sugar fermentation, esterification, and transesterification	Biobutanol, bioethanol, biodiesel
Second [16,17]	Waste cooking oil, nonedible oilseeds, forest residue, sugarcane bagasse, cereal straw	Thermochemical process, fermentation, physical, chemical, biological pretreatment of feedstock	Biobutanol, bioethanol, biodiesel, syngas
Third [18,19]	Algae	Cultivation, harvesting, oil extraction, transesterification, or thermochemical process, or fermentation	Biobutanol, bioethanol, biodiesel, biohydrogen, methane, syngas
Fourth [20]	Algae and other microbes	Metabolic engineering of algae, cultivation, harvesting, oil extraction, transesterification, or thermochemical process, or fermentation	Biobutanol, bioethanol, biodiesel, biohydrogen, methane, syngas

are also a source of food for the human population, which would start a competition between food and fuel, increasing food prices. To minimize this issue, forest and agricultural wastes and nonedible crops can be used for producing biofuel. These feedstocks are considered second-generation feedstock. Vegetative grasses such as switchgrass and *Miscanthus* are also considered as second-generation feedstock as these energy-rich crops can grow on degraded land. However, most of the second-generation feedstock requires fertile land which may compete with the food crops. Therefore, it is necessary to explore alternatives so that there is no competition between nonfood and food crop [22]. Algae are considered as the third-generation feedstock as it has the potential to produce biofuels, and it can grow in an aquatic environment (brackish/saline water/coastal seawater) throughout the year. Photosynthesis and growth rate of algae is faster than any other terrestrial plant [23]. It contains nearly 70% lipid on a dry weight basis [24].

Some of the algae species with high potential of biofuel production are *Chaetocero scalcitrans*, *Botryococcus braunii*, several Chlorella species, *Schizochytrium limacinum*, *Isochrysis galbana*, Scenedesmus, and Nanochloropsis [18,19,25]. The scientist has observed that the oil content of the fast-growing algae is low, whereas slow-growing organisms have high lipid content. For commercializing algae biofuel, identification of the correct species that have high biomass and lipid content is necessary. Phototropic and heterotrophic type of cultivation also affects the yield of biomass in the same strain [18,19]. Scientists, therefore, are looking for a more sustainable and cost effective cultivation method for maximizing the biofuel yield. Genetic engineering of the algae that would give high biomass yield is the solution to this problem, and this type of feedstock is considered as the fourth-generation feedstock [23].

For the production of biofuels in India, the focus is mainly concentrated on nonedible feedstock. Since India has a huge population, therefore, the country cannot afford to grow edible crops on the fertile land for biofuel production. Moreover, the wastelands which are unfit for the production of food crops can be utilized for growing the nonedible feedstock. India can also concentrate on wastes for biofuel production. Some of the feedstock which can be utilized for the biofuel production and their utilization in a compression ignition engine is discussed here.

1.1 Karanja oil

Karanja also called as *Pongamia pinnata* is a medium-sized tree. It can survive in saline, drought, heat, and frost conditions. It can grow along riverbanks, coasts, and can reclaim marginal lands. In India, it is found from Kanyakumari to the foothills of the Himalayas. The fruits of Karanja have a viability period of 1 year, and the yield of seeds is 9–90 kg per tree. The Karanja pods contain single seed which is elliptical. They are thick-walled, 3–6 cm long, and 2–3 cm wide. The seeds are 10–20 mm long and brown. The seeds contain 27%–39% oil. The oil is inedible as it contains toxic diketone pongamol and Karanjin.

Oleic acid is the primary fatty acid in the Karanja oil, followed by linoleic and palmitic acid. The oil is used for making soaps, in the leather industry for tanning and finishing, and in veterinary medicines. The Karanja wood is also used for brushing of teeth [26–29].

The direct use of Karanja oil in a diesel engine is not feasible due to its high kinematic viscosity (nearly 46 mm^2/s). There are many ways of using Karanja oil in the diesel engine. First, the straight oil can be blended with diesel; second, it can be preheated to reduce its viscosity; third, the vegetable oil can be esterified (Eq. 1) and transesterified (Eq. 2) using a homogeneous catalyst, fourth, the vegetable oil can be esterified and transesterified using a heterogeneous catalyst, and lastly, a non-catalytic transesterification can be carried out in supercritical conditions. Many researchers have used these methods to operate the diesel engine with the modified Karanja oil. On the basis of the free fatty acid content of the Karanja oil, the biodiesel (K100) can be produced either by transesterification or a combination of esterification and transesterification process. Table 2 compares the properties of Karanja oil methyl ester and diesel.

$$\underset{\text{FFA}}{R_1 - COOH} + \underset{\text{Alcohol}}{ROH} \overset{H^+}{\longleftrightarrow} \underset{\text{Fatty Acid Ester}}{R - O - CO - R_1} + \underset{\text{Water}}{H_2O} \tag{1}$$

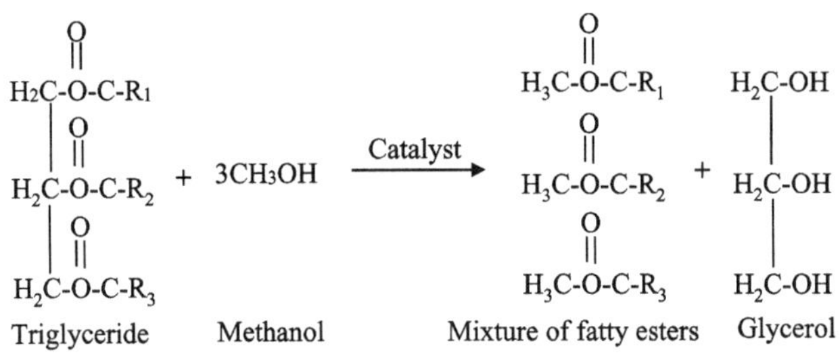

(2)

TABLE 2 Properties of neat diesel and Karanja oil methyl ester.

Property	Diesel	Karanja oil methyl ester (K100)
Average molecular formula	$C_{14}H_{28}$	$C_{20}H_{36}O_2$
Kinematic viscosity at 40°C, cSt	3.6	4.9
Density at 15°C, g/cm^3	0.840	0.858
Lower heating value, kJ/kg	42,700	41,200
Cetane index	45–55	49
Flashpoint, °C	74	135

Many researchers have investigated the effect of Karanja oil biodiesel in a CI engine [30–36]. A vehicle having medium-duty, variable speed, and the four-cylinder diesel engine was tested with K100 and its blends with diesel [32]. The authors studied the engine's performance and emission characteristics at 1800 and 2600 rpm at different load conditions. The CO emissions reduced with K100 and its blends at high load and speed conditions. A decrease in HC emissions and smoke opacity was observed with an increase in load, whereas NOx emission increased. The methyl esters of Jatropha, Karanja, and Sesame were compared for their engine performance and emission characteristics by Banapurmath et al. [33]. The authors found slightly higher HC and CO emissions with the neat methyl esters, in comparison to diesel.

The thermal efficiency of the engine with the esters was close to diesel at 80% load, and Karanja oil ester gave higher thermal efficiency than Jatropha oil ester at the same load. The ethyl and methyl ester of Karanja oil was prepared by Baiju et al. [34] using ethanol and methanol as the catalyst for transesterification. The viscosity of the ethyl ester of Karanja oil was slightly higher than its methyl ester. Moreover, both the viscosities were within the biodiesel standards. The authors used blends of both the esters with diesel in a CI engine. The thermal efficiency of the engine, at part load condition, with 20% ethyl ester was lower than 20% methyl ester. The HC and CO emissions were lower with methyl ester in comparison to ethyl ester. For all blending proportions of both the esters, the smoke opacity was lower than diesel. The NOx emission with diesel was higher at full load, whereas at part load conditions, the NOx emission with esters was 10%–25% higher than that of diesel. The effect of different combustion chambers (Toroidal, Shallow depth, and Hemispherical) on a diesel engine fueled with 20% blend of Karanja oil methyl ester and diesel was studied by Dwivedi and Sharma [35]. The authors found that Toroidal combustion chamber resulted in higher thermal efficiency and lower fuel consumption as compared to others. Chauhan et al. [36] compared the performance of a CI engine fueled with Karanja biodiesel, its blend with diesel, and baseline diesel. The CO_2 emissions with the blends and neat Karanja biodiesel were higher than diesel, and maximum emissions were observed with neat biodiesel. Higher carbon and oxygen content of the biodiesel was considered as the major reason for the observation. The thermal efficiency of the engine with biodiesel was lower than diesel. The NO emissions with biodiesel were higher than diesel, whereas smoke was lower compared to diesel.

All the studies discussed, show that the CO_2 emissions are higher with Karanja oil methyl ester, along with small penalty on thermal efficiency and fuel consumption. If the methyl ester is to be used as a substitute to fossil diesel, then these issues should be addressed [28]. Some of the techniques identified from literature for reducing engine emission and improving its performance are blending with low carbon fuel and oxygenates and inducting carbon-free gaseous fuels.

1.2 Low carbon fuels

Some of the low carbon fuels that have been extensively studied are discussed here. Table 3 presents its properties.

1.2.1 Pine oil

The pine tree is widely grown for its bark, essential oil, and turpentine. The height of the tree can be 40m. The oleoresin obtained from the pine tree has a fresh forest smell and is pale yellow. A living pine tree can give nearly 2.75kg of oleoresin. The oleoresin obtained from the pine tree is subjected to the steam distillation process. The condensed lower boiling fractions of the oleoresin are treated with orthophosphoric acid to formulate the pine oil. The major constituents of pine oil are pinene ($C_{10}H_{16}$) and terpineol ($C_{10}H_{18}O$). The molecule formula indicates that the pine oil has oxygen content in it, and it has a distinct composition in comparison to biodiesel. The carbon to hydrogen ratio of the oil is low, which is an inherent advantage in terms of particulate emission. The properties of pine oil, i.e., boiling point, flash point, and viscosity are similar to alcohols, whereas its calorific value is close to diesel [37–40].

The effect of pine oil and diesel blends were studied in a multicylinder CI engine with varying exhaust gas recirculation (EGR) rates [41]. The authors observed improvement in engine performance with pine oil due to its low boiling point and viscosity. The NO emissions were found to increase, whereas the soot emissions decreased. With the same rate of EGR, an increase in the proportion of the pine oil lead to an increase in NO emission without much effect on NO emissions. The authors also found that the solubility of pine oil in diesel is excellent, resulting in the formation of homogenous mixtures.

TABLE 3 Properties of biofuels [4].

Property	Eucalyptus oil	Pine oil	Orange oil	Camphor oil
Source of oil extraction	Leaves	Leaf needles	Seed	Branches
Molecular formula	$C_{10}H_{18}O$	$C_{11}H_{10}BrN_5$	$C_{10}H_{20}O$	$C_{10}H_{16}O$
Kinematic viscosity at 40°C, cSt	2.0	1.3	3.5	1.9
Density at 15°C, g/cm^3	0.895	0.875	0.816	0.894
Lower heating value, MJ/kg	43.27	42.8	34.65	38.2
Cetane index	28	11	47	5
Flash point, °C	54	52	74	50

Vallinayagam et al. [42] blended pine oil with diesel in various proportions and tested it on a stationary diesel engine. The engine performance improved with the increase in pine oil proportion in the blend. The authors found that engine operation with neat pine oil resulted in a significant reduction in HC, CO, and smoke emissions. In contrast, NOx emissions were higher than diesel at full load. Also, the thermal efficiency of the engine with neat pine oil improved slightly in comparison to diesel. To completely replace diesel for operating a diesel engine Vallinayagam et al. [37] prepared blends of Kapok methyl ester and pine oil. The authors found that equal blending of the two fuels had thermal efficiency similar to diesel and HC, CO, and smoke emissions were found to decrease by 8.1%, 18.9%, and 12.5%, respectively. Higher content of pine oil in the blend resulted in improved engine performance, but due to lower cetane number of pine oil, the engine started knocking. When the biodiesel content was higher in the blend, the power loss of the engine was observed because of the high viscosity of biodiesel that affects the combustion process.

1.2.2 Eucalyptus oil

The eucalyptus tree can grow in nearly all climatic conditions. The eucalyptus genus comprises more than 900 species, and it is a native of Australia [43]. It was introduced in India in the 1790s mainly for meeting the firewood demand as its growth rate is very high. Nearly 7.5 million hectares of land is under eucalyptus plantation in India [44]. Eucalyptus oil is extracted from the leaves of the tree using a steam distillation process [45]. The leaves are available throughout the year [46]. The main component of eucalyptus oil is 1,8-cineol, and the leaves contain 6%–9% liquid and semisolid type hydrocarbons. The chemical structure (Fig. 1) of the oil is similar to naphthenic bases; hence it is also used as an effective cosolvent [47]. The oil has many other uses such as biopesticide, repellant, and antiseptic, etc. [48]. The price of the eucalyptus oil is exorbitant, making it economically unfeasible for use as an alternate fuel.

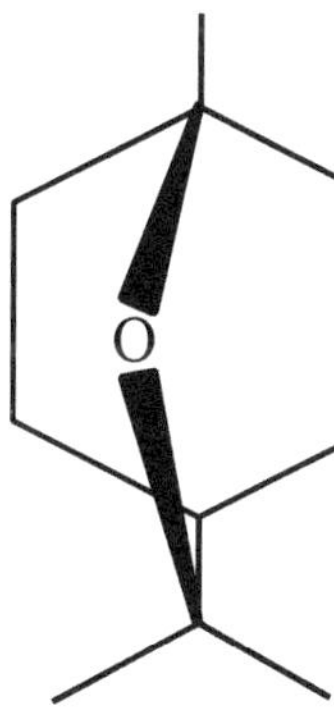

FIG. 1 Chemical structure of eucalyptus oil.

Nevertheless, studies have been conducted on using eucalyptus oil in a diesel engine. The oil has low viscosity and boiling point as compared to diesel which is helpful in atomization, evaporation, and the combustion process. Moreover, the calorific value of eucalyptus oil is close to diesel which makes it a viable alternative to diesel fuel. However, the cetane number of the fuel is low, which prolongs the ignition delay resulting in poor ignition quality.

To eliminate diesel as a fuel, paradise oil methyl ester, and eucalyptus oil were blended in various proportions by Devan et al. [49] and tested in a diesel engine. An equal blending of the two fuels resulted in 34.5%, 37%, and 49% reduction in HC, CO, and smoke emission, respectively and NOx emission increased by 2.7% as compared to diesel. The thermal efficiency of the engine was also comparable. Eucalyptus oil was blended with fossil diesel by Anandevelu et al. [50] in various proportions. The authors found that 40% blend of the oil in diesel resulted in better performance and reduction in CO, HC, and smoke emissions.

Further, due to longer ignition delay and sudden rise in heat release, the peak in cylinder pressure was higher than diesel resulting in higher NO_x emission. To counteract the high NOx emission, the same research group studied the effect of EGR technique on the emission [51]. The authors observed that even with 15% EGR rate, the thermal efficiency was 3.8% higher than diesel. With the increase in the percentage of eucalyptus oil, the smoke emissions were lower, and the NOx emissions reduced by 65%.

1.2.3 *Orange oil*

Orange is a citrus fruit, and it is consumed all over the world in the form of juice and natural and peeled forms. It contains many nutrients such as vitamin A, B, and C, dietary fiber, minerals, and phytochemicals like amino acids, phenolic acids, triterpenes, flavonoids, and carotenoids [52,53]. During the production of juice from orange nearly half of the orange weight is converted into juice and the rest half is considered as a waste which includes pulp, seeds, peel, orange leaves, and whole fruits that do not match the quality requirements [54]. These residues can be utilized for the production of essential oils, bio-oils, pectin, organic fertilizers, and antioxidant compounds. They can also be used as a substrate for producing high-value compounds such as ethanol, enzymes, organic acids, adsorbent materials, microbial proteins, and biologically active secondary metabolites. The essential oil obtained from the peel is used as an ingredient in beverages, foods, perfumes, and cosmetics, aromatic essences. The D-limonene obtained from the peel is used for the production of solvent and ink. The citric pulp bran is used for producing animal feed, and the pulp is also used in the beverage and food industry [55].

Researchers have also used the orange peel to produce oil using the solvent extraction technique. The obtained oil has high viscosity and cetane index in comparison to diesel, whereas its calorific value is lower than diesel. The oil has been used by the researchers to operate a diesel engine. Purushothaman and Nagarajan [56] operated a diesel engine with neat orange oil and compared its performance with diesel. The authors found that the thermal efficiency of the engine was higher than diesel. The CO and HC emissions with orange oil were lower than diesel. In contrast, NO_x emission with orange oil was higher than diesel. Due to slightly higher ignition delay, the peak pressure and heat release with orange were higher than diesel. The authors attributed the rise in ignition delay to lower volatility characteristics of the oil which affects the physical delay. Methyl ester of the orange peel oil was blended with diesel and tests were carried out on a single-cylinder diesel engine by Deep et al. [57]. The authors observed that 20% blending of the methyl ester in diesel resulted in higher thermal efficiency than diesel. The HC and CO emissions were lower than diesel, whereas the NO_x emissions increased substantially. Rahman et al. [58] operated a multicylinder diesel engine with 10% orange peel oil blended in diesel. The authors observed that the brake specific particulate number and mass higher for the blend in comparison to diesel, similarly fuel consumption was also higher than diesel. Whereas CO and NOx emission was lower for the blend. The authors also studied the properties of the blend and found lower viscosity and density and higher heating value in comparison to diesel. The surface tension of the blend was also lower than diesel. To reduce NO_x emission, Dinesh et al. [59] prepared an emulsion of the orange peel oil, water, and used Span 80 as the nonionic surfactant. The authors observed that the NOx and smoke emission was lower than neat orange oil engine operation and diesel, as the water lowers the combustion temperature during combustion. Moreover, the HC and CO emissions with the emulsions were also lower than neat orange oil engine operation. Further, the thermal efficiency with the emulsions was also higher than diesel and neat orange oil engine operation. For the reduction of NOx emission with orange oil, Mahesh Kumar et al. [60] prepared an emulsion of orange peel oil methyl ester with titanium oxide nanoparticle dissolved in water and Span 80 as the surfactant. The authors found that the thermal efficiency with the neat methyl ester was lower than diesel and the thermal efficiency increased with the emulsions, but it was still lower than diesel. The HC, CO, and smoke emissions with the neat methyl ester and the emulsions were lower than diesel. Also, the emissions with the emulsions were lower than neat methyl ester operation. The oxides of nitrogen emission was higher than diesel with neat methyl ester and emulsions, However, the emission with the emulsions was observed to be lower than neat methyl ester engine operation.

1.2.4 Camphor oil

The genus *Cinnamomum* Schaeff. (Lauraceae) consists of 250 species. The tree is evergreen and grows in tropical and subtropical South America and Asia-pacific. Nearly 45 species are found in India out of which 21 are reported in the Western Ghats. Sixteen of these species is confined only to Southern

Western Ghats. *Cinnamomum* is used as a spice and as a medicine. The obtained essential oils have many biological activities such as anti-inflammatory, antiparasitic, antimicrobial, antihyperglycemic, antioxidant, antitumor, antidiabetic, and so on [61]. The wood of camphor laurel *Cinnamomum camphora* tree is used for the extraction of camphor oil. The molecular formula of camphor oil is $C_{10}H_{16}O$, and its chemical name is 1,7,7-trimethylbicyclo[2.2.1]heptane-2-one [62,63]. The oil is available in four grades: yellow, brown, white, and blue. The commonly available grade is the white one which has 1,8 cineol (11.5%), linalool (23%), camphor (50%), and some terpene derivatives as the primary components. The camphor oil on recrystallization forms solid camphor which contains secondary components which are excluded from the primary components [64,65]. Due to its low viscosity, it readily ignites without leaving any residue. Also, camphor is a member of the terpenoid ketone family which are naturally occurring organic compounds. As camphor has a bicyclic ring structure and a keto group, the nanostructure of the soot particles can be modified such that their oxidation reactivity is improved thereby improving the performance of the diesel particulate filter [66].

Camphor oil was blended with cashew nut shell oil and tested in a CI engine by Kasiraman et al. [62]. Due to the low cetane index of the camphor oil, blending with cashew nut shell oil was limited to 30% by volume. The authors observed that the addition of camphor oil to cashew nut shell oil resulted in the improved engine efficiency in comparison to neat cashew nut shell oil and its performance was at par with diesel-fueled engine. The authors also observed that smoke, CO, and HC emissions were lower, and NO emissions were higher in comparison to diesel. Mohamed Shameer and Ramesh [63] compared the NOx concentration of 20% blend of animal fat biodiesel, waste cooking oil biodiesel, and camphor oil in diesel with baseline diesel. The authors found that NOx emission was lower with camphor oil blend than animal fat and waste cooking oil biodiesel, but it was still higher than diesel. The authors attributed the trend to lower cetane number of the oil, which reduced the combustion temperature. To correlate the combustion temperature with NOx emission, the authors used a thermal imager to measure the cylinder head temperature. The authors found a direct relation between NOx emission and cylinder head temperature, i.e., highest NOx emission was observed with animal fat biodiesel blend and the cylinder head temperature was also observed to be highest with animal fat biodiesel. Similarly, camphor oil blend resulted in lowest emission among the blends and the head temperature was also found to be the lowest. For improving the diesel particulate filter regeneration efficiency, the soot oxidative reactivity of 10% blend of camphor oil in diesel was studied by Morajkar et al. [66]. The authors found that the smoke point of the blend was lower than diesel which shows that the soot formation tendency during combustion is lower with the blend in comparison to diesel. XRD, Raman, and HRTEM studies reveal that the primary particle diameter with core-shell structure was smaller with the blend. A high percentage of lattice fringes were short, and the fringe tortuosity was increased with the blend in comparison to diesel soot resulting in an enhanced degree of disorder in the crystal structure thereby making it easy for oxygen to attack during oxidation. The EDX measurement shows that the soot produced by the blend had higher aliphatic and oxygen content which enhances the oxidation kinetics of the soot.

1.2.5 Di-ethyl ether (DEE)

Oxygenated organic compounds such as ethers with chemical formula R-O-R are potential fuel for diesel engines as they have low molecular weight, and the activation energy required for breakage of bonds is low thereby increasing their cetane number. The presence of oxygen in ethers further help in the reduction of soot, thereby making them a favorable fuel for diesel engines [67]. Di-ethyl ether (DEE) is one such ether which has been used by many researchers as a fuel for diesel engines. The dehydration of ethanol can produce DEE in the presence of acid catalysts. Since ethanol is produced from biomass by the fermentation process, DEE can be considered as a renewable fuel [68]. DEE has a chemical formula of $C_4H_{10}O$ and contains 21.6% oxygen by mass. It is a liquid at atmospheric conditions. It has low autoignition temperature, high cetane number, and volatility. Further, its miscibility in diesel is high, and its flammability limit is broad [69].

DEE was first used as neat fuel for operating the diesel engine as it aids in cold-starting due to its low autoignition temperature. However, it was found that the fuel injection system started wearing out due to its low viscosity. Therefore, it is recommended that the DEE can be used as a blend in limited proportions. It was also found that during storage due to high volatility, air pollution was caused and peroxides were formed because of oxidation [70].

In neem biodiesel, Sivalakshmi and Balusamy [71] added 5%, 10%, and 15% DEE on a volume basis. With 5% DEE addition in biodiesel, the brake thermal efficiency improved and with a significant reduction in HC, CO, and smoke emissions. In contrast, higher NOx emissions were observed with all the blends in comparison to diesel. The peak pressure and heat release with 5% blend of DEE were higher than other blends and biodiesel. Due to higher volatility of DEE, a large amount of air-fuel mixture is prepared for ignition resulting in higher heat release. With the higher percentage of DEE in the blend, the ignition delay is reduced, and the less air-fuel mixture is prepared in the initial stage of combustion, thereby reducing the peak heat release. Ibrahim [72] studied the oxygen enhancing effect of DEE in a diesel by blending it up to 15%. The authors noted that 15% DEE addition in diesel resulted in 7.4% improvement in thermal efficiency as compared to diesel. Improvement in engine combustion characteristics was also observed with the 15% blend of DEE in diesel. However, engine stability was reduced during the experiments. In the ternary blends of ethanol, jatropha biodiesel and diesel, and methanol jatropha biodiesel and diesel, 5% and 10% DEE were blended was added as an ignition enhancer by Venua and Madhavan [73]. Addition of DEE to the ethanol biodiesel blend increased BSFC, combustion pressure and duration with the reduction in smoke and NOx due to high latent heat of evaporation and reduced ignition delay. Whereas, DEE addition in methanol biodiesel blend resulted in a decrease in combustion pressure, duration, and fuel consumption and an increase in carbon dioxide, carbon monoxide, particulate matter emission. Tudu et al. [74] blended 40% of the light fraction of tyre pyrolysis oil with diesel and found that the cetane number of the blend was lower than diesel. To improve the ignition quality of the blend, the authors added 1%–4% DEE and tested the blends on a single-cylinder diesel engine. The authors found a reduction in ignition delay with the increment in DEE in the blend. The engine-out emissions and its performance were better with 4% DEE addition in the blend at full load. A reduction of 6% in fuel consumption and 25% lower NO emission was observed with 4% DEE blend in comparison to diesel at full load.

1.2.6 Acetone

The fermentation of biomass using suitable bacteria can result in the formation of a mixture of acetone, butanol, and ethanol (ABE). The three liquids are typically in the ratio of 3:6:1 on a volume basis [75]. The mixture can then be purified to obtain the three liquids separately. Acetone is a major organic solvent, and many modern industrial processes rely heavily on it. It is used extensively as a solvent for synthetic materials and polypropylene, polycarbonate, polystyrene, and others. It is also an essential ingredient of varnishes and paints [76].

Many studies show that acetone can be used as an additive for operating the diesel engine as its properties are similar to alcohols with the high latent heat of vaporization. ABE was emulsified in diesel and used to operate a CI engine [77]. In all, 20% by volume of ABE in 80% of diesel resulted in better engine efficiency, reduction in NOx, polycyclic aromatic hydrocarbons, unburnt hydrocarbon, and near-zero soot emission in comparison to diesel. Chang et al. [78] blended water containing ABE with soybean biodiesel and diesel in various proportions and tested the blends on a multicylinder diesel engine. The blends containing ABE, biodiesel, and diesel resulted in better thermal efficiency that diesel and biodiesel-diesel blends. These blends reduced both particulate matters and NOx emission and the reduction was nearly 4.3%–30.7% and 10.9%–63.1%, respectively due to lower aromatic content of the blends, the polycyclic aromatic hydrocarbon emission reduced by 26.7%–67.6%. Lin et al. [79] formed blends by adding 1–3 wt% of water containing acetone with diesel (95%–97%) and also used 1 wt% of isopropyl alcohol and neat and biodiesel of soybean oil as a stabilizer. Trial results indicated that blend of 3% water containing acetone, 1% propyl alcohol, and neat soybean oil had excellent fuel stability and the BSFC was reduced by 1.4%–5.5% at maximum load condition. The blend emitted less NO, particulates and HC due to acetone cooling effect and higher fuel oxygen content. The blend also emitted 4.92% less CO_2 in comparison to diesel. The particulates, carbon and hydrocarbons emitted from CI engine fueled with waste cooking oil biodiesel, acetone, and isopropyl alcohol were examined by Tsai et al. [80]. The tested blends were diesel with 1%–3% volume of pure acetone and 1% volume isopropyl alcohol and remaining biodiesel. The experimental results showed that the addition of acetone aided in the reduction of particulate matter and particulate organic carbon emission. With all the blends, total hydrocarbon emissions were reduced in comparison with diesel. They also established that the blend containing acetone reduced emissions of particulates, carbon, and hydrocarbons. A single-cylinder diesel engine running at different engines was fueled with various mixtures of butanol, acetone, and diesel [81]. The authors found a decrease in ignition delay with increase in the butanol-acetone (BA) mixture in the blend with diesel which improves the combustion quality. The brake power with 10% BA blend showed a 5% improvement whereas with 20% and 30% BA blend the brake power was comparable with diesel at all engine speeds. The thermal efficiency with 10% BA blend was comparable with diesel at all the speeds whereas it was higher with 20% and 30% BA blend in comparison to diesel.

1.2.7 Monoethanolamine (MEA)

Monoethanolamine is a solvent used in the coal-fired power stations for absorbing post-combustion CO_2 emissions. The solvent is readily available, its cost is low, and its absorption rate is fast. However, the parasitic energy consumption and the capital cost involved in an MEA-based post-combustion capture system is a hindrance to its use [82]. MEA can be produced by reacting ammonia/water with ethylene oxide at a pressure of 50–70 bar to keep ammonia in the liquid phase. The process is exothermic and does not require any catalyst. The ratio of ammonia and ethylene oxide plays a significant role in deciding the composition of the resulting mixture. If ammonia reacts with one mole of ethylene oxide, monoethanolamine is formed, with two molecules of ethylene oxide, diethanolamine is formed whereas with three moles of ethylene oxide triethanolamine is formed (Fig. 2). After the reaction, the distillation of the resulting mixture is carried out first to remove excess ammonia and water. Then the amines are separated using a three-step distillation setup [83].

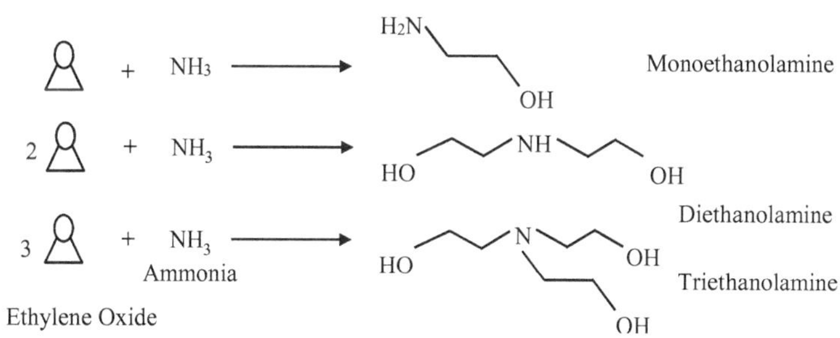

FIG. 2 Chemical reaction for the formation of monoethanolamine.

Amines such as *N,N′*-diphenyl-1,4-phenylenediamine (DPPD) and *N*-phenyl-1,4-phenylenediamine (NPPD), p-phenylenediamine and ethylenediamine have good antioxidation characteristics [84,85]. When these amines were used as an additive in a diesel engine, reduction in NOx emissions was observed without affecting the engine performance or other emissions [86]. However, monoethanolamine has not been used as a fuel additive until now. Therefore, the authors have explored the potential of the amine as an additive in this novel study.

1.3 Gaseous fuels

Apart from operating the diesel engine with blends of biofuels, they can be run on gaseous fuels as well. Some of the gaseous fuels used by researchers are liquefied petroleum gas (LPG) [87], compressed natural gas (CNG) [88], biogas [89], producer gas [90], hydrogen [91], and Brown's gas [92]. However, these gases have a low cetane number, and their autoignition temperature is high. Therefore, another fuel which has a high cetane number (diesel or biofuel) is direct-injected into the cylinder that acts as a source of ignition for the gaseous fuels, thereby starting the combustion process. Among these fuels, hydrogen and Brown's gas are free of carbon and are discussed in detail in this section.

1.3.1 Hydrogen

Hydrogen is used for the production of various chemicals and as a feedstock for producing refined petroleum products. It is also used in the pharmaceutical, electronic, and metallurgical industries. It finds application as a fuel for propelling space shuttles and rockets. Approximately, 140 MToe of hydrogen is annually produced, which is nearly 2% of the energy requirement. Presently, hydrogen is produced from oil, natural gas, and coal, which is nearly 96% of the total hydrogen production. Renewable sources such as reforming of biomass or splitting of water by electrolysis have also been investigated for hydrogen production, wherein the energy is also supplied by renewable sources. Fig. 3 shows the various pathways through which hydrogen can be produced. Moreover, for making hydrogen as the primary source of energy, extensive is required for establishing distribution and storage infrastructure [93–95].

Recently, hydrogen gas has been explored as a fuel for both the SI engine and the CI engine [96–98]. Its heating value is 120 MJ/kg, about 2.75 times of hydrocarbon-based fuels [99]. However, its volumetric energy density is very low, i.e., a separate tank is required for carrying hydrogen in a vehicle. Changes in engine similar to the use of CNG for operating it with hydrogen [100]. The engine can run on very lean mixtures as its flammability limit is very wide. Its ignition energy is low, but its autoignition temperature is high. Its octane number and flame speed are very high [101].

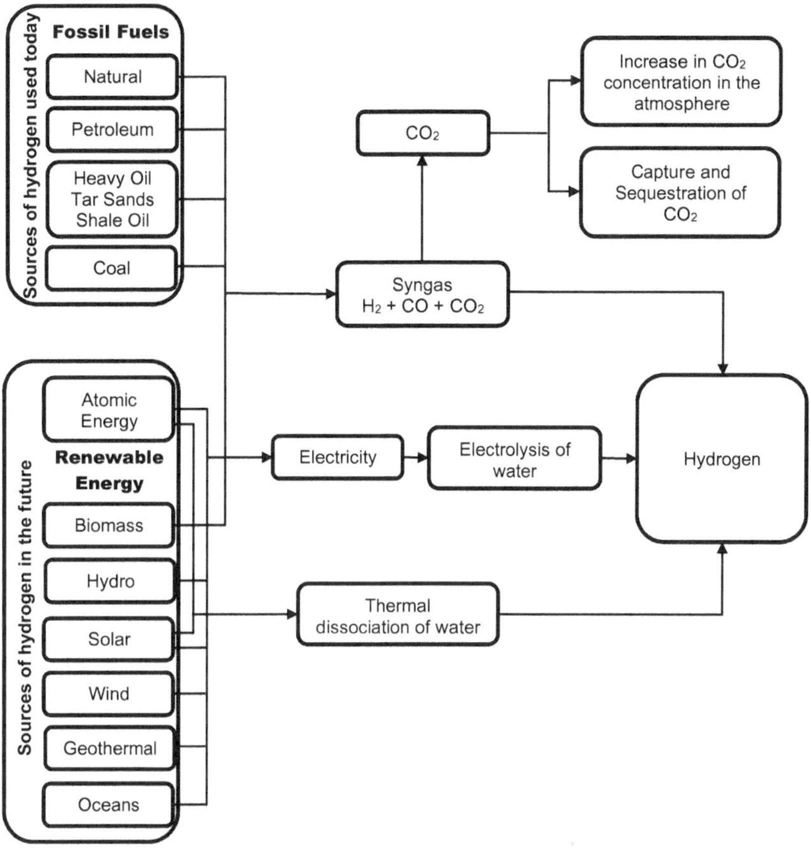

FIG. 3 Different routes for hydrogen production.

The direct addition of hydrogen in a CI engine is limited due to low operating range and improved by using low ignition temperature fuels such as diesel or biofuels. Many researchers have operated a diesel engine by either inducing hydrogen through a carburettor or injecting hydrogen through a solenoid injector [102,103]. Waste cooking oil and its emulsion were used as the direct-injected fuel in a dual fuel engine with hydrogen being inducted in small quantities [91]. Improvement in engine performance and reduction in emissions was observed, but the authors found no advantage in adding hydrogen at low power output from the engine. Senthil Kumar et al. [104] also observed similar output with Jatropha as a primary fuel. Rubber seed oil as the direct-injected fuel by Edwin Geo et al. [105] HC, CO, and smoke level were found to decrease at all loads as the hydrogen energy content increased. To counter NOx emissions rise in hydrogen inducted dual fuel engine Bose and Maji [106] used exhaust gas recirculation technique. However, the HC and CO emission was higher with this technique.

Automobile manufacturers have built several prototype vehicles which use hydrogen for powering their internal combustion engines. Some of them include BMW, Mazda, and Ford. Toyota Prius hybrid was also modified to operate on hydrogen [93]. Fuel cell-powered vehicles that utilize compressed hydrogen are also being commercially produced. Some of the vehicles available on sale and lease in the United States, South Korea, and Japan are Hyundai Tucson FCEV, Nexo, Honda Clarity, and Toyota Mirai [107].

1.3.2 Oxyhydrogen (HHO)

Studies show that hydrogen needs to be stored onboard a vehicle in high-pressure vessels for its use in an internal combustion engine due to its low energy density. Also, there are many safety-related issues with storing hydrogen in a vehicle. This problem can be countered by onboard production of hydrogen using electrolyzers, thereby reducing the storage issue. The electrolyzers split water into hydrogen and oxygen by passing a voltage through it; this process is known as electrolysis [108,109]. The electrolyzer units are of two types:

(1) Water is split into hydrogen and oxygen gas and released at anode and cathode, respectively. These gases are then collected from the respective electrodes.

(2) Production of HHO which is a combustible mixture of hydrogen and oxygen.

In an electrolyzer, when the electrolysis of water takes place at atmospheric pressure and a temperature of 25°C, the potential difference required between the anode and the cathode is nearly 1.23 V [110]. If the electrolyte contains potassium hydroxide, then at an electrolytic temperature of 80°C, the current density and voltage requirement shoot up to 2000 A/m^2 and 2–2.5 V, respectively [111]. The electrolyzer contains an ion conductive membrane placed between the two electrodes that separate the oxygen and hydrogen gases produced at the electrodes and allows only OH^- ions to pass through. The equations for the electrolysis processes are given by Eqs. (3)–(6) [112].

$$\text{Electrolyte } 2H_2O \rightarrow 2H^+ + 2OH^- \tag{3}$$

$$\text{Cathode } 2H^+ + 2e^- \rightarrow H_2 \tag{4}$$

$$\text{Anode } 2OH^- \rightarrow H_2O + {}^1/{}_2O_2 + 2e^- \tag{5}$$

$$\text{Result } H_2O \rightarrow H_2 + {}^1/{}_2O_2\ \Delta Hr = 240\ \text{kJ/mol} \tag{6}$$

where ΔH_r is the enthalpy of the reaction, i.e., 240 kJ/mol of energy is required for the production of 2 g of hydrogen. The theoretically same amount of energy should be released on the combustion of 2 g of hydrogen. When the conductive membrane present between the electrodes is removed, the hydrogen and oxygen produced during electrolysis are allowed to blend. It should be noted here that the electrolysis of water is such that the oxygen and hydrogen did not form O_2 and H_2 molecules and they are still in a monoatomic state, i.e., a single atom per molecule [92]. The mixture is known as oxyhydrogen gas (HHO) or Brown's gas, and on a volume basis, hydrogen and oxygen present in the gas are in the ratio of 2:1 (Eq. 4). From 1 kg of water, nearly 1860 L of HHO can be produced by consuming 22.9 MJ of electric power. The HHO produced contains nearly 1240 L of hydrogen and 620 L of oxygen. At 0.1 MPa and 273 K, the density of HHO gas mixture, H_2 and O_2 is 0.54, 0.09, and 1.43 g/m^3, respectively [112].

Yull Brown first produced the gas in a single cell electrolyzer and used it for welding purpose. Hence the gas was named as Brown's gas. As the gas does not contain carbon, its burning is sootless, and it finds use in soldering, brazing, acryl process, glass process, and so on [113]. Studies were also conducted in using the gas as a fuel for both SI and CI engine. The Brown's gas is lighter than air and is an odorless and colorless gas. Its flammability limit is much more in comparison to diesel [114]. The calorific value of the gas is very high, and it has a potency of three times that of gasoline and eight times that of diesel [115]. On the ignition of the combustible mixture of HHO, both implosion and explosion occur resulting in the formation of water and releasing the bond energy in the form of heat. In monoatomic molecules such as HHO, no interatomic bonds are broken (such as H_2 and O_2) only their reshuffling takes place resulting in the formation of water again. Since oxygen and hydrogen do not occur in a diatomic state, the energy in the HHO gas is more as these bonds were not made. The water vapor in this state is unstable and releases more energy in comparison to that of hydrogen burning with oxygen. When a mixture of HHO and fuel is combusted, the formed water collides with the fuel molecules and water forms the core and the fuel molecules transform into water shell due to differences in their densities. In the combustion stroke, water explodes into steam due to high heat and pressure, thereby atomizing the fuel when the combustion starts due to rapid release of heat, the cylinder temperature increases resulting in the splitting of water into hydrogen and oxygen which once again reignites increasing combustion efficiency [92].

Many researchers have used HHO as a fuel in a CI engine. The HHO can be injected through an injector or inducted through a venturi due to pressure difference in the intake manifold. However, due to lower cetane number of HHO, the ignition of fuel does not take place; hence a higher cetane number fuel such as diesel or biodiesel is directly injected in the cylinder to start the combustion. Yilmaz et al. [92] used HHO as an additive for operating a four-cylinder diesel engine. An in-house electrolyzer unit generated the HHO and its injection in the intake manifold was controlled by an ECU. The engine was tested at various speeds, and the amount of HHO injection was controlled accordingly. The authors observed that the engine torque on an average increased 19.1% and the specific fuel consumption decreased 14%. CO and HC emissions were decreased by 13.5% and 5%, respectively. Bari et al. [116] also studied the efficiency and emissions of an unmodified four-cylinder diesel engine fueled with different amounts of HHO. The authors carried out the tests at three different power output of the engine at a constant speed of 1500 rpm. A commercial HHO generator was used for producing HHO. The authors observed a maximum increment of 8.8% in brake thermal efficiency and a reduction of 8.1% in specific fuel consumption. Moreover, the CO_2, CO, and HC emissions decreased while the NOx emissions increased because of high peak pressure and temperature as HHO has a high flame velocity and energy content. A single-cylinder diesel engine was used by Premkaritkumar [117] to study the effects of HHO, which was produced by an in-house made generator powered by a 12 V source. The engine was tested with 1 1 pm to 3.31 pm flow rate of HHO under varying load conditions. The authors, in their subsequent work, also varied the injection timing of diesel to study its effect on the diesel engine [118]. The authors found an 11.6% increase in thermal efficiency at full load and 3.31 pm HHO flow rate whereas a 4.15% reduction in the efficiency was observed at 25% load and 1 1 pm. At high HHO flow rate and load, smoke, HC, and CO emission decreases whereas CO_2 and NO_x emission increases and at low loads and low flow rate of HHO, the CO_2 and NO_x emission was low whereas the other regulated emissions were higher than neat diesel engine operation.

1.4 Present study

In the present scenario, it is essential to reduce the carbon dioxide emission from the exhaust of a CI engine. This can be achieved either by reducing fuel consumption, i.e., increasing the engine's thermal efficiency or by using fuels that have low carbon content. The literature review shows that renewable Karanja oil methyl ester has the potential to completely replace diesel as it is inedible oil and is abundantly available in many parts of the world. However, the studies carried out on engines fueled with the methyl ester reveal that there are shortcomings in completely replacing diesel. These shortcomings are high fuel consumption and high CO_2 and NO_x emissions. The review further shows that if the low carbon content fuels such as pine oil, eucalyptus oil, orange oil, camphor oil, di-ethyl ether, MEA, and acetone are blended with other fuels, they increase the performance of the engine and facilitate emission reduction.

Similarly, the inducted gaseous fuels namely hydrogen and HHO having no carbon content, also improve the engine performance along with lower emissions. Therefore, it is imperative to carry out a study for comparing the emission reduction potential of the blends of Karanja oil methyl ester and the low carbon content fuels. The engine performance and emission with hydrogen and HHO as the inducted fuel and Karanja oil methyl ester as the direct-injected fuel is also compared in this study.

2 Experimental setup

A water-cooled mono-cylinder compression ignition engine was used in the present study. At a constant speed of 1500 rpm, the engine generated a maximum power of 5.2 kW. For loading the engine, it was coupled to an eddy current dynamometer. The specification of the test engine is presented in Table 4. No modifications were made in the engine's fuel injection system and its structure for carrying out the tests. The engine's fuel consumption with different test fuels was measured on a volume basis, and the flow rate of air was measured using U-tube manometer. AVL five gas analyzer was used to measure the emissions, namely CO_2, CO, NO, and HC. Based on air and fuel flow rate the absolute values of emissions were converted to specific emissions. An AVL smoke meter working on the principle of light extinction technique was used to measure the smoke opacity of the engine exhaust. The schematic diagram of the experimental setup is shown in Fig. 4. The results in this study are shown at the rated power of 5.2 kW.

2.1 Gaseous fuel injection system

In the present study, hydrogen and oxyhydrogen gas were inducted in the intake manifold separately. The hydrogen cylinder (1) shown in Fig. 2 stored the gas at nearly 220 bar pressure, and with the help of two-stage regulator, the pressure of the gas was brought down to 1.5 bar. The hydrogen was first bubbled through a flame trap (4), and then it was passed through a one-way flame arrester (5). The flame arrester and flame trap are used in the present study as a preventive measure. Since there is a chance of backfiring, i.e., when the intake valve is open, and the mixture of hydrogen and air starts entering the cylinder, due to high temperature inside the cylinder, the mixture may start burning. The flame may travel through the pipe (hydrogen has high flame speed) and reach the hydrogen

TABLE 4 Engine specifications.

Make and model	Kirloskar AV1
Engine type	Single-cylinder, water-cooled, direct-injection, constant speed
Bore (mm)* Stroke (mm)	87.5*110
Compression ratio	17.5
Rated power @ 1500 rpm in kW	5.2
Injection pressure (bar)	200
Injection timing	23 deg. before top dead centre

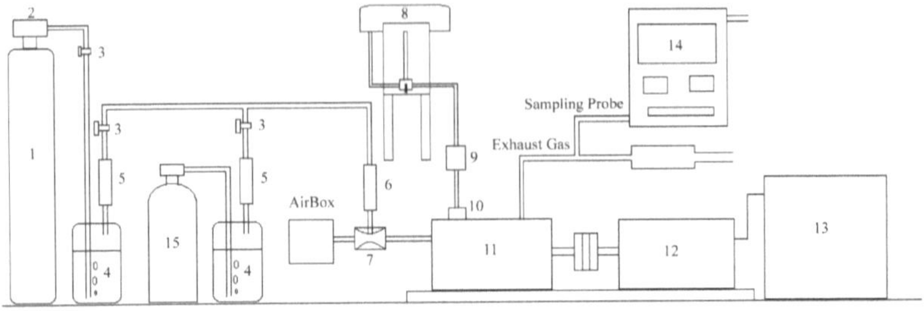

FIG. 4 Engine experiment test setup. 1. Hydrogen cylinder; 2. Regulator; 3. Needle valve; 4. Flame trap; 5. Flame arrester; 6. Rotameter; 7. Venturi; 8. Fuel tank; 9. Fuel pump; 10. Fuel injector; 11. Engine; 12. Eddy current dynamometer; 13. Control panel; 14. Emission analyzer; 15. Oxyhydrogen generator.

cylinder which may cause an explosion. After the flame arrester, the gas is passed through a needle-valve (3) and then to a rotameter (6) which was used to control the flow rate of hydrogen. Finally, it was inducted in the air intake pipe through a venturi (7). During the suction stroke, a partial vacuum is created inside the cylinder due to this difference in pressure the hydrogen flows through the venturi and forms a homogeneous mixture with air before entering the cylinder.

The oxyhydrogen was also inducted into the intake manifold through a flame trap, flame arrester, rotameter, and venturi. In the present study, the oxyhydrogen was produced in real time using an in-house made electrolyzer (15) which is a three-cell chamber and requires 12 V power input. Distilled water was poured in the chamber for producing HHO.

2.2 Test fuels

Table 3 shows that the viscosity of the low carbon fuels, namely, pine oil, eucalyptus oil, and camphor oil, except for orange oil, is lower than diesel. Therefore, it is not suitable to use them as a sole fuel for operating the diesel engine. All the fuels have nearly half the carbon content in comparison to Karanja oil. Therefore, to reduce CO_2 emissions equal blends (on a volume basis) of the low carbon content fuels and Karanja oil methyl ester were prepared. Table 5 shows that the oxygenates, such as DEE, acetone, and MEA have a low number of carbon atoms and high latent heat of vaporization. Their use as an additive for enhancing the ignition is already known. Therefore, they were used as a 20% blend in Karanja oil methyl ester. The nomenclature of the test blends is presented in Table 6. Similarly, it is not feasible to operate the diesel engine only on the gaseous fuels. Therefore, they were operated in the dual fuel mode with Karanja oil methyl ester as the direct-injected fuel. The energy share of hydrogen was fixed at 12% throughout the experiments. Since the thermal efficiency of the engine was found to be maximum at this energy share. Furthermore, the energy share of HHO was fixed at 5% as the engine efficiency was found to be the maximum.

TABLE 5 Properties of DEE, acetone and MEA.

Property	DEE	Acetone	MEA
Molecular formula	$(C_2H_5)_2O$	C_3H_6O	C_2H_7NO
Kinematic viscosity at 40°C, cSt	0.23	0.34	–
Density at 15°C, g/cm^3	0.715	0.791	0.997
Lower heating value, MJ/kg	33.89	29.6	–
Cetane index	~125	–	–
Latent heat of vaporization, kJ/kg	350	551	175

TABLE 6 Test fuel composition and nomenclature.

Test fuel composition	Nomenclature
100% Diesel	Diesel
100% Karanja oil methyl ester	K100
50% Karanja oil methyl ester and 50% Eucalyptus oil	K50-EU50
50% Karanja oil methyl ester and 50% Camphor oil	K50-CMO50
50% Karanja oil methyl ester and 50% Pine oil	K50-PO50
50% Karanja oil methyl ester and 50% Orange oil	K50-O50
80% Karanja oil methyl ester and 20% Acetone	K80-Acetone20
80% Karanja oil methyl ester and 20% MEA	K80-MEA20
80% Karanja oil methyl ester and 20% DEE	K80-DEE20
100% Karanja oil methyl ester and 5% energy share of HHO	K100+5%HHO
100% Karanja oil methyl ester and 12% energy share of H_2	K100+12%H_2

2.3 Experiment

The experiments were first carried out with the test fuels without induction of hydrogen and oxyhydrogen. The fuel sample with which the engine was operated was placed in the fuel tank and the engine was cranked with the help of a starter motor until it gained enough speed to sustain combustion. The engine was first allowed to stabilize at the no-load condition and then the engine was slowly loaded through the eddy current dynamometer till it reached full load condition. The engine was then stabilized at this load. The data such as fuel flow rate, exhaust gas temperature, and engine-out emissions were recorded. The readings were taken thrice and averaged. After operating the engine with one sample, the tank was filled with diesel and the engine was operated with diesel at no-load condition. The same method was applied to all the test samples. When the engine was operated in dual fuel mode with the induction of hydrogen/HHO, the needle valve (3) placed after the flame trap (4) was opened so that the hydrogen/HHO could be inducted through the venturi (7) in the intake manifold due to difference in pressure. At the engine exhaust end, the sampling probes connected to the emission analyzer were placed in the exhaust pipe for measuring the emissions.

3 Results and discussion

The CO_2 emission for the test fuels at full load is depicted in Fig. 5. Engine operation with K100 resulted in a 26.2% increase in CO_2 emission in comparison to diesel. As the number of carbon atoms in the methyl ester is higher than that in diesel, the increase in emissions is observed. Also, the methyl ester has a higher energy density in comparison to diesel, hence the mass of the injected fuel is higher, thereby increasing the emission. As the Karanja oil methyl is renewable, the net CO_2 emission is negative since the *Pongamia pinnata* plant will absorb the CO_2 from the atmosphere for the photosynthesis process. Equal blends of eucalyptus oil, pine oil, camphor oil, and orange oil with Karanja oil methyl ester resulted in a reduction in CO_2 emissions, as the total amount of carbon content present in the charge reduced. As compared to K100, the percentage reduction in the CO_2 emission with K50EU50, K50CMO50, K50PO50, and K50O50 is 7.54%, 9%, 9.4%, and 27.4%, respectively. The maximum CO_2

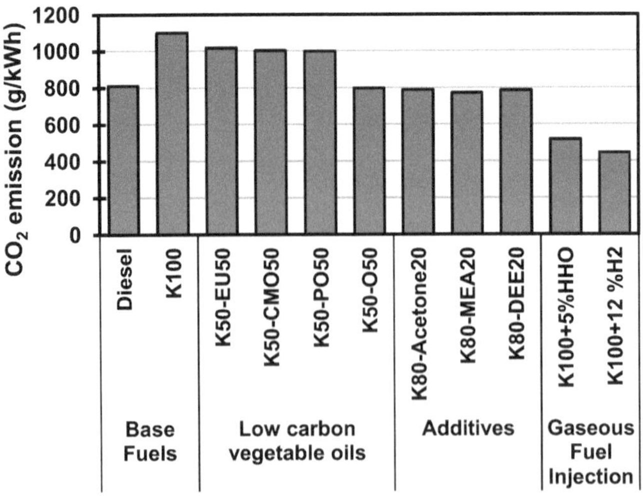

FIG. 5 CO_2 emission for various fuels at maximum load.

reduction is observed with the blend of orange oil as it has the minimum density in comparison to the other blends. The oxygenated additives such as acetone, MEA, and DEE were also blended with Karanja oil methyl ester and their effect on CO_2 emission is also shown in Fig. 5. It is seen that as the number of carbon atoms in the additives increases the CO_2 emission, however, the CO_2 emission was lower than K100 and the blends of Karanja oil methyl ester and low carbon vegetable oils. The reduction is mainly caused by the low density of the additives and less number of carbon atoms. Another reason is the decrement in combustion efficiency, which affects the conversion of the intermediate carbon monoxide to carbon dioxide. The CO_2 emission with K80Acetone20, K80MEA20, and K80DEE20 is 785.8, 769.66, and 784.24 g/kWh, respectively. The engine was also operated in dual fuel mode with hydrogen and oxyhydrogen as the inducted fuel and K100 as the direct-injected fuel. The results show a drastic reduction in CO_2 emission with both gases. With hydrogen induction, the CO_2 emission reduced by 59.48% whereas with HHO, CO_2 emission reduced by 53% in comparison to K100. As both the gases are carbon-free, and on their addition, the amount of K100 injected in the combustion chamber is reduced. Therefore, the amount of carbon content entering the cylinder reduces, thereby reducing the carbon dioxide emission.

The NO emission for the blends of Karanja oil methyl ester with low carbon fuels, additives and H_2 and HHO in dual fuel mode is compared with base diesel and K100 in Fig. 6. The NO emission with K100 is 13% higher than diesel since the oxygen content of the charge increases as K100 itself contains oxygen. Also, the cylinder temperature is higher when the engine is operated with K100, which also favors NO formation. The engine operation with equal blends of low carbon vegetable fuels and Karanja biodiesel resulted in an increase in

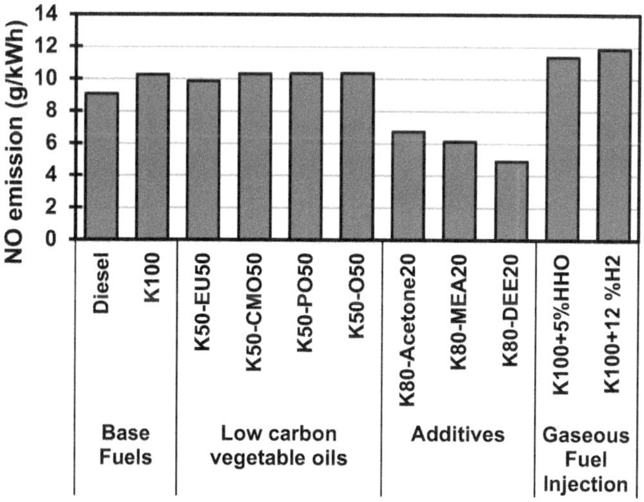

FIG. 6 NO emission for various fuels at maximum load.

NO emission except for K50EU50. The NO emission with K50EU50, K50CMO50, K50PO50, and K50O50 is 9.8, 10.32, 10.3, and 10.35 g/kWh, respectively. The highest emission was observed with orange oil blend due to improvement in combustion caused by better fuel properties. As CMO, PO, and O have low cetane number, the start of combustion is postponed resulting in more fuel being prepared for combustion, thereby increasing the emission, whereas with eucalyptus oil blend, the combustion is improper, resulting in lower NO emission. The lowest NO emission is observed with diethyl ether blend since the cetane number of DEE is very high, resulting in lower ignition delay. Hence, less fuel is prepared for burning in the initial phase of combustion, thereby lowering the heat release and the combustion temperature. The NO emission with acetone and MEA additive is low since the latent heat of vaporization of the two is high. Hence, they absorb some heat produced during the compression stroke for evaporation, thereby decreasing the cylinder temperature and in turn, decreasing the combustion temperature. When the engine is operated in dual fuel mode with hydrogen and oxyhydrogen the NO emission increases in comparison to K100 and the maximum emission is observed with hydrogen induction. The calorific value of both hydrogen and oxyhydrogen is very high hence large amount of heat is released during their combustion resulting in higher cylinder temperature. Also, the presence of oxygen-containing biodiesel provides the extra oxygen for the nitrogen radical to be converted into NO.

The smoke opacity with different test fuels at full load condition is displayed in Fig. 7. The smoke opacity with K100 and diesel operation is 49.4% and 51%, respectively. The presence of fuel-borne extra oxygen is helpful in soot

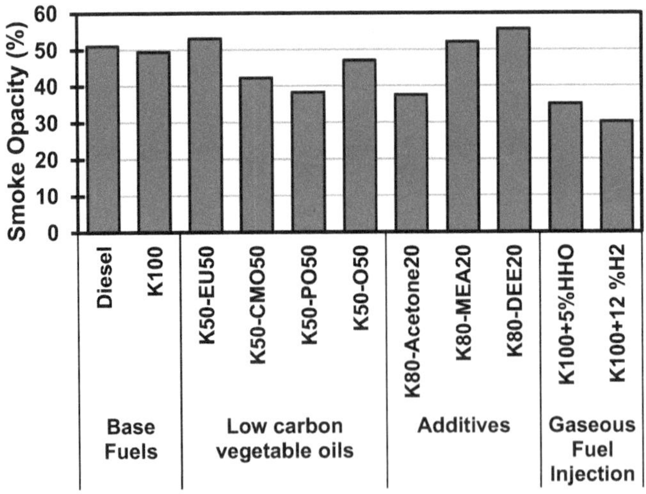

FIG. 7 Smoke opacity for various fuels at maximum load.

reduction as the particles are easily oxidized at higher temperatures. Among the equal blends of low carbon vegetable oils and biodiesel, the lowest emission is observed with pine oil (38.2%), followed by camphor oil (42.2%), orange oil (47%), and eucalyptus oil (53%). The viscosity of pine oil is lowest among the blends hence the atomization of the fuel molecules, their vaporization, and mixing with air will be better than the other blends along with the presence of extra oxygen will lead to an improvement in combustion thereby reducing the soot emission. Addition of additives (DEE and MEA) resulted in higher emissions than K100 at full load condition. The smoke opacity with K80MEA20 and K80DEE20 is 52.1% and 55.6%, respectively. The increase in emission with the two additives is due to lower thermal efficiency of the engine, which increases the amount of fuel injected per cycle. As more amount of fuel enters the cylinder, and due to improper combustion, the in-cylinder temperature is lower, thereby increasing the soot emission. The operation of the engine in dual fuel mode resulted in a decrease in soot emission for both HHO and H_2. As compared to K100, the soot emission with H_2 and HHO reduced by 38.84% and 28.94%, respectively. The reduction in emission with the gaseous fuels is due to the reduction in the carbon content entering into the cylinder per cycle. Also, due to high cylinder temperature and flame velocity of both the fuels, improvement in the combustion process is observed, which improves the reduction in soot emission.

The HC emission at full load for the test fuels compared to base fuels is shown in Fig. 8. 0.34 and 0.39 g/kWh is the unburned hydrocarbon emission with diesel and K100, respectively. Due to poor physical properties of K100 and higher C/H ratio, the emission with K100 tends to increase. The HC

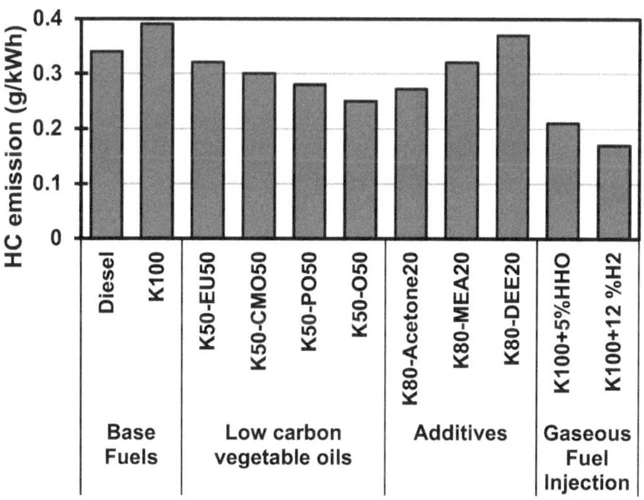

FIG. 8 HC emission for various fuels at maximum load.

emission is found to decrease for all the blends of low carbon content vegetable oil in comparison to the biodiesel due to their lower carbon to hydrogen ratio. At full load condition, the HC emission with K50EU50 blend is 0.32 g/kWh, K50CMO50 blend is 0.3 g/kWh, and with K50PO50 blend is 0.28 g/kWh. The lowest HC emission among the vegetable oil blends is found to be 0.25 g/kWh with K50O50 blend. The reason for the lowering of the emission is a reduction in the amount of fuel entering the cylinder, enhancement in combustion, and the presence of excess oxygen compared to other blends. The HC emission with the additives blended in the biodiesel was found to be lower than the neat biodiesel engine operation. Moreover, the highest HC emission is observed with DEE (0.37 g/kWh), followed by MEA (0.32 g/kWh) and acetone (0.27 g/kWh). Due to the cooling effect of the additives, the HC emissions may increase as the cylinder temperature reduces. However, the carbon to hydrogen ratio of the additives is very low; hence the amount of unburned hydrocarbons reduces. The HC emissions were found to decrease when the engine was operated in dual fuel mode with H_2 and HHO. The HC emission with HHO and H_2 reduced by 46.15% and 56.4%, respectively as compared to K100 engine operation. The drastic reduction in the HC emission with the gaseous fuel is caused by higher combustion temperature and high flame speed of the two fuels, thereby reaching the quench layers of the combustion chamber. Another reason for the reduction in the emission is the substitution of carbon-rich biodiesel with carbonless gaseous fuels. Further, the reduction in emission with hydrogen is more than oxyhydrogen as the calorific value, and flame speed of the former gas is more.

Fig. 9 shows the CO emission for various test fuels at full load. The carbon monoxide emission with K100 is 49.5% higher as compared to diesel. Due to

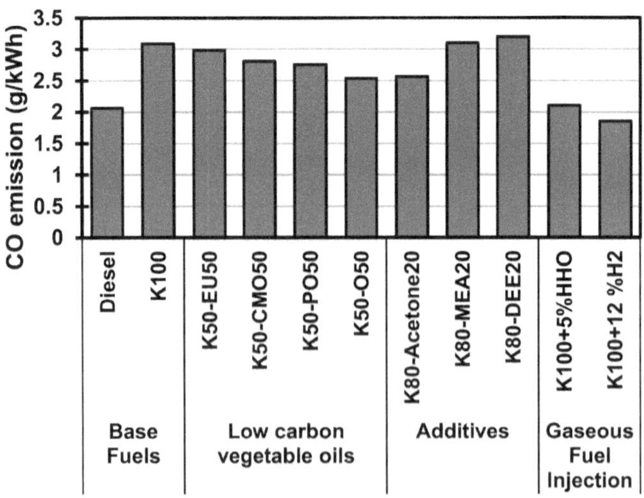

FIG. 9 CO emission for various fuels at maximum load.

inferior combustion caused by poor physical properties of the biodiesel, the conversion of carbon monoxide to carbon dioxide is reduced. The CO emission for blends of the methyl ester with eucalyptus oil, camphor oil, pine oil, and orange oil is 2.98, 2.8, 2.75, and 2.5 g/kWh, respectively, which is found to be lower than the engine operation with neat biodiesel. The minimum CO emission among the blends is observed with orange oil blend due to lower carbon content and improved combustion. The emission was observed to increase with DEE and MEA blend and was found to be higher than K100. Although the oxygen content with the additives increases, due to the reduction in combustion efficiency, the emission was found to increase. Similar to HC emissions, the CO emission decreases in dual fuel mode with both hydrogen and oxyhydrogen. The CO emission at full load with H_2 and HHO is 1.85 g/kWh and 2.1 g/kWh. As the combustion improves with the two fuels, the cylinder temperature is higher, so the probability of conversion of CO to CO_2 is more. Furthermore, the reduction in the hydrocarbon content further increases the CO reduction.

The brake thermal efficiency at full load condition for the test fuels is shown in Fig. 10. A marginal reduction of 4% in the thermal efficiency is observed with K100 in comparison to diesel, owing to its poor physical properties. The thermal efficiency of the engine on operating with the vegetable oil blends was slightly better, except K50EU50 blend. The maximum thermal efficiency was observed among the blends to be 32.36% with K501O50. The improved properties of the blend lead to an improvement in combustion, thereby producing more useful work. The thermal efficiency with the additives was lower than the base Karanja oil methyl ester. At full load, the efficiency with K80Aectone20, K80MEA20, and K80DEE20 is 26.93%, 24.65%, and

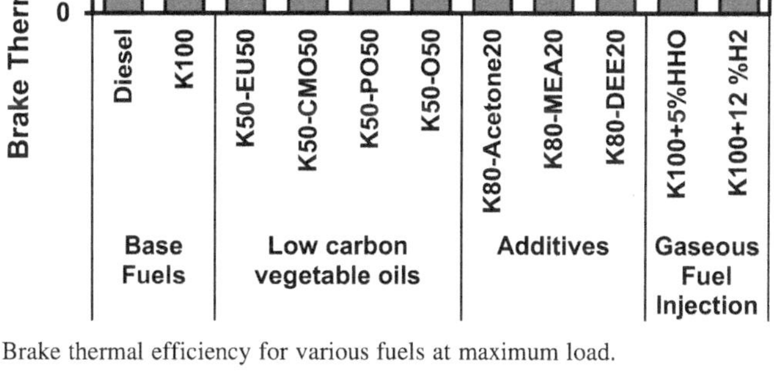

FIG. 10 Brake thermal efficiency for various fuels at maximum load.

25.95%, respectively. The reduction in efficiency with the acetone and MEA is due to their cooling effect that reduces the combustion temperature, thereby reducing the combustion efficiency. The reduction in efficiency with DEE is due to its higher cetane number that causes less fuel to burn in the premixed phase thereby reducing the peak heat release and the heat released in the second phase of combustion is not fully utilized for producing useful work which decreases the thermal efficiency. With the induction of both HHO and H_2 in the intake manifold, the thermal efficiency was found to increase in comparison to both diesel and K100. At full load, the thermal efficiency with HHO and H_2 increased by 12.67% and 16.84%, respectively, in comparison to K100 and 8.53% and 12.54%, respectively in comparison to diesel. The increase in efficiency is caused by the high calorific value of the two fuels, which releases a large amount of energy in the form of heat in the initial phase of combustion. This energy can be utilized in producing useful power, thereby increasing the thermal efficiency.

The brake specific energy consumption with blends of biodiesel and low carbon vegetable oils, blends of biodiesel and oxygenated additives and gaseous fuel in dual fuel mode is compared with base diesel and K100 in Fig. 11. The energy consumption with K100 and diesel is 12.48 and 12.03 MJ/kWh. The slight increase in energy consumption with K100 is due to its lower heating value as compared to diesel which increases its consumption. The energy consumption of the K50EU50 blend was higher than the K100. Whereas, the energy consumption with the other vegetable oil blends, namely, K50CMO50, K50PO50, and K50O50, was found to be lower than K100. Among these blends, the lowest energy consumption was observed with orange oil blend

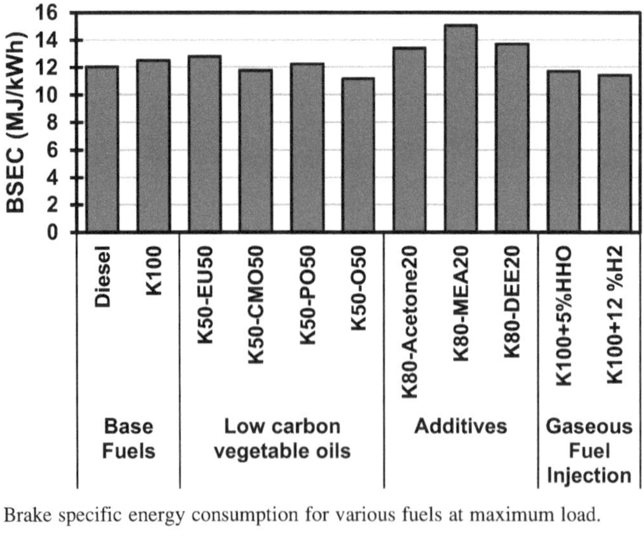

FIG. 11 Brake specific energy consumption for various fuels at maximum load.

(11.16 MJ/kWh). Due to superior combustion and better fuel properties, the thermal efficiency of the engine was improved, which in turn reduced the energy consumption. The energy consumption with the oxygenated additives was observed to increase at full load condition, and the maximum energy consumption of 15.03 MJ/kWh was observed with K80MEA20 blend. The energy consumption in the dual fuel mode with H_2 and HHO is observed to be in the range of 11.42–11.69 MJ/kWh. The energy consumption with the two gaseous fuels is found to be lower than K100. The decrease in energy consumption is caused by the higher energy content of the two fuels, which improves the combustion process, thereby improving the thermal efficiency and hence less heat is wasted.

4 Conclusion

This study intends to reduce CO_2 emissions from a CI engine. The techniques used to achieve the same were; blending of biodiesel with low carbon content vegetable oil in equal proportions, blending of 20% oxygenated additives, and induction of carbon-free gaseous fuels. The following conclusions were drawn from the experimental study:

1. The CO_2 emission with K100 was higher than diesel as a result of its higher carbon content. Hydrogen induction emitted minimum CO_2 due to zero carbon content in hydrogen.
2. The NO emission with K100 was higher than diesel due to excess oxygen in K100. Hydrogen addition emitted maximum NO, while DEE blend emitted minimum NO.
3. The smoke opacity with the methyl ester was marginally reduced compared to diesel. Hydrogen emitted minimum smoke due to high flame velocity and absence of carbon atoms.
4. K100 emitted higher HC and CO compared to diesel. Hydrogen induction emitted minimum HC and CO emission among the tested techniques.
5. The thermal efficiency of the engine with the methyl ester was slightly lower than diesel. The operation of the engine in dual fuel mode with hydrogen resulted in highest thermal efficiency among the blends as well as diesel.

The use of hydrogen in dual fuel mode simultaneously reduced both CO_2 and smoke by 59.48% and 38.84%, respectively. However, NO emission was found to increase. After comparing various methods of reducing CO_2, it is found that hydrogen and Karanja oil methyl ester in dual fuel mode can reduce the emissions without affecting the engine efficiency. However, the hydrogen presently is produced using fossil fuels, but whenever in the future the hydrogen is produced from renewable sources, this combined system can be an effective way of eliminating fossil diesel and operating the CI engine only with renewable fuels.

Abbreviations

ABE	acetone-butanol-ethanol
BSFC	brake specific fuel consumption
BTE	brake thermal efficiency
CI	compression ignition engine
CNG	compressed natural gas
CO	carbon monoxide emission
CO_2	carbon dioxide emission
DEE	diethyl ether
EGR	exhaust gas recirculation
H_2	hydrogen gas
HC	unburned hydrocarbon emission
HHO	oxyhydrogen
HRTEM	high-resolution tunneling electron microscope
LPG	liquified petroleum gas
MEA	monoethanolamine
MToe	million tons of oil equivalent
NOx	oxides of nitrogen emission
OPEC	organization of the petroleum exporting countries
XRD	X-ray diffraction technique

References

[1] Darda S, Papalas T, Zabaniotou A. Biofuels journey in Europe: currently the way to low carbon economy sustainability is still a challenge. J Clean Prod 2019;208:575–88.

[2] Azad AK, Rasul MG, Khan MMK, Sharma SC, Hazrat MA. Prospect of biofuels as an alternative transport fuel in Australia. Renew Sustain Energy Rev 2015;43:331–51.

[3] Geo VE, Sonthalia A, Aloui F, Femilda Josephin JS. Study of engine performance, emission and combustion characteristics fueled with diesel-like fuel produced from waste engine oil and waste plastics. Front Environ Sci Eng 2018;12(4):8–17.

[4] Thiyagarajan S, Edwin Geo V, Martin LJ, Nagalingam B. Comparative analysis of various methods to reduce CO_2 emission in a biodiesel fueled CI engine. Fuel 2019;253:146–58.

[5] Christensen J, Hewitson B, Busuioc A, Chen A, Gao X, Held I, et al. Regional climate projections. Climate change 2007: The physical science basis. In: Contribution of working group I to the fourth assessment report of the intergovernmental panel on climate change; 2007. p. 847–940.

[6] Cheah WY, Ling TC, Juan JC, Lee DJ, Chang JS, Show PL. Biorefineries of carbon dioxide: from carbon capture and storage (CCS) to bioenergies production. Bioresour Technol 2016;215:346–56.

[7] Gaurav N, Sivasankari S, Kiran GS, Ninawe A, Selvin J. Utilization of bioresources for sustainable biofuels: a review. Renew Sustain Energy Rev 2017;73(November 2016):205–14.

[8] British Petroleum. British petroleum statistical review of world energy [Internet]. BP Statistical Review of World Energy; 2019.

[9] Demirbas M. Biorefineries for biofuel upgrading: a critical review. Appl Energy 2009;86: S151–61.
[10] Naik SN, Goud VV, Rout PK, Dalai AK. Production of first and second generation biofuels: a comprehensive review. Renew Sustain Energy Rev 2010;14(2):578–97.
[11] Bhuiya MMK, Rasul MG, Khan MMK, Ashwath N, Azad AK. Prospects of 2nd generation biodiesel as a sustainable fuel—part: 1 selection of feedstocks, oil extraction techniques and conversion technologies. Renew Sustain Energy Rev 2016;55:1109–28.
[12] Lee RA, Lavoie JM. From first-to third-generation biofuels: challenges of producing a commodity from a biomass of increasing complexity. Anim Front 2013;3(2):6–11.
[13] Kumar N, Sonthalia A, Pali HS, Sidharth, Gupta AK, De A, Aggarwal SK, Kushari A, Runchal A. Next-generation biofuels—opportunities and challenges. In: Innovations in sustainable energy and cleaner environment. Singapore: Springer; 2020. p. 171–91.
[14] Chavez-Rodriguez MF, Nebra SA. Assessing GHG emissions, ecological footprint, and water linkage for different fuels. Environ Sci Technol 2010;44(24):9252–7.
[15] Ajanovic A, Haas R. Economic challenges for the future relevance of biofuels in transport in EU countries. Energy 2010;35(8):3340–8.
[16] Sims REH, Mabee W, Saddler JN, Taylor M. An overview of second generation biofuel technologies. Bioresour Technol 2010;101(6):1570–80.
[17] Tye Y, Lee KT, Wan Nadiah W, Peng LC. Potential of *Ceiba pentandra* (L.) Gaertn. (kapok) fiber as a resource for second generation bioethanol: parametric optimisation and comparative study of various pretreatments prior enzymatic saccharification for sugar production. Bioresour Technol 2013;140C:10–4.
[18] Singh J, Gu S. Commercialization potential of microalgae for biofuels production. Renew Sustain Energy Rev 2010;14(9):2596–610.
[19] Kovacevic V, Wesseler J. Cost-effectiveness analysis of algae energy production in the EU. Energy Policy 2010;38(10):5749–57.
[20] Rodolfi L, Zittelli GC, Bassi N, Padovani G, Biondi N, Bonini GTM. Microalgae for oil: strain selection, induction of lipid synthesis and outdoor mass cultivation in a low-cost photobioreactor. Biotechnol Bioeng 2009;102:100–12.
[21] OECD/IEA. From 1st to 2nd generations of biofuel technologies, overview of current industry and RD&D activities; 2008.
[22] Sims REH, Mabee W, Saddler JNTM, Sims REH, Mabee W, Saddler JN, Taylor M, Sims REH, Mabee W, Saddler JN, Taylor M. An overview of second generation biofuel technologies. Bioresour Technol 2010;101(6):1570–80.
[23] Dutta K, Daverey A, Lin JG. Evolution retrospective for alternative fuels: first to fourth generation. Renew Energy 2014;69:114–22.
[24] Suali E, Sarbatly R. Conversion of microalgae to biofuel. Renew Sustain Energy Rev 2012; 16(6):4316–42.
[25] Rodolfi L, Chini Zittelli G, Bassi N, Padovani G, Biondi N, Bonini G, et al. Microalgae for oil: strain selection, induction of lipid synthesis and outdoor mass cultivation in a low-cost photobioreactor. Biotechnol Bioeng 2009;102(1):100–12.
[26] Patel RL, Sankhavara CD. Biodiesel production from Karanja oil and its use in diesel engine: a review. Renew Sustain Energy Rev 2017;71(April 2015):464–74.
[27] Vivek GAK. Biodiesel production from Karanja oil. J Sci Ind Res (India) 2004; 63(1):39–47.
[28] Sharma YC, Singh B, Upadhyay SN. Advancements in development and characterisation of biodiesel: a review. Fuel 2008;87(12):2355–73.

[29] Kesari V, Rangan L. Genetic diversity analysis by RAPD markers in candidate plus trees of Pongamia pinnata, a promising source of bioenergy. Biomass Bioenergy 2011;35(7): 3123–8.

[30] Srivastava PK, Verma M. Methyl ester of Karanja oil as an alternative renewable source energy. Fuel 2008;87(8):1673–7.

[31] Sahoo PK, Das LM. Combustion analysis of Jatropha, Karanja and Polanga based biodiesel as fuel in a diesel engine. Fuel 2009;88(6):994–9.

[32] Dhar A, Agarwal AK. Performance, emissions and combustion characteristics of Karanja biodiesel in a transportation engine. Fuel 2014;119:70–80.

[33] Banapurmath NRR, Tewari PGG, Hosmath RSS. Performance and emission characteristics of a DI compression ignition engine operated on Honge, Jatropha and sesame oil methyl esters. Renew Energy 2008;33(9):1982–8.

[34] Baiju B, Naik MK, Das LM. A comparative evaluation of compression ignition engine characteristics using methyl and ethyl esters of Karanja oil. Renew Energy 2009;34(6): 1616–21.

[35] Dwivedi G, Sharma MP. Investigation and improvement in cold flow properties of Pongamia biodiesel. Waste Biomass Valoriz 2015;6(1):73–9.

[36] Chauhan BS, Kumar N, Cho HM, Lim HC. A study on the performance and emission of a diesel engine fueled with Karanja biodiesel and its blends. Energy 2013;56:1–7.

[37] Vallinayagam R, Vedharaj S, Yang WM, Lee PS, Chua KJE, Chou SK. Pine oil–biodiesel blends: a double biofuel strategy to completely eliminate the use of diesel in a diesel engine. Appl Energy 2014;130:466–73.

[38] Vallinayagam R, Vedharaj S, Yang WM, Saravanan CG, Lee PS, Chua KJE, et al. Impact of pine oil biofuel fumigation on gaseous emissions from a diesel engine. Fuel Process Technol 2014;124:44–53.

[39] Vallinayagam R, Vedharaj S, Yang WM, Saravanan CG, Lee PS, Chua KJE, et al. Impact of ignition promoting additives on the characteristics of a diesel engine powered by pine oil–diesel blend. Fuel 2014;117:278–85.

[40] Vallinayagam R, Vedharaj S, Yang WM, Saravanan CG, Lee PS, Chua KJE, et al. Emission reduction from a diesel engine fueled by pine oil biofuel using SCR and catalytic converter. Atmos Environ 2013;80:190–7.

[41] Huang H, Liu Q, Shi C, Wang Q, Zhou C. Experimental study on spray, combustion and emission characteristics of pine oil/diesel blends in a multi-cylinder diesel engine. Fuel Process Technol 2016;153:137–48.

[42] Vallinayagam R, Vedharaj S, Yang WM, Lee PS, Chua KJE, Chou SK. Combustion performance and emission characteristics study of pine oil in a diesel engine. Energy 2013;57: 344–51.

[43] Cavaleri MA, Ostertag R, Cordell S, Sack L. Native trees show conservative water use relative to invasive trees: results from a removal experiment in a Hawaiian wet forest. Conserv Physiol 2014;2(1), cou016.

[44] Durai MV, Ravi N, Rishi RR, Shettapanavar V, Karnat NM. Impacts of eucalyptus plantations on ground water resources. Int J Sci Nat 2019;10(2):75–81.

[45] Vallinayagam R, Vedharaj S, Yang WM, Roberts WL, Dibble RW. Feasibility of using less viscous and lower cetane (LVLC) fuels in a diesel engine: a review. Renew Sustain Energy Rev 2015;51:1166–90.

[46] Tarabet L, Loubar K, Lounici MS, Hanchi S, Tazerout M. Eucalyptus biodiesel as an alternative to diesel fuel: preparation and tests on di diesel engine. J Biomed Biotechnol 2012; 2012:235485.

[47] Poola RB, Nagalingam B, Gopalakrishnan KV. Performance studies with biomass-derived high-octane fuel additives in a two-stroke spark-ignition engine. Biomass Bioenergy 1994; 6(5):369–79.

[48] Prabuseenivasan S, Jayakumar M, Ignacimuthu S. In vitro antibacterial activity of some plant essential oils. BMC Complement Altern Med 2006;6(1):39.

[49] Devan PK, Mahalakshmi NV. A study of the performance, emission and combustion characteristics of a compression ignition engine using methyl ester of paradise oil-eucalyptus oil blends. Appl Energy 2009;86(5):675–80.

[50] Anandavelu K, Alagumurthi N, Saravannan CG. Experimental investigation of using eucalyptus oil and diesel fuel blends in Kirloskar TV1 direct injection diesel engine. J Sustain Energy Environ 2011;2:93–7.

[51] Anandavelu K, Alagumurthi N, Saravanan CG. Performance and emission studies on biofuel-powered Kirloskar TV-1 direct-injection diesel engine with exhaust gas recirculation. Int J Sustain Energy 2011;30(Supp 1):S66–75.

[52] Meléndez-Martínez AJ, Britton G, Vicario IM, Heredia FJ. The complex carotenoid pattern of orange juices from concentrate. Food Chem 2008;109(3):546–53.

[53] Roussos PA. Phytochemicals and antioxidant capacity of orange (*Citrus sinensis* (l.) Osbeck cv. Salustiana) juice produced under organic and integrated farming system in Greece. Sci Hortic (Amsterdam) 2011;129(2):253–8.

[54] Braddock R. By-products of citrus fruits. Food Technol 1995;49(9):74–7.

[55] Rezzadori K, Benedetti S, Amante ER. Proposals for the residues recovery: orange waste as raw material for new products. Food Bioprod Process 2012;90:606–14.

[56] Purushothaman K, Govindan N. Performance, emission and combustion characteristics of a compression ignition engine operating on neat orange oil. Renew Energy 2009;34:242–5.

[57] Deep A, Singh A, Vibhanshu V, Khandelwal A, Kumar N. Experimental investigation of orange peel oil methyl ester on single cylinder diesel engine. SAE International; 2018. Technical Paper 2013-24-01.

[58] Rahman SMA, Hossain FM, Van TC, Dowell A, Islam MA, Rainey TJ, et al. Comparative evaluation of the effect of sweet orange oil-diesel blend on performance and emissions of a multi-cylinder compression ignition engine. In: 7th BSME international conference on thermal engineering; 2017. p. 1–7.

[59] Dinesh RS, Munuswamy B, Devarajan Y. Emission and performance study emulsified orange peel oil biodiesel in an aspirated research engine. Pet Sci 2019;16(1):180–6.

[60] Mahesh AR, Kannan M, Nataraj G. A study on performance, emission and combustion characteristics of diesel engine powered by nano-emulsion of waste orange peel oil biodiesel. Renew Energy 2020;146:1781–95.

[61] Sriramavaratharajan V, Stephan J, Sudha V, Murugan R. Leaf essential oil of Cinnamomum agasthyamalayanum from the Western Ghats, India—a new source of camphor. Ind Crop Prod 2016;86:259–61.

[62] Kasiraman G, Nagalingam B, Balakrishnan M. Performance, emission and combustion improvements in a direct injection diesel engine using cashew nut shell oil as fuel with camphor oil blending. Energy 2012;47(1):116–24.

[63] Shameer PM, Ramesh K. Experimental evaluation on performance, combustion behavior and influence of in-cylinder temperature on NOx emission in a D.I diesel engine using thermal imager for various alternate fuel blends. Energy 2017;118:1334–44.

[64] Chen HP, Yang K, You CX, Lei N, Sun RQ, Geng ZF, et al. Chemical constituents and insecticidal activities of the essential oil of *Cinnamomum camphora* leaves against *Lasioderma serricorne*. J Chem 2014;2014:963729.

[65] Guo S, Geng Z, Zhang W, Liang J, Wang C, Deng Z, et al. The chemical composition of essential oils from *Cinnamomum camphora* and their insecticidal activity against the stored product pests. Int J Mol Sci 2016;17:1836.

[66] Morajkar PP, Guerrero Penã GDJ, Raj A, Elkadi M, Rahman RK, Salkar AV, et al. Effects of camphor oil addition to diesel on the nanostructures and oxidative reactivity of combustion-generated soot. Energy Fuel 2019;33(12):12852–64.

[67] Nanthagopal K, Ashok B, Garnepudi RS, Tarun KR, Dhinesh B. Investigation on diethyl ether as an additive with *Calophyllum inophyllum* biodiesel for CI engine application. Energ Conver Manage 2019;179:104–13.

[68] Barik D, Murugan S. Effects of diethyl ether (DEE) injection on combustion performance and emission characteristics of Karanja methyl ester (KME)–biogas fueled dual fuel diesel engine. Fuel 2016;164:286–96.

[69] Ibrahim A. Investigating the effect of using diethyl ether as a fuel additive on diesel engine performance and combustion. Appl Therm Eng 2016;107:853–62.

[70] Bailey B, Eberhardt J, Goguen S, Erwin J. Diethyl ether (DEE) as a renewable diesel fuel; 1997. SAE Tech Pap 972978.

[71] Sivalakshmi S, Balusamy T. Effect of biodiesel and its blends with diethyl ether on the combustion, performance and emissions from a diesel engine. Fuel 2013;106:106–10.

[72] Ibrahim A. Investigating the effect of using diethyl ether as a fuel additive on diesel engine performance and combustion. Appl Therm Eng 2016;107:853–62.

[73] Venu H, Madhavan V. Influence of diethyl ether (DEE) addition in ethanol-biodiesel-diesel (EBD) and methanol-biodiesel-diesel (MBD) blends in a diesel engine. Fuel 2017;189:377–90.

[74] Tudu K, Murugan S, Patel SK. Effect of diethyl ether in a DI diesel engine run on a tyre derived fuel-diesel blend. J Energy Inst 2016;89(4):525–35.

[75] Bellido C, Infante C, Coca M, González-Benito G, Lucas S, García-Cubero MT. Efficient acetone–butanol–ethanol production by *Clostridium beijerinckii* from sugar beet pulp. Bioresour Technol 2015;190:332–8.

[76] Zhou M, Qiu X, Yang D, Wang W. Synthesis and evaluation of sulphonated acetone–formaldehyde resin applied as dispersant of coal–water slurry. Energ Conver Manage 2007; 48(1):204–9.

[77] Chang Y-C, Lee W-J, Lin S-L, Wang L-C. Green energy: water-containing acetone–butanol–ethanol diesel blends fueled in diesel engines. Appl Energy 2013;109:182–91.

[78] Chang Y-C, Lee W-J, Wu TS, Wu C-Y, Chen S-J. Use of water containing acetone–butanol–ethanol for NOx-PM (nitrogen oxide-particulate matter) trade-off in the diesel engine fueled with biodiesel. Energy 2014;64:678–87.

[79] Lin S-L, Lee W-J, Lee C-F, Chen S-J. Energy savings and emission reduction of nitrogen oxides, particulate matter, and polycyclic aromatic hydrocarbons by adding water-containing acetone and neat soybean oil to a diesel-fueled engine generator. Energy Fuel 2010;24(8): 4522–33.

[80] Tsai J-H, Chen S-J, Huang K-L, Lin W-Y, Lee W-J, Lin C-C, et al. Emissions from a generator fueled by blends of diesel, biodiesel, acetone, and isopropyl alcohol: analyses of emitted PM, particulate carbon, and PAHs. Sci Total Environ 2014;466–467:195–202.

[81] Algayyim SJM, Wandel AP, Yusaf T, Al-Lwayzy S, Hamawand I. Impact of butanol-acetone mixture as a fuel additive on diesel engine performance and emissions. Fuel 2018;227:118–26.

[82] Li K, Leigh W, Feron P, Yu H, Tade M. Systematic study of aqueous monoethanolamine (MEA)-based CO_2 capture process: techno-economic assessment of the MEA process and its improvements. Appl Energy 2016;165:648–59.

[83] Luis P. Use of monoethanolamine (MEA) for CO_2 capture in a global scenario: consequences and alternatives. Desalination 2016;380:93–9.
[84] Varatharajan K, Cheralathan M, Velraj R. Mitigation of NOx emissions from a jatropha biodiesel fuelled DI diesel engine using antioxidant additives. Fuel 2011;90(8):2721–5.
[85] Varatharajan K, Cheralathan M. Effect of aromatic amine antioxidants on NOx emissions from a soybean biodiesel powered DI diesel engine. Fuel Process Technol 2013;106:526–32.
[86] Palash SM, Kalam MA, Masjuki HH, Arbab MI, Masum BM, Sanjid A. Impacts of NOx reducing antioxidant additive on performance and emissions of a multi-cylinder diesel engine fueled with Jatropha biodiesel blends. Energ Conver Manage 2014;77(x):577–85.
[87] Guan C, Cheung CS, Ning Z, Wong PK, Huang Z. Comparison on the effect of using diesel fuel and waste cooking oil biodiesel as pilot fuels on the combustion, performance and emissions of a LPG-fumigated compression-ignition engine. Appl Therm Eng 2017;125:1260–71.
[88] Vávra J, Bortel I, Takáts M, Diviš M. Emissions and performance of diesel–natural gas dual-fuel engine operated with stoichiometric mixture. Fuel 2017;208:722–33.
[89] Barik D, Murugan S. Experimental investigation on the behavior of a DI diesel engine fueled with raw biogas–diesel dual fuel at different injection timing. J Energy Inst 2016;89(3): 373–88.
[90] Dhole AE, Yarasu RB, Lata DB, Priyam A. Effect on performance and emissions of a dual fuel diesel engine using hydrogen and producer gas as secondary fuels. Int J Hydrogen Energy 2014;39(15):8087–97.
[91] Senthil Kumar M, Jaikumar M. Studies on the effect of hydrogen induction on performance, emission and combustion behaviour of a WCO emulsion based dual fuel engine. Int J Hydrogen Energy 2014;39(32):18440–50.
[92] Yilmaz AC, Uludamar E, Aydin K. Effect of hydroxy (HHO) gas addition on performance and exhaust emissions in compression ignition engines. Int J Hydr 2010;35:11366–72.
[93] Olah GA, Goeppert APG, Surya Prakash GK. Beyond oil and gas: the methanol economy. 2nd ed. Weinheim: Wiley-VCH; 2009. p. 1–352.
[94] Holladay JD, Hu J, King DL, Wang Y. An overview of hydrogen production technologies. Catal Today 2009;139(4):244–60.
[95] Alves HJ, Bley Junior C, Niklevicz RR, Frigo EP, Frigo MS, Coimbra-Araújo CH. Overview of hydrogen production technologies from biogas and the applications in fuel cells. Int J Hydrogen Energy 2013;38(13):5215–25.
[96] Verhelst S, Wallner T. Hydrogen-fueled internal combustion engines. Prog Energy Combust Sci 2009;35(6):490–527.
[97] Saravanan N, Nagarajan G, Narayanasamy S. An experimental investigation on DI diesel engine with hydrogen fuel. Renew Energy 2008;33(3):415–21.
[98] Sonthalia A, Rameshkumar C, Sharma U, Punganur A, Abbas S. Combustion and performance characteristics of a small spark ignition engine fuelled with HCNG. J Eng Sci Technol 2015;10(4):404–19.
[99] Ogden JM, Williams RH, Larson ED. Societal lifecycle costs of cars with alternative fuels/engines. Energy Policy 2004;32(1):7–27.
[100] Balat M. Current alternative engine fuels. Energy Source 2005;27(6):569–77.
[101] Balat M. Potential importance of hydrogen as a future solution to environmental and transportation problems. Int J Hydrogen Energy 2008;33(15):4013–29.
[102] Boopathi D, Sonthalia A, Devanand S. Experimental investigations on the effect of hydrogen induction on performance and emission behaviour of a single cylinder diesel engine fuelled with palm oil methyl ester and its blend with diesel. J Eng Sci Technol 2017;12(7):1972–87.

[103] Sandalci T, Karagöz Y. Experimental investigation of the combustion characteristics, emissions and performance of hydrogen port fuel injection in a diesel engine. Int J Hydrogen Energy 2014;39(32):18480–9.
[104] Senthil Kumar M, Ramesh A, Nagalingam B. Use of hydrogen to enhance the performance of a vegetable oil fuelled compression ignition engine. Int J Hydrogen Energy 2003;28(10): 1143–54.
[105] Edwin Geo V, Nagarajan G, Nagalingam B. Studies on dual fuel operation of rubber seed oil and its bio-diesel with hydrogen as the inducted fuel. Int J Hydrogen Energy 2008; 33(21):6357–67.
[106] Bose PK, Maji D. An experimental investigation on engine performance and emissions of a single cylinder diesel engine using hydrogen as inducted fuel and diesel as injected fuel with exhaust gas recirculation. Int J Hydrogen Energy 2009;34(11):4847–54.
[107] DOE U. Compare fuel cell vehicles [Internet]; 2018.
[108] Polverino P, D'Aniello F, Arsie I, Pianese C. Study of the energetic needs for the onboard production of oxy-hydrogen as fuel additive in internal combustion engines. Energ Conver Manage 2019;179:114–31.
[109] Sapountzi FM, Gracia JM, Weststrate CJ(K-J), Fredriksson HOA, Niemantsverdriet JW(H). Electrocatalysts for the generation of hydrogen, oxygen and synthesis gas. Prog Energy Combust Sci 2017;58:1–35.
[110] Gupta R. Hydrogen fuel: production, transport and storage. Taylor & Francis Group: CRC Press; 2009.
[111] Srinivasan S, Salzano FJ. Prospects for hydrogen production by water electrolysis to be competitive with conventional methods. Int J Hydrogen Energy 1977;2(1):53–9.
[112] Rimkus A, Matijosius J, Bogdevicius M, Bereczky A, Torok A. An investigation of the efficiency of using O_2 and H_2 (hydrooxile gas -HHO) gas additives in a CI engine operating on diesel fuel and biodiesel. Energy 2018;152:640–51.
[113] Falahat A, Hamdan M, Yamin J. Engine performance powered by a mixture of hydrogen and oxygen fuel obtained from water electrolysis. Int J Automot Technol 2014;15(1):97–101.
[114] Uludamar E, Tosun E, Tuccar G, Yildizhan S, Calik A, Yildirim S, et al. Evaluation of vibration characteristics of a hydroxyl (HHO) gas generator installed diesel engine fuelled with different diesel-biodiesel blends. Int J Hydrogen Energy 2017;42:23352–60.
[115] Arat HT, Baltacioglu MK, Özcanli M, Aydin K. Effect of using hydroxy—CNG fuel mixtures in a non-modified diesel engine by substitution of diesel fuel. Int J Hydrogen Energy 2016;41:8354–63.
[116] Bari S, Mohammad EM. Effect of H_2/O_2 addition in increasing the thermal efficiency of a diesel engine. Fuel 2010;89(2):378–83.
[117] Premkartikkumar SR, Annamalai K, Pradeepkumar AR. Effectiveness of oxygen enriched hydrogen-HHO gas addition on DI diesel engine performance, emission and combustion characteristics. Therm Sci 2014;18:259–68.
[118] Premkartikkumar SR. Influence of oxygen enriched hydrogen gas as a combustion catalyst in a DI diesel engine operating with varying injection time of a diesel fuel. RSC Adv 2015; 5(39):30583–91.

Chapter 11

Life cycle assessment of photosynthetic microalgae for sustainable biodiesel production

Neha Arora and George P. Philippidis
Patel College of Global Sustainability, University of South Florida, Tampa, FL, United States

1 Microalgae production and processing

The growing energy demand worldwide and the climate change crisis have prompted the development and utilization of renewable and sustainable energy sources. Among renewable energy sources, biomass to biofuel conversion technologies are being developed and expected to make a significant contribution in fulfilling the world's future energy needs. In this regard, microalgal biomass has received considerable attention, particularly for biodiesel production because of high oil yields (5000–100,000 L/ha-year), high photosynthetic efficiency, and sequestration of atmospheric and industrial CO_2 [1,2]. They also have the ability to grow on waste streams (industrial and municipal) and saline or brackish water, and can be cultivated on unproductive or barren land [1,2]. Once harvested, microalgal biomass can be converted into an array of commodity products (biodiesel, bioethanol, biogas, biofertilizer) and high-value specialty products (omega-3 fatty acids, vitamins, carotenoids, proteins), which offer algal derivatives multiple routes for market penetration and expansion [3].

However, near-term commercial viability of algal biodiesel production is rather uncertain [4]. The major drawback is the current high production and downstream processing cost [5]. Typically, the process begins with the culturing of algae followed by harvesting, drying of the microalgal biomass, oil extraction, and transesterification of the algal oil to biodiesel [6]. Moreover, valuable coproducts and energy can be also produced, as depicted in Fig. 1. The cultivation system, the medium employed, and the pH, temperature, and light exposure determine the overall energy requirements and the cost of production for the cultivation step [7]. Large-scale algae cultivation is generally done in open raceway ponds (ORPs) or photobioreactors (PBRs). The OPRs are outdoor, closed-loop, oval, and shallow (~0.25–0.4 m) cultivation systems

Bioenergy Resources and Technologies. https://doi.org/10.1016/B978-0-12-822525-7.00004-4
Copyright © 2021 Elsevier Inc. All rights reserved.

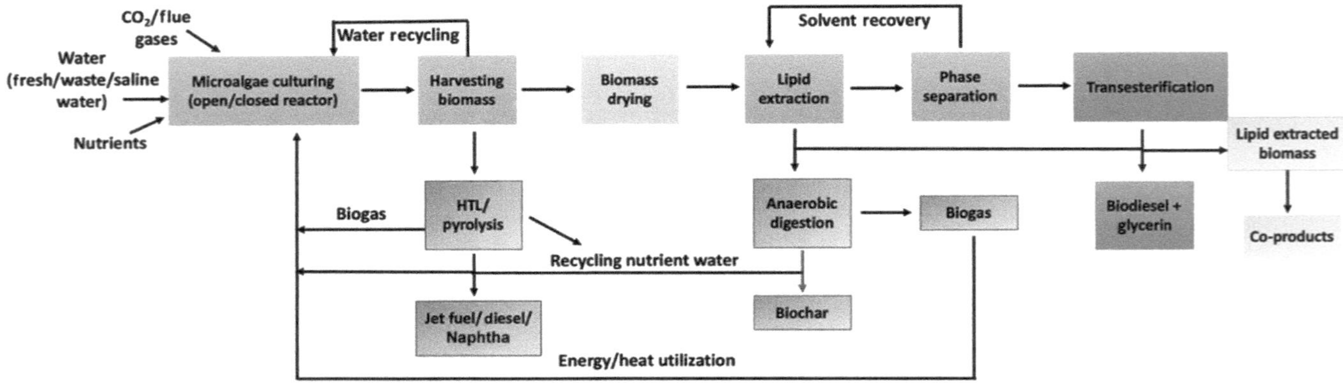

FIG. 1 Schematic representation of microalgal biodiesel production pathways.

in which the algae culture can be stirred using a paddle wheel or by sparging air [8]. On the other hand, PBRs are closed transparent well-controlled systems of various designs, such as a tubular, flat panel, or horizontal [8,9]. Although PBRs achieve more than 10 times higher volumetric productivity of algae than ORPs, they are expensive and require significantly more energy to operate [10]. In contrast, ORPs are rather inexpensive and less energy-intensive compared to PBRs but are more prone to contamination by invading species that lead to culture crushes. The major costs associated with algal cultivation are pumping of water and CO_2 and providing nutrients (nitrogen and phosphorous) given that to produce 1 kg of algal biomass, approximately 876.36 kg of H_2O, 0.08 kg of nitrogen, and 0.017 kg of phosphate are required, along with 1.83 kg of CO_2 fixation [11,12]. To reduce freshwater and nutrient requirements, several studies have highlighted the importance of integrating algae cultivation with wastewater treatment [13]. Utilization of wastewater not only reduces water (by ~90%) and nutrient (by ~100%) requirements but also makes the process more sustainable [12].

Algae harvesting and drying and lipid extraction account for 60% of total energy cost and thus are considered major challenges for economically viable algal biodiesel production [14,15]. The commonly used harvesting methods are centrifugation, flocculation, gravitational settling, dissolved air filtration, and microfiltration, some of which are often combined to optimize biomass production per unit of energy used [16]. Similarly, two or more techniques, including thermal, spray, and solar, are utilized to dry the algal paste [6]. Several reports have suggested the utilization of wet or semidry microalgal biomass to reduce the cost of biodiesel production. To this end, two techniques, hydrothermal liquefaction (HTL) and pyrolysis, have been extensively studied to directly convert algal biomass to oils (Fig. 1). In HTL, the wet algal biomass (85%–90% water content) is processed using high pressure (10–25 MPa) and moderate temperature (200–400°C) and converted into a nonaqueous biocrude (20%–60% of weight), an aqueous phase (30%–50% of weight), a gas phase (1%–8% of weight), and a solid phase (2%–3% of weight) [17–21]. The biocrude phase consists of the algal oil that is subsequently upgraded to renewable diesel or to biodiesel via transesterification, while the aqueous phase rich in organic acids, nitrogen, and phosphorous can be recycled as a nutrient source in the algae cultivation system [15,19–21]. The biogas phase contains CO_2, CH_4, and other volatile organic compounds, which can be utilized for combined heat and power generation (CHP), while the solid phase is biochar that can serve as soil amendment or be converted to high-value products [17]. Although HTL is compatible with existing petroleum refineries, it causes acid catalyst contamination during oil upgrading and also emits higher NO_X and N_2O emissions than conventional lipid extraction methods [22]. To reduce such nitrogen emissions, an additional pretreatment step is recommended, thereby converting the algal proteins to peptides and amino acids, which are retained in the aqueous phase [22]. Pyrolysis involves the thermal conversion of algae biomass with about 20% water content under high temperatures (400–1200°C) to biocrude, biochar, and biogas phases [3,19,20]. In comparison to HTL, pyrolysis is more expensive and requires ~80% dry algal biomass.

The alternative to using HTL or pyrolysis on algal biomass is lipid extraction from algal biomass using conventional solvents, supercritical CO_2, supercritical water or expeller press [14]. Conventional solvent extraction involves a combination of organic solvents, such as chloroform, methanol, and hexane. The main disadvantages of using organic solvents are their low selectively, cost, and environmental toxicity. Supercritical extraction using either water or CO_2 are better alternatives, as they can convert wet algae biomass to biodiesel without using solvents, thus making the process nontoxic and environmentally friendly [14,23]. The algal oil is subsequently converted to biodiesel using the well-established transesterification process operated at low temperatures (80–120°C) using an acid or base catalyst and methanol or ethanol (Fig. 1). The resultant mixture consists of fatty acid methyl esters (FAMEs) or fatty acid ethyl ester (FAEEs), with glycerine being a coproduct [24]. Furthermore, by integrating into the process of biogas production energy demand can be significantly reduced. Biogas is generated via anaerobic digestion (AD) of the lipid-extracted microalgal biomass residue (LEA) and can be utilized to generate heat and electricity, thus powering algae biodiesel production (Fig. 1).

During the last decade, numerous promising algae conversion technologies have been developed and investigated to scale up algal-based biodiesel. However, before commercialization can be achieved, the environmental impacts of those technologies need to be thoroughly assessed [25]. Life cycle assessment (LCA) is the commonly used method to analyze the environmental impact associated with all the steps of a production process [14]. It enables researchers to identify key bottlenecks in the algae to biodiesel conversion pathway, which should be improved to augment the environmental profile of algal biodiesel and facilitate its commercial development [8]. Next, we review important LCA terms and analyze and compare LCA studies of microalgal biodiesel production reported in the literature.

2 Life cycle assessment (LCA)

The LCA is a tool that helps assess the environmental impact and resource utilization during the manufacture of a product starting from the acquisition of raw materials and including product development and waste management [26]. The principles and framework of LCA are described in ISO 14040-43 and comprise four phases [7]:

1. Goal and scope definition that includes identifying the functional unit (FU) and system boundaries for the target products.
2. Life cycle inventory analysis focusing on the raw material (input) procurement and output flows (product, co-product, and energy) of the production process.

3. Life cycle impact assessment, identifying key impact categories (such as energy, greenhouse gas emissions, cost), classifying the impact categories, linking with the inventory data, and calculating indicator values.
4. Interpretation of the results followed by conclusions and recommendations.

Functional unit (FU) is the quantitative description of the production function to be assessed and provides a reference to which all the inputs and outputs can be normalized [27]. Commonly used FUs in algae LCA analysis are 1 MJ of biodiesel, 1 kg of algal biomass, 1 ha of cultivation area or 1 ton of biodiesel (Table 1).

The system boundary defines the processes that are included in the LCA assessment and its selection is based on the scope of the study [69]. Various types of system boundaries have been used to study the algae-to-biodiesel process, such as cradle to grave (also known as cradle to wake, pond to wheels or well to wheels), cradle to gate (also known as well to tank or well to pump), and tank to wheels, as listed in Table 1 (5.15). Cradle to grave is a complete assessment starting with algae cultivation and ending with biodiesel combustion in diesel engines, while cradle to gate is a partial assessment from algal cultivation to biodiesel transportation from the manufacturing site [36]. Tank to wheel only includes the algal biodiesel combustion and is generally combined with well to wheel [28].

Life cycle inventory (LCI) accounts for all the product pathways inclusive of both input and output resources. For example, the LCI of algal biodiesel production includes the water, carbon, land, and nutrient footprints for the raw materials, energy requirement along with the credit for the sequestered CO_2, electricity/heat generation, combustion of biodiesel, and any coproducts formed [43]. Net energy value (NEV) is the difference between the total energy produced and the total energy consumed [5]:

$$\text{NEV} = \text{Output Energy} - \text{Input Energy}$$

A positive NEV indicates that the process produces more energy than it consumes, which is highly desirable. Net energy ratio (NER) is defined as the ratio of energy contained in the final product to the cumulative energy consumed (direct and indirect) for manufacturing the product. It is a vital parameter used to estimate the net energy income of the system and also to determine the economic viability of the process [29]. If NER >1, the process generates energy and is viable, whereas if NER < 1 the process is not viable [49]. NER is calculated as $\text{NER} = \Sigma\, E_{output}/\Sigma\, E_{input}$. Energy return on investment (EROI), also known as energy returned on energy invested (ERoEI), is the ratio of the energy output to the energy input per defined functional unit [16].

Life cycle-greenhouse gases (LC-GHG) quantify the total emissions of GHGs (CO_2, CH_4, and N_2O) during the product life cycle. It is the principal metric used by the US Environmental Protection Agency to determine if a biofuel qualifies for the Renewable Fuel Standard [56]. The RFS program

TABLE 1 Key data from life cycle assessment studies of algal biofuel production.

Microalgae	Cultivation system	Biomass productivity (g/m^2-d)	Lipid content (%)	Functional unit	System boundary	Process	Type of fuel	Coproducts	NER (output/input)	LC-GHG (kg CO_2 eq/FU)	Reference
–	ORP	–	–	1 GJ	Cradle to wake	Refinery HTL	Bio-Jet	–	–	54.2	[15]
						Wastewater HTL				21.2	
Chlorella vulgaris	Tubular PBR	–	32.76	1 kg algae biomass	Cradle to gate	Transesterification	Biodiesel	LEA	0.2–0.3	–	[5]
	ORP	–	25	1 kg biodiesel				Protein and succinic acid	–	0.33	[28]
		15	40	1 MJ		Anaerobic digestion with pretreatment		Biogas	0.7[a]	−0.061	[29]
						HTL			1.25[a]	−0.0068	
						Anaerobic digestion without pretreatment			1.27[a]	−0.173	
						Transesterification			1.52[a]	−0.10	
						Pyrolysis			2.46[a]	0.265	
		–	23	1 ha		Transesterification		–	0.20	–	[30]
	Flat plate PBR with municipal wastewater	25	30	Algal biomass produced in year				–	–	40	[31]
	ORP	40 tons/ha-yr	–	1-ton biodiesel		Transesterification + Anaerobic digestion		Biogas	–	0.713	[32]
	Air-lift tubular PBR		–						–	1.19	
	ORP	–	–	1 MJ	Well to tank	Transesterification		Glycerin + lipid extracted biomass	1.70[b]	1.82	[33]
	OPR with wastewater	24.75	–	1 kg of algal biomass	–			–	–	0.306	[34]
	ORP (semicontinuous)	11.48	–		Cradle to gate			–	–	0.027	[35]
Nannochloropsis salina	ORP	–	24.3	1 MJ		Subcritical water extraction		Biogas	–	3.65	[14]
	PBR + Glycerine					Algae sugarcane biorefinery	Biodiesel	Biochar	–	0.021	[36]

	ORP (high volume)	50	50		Well to pump	Transesterification + anaerobic digestion_		–	0.68[a]	−0.042	[37]
	PBR	150	60			Transesterification		Glycerin	0.93[a]	−0.075	[38]
Isochyrsis sp. + *Chlorella* sp.	ORP	–	–		Well to wake	Pyrolysis	Bio-Jet	–	1	0.068	[39]
–				1-ton biodiesel		Anaerobic digestion	Biodiesel	Biogas	0.74	0.16	[6]
Nannochloropsis sp.		3	–	1 MJ	Pond to wheels	Transesterification		–	1.37[a]	1.8	[40]
	ORP + wastewater	16	25	1 kg of algal oil				Biogas	0.48	–	[2]
	Flat Panel PBR	–	29	1 MJ		Supercritical CO_2 extraction		Glycerol	–	0.1	[41]
	Integrated PBR + ORP	–	45		Cradle to gate	Transesterification		–	–	0.2	[42]
–	ORP	27	40		Pond to Wheels	Wet lipid extraction + hydrothermal gasification		Biogas	–	0.041	[43]
–	ORP + Dairy manure	–	–	1 GJ	Cradle to grave	Anaerobic digestion		Biogas + Biochar	0.77	19	[44]
Scenedesmus dimorphus	ORP	15	25	1 kg biodiesel	Cradle to gate		Biodiesel	Biogas	–	0.122	[45]
	ORP + wastewater	6	20	1 MJ		Transesterification + Anaerobic digestion			0.125[a]	1	[46]
	ORP	13	–		Well to pump	HTL		–	1.24	−0.011	[21]
						Pyrolysis			2.28	−0.210	
–	ORP	–	–	1 tkm	Cradle to grave	Transesterification		Glycerol + animal feed	–	−0.46	[7]
–	ORP	15	–					–	–	0.063	[47]
		30	–						–	0.108	
Neochloris oleoabundans		–	–	1 kg algal biomass	Cradle to gate	Harvesting with Chitosan		–	–	0.3	[16]
–		–	–	1 MJ		HTL + aqueous phase recycling		Biochar	2.5[b]	0.03	[17]
–	ORP	–	–	per million $	Cradle to grave	Transesterification		–	–	83	[48]
–	Rotating algal biofilm reactor + wastewater	30	42	1 MJ	Well to product	HTL	Renewable diesel	–	0.33	−0.043	[49]

Continued

TABLE 1 Key data from life cycle assessment studies of algal biofuel production.—cont'd

Microalgae	Cultivation system	Biomass productivity (g/m^2-d)	Lipid content (%)	Functional unit	System boundary	Process	Type of fuel	Coproducts	NER (output/input)	LC-GHG (kg CO_2 eq/FU)	Reference
–	ORP	23.7	14			Wet lipid extraction	Biodiesel	–	–	3.75	[22]
						HTL				3.01	
						Pretreatment + HTL				2.95	
Alexandrium minutum	Indoor ORP	–	–		Cradle to grave	Transesterification		–	–	146.3	[10]
	Outdoor ORP	–	–					–	–	23.24	
Tetraselmis chui	PBR	–	–	1 MWh electricity				Biogas + biochar	–	0.42	[3]
Tetraselmis sp.	ORP + wastewater	–	–	1 vehicle km traveled		Transesterification + direct combustion of residual algal biomass			1.99^b	0.2	[50]
						Direct combustion of algal biomass	–	Electricity	4.09^b	0.1	
Desmodesmus sp.	Low-cost pond	23	38	1 MJ		HTL	Biodiesel	–	$6–7^b$	–	[51]
Mixed algae	HRAP	8–35	–	MJ/year		HTL		Biogas	–	−2.67	[52]
				1 MJ		Pyrolysis				29.48	
–	ORP	30	41		Well to wheels	Transesterification	Renewable diesel	Naphtha	–	1.32	[53]
						Hydrolysis + wet lipid extraction			–	0.06	
Galdieria sulfuraina	ORP + CO_2 from waste streams (Coal)	25	35		Cradle to grave	HTL			0.15^a	0.02	[54]
	ORP + cellulosic sugar								$-ve^a$	−ve	
	ORP + CO_2 from hydrogen carbonate								0^a	0.15	
Scendesmus armatus	ORP	–	–		Cradle to gate	Transesterification	Biodiesel	–	0.34^a	0.012	[55]
–	ORP + CO_2 from ethanol production	–	–		Cradle to grave	HTL	Jet fuel	–	–	0.04	[56]

	ORP + CO_2 from an industrial source							–	–	0.058	
–	ORP	–	–		Well to tank	Transesterification + anaerobic digestion	Biodiesel	Biogas	–	0.029	[57]
–	ORP + source separated urine	–	–	317 MJ	Cradle to grave	Transesterification		–	–	-1.9×10^4	[58]
–	Flat pane; PBR	–	–	10 GJ biodiesel		Transesterification with supercritical methanol + anaerobic digestion		Glycerin + biofertilizer	–	0.81	[59]
–	ORP + centrate wastewater	13.7	20	1 vehicle km traveled	Well to wheel	Wet lipid extraction		–	–	−6.9	[13]
–	ORP	25	25	1 MJ		HTL	Bio-oil	Biogas	–	31	[60]
Thraustochytrids sp.	Bioreactor + molasses	71 g/L	52			Transesterification	Hydro processed biodiesel	–	1.35^b	0.017	[61]
	Bioreactor + glycerol							–	0.49^b	0.072	
–	ORP	–	–	1 metric ton of Biodiesel	Cradle to grave	Spiral plate centrifuge + Transesterification	Biodiesel	Glycerin	–	0.69	[56]
–	High rate algal pond	25	–	1 MJ		Transesterification + Anaerobic digestion		Biogas	1.25	2	[62]
Chlorella protothecoides	Mixotrophic (ORP + reactor) using sugarcane	–	50						1.6	1.8	
Nostoc sp.	Reactor using sugarcane	80 g/L	50						1.7	1.5	
–	ORP	25	25						–	0.071	[63]
–	ORP	–	–			Transesterification	Renewable diesel	Biogas	2.01^b	0.04	[64]
–	ORP + digestate	–	–	20,000 MJ/year			Biodiesel	Biogas	0.54^b	–	[65]
–	ORP + wastewater	–	–	1 MJ	Pond to pump			–	–	67	[66]
Phaeodactylum tricornutum	ORP + Sea water	–	30	1 kg of algal biomass	Cradle to gate		Omega-3 fatty acids	–	–	0.0257	[67]
Chlamydomonas reinhardtii	ORP	25	–	1 MJ		HTL	Renewable biodiesel	Biogas	1.15	–	[68]

–, no value reported; *LEA*, lipid extracted algal biomass; HTL, hydrothermal liquefaction.
[a] *NER = input/output.*
[b] *EROI values.*

mandates that biomass-based biofuels, including algal biofuels, reduce GHG emissions by at least 50% relative to a 2005 petroleum baseline [54]. Global warming potential (GWP) is a measure of global warming in 100 years (GWP 100) for CO_2 equivalent gases expressed in kg [3]. Eutrophication potential is expressed in terms of nitrogen (NH_3 and NO_x) and phosphate (PO_4^-) equivalents discharged into the environment, including wastewater discharges and air emissions, during the product life cycle [3]. Finally, the life cycle cost (LCC) determines the economic feasibility of the production process.

3 LCA studies of algal biodiesel production

To systematically assess the environmental impact of algae-based biodiesel production, it is essential to compare the reported studies in terms of NER and LC-GHG emissions. This section discusses and compares 51 LCA studies conducted on algal biodiesel and jet biofuel to provide a comprehensive overview of the feasibility of using algae as a renewable fuel feedstock (Table 1). The studies utilize various cultivation systems (ORP and PBR), culturing media (seawater, wastewater, freshwater), system boundaries, functional units, and most importantly, various processes to convert algal oil to biodiesel and coproducts. Most of the studies are based on projecting lab or pilot-scale data to large scale since there is no actual commercial algal biodiesel production data publicly available. The NER and LC-GHG emissions from conventional diesel and various plant biodiesel sources (soybean, rapeseed, jatropha, and canola) are listed in Table 2 for comparison with algal biodiesel.

The first LCA studies focused on the conversion of algal lipids to biodiesel via transesterification and comparison of ORPs vs PBRs [36–42]. A brief comparison of ORP and PBR based on the key parameters is presented in Table 3.

Most of these LCA studies reported ORPs to be better cultivation systems than PBRs due to their higher NER since PBRs incur higher energy cost and capital cost. Moreover, the GWP emissions of ORPs are lower (713 kg CO_2-

TABLE 2 Comparison of NER and LC-GHG among conventional diesel, plant-based biodiesel, and algal biodiesel.

Type of diesel	NER	LC-GHG (g CO_2-eq/MJ)	Reference
Conventional diesel	0.19	93.08	[38]
Soybean biodiesel	1.64	5.01	
Rapeseed biodiesel	2.28	40.7	[70]
Jatropha biodiesel	1.73	56.7	[71]
Microalgae biodiesel	0.2–7.8	(−75.29)–534	[37]

TABLE 3 Comparison of open raceway ponds (ORP) and photobioreactor (PBR) based on the operating parameters [57,58].

Operating parameters	Open raceway ponds (ORP)	Photobioreactor (PBR)
Location	Outdoor	Indoor
Cultivation area	Very large	Small
Cultivation capacity per unit land area	Moderate	Very large
Illumination	Solar light	Artificial/solar light
Light utilization	Low	High
Photoperiod flexibility	Cannot be altered Seasonal light/dark cycle	Very flexible
Water loss	High	Practically none
Energy requirement	Low	High
Contamination	High	Low
Capital cost	Low	High
Labour requirement	High	Low
Maximum production	Low; 0.1–0.2 g/L	High; 2–8 g/L
Ease of operation	Simple	Difficult
Degree of process control	Very low	High
Downstream processing energy	Very high	Low
Cleaning	Simple	Difficult

eq/ton of biodiesel) compared to PBRs (11,900 kg CO_2-eq/ton of biodiesel) [32]. Similar results were reported in another study, where ORPs and PBRs were compared, and it was suggested that a minimum lipid content of 60% (dry cell weight) and biomass productivity of 150 g/m^2-day are required to achieve a NER of 0.93 [38]. Various PBR building materials were compared to decrease the cost of those reactors, including glass, polyvinylchloride (PVC), polycarbonate (PC), polymethyl methacrylate (PMMA), and high-density polyethylene (HDPE) over a 20-year life span [31]. Comparison of these construction materials based on their cost, life span, energy content, light transmittance, etc. is summarized in Table 4.

Among those materials, the lowest environmental impacts were reported for HDPE, but GHG emissions were slightly higher than those of flat-panel glass, which had the lowest eutrophication potential. Furthermore, the utilization of

TABLE 4 Comparison of different construction material for photobioreactor [72].

Material characteristics	Flat glass	PVC	PC	PMMA	HDPE
Cost (kg/PBR)	260	140	120	119	95
Material energy content (mg/kg)	25–28	–	–	–	131–135
Life of the material (year)	18–20	10–15	5–10	10	20–22
Energy content (MJ/m^2)	310		–	131.4	1456
Life span (year)	20	–	–	20	73–75
Light transmittance (%)	80–90	75	83	95	90

various wastewater streams was assessed for cultivating algae [13,50]. When three wastewater sources were compared, namely conventional activated sludge, biological nitrogen removal, and source-separated urine, the latter resulted in negligible energy requirements and GHG emissions of -1.9×10^4 kg CO_2-eq/MJ [58]. The authors also evaluated four different algae biomass conversion pathways: (a) anaerobic digestion (AD) for bioelectricity generation; (b) biodiesel from lipids with AD; (c) biodiesel with direct combustion of residual algal biomass for bioelectricity; and (d) direct combustion of algal biomass [50]. They also assessed various CO_2 sources, including pure CO_2, CO_2 captured from a coal-fired plant, and flue gas (12.5% CO_2) along with utilizing wastewater in the algae cultivation. The best conversion pathways with EROI >1 was achieved with direct combustion of residual algal biomass for bioelectricity and with direct combustion of algal biomass utilizing wastewater for nutrients [50]. Further, comparison of HTL and pyrolysis for the production of renewable diesel showed that the latter conversion process showed higher NER (2.28) as compared to the former (1.24). However, the GHG emissions were higher in the pyrolysis process (210 g CO_2-eq/MJ) while HTL showed negative GHG (-11.4 g CO_2-eq (Table 1) [21]. Contrary to this study, Hognon et al. [68] reported higher NER (1.15) using HTL process combined with belt filter press and heat exchange as compared to pyrolysis (NER $=0.91$) A comparison of freshwater (FW) and seawater (SW) in combination with ORP or PBR showed that the EROI for ORP-SW was close to 1 and decreased in the order ORP-FW $>$ PBR-SW $>$ PBR-FW [73]. A similar trend was observed for GHG emissions. The comparison of different sources of CO_2 namely CO_2 obtained as a co-product of NH_3 production, CO_2 recovered from flue gas from a cement plant, and direct flue gas injection (22% CO_2) into the microalgal cultivation system has also been reported [34]. The minimum environmental effects including GWP were obtained when direct flue gas was utilized to cultivate *C. vulgaris* (Table 1). The GWP was reduced to 0.306 kgCO_2-eq when freshwater was replaced by wastewater. A recent study also reported a substantial reduction (15%) in the GWP when waste CO_2 obtained from biogas upgrading plant was utilized instead of pure CO_2 [67].

Utilization of supercritical CO_2 (sCO_2) has been proposed to replace the use of organic solvents for lipid extraction and recovery [7,37,41,63,74]. However, the energy utilized was comparatively higher than conventional lipid extraction, thus resulting in a decreased NER (Table 1). Supercritical CO_2 extraction reportedly has an NER of 3.07 (input/output) with energy consumption of 0.03 kWh/g-oil [37]. Similar results have been reported by others, who concluded that the worst LCA was observed when algal oil was extracted using SCO_2, which emitted 597.7 kg CO_2-eq/MJ [7]. Harvesting methods were also evaluated by comparing flocculation (chitosan, ferric sulfate, alum), centrifugation, and filter press separation [16]. The best harvesting method was chitosan-based flocculation, as it required low energy input (5 MJ) and resulted in low GHG emissions (0.3 kg CO_2-eq/MJ) [16].

Interestingly, a combination of flat-panel PBR cultivation (algal oil productivity of 53,200 L/ha-year), harvesting by flocculation with chitosan, lipid extraction with supercritical methanol, and anaerobic digestion of the lipid-extracted biomass resulted in minimum GHG emissions (805 kg CO_2-eq/MJ) and eutrophication potential (615 kg N-eq/MJ), as compared to the base case of 5340 kg CO_2-eq/MJ and 2820 kg N-eq/MJ, when using OPR, harvesting by centrifugation, thermal drying, hexane extraction, and transesterification along with landfilling of residual algal biomass [59]. Indeed, numerous studies have highlighted the importance of coupling biodiesel and biogas production, which decreases the NER and GHGs. For example, the LCA of utilizing dairy manure to cultivate algae for the production of biodiesel and biogas was assessed and the authors reported that AD of lipid extracted biomass combined with AD of dairy manure processed via pyrolysis and recycling of nutrients resulted in NER of 0.77 [44]. Moreover, integration of biodiesel and biogas with coproduct manufacture, particularly high value-added ones, can be economically beneficial. For instance, the best-case LCA scenario was the cultivation of algae in OPRs with flue gas, flocculation with aluminium sulfate and centrifugation, homogenization to break the cells followed by hexane lipid recovery and transesterification, along with AD of residual biomass and omega-3 fatty acid and glycerine production [75].

Additionally, HTL to produce biocrude has been evaluated by several authors and could prove to be a promising algae processing technology (Table 1). It has been reported that algae biofuels produced using HTL incur more than 50% reduction in GHG emissions and therefore can qualify under RSF mandates [56]. Moreover, by incorporating brackish water in OPRs, CO_2 from acid gas striping, nitrogen, and phosphate from commercial fertilizers, dewatering using DAF, decanter centrifuging, and HTL for biocrude conversion to biodiesel, along with recycling the biogas and aqueous phase, an EROI of 2.3 with 30g CO_2-eq/MJ emissions is achieved [17]. Another study assessed the LCA for production of jet biofuel from algae using HTL by comparing two production sites: a petroleum refinery and a wastewater treatment plant (WWTP) [22]. Algae cultivation at the WWTP using wastewater, harvesting with gravity sedimentation and centrifugation, and dewatering with solar drying (10%) followed by HTL and upgrading to jet fuel resulted in 76% lower GHG emissions as compared to conventional jet fuel (Table 1). Furthermore, the same study assessed the LCA of various pathways (HTL, HTL with pretreatment, and wet lipid extraction) for algal biodiesel production using municipal wastewater in OPRs and concluded that HTL generates the highest GHG emissions.

The potential of attached growth of algae combined with HTL was assessed using rotating algal biofilm reactor (RABR) and wastewater, followed by mechanical harvesting and HTL for production of algal biodiesel [49]. At biomass productivity of 30 g/m^2-d and an algae oil content of 42%, the process had a NER of 0.33 and negative GHG emissions of −43.6 g CO_2-eq/MJ (Table 1). In a recent study, various carbon sources (waste CO_2 from coal- and gas-powered plants, captured CO_2, hydrogen carbonate, and cellulosic sugar) were evaluated for conversion of algal biomass to renewable diesel using HTL with naphtha as coproduct [54]. Among the tested conditions, the lowest NER was achieved when using CO_2 from a colocated coal power plant, pure CO_2 from a pipeline, hydrogen carbonate, and cellulosic sugar (Table 1).

Thermochemical conversion using pyrolysis is another promising algae technology for conversion of semidried algal biomass to bio-oil. An LCA comparison was conducted among wastewater-cultivated algae combined with pyrolysis, saline water-cultivated algae with oil extraction followed by hydroprocessing, and saline water-cultivated algae combined with pyrolysis [76]. While all three processes resulted in negative GHG emissions, the latter process (saline water and pyrolysis) achieved the highest reduction in GHG emission of 81% compared to petroleum gasoline (91.3 g CO_2-eq/MJ). When the feasibility of slow pyrolysis of algae for the production of biodiesel, electricity, and biochar was assessed, direct solvent conversion of lipids from dried algae resulted in significantly lower GWP compared to slow pyrolysis [3]. In another LCA study, comparison of HTL, pyrolysis, AD, and transesterification for the production of algal diesel showed that the lowest NER (input/output) was achieved by AD with pretreatment, followed by HTL, AD without pretreatment, transesterification, and pyrolysis [29]. The GWP was negative for all those processes except for pyrolysis, which had the maximum GWP emissions (265.26 g CO_2-eq/MJ), putting it at a significant disadvantage compared to AD, HTL, and transesterification.

Interestingly, integration of algae cultivation with a sugarcane biorefinery for the production of both ethanol and biodiesel was considered [36]. The authors assessed mixotrophic mode of algae cultivation in a PBR using glycerine available on-site and CO_2 from the sugarcane juice fermentation. The algae-sugarcane biorefinery system showed a 50% reduction in fossil fuel consumption compared to traditional sole ethanol production with 21 g CO_2-eq/MJ of GHG emissions [36]. A comparison among phototrophic (sunlight), heterotrophic (sugars), and mixotrophic (sunlight + sugars) modes of algae cultivation for the production of biodiesel using transesterification followed by AD of lipid-extracted biomass showed that mixotrophic cultivation has the best NER (1.7) and minimum GWP (1.5 kg CO_2-eq/MJ), followed by heterotrophic and autotrophic cultivation [62] (Table 1). However, when autotrophic and heterotrophic modes were compared in ORPs and PRBs using freshwater or wastewater, the lowest energy demand (1.26 MJ) and GHG emissions (67 g CO_2-eq/MJ) were reported for the phototrophic mode in ORP using wastewater followed by heterotrophic mode in PBR [66].

4 Conclusions and recommendations

Algae-derived biodiesel represents a promising renewable alternative to fossil fuels. It can be produced using a range of algae processes, whose sustainability profile needs to be evaluated using life cycle assessment. The LCA adopts a cradle to grave methodology that encompasses resource (materials and energy) procurement, utilization, and depletion, GHG emissions, and eutrophication potential to rate sustainability performance. The LCA studies presented in this chapter rated the environmental and long-term economic viability for algal-based biofuels. Wide variability in key LCA metrics, such as NER and LC-GHG, is reported among processes as a result of different system boundaries, sources of fossil energy, technology features, and coproduct formation. Overall, it appears that significant improvements are still required along with the entire algae production and conversion chain before algal biodiesel can be considered sustainable. These include attainment of higher biomass productivity (>25 g/m^2-d), lower energy consumption for water pumping and algae drying, integration of biogas generation with biodiesel production, recycling of nutrients, and use of saline or low-quality water, as well as CO_2 from flue gases and even using mixotrophic cultivation mode. The LCAs also showed that integrating algal biodiesel production with value-added coproducts, such as nutraceuticals from omega-3 oils and animal feed from protein-rich residual algal biomass, can further reduce the cost and carbon footprint of algal biodiesel.

References

[1] Singh A, Olsen SI. A critical review of biochemical conversion, sustainability and life cycle assessment of algal biofuels. Appl Energy 2011;88(10):3548–55.

[2] Razon LF, Tan RR. Net energy analysis of the production of biodiesel and biogas from the microalgae: *Haematococcus pluvialis* and Nannochloropsis. Appl Energy 2011;88(10):3507–14.

[3] Grierson S, Strezov V, Bengtsson J. Life cycle assessment of a microalgae biomass cultivation, bio-oil extraction and pyrolysis processing regime. Algal Res 2013;2(3):299–311.

[4] Medeiros DL, Sales EA, Kiperstok A. Energy production from microalgae biomass: carbon footprint and energy balance. J Clean Prod 2015;96:493–500.

[5] Dasan YK, Lam MK, Yusup S, Lim JW, Lee KT. Life cycle evaluation of microalgae biofuels production: effect of cultivation system on energy, carbon emission and cost balance analysis. Sci Total Environ 2019;688:112–28.

[6] Yanfen L, Zehao H, Xiaoqian M. Energy analysis and environmental impacts of microalgal biodiesel in China. Energy Policy 2012;45:142–51.

[7] Valente A, Iribarren D, Dufour J. How do methodological choices affect the carbon footprint of microalgal biodiesel? A harmonised life cycle assessment. J Clean Prod 2019;207:560–8.

[8] Slade R, Bauen A. Micro-algae cultivation for biofuels: cost, energy balance, environmental impacts and future prospects. Biomass Bioenergy 2013;53:29–38.

[9] Dogaris I, Brown TR, Loya B, Philippidis G. Cultivation study of the marine microalga *Picochlorum oculatum* and outdoor deployment in a novel bioreactor for high-density production of algal cell mass. Biomass Bioenergy 2016;89:11–23.

[10] Sevigné Itoiz E, Fuentes-Grünewald C, Gasol CM, Garcés E, Alacid E, Rossi S, et al. Energy balance and environmental impact analysis of marine microalgal biomass production for biodiesel generation in a photobioreactor pilot plant. Biomass Bioenergy 2012;39:324–35.

[11] Chisti Y. Biodiesel from microalgae. Biotechnol Adv 2007;25(3):294–306.

[12] Yang J, Xu M, Zhang X, Hu Q, Sommerfeld M, Chen Y. Life-cycle analysis on biodiesel production from microalgae: water footprint and nutrients balance. Bioresour Technol 2011;102 (1):159–65.

[13] Mu D, Min M, Krohn B, Mullins KA, Ruan R, Hill J. Life cycle environmental impacts of wastewater-based algal biofuels. Environ Sci Technol 2014;48(19):11696–704.

[14] Ponnusamy S, Reddy HK, Muppaneni T, Downes CM, Deng S. Life cycle assessment of biodiesel production from algal bio-crude oils extracted under subcritical water conditions. Bioresour Technol 2014;170:454–61.

[15] Fortier M-OP, Roberts GW, Stagg-Williams SM, Sturm BSM. Life cycle assessment of bio-jet fuel from hydrothermal liquefaction of microalgae. Appl Energy 2014;122:73–82.

[16] Beach ES, Eckelman MJ, Cui Z, Brentner L, Zimmerman JB. Preferential technological and life cycle environmental performance of chitosan flocculation for harvesting of the green algae *Neochloris oleoabundans*. Bioresour Technol 2012;121:445–9.

[17] Liu X, Saydah B, Eranki P, Colosi LM, Greg Mitchell B, Rhodes J, et al. Pilot-scale data provide enhanced estimates of the life cycle energy and emissions profile of algae biofuels produced via hydrothermal liquefaction. Bioresour Technol 2013;148:163–71.

[18] Jena U, Das KC. Comparative evaluation of thermochemical liquefaction and pyrolysis for bio-oil production from microalgae. Energy Fuel 2011;25(11):5472–82.

[19] Hognon C, Delrue F, Texier J, Grateau M, Thiery S, Miller H, et al. Comparison of pyrolysis and hydrothermal liquefaction of *Chlamydomonas reinhardtii*. Growth studies on the recovered hydrothermal aqueous phase. Biomass Bioenergy 2015;73:23–31.

[20] Chiaramonti D, Prussi M, Buffi M, Rizzo AM, Pari L. Review and experimental study on pyrolysis and hydrothermal liquefaction of microalgae for biofuel production. Appl Energy 2017;185:963–72.

[21] Bennion EP, Ginosar DM, Moses J, Agblevor F, Quinn JC. Lifecycle assessment of microalgae to biofuel: comparison of thermochemical processing pathways. Appl Energy 2015;154:1062–71.

[22] Mu D, Ruan R, Addy M, Mack S, Chen P, Zhou Y. Life cycle assessment and nutrient analysis of various processing pathways in algal biofuel production. Bioresour Technol 2017;230:33–42.

[23] Taher H, Al-Zuhair S, Al-Marzouqi AH, Haik Y, Farid M, Tariq S. Supercritical carbon dioxide extraction of microalgae lipid: process optimization and laboratory scale-up. J Supercrit Fluids 2014;86:57–66.
[24] Arora N, Patel A, Mehtani J, Pruthi PA, Pruthi V, Poluri KM. Co-culturing of oleaginous microalgae and yeast: paradigm shift towards enhanced lipid productivity. Environ Sci Pollut Res 2019;26(17):16952–73.
[25] Collet P, Hélias A, Lardon L, Steyer J-P, Bernard O. Recommendations for life cycle assessment of algal fuels. Appl Energy 2015;154:1089–102.
[26] Finnveden G, Hauschild MZ, Ekvall T, Guinée J, Heijungs R, Hellweg S, et al. Recent developments in life cycle assessment. J Environ Manage 2009;91(1):1–21.
[27] Sills DL, Van Doren LG, Beal C, Raynor E. The effect of functional unit and co-product handling methods on life cycle assessment of an algal biorefinery. Algal Res 2020;46:101770.
[28] Gnansounou E, Kenthorai RJ. Life cycle assessment of algae biodiesel and its co-products. Appl Energy 2016;161:300–8.
[29] Sun C-H, Fu Q, Liao Q, Xia A, Huang Y, Zhu X, et al. Life-cycle assessment of biofuel production from microalgae via various bioenergy conversion systems. Energy 2019;171:1033–45.
[30] Pragya N, Pandey KK. Life cycle assessment of green diesel production from microalgae. Renew Energy 2016;86:623–32.
[31] Soratana K, Landis AE. Evaluating industrial symbiosis and algae cultivation from a life cycle perspective. Bioresour Technol 2011;102(13):6892–901.
[32] Stephenson AL, Kazamia E, Dennis JS, Howe CJ, Scott SA, Smith AG. Life-cycle assessment of potential algal biodiesel production in the United Kingdom: a comparison of raceways and air-lift tubular bioreactors. Energy Fuel 2010;24(7):4062–77.
[33] Shirvani T, Yan X, Inderwildi OR, Edwards PP, King DA. Life cycle energy and greenhouse gas analysis for algae-derived biodiesel. Energ Environ Sci 2011;4(10):3773.
[34] Collotta M, Champagne P, Mabee W, Tomasoni G, Leite GB, Busi L, et al. Comparative LCA of flocculation for the harvesting of microalgae for biofuels production. Procedia CIRP 2017;61:756–60.
[35] Yadav G, Dubey BK, Sen R. A comparative life cycle assessment of microalgae production by CO_2 sequestration from flue gas in outdoor raceway ponds under batch and semi-continuous regime. J Clean Prod 2020;258:120703.
[36] Souza SP, Gopal AR, Seabra JEA. Life cycle assessment of biofuels from an integrated Brazilian algae-sugarcane biorefinery. Energy 2015;81:373–81.
[37] Quinn JC, Davis R. The potentials and challenges of algae based biofuels: a review of the techno-economic, life cycle, and resource assessment modeling. Bioresour Technol 2015;184:444–52.
[38] Batan L, Quinn J, Willson B, Bradley T. Net energy and greenhouse gas emission evaluation of biodiesel derived from microalgae. Environ Sci Technol 2010;44(20):7975–80.
[39] Guo M, Cheng S, Chen G, Chen J. Improvement of lipid production in oleaginous yeast *Rhodosporidium toruloides* by ultraviolet mutagenesis. Eng Life Sci 2019;19(8):548–56.
[40] Passell H, Dhaliwal H, Reno M, Wu B, Ben Amotz A, Ivry E, et al. Algae biodiesel life cycle assessment using current commercial data. J Environ Manage 2013;129:103–11.
[41] Monari C, Righi S, Olsen SI. Greenhouse gas emissions and energy balance of biodiesel production from microalgae cultivated in photobioreactors in Denmark: a life-cycle modeling. J Clean Prod 2016;112:4084–92.
[42] Khoo HH, Sharratt PN, Das P, Balasubramanian RK, Naraharisetti PK, Shaik S. Life cycle energy and CO_2 analysis of microalgae-to-biodiesel: preliminary results and comparisons. Bioresour Technol 2011;102(10):5800–7.

[43] Azadi P, Brownbridge G, Mosbach S, Smallbone A, Bhave A, Inderwildi O, et al. The carbon footprint and non-renewable energy demand of algae-derived biodiesel. Appl Energy 2014;113:1632–44.

[44] Chowdhury R, Freire F. Bioenergy production from algae using dairy manure as a nutrient source: life cycle energy and greenhouse gas emission analysis. Appl Energy 2015;154:1112–21.

[45] Togarcheti SC, Kumar MM, Chauhan VS, Mukherji S, Ravi S, Mudliar SN. Life cycle assessment of microalgae based biodiesel production to evaluate the impact of biomass productivity and energy source. Resour Conserv Recycl 2017;122:286–94.

[46] Mediboyina MK, Banuvalli BK, Chauhan VS, Mudliar SN. Comparative life cycle assessment of autotrophic cultivation of Scenedesmus dimorphus in raceway pond coupled to biodiesel and biogas production. Bioprocess Biosyst Eng 2020;43(2):233–47.

[47] Campbell PK, Beer T, Batten D. Life cycle assessment of biodiesel production from microalgae in ponds. Bioresour Technol 2011;102(1):50–6.

[48] Malik A, Lenzen M, Ralph PJ, Tamburic B. Hybrid life-cycle assessment of algal biofuel production. Bioresour Technol 2015;184:436–43.

[49] Barlow J, Sims RC, Quinn JC. Techno-economic and life-cycle assessment of an attached growth algal biorefinery. Bioresour Technol 2016;220:360–8.

[50] Clarens AF, Nassau H, Resurreccion EP, White MA, Colosi LM. Environmental impacts of algae-derived biodiesel and bioelectricity for transportation. Environ Sci Technol 2011;45 (17):7554–60.

[51] Beal CM, Gerber LN, Sills DL, Huntley ME, Machesky SC, Walsh MJ, et al. Algal biofuel production for fuels and feed in a 100-ha facility: a comprehensive techno-economic analysis and life cycle assessment. Algal Res 2015;10:266–79.

[52] Roostaei J, Zhang Y. Spatially explicit life cycle assessment: opportunities and challenges of wastewater-based algal biofuels in the United States. Algal Res 2017;24:395–402.

[53] Dutta S, Neto F, Coelho MC. Microalgae biofuels: a comparative study on techno-economic analysis & life-cycle assessment. Algal Res 2016;20:44–52.

[54] Somers MD, Quinn JC. Sustainability of carbon delivery to an algal biorefinery: a techno-economic and life-cycle assessment. J CO_2 Util 2019;30:193–204.

[55] Wibul P, Malakul P, Pavasant P, Kangvansaichol K, Papong S. Life cycle assessment of biodiesel production from microalgae in Thailand: energy efficiency and global warming impact reduction. Chem Eng Trans 2012;29:1183–8.

[56] Connelly EB, Colosi LM, Clarens AF, Lambert JH. Life cycle assessment of biofuels from algae hydrothermal liquefaction: the upstream and downstream factors affecting regulatory compliance. Energy Fuel 2015;29(3):1653–61.

[57] Woertz IC, Benemann JR, Du N, Unnasch S, Mendola D, Mitchell BG, et al. Life cycle GHG emissions from microalgal biodiesel – a CA-GREET model. Environ Sci Technol 2014;48 (11):6060–8.

[58] Clarens AF, Resurreccion EP, White MA, Colosi LM. Environmental life cycle comparison of algae to other bioenergy feedstocks. Environ Sci Technol 2010;44(5):1813–9.

[59] Brentner LB, Eckelman MJ, Zimmerman JB. Combinatorial life cycle assessment to inform process design of industrial production of algal biodiesel. Environ Sci Technol 2011;45 (16):7060–7.

[60] Frank ED, Elgowainy A, Han J, Wang Z. Life cycle comparison of hydrothermal liquefaction and lipid extraction pathways to renewable diesel from algae. Mitig Adapt Strateg Glob Change 2013;18(1):137–58.

[61] Chen B, Wan C, Mehmood MA, Chang J-S, Bai F, Zhao X. Manipulating environmental stresses and stress tolerance of microalgae for enhanced production of lipids and value-added products–a review. Bioresour Technol 2017;244:1198–206.

[62] Orfield ND, Levine RB, Keoleian GA, Miller SA, Savage PE. Growing algae for biodiesel on direct sunlight or sugars: a comparative life cycle assessment. ACS Sustain Chem Eng 2015;3 (3):386–95.

[63] Yuan J, Kendall A, Zhang Y. Mass balance and life cycle assessment of biodiesel from microalgae incorporated with nutrient recycling options and technology uncertainties. GCB Bioenergy 2015;7(6):1245–59.

[64] Zaimes GG, Khanna V. Microalgal biomass production pathways: evaluation of life cycle environmental impacts. Biotechnol Biofuels 2013;6(1):88.

[65] Bessette AP, Stuart BJ, Resurreccion EP, Kumar S. Algae-powered sustainable community design life cycle assessment and techno-economic analysis. ACS Sustain Chem Eng 2020;8 (4):1916–22.

[66] Azari A, Noorpoor AR, Bozorg-Haddad O. Carbon footprint analyses of microalgae cultivation systems under autotrophic and heterotrophic conditions. Int J Environ Sci Technol 2019;16(11):6671–84.

[67] Porcelli R, Dotto F, Pezzolesi L, Marazza D, Greggio N, Righi S. Comparative life cycle assessment of microalgae cultivation for non-energy purposes using different carbon dioxide sources. Sci Total Environ 2020;721:137714.

[68] Hognon C, Delrue F, Boissonnet G. Energetic and economic evaluation of *Chlamydomonas reinhardtii* hydrothermal liquefaction and pyrolysis through thermochemical models. Energy 2015;93:31–40.

[69] Tillman A-M, Ekvall T, Baumann H, Rydberg T. Choice of system boundaries in life cycle assessment. J Clean Prod 1994;2(1):21–9.

[70] Larson ED. A review of life-cycle analysis studies on liquid biofuel systems for the transport sector. Energy Sustain Dev 2006;10(2):109–26.

[71] Mohammad BT, Al-Shannag M, Alnaief M, Singh L, Singsaas E, Alkasrawi M. Production of multiple biofuels from whole camelina material: a renewable energy crop. BfioResources 2018;13:14.

[72] Johnson TJ, Katuwal S, Anderson GA, Gu L, Zhou R, Gibbons WR. Photobioreactor cultivation strategies for microalgae and cyanobacteria. Biotechnol Progress 2018;34(4):811–27.

[73] Resurreccion EP, Colosi LM, White MA, Clarens AF. Comparison of algae cultivation methods for bioenergy production using a combined life cycle assessment and life cycle costing approach. Bioresour Technol 2012;126:298–306.

[74] Shimako AH, Tiruta-Barna L, Pigné Y, Benetto E, Navarrete Gutiérrez T, Guiraud P, et al. Environmental assessment of bioenergy production from microalgae based systems. J Clean Prod 2016;139:51–60.

[75] Sawaengsak W, Silalertruksa T, Bangviwat A, Gheewala SH. Life cycle cost of biodiesel production from microalgae in Thailand. Energy Sustain Dev 2014;18:67–74.

[76] Handler RM, Shonnard DR, Kalnes TN, Lupton FS. Life cycle assessment of algal biofuels: influence of feedstock cultivation systems and conversion platforms. Algal Res 2014;4:105–15.

Section E

Socio-economic aspect of bioenergy

Chapter 12

Social, economic, and environmental aspects of bioenergy resources

Zobaidul Kabir
School of Environmental and Life Sciences, University of Newcastle, Ourimbah, NSW, Australia

1 Introduction

The importance of the production and the use of renewable energy as a substitute of fossil fuel are increasing. The essential purpose of the use of renewable energy is to decrease the reliance on the limited stock of fossil fuel and to lessen the release of greenhouse gases (GHGs). The ultimate goal of using renewable energy is to ensure energy security in future and to control the global warming caused by GHGs [1, 2]. Undeniably, the adversarial effect of the use of nonrenewable energy on global warming is forcing the global society to move to renewable energy, where the use of renewable energy emits less GHGs [3]. Globally, the demand for total energy usage will increase to nearly 50% in the next 25 years [4, 5] due to rapid economic growth particularly in Asia [6]. The increasing demand for finite fossil fuel has prompted an urgent need for alternative energy that is renewable [7]. Both developed and developing counties are realizing the potentials of renewable energy and, therefore, have adopted necessary policy strategies and technologies to produce renewable energy. The sources of renewable energy include solar, wind, biomass, geothermal, and tidal wave.

Among the renewable energies, bioenergy (hereafter bioenergy and biomass will be used interchangeably in this chapter) is receiving consideration increasingly, while the production of solar and wind energies is more advanced at this moment [8]. In the past, the traditional use of bioenergy was only limited to domestic cooking and heating. Due to the development of technology to transform the bioenergy sources to bio-power such as ethanol or biodiesel, the advantages and scope of the use of bioenergy are progressively acknowledged globally compared to the use of fossil fuel. One of the key advantages of bioenergy is that there is a large number of sources available to generate bio-power

Bioenergy Resources and Technologies. https://doi.org/10.1016/B978-0-12-822525-7.00003-2
Copyright © 2021 Elsevier Inc. All rights reserved.

[9]. The possibilities of bioenergy production, therefore, are substantial and may play an essential role in protecting energy security globally [10]. There is a diversity of feedstock, particularly in the forest and agricultural sectors. These feedstocks are used to make useful end products. These include, for example, pellets, wood chips, bioethanol, biogas, and biodiesel and these end products are used for generating electricity, heat, and bioenergy [11]. In 2016, the total primary energy supply of biomass was 56.5 EJ, of which 86% of the supply was solid biomass including wood chips, wood pellets, and fuelwood for cooking and heating, 7% of the supply was waste sources, and the share of biofuels and biogas share was 6% and 2%, respectively [12]. Certainly, technological improvement has enhanced the generation of electricity and heat from various sources of bioenergy [13]. The trend shows that the supply of bioenergy is increasing overtime. The graph shows that the total supply of biomass has increased by 56.6 EJ compared to the total supply of 42.8 EJ in 2000 (Fig. 1).

Primarily, bioenergy in the form of residues of crops and forest wood is used for cooking and heating, mostly in the developing countries [14–16]. However, due to the development of technologies, bioenergy is increasingly being used in both developing (e.g., Brazil and Thailand) and developed countries for the industrial application and transportation rather than simply cooking or heating. Given the technological development overtime, the utilization of bioenergy is now more efficient, and the application of bioenergy is more diversified than before. In the European Union, Australia, and the United States, for example, bioenergy is being used in different industries, both power and non-power generation industries [17, 18]. Among the developed countries, Sweden and Finland are using bioenergy for industry application 16% and 19%, respectively,

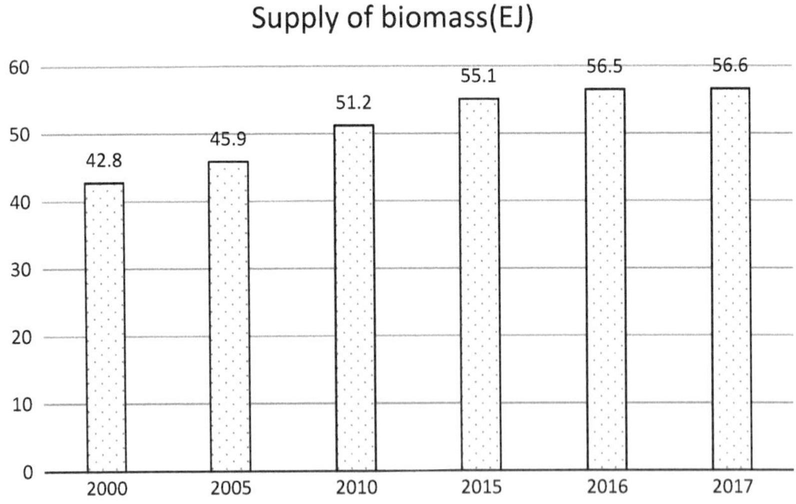

FIG. 1 Trend of the global supply of biomass. *(Source: World Business Association, 2019.)*

the highest rates. The developing counties are also using technologies for bioenergy production not only to meet the increasing demand of fuel globally but also to meet domestic demand due to population growth and industrialization [2]. For example, Brazil produced 34.45 billion liters ethanol in 2019; this was 4% more than the total ethanol produced in 2018. Most of the ethanol (33.93 billion liters) is used to meet the domestic demand, while only 0.52 billion ethanol was exported. Similarly, the country produced 5.8 billion liters ethanol in 2019, an increase of 8% of production compared to 2018 [19].

There has been substantial development in electricity generation using renewable energy technologies [14, 20]. In 2017, the contribution of electricity generation from renewable energy sources was about 25% of the total electricity generated globally, and the total generation was 6191 TWh [21]. Among the renewable energy sources, biomass is the third largest source for electricity generation following the hydropower and wind, where the biomass accounts for the generation of 596 TWh [11]. In the transport sector, the use of biofuels in the form of ethanol and biodiesel is increasing, and biofuel is found to be one of the most effective alternatives for reducing dependency on fossil fuel [9, 22]. In 2017, the estimated contribution of bioenergy in the transportation sector was almost 3%, where the total production was about 3.5 EJ or 138 billion liters. Furthermore, the domination of the use of biomass in the direct heating is widespread. In 2017, the residential, commercial, and agricultural sectors consumed 40 EJ of biomass for heating and cooking purposes [11, 21].

In 2017, about 70% of the total renewable energy consumed was bioenergy [21, 23], and this was consumed by sectors including residence, industry, electricity, and transportation. For the heating purpose, almost half (42%) of the total bioenergy is consumed [11] by residential and commercial sector for space heating, and 12% is consumed by industrial sector for industrial process. Furthermore, the electricity and transport sectors share 11% and 5% of bioenergy sources, respectively. It is to be noted that the transport sector uses liquid bioenergy in the form of bioethanol and biodiesel, etc.; a good option alternative to fossil fuel. According to the World Bioenergy Council (2013), the global use of consumption of biomass for heating and industrial application is expected to be almost double by 2050. Furthermore, the production of electricity from biomass will continue to increase. It is estimated that the production of electricity from biomass will increase from 1.3% to 2.4%–3.3% by 2030 [24]. One of the key prospects of bioenergy use is the use of biofuel by the transport sector [9, 22]. Due to the increasing demand for biofuel,[a] particularly in the transport sector, the growing of biofuel is faster than before [15]. Today, they represent 6% of total bioenergy, and it is expected that the consumption of befouling will grow 3% annually over the next 5 years [15, 24].

a. The advantages of biofuel are that biofuel is a clean energy, emit less greenhouse gases than fossil fuel and there are ample opportunities to produce biofuel due to available resources and technological development.

Furthermore, the trade of biomass feedstock is rapidly growing globally [25, 26]. The export and import of bioenergy feedstock include vegetable oil, wood chips, vegetable oils and agricultural residues, and processed bioenergy carriers, for example, ethanol, biodiesel, and wood pellets are growing rapidly. While present estimates indicate that bioenergy trade is modest (about 2% of current bioenergy use), much larger quantities of these products will be traded internationally in the long run. Notably, the trade will be an essential component of the sustained growth of the bioenergy sector [25, 26]. The main exporter of pallets is North America. The major quantity of these pallets is imported by Europe. The consumption of wood pallets will continue to increase sharply. According to the European Biomass Association (AEBIOM), the consumption of wood pallets will increase from 105 PJ in 2008 to 1340 PJ in 2030 [27]. On the other hand, Canada is the leading exporter who also supplies pellets to the United States [25]. Brazil is extensively producing bioenergy resources with the advantage of adequate land for farming bioenergy and forestland. The production of bioenergy sources is possibly sufficient to meet the demand of total primary energy in Brazil by 2030. Also, because of the surplus production of bioenergy, Brazil, to gradually going to be a key exporter [25].

For bioenergy supply, the residues from the agricultural sector are one of the key potential sources. The present share of agricultural residues to total bioenergy production is 3% although the quantity of agricultural residues is increasing [25]. It is estimated that the amount of agricultural residues from all main crops for energy has potential to generate from 4.3 to 9.4 billion tonnes per year and subsequently the prospect of energy generation this range of the amount of residue is projected from 17.8 to 82.3 EJ [27]. The amount of energy to be generated from agricultural residues has the potential to meet about 3%–14% of the total energy supply globally [11].

Another important source with potential involvement in the expansion of bioenergy production is forest sector [25]. Forest sector accounts for more than 85% of total biomass utilized for energy generation. Woodfuel is one of the main products from forests that are used for biofuel production [9, 22]. Currently, most of the woodfuel is utilized for conventional cooking and heating in rural areas of the developing countries, particularly in Asia and Africa [3]. However, due to the development of various technologies for bioenergy production, it is expected that it will be possible to generate energy efficiently from these woodfuel or forest resources. There are 1.9 billion m^3 of woodfuel produced from forests and used for energy purpose globally [28].

For the supply of bioenergy, the third type of sources is municipal solid waste (MSW), where the MSW is used for bioenergy production, mainly in city areas in the developed and developing countries [29]. In 2017, the supply of waste to energy for the household purpose was 2.51 EJ, where the share of the energy produced from MSW was 58% and the rest 42% was sourced from industrial waste [11]. It has been assessed that the production of MSW and industrial waste amount to 1.3 billion tonnes per year worldwide and the amount would grow up to 2.2 billion tons per year by 2025 [13]. The key sources of MSW are significant cities both in the developed and developing countries, and the generation of energy is on the rise [29, 30]. In fact, the increase of energy generation from MSW can be attributed to rapid urbanization, industrialization, population growth, and economic progress in many developing countries [29–31]. There are available technologies developed for the generation of energy using MSW, and now many countries are producing energy using these technologies with the appropriate management [29, 30].

The potential welfare of bioenergy use is now clearly acknowledged, and the portion of bioenergy of total energy supply is increasing [32]. However, there are impacts of bioenergy development from socioeconomic and environmental aspects, where the impacts are either positive or negative. The scale of these impacts, however, may vary from one country to another due to the different socioeconomic and political contexts. In addition, the type and size of technologies for the production of bioenergy and its utilization, land-use policy, and market share may depend on the socioeconomic and environmental aspects of a country, and therefore, the impacts of bioenergy development may vary substantially. For example, the impacts of bioenergy in the developed countries will differ from the developing countries given the socioeconomic context and the extent and pattern of the use of technology. Even among the developing countries, the impacts of bioenergy may vary given the national and local context, including land-use rights, policy for bioenergy development, and availability of land and population growth. The aim of this chapter is to provide an overview of the impacts of bioenergy from social, economic, and environmental aspects. In this chapter, Section 2 analyzed the impacts of bioenergy from environmental aspects after an introduction. Section 3 focuses on the impacts of bioenergy from social aspects. Section 4 analyzed the impacts of bioenergy from an economic point of view. Finally, the conclusion was made after Section 4.

2 Impacts of bioenergy: Environmental aspects

While bioenergy is one of the potential renewable energies, the production of bioenergy also has a variety of environmental effects. The extent of impacts, however, depends on the sources of bioenergy, the types of biomass, local context including policies and regulations, the use of technology and efficient use of technology, and the intensity and efficiency of biomass use. The overall environmental impacts are related to the bioenergy production on land use, soil quality and erosion, biodiversity, climate change impact, and water quality and quantity [24]. The impacts are discussed as follows.

2.1 Water quantity and quality

One of the key environmental aspects of bioenergy resources is water availability and competition [33]. The use of bioenergy has impacts on both the quality and quantity of water. Water is a vital source for agricultural production, including feedstock. However, the extent of the use of water may depend on the type of the crops for biomass production and availability of water in addition to production system technologies and locations [12]. Also, the impact on water use depends on the land use, landscape shape, and availability of water, topological factors, and management practice in a country [33]. In general, there is a short supply of water in many regions of the world and the availability of water is gradually decreasing. This is due to the use of water for more food production to meet the demand of an increasing population [34, 35]. The extension of bioenergy crop production at a large scale may lead to the use of a huge amount of water and an increase in evapotranspiration. This could make worse in some countries, where stress on the availability of water already exists [36]. The demand for crops for bioenergy associated with the availability of land varies geographically and so the use of water.

The production of bioenergy requires water consumption mainly due to the production of crops as well as the use of land. It is found that the production of ethanol from corn causes potential stress on the use of water in the United States [2, 37, 38]. Although other sources of ethanol require the use of water, the study shows that the production of corn consumes more water than other crops such as wheat and soybean [2, 38]. Furthermore, the use of land for crops other than agricultural production or the conversion of the use local agricultural land for the production of grasses such as sugarcane, switchgrass, and Miscanthus can also have significant effects on water. This may cause quick runoff surface water, loss of capacity of regeneration of water, and reduction of the storage of soil water both locally and regionally [39]. In addition, the removal of corn stover (leaves and stalk, leaves, sheaths, and husks) from cornfield has potential impacts on water quantity, where the removal of corn stover stops the reduction of water yield [40, 41]. Studies on groundwater demonstrate that the extensive plantation of bioenergy crops would reduce the availability of surface water. In general, the alteration of land use for the production of bioenergy crops can reduce the sources of water as well as the number of water resources [41, 42].

The production of bioenergy crops also affects not only the quantity of water but also the quality of water. The worsening of quality of water is significantly associated with increasing farming of bioenergy crops. The quality of water is deteriorated due to the release of nutrients caused by the leaching and surface runoff. For example, any agricultural or bioenergy crops require the use of fertilizer, and it evidenced that among the bioenergy crops, the production of corn has higher fertilizer use and lower efficiency of nutrient use than any other bioenergy crops. The increase of the frequency of corn plantation along with the use of fertilizer may release a significant amount of nitrate to waterways from farmland and thereby pollute the water of surface and ground levels [36, 42]. For example, the pH of the groundwater was found to be alkaline in some of the water samples. Leaching and surface are runoff often occurs from agricultural lands due to the use of chemical fertilizers. Therefore, both surface and groundwater tend to be polluted by the release of nutrients [42].

However, land use for bioenergy production may also offer some environmental benefits. In particular, the transition of land use from arable to perennials may have substantial benefits. It was estimated that the use for arable land for growing recurrent grasses might save the loss of total nitrogen from 30% to 40%, where the loss of nitrogen may occur from the traditional system of cotton production [43]. Study indicates that the firming of bioenergy crops in marginal land or erodible areas may reduce not only the flow of water but also the loss of nutrients [42]. Moreover, increasing recurrent grasses requires approximately no pesticide, and this is clearly useful for the amelioration of the quality of water [2]. In addition, the pollution of water may be possible to reduce throughout the selection appropriate species of bioenergy crops and managing good practice such as production rate, the use of proper amount, and type of fertilizer [35, 44].

2.2 GHG emissions

Another key impact of bioenergy development from environmental points of view is the reduction of greenhouse gases (GHGs) emissions. The two main components of GHGs are CO_2 and N_2O because of their large amount to be released by industries or any similar kind of activities. The release of these gases is a larger quantity than other GHGs, for example, SO_2 [35, 44]. It is well recognized that the emissions of GHGs that may result from the use of biofuel are far less than the use of fossil fuel [18, 44–46]. Results from the study, e.g. [47], show that the use of ethanol for the transportation may reduce 40%–85% of GHG emissions compared to the use of fossil fuel (e.g., gasoline) on a per megajoule (MJ) energy basis despite the fact that the quantity of GHGs reduction varies based on the nature of the feedstock. For the reduction of GHGs, it is recommended to the use of marginal land where available and possible to use. The study [47] finds that the farming of bioenergy crops on marginal land may release less GHGs and thereby has a positive impact on the environment [48]. The substitution of biomass for gasoline or diesel used for transport also contributes to the more reduction of emission of GHGs [24, 49].

However, there are also negative impacts of bioenergy production. From the conversion of land to another use may emit direct and indirect emission of GHGs and this may substantially reduce the climate benefits [24, 50], and this needs to be taken into account with due importance. Furthermore, while the use of biomass in the developed counties is for the generation of heat or energy such as biofuel, the use of biomass in many developing countries is still traditional. In many developing countries, burning of traditional primary biomass for cooking using low-efficiency stoves produce a large amount of incomplete combustion products including suspended particular matters. This may have negative consequences on climate change, women health, and local air pollution [51]. Furthermore, the extension of the cultivation of bioenergy crops; for example, corn can be driven by the increasing demand for ethanol, may stimulate the N_2O emission. Some of the bioenergy crops, for example, the cultivation of corn requires the use of more fertilizer compared with other crops. Nitrogen fertilizer is used by farmers when they produce corn. This fertilizer is responsible for soil denitrification process and aggravating N_2O emission directly. The emission of N_2O from nitrogen fertilizer accounts for 40% of the total GHG emissions, with an additional 17% from trash burning [52, 53]. Therefore, the reasonable choice of bioenergy crops and the use of fertilizer are essential of controlling the N_2O emission.

Biogas can significantly contribute to abate greenhouse gas emission. Particularly, the upgrading of biomethane can generally improve air quality and reduce GHG emissions [54]. In the developing countries, biogas has played a critical role in curtailing indoor air pollution by replacing highly polluting firewood and cow dung as a cooking fuel, in addition to providing much-needed energy resources for cooking. In addition, bioenergy has also provided an opportunity for improving health and sanitation by effective stabilization of biowastes, especially during anaerobic digestion [55].

2.3 Biodiversity

The rising demand for bioenergy feedstock also has led to the increased clearing of forestlands for growing more grains and oilseeds for biofuels, thereby having impacts on biodiversity and ecological balance. This reduction in biodiversity could also be an issue with second-generation biofuel crops such as energy crops [56]. The use of fertilizers and pesticides for bioenergy crops can contaminate water and thereby affect aquatic biodiversity. The outflow of phosphorus and nitrogen from crop field into surrounding water bodies can lead to a loss of a variety of plants and animals, as well as an increase in unwanted algae [57, 58]. The aquatic biodiversity affected by polluted water may disrupt the food chain for some animals.

The impact of bioenergy production on biodiversity is one of the critical environmental issues associated with the use of land for food production and the ecosystem services [59]. The extent of effects of the production of bioenergy crops depends on the land-use condition to be used for bioenergy crop, the types of bioenergy to be produced, and the landscape profile or place [60]. It is well recognized that the conversion of land use is one of the key factors that may affect biodiversity. The direct change of land-use pattern and production system of crops associated with bioenergy may affect the biodiversity [60].

For the long term, the stress to the conversion of landscape ecosystems to energy crop farming may affect the biodiversity adversely. The extension of rigorous cultivation of bioenergy production using fertilizers and chemicals may affect biodiversity [60]. The discharge of nutrients and chemicals used for bioenergy crop production may lead to the transformation of the composition of species be it animals or plants or in the adjacent ecosystems. Importantly, arable monocultures are usually worse than mixed cropping systems and perennial crops and may have additional negative effects on biodiversity, for example, animal-human conflicts or loss of habitat for a particular species [13]. The practice of monocultures may cause the growth and extension of invasive species, and this is harmful to agricultural land [61]. Countries, especially ones with environmental degradation and increased land pressure from biofuels are likely to affect already fragile ecosystems.

Furthermore, the rising demand for biofuel in tropical countries may cause the clearance of natural habitats particularly tropical rainforests. For example, the expansion of soy plantation into the Amazon forest for bioenergy production [61]. This has caused a significant loss of biodiversity and decreases the quality of natural vegetation. In this situation, better management practices may reduce the risk of biodiversity loss at locations.

While the production of bioenergy sources such as crops, the feedstock can have negative impacts on biodiversity, there are also positive impacts. There is land available in many countries that may be arid or degraded. The use of degraded land for bioenergy crops may stimulate the enhancement of the variety of species. The growing crops for bioenergy on either low-productive or marginal lands available in a country may improve the landscape design [62, 63]. Study finds that the use of grassland for biofuels crops may increase the local productivity, in addition, to support the maintenance of ecosystem functions due to good practice production system [60].

The growth of Miscanthus on the arable land has the capacity to sequester inorganic carbon from the air [64]. In addition, the application of biochar is significant for the enrichment of organic carbon in the soil. This is because the aggregation of organic carbon with electronegativity can absorb CO_2 from the air. Therefore, this process can be useful to both the carbon sink of agricultural soil and the protection of air quality, such as mitigation in nitrogen oxides, methane, and $PM_{2.5}$ [65]. In summary, the development bioenergy requires to select the most suitable areas, plant types, and mitigation measures for the biodiversity and carbon sequestration.

2.4 Soil quality and erosion

One of the major issues of concern in bioenergy production is soil erosion. The erosion reduces the soil quality, where the bioenergy crops are produced and consequently diminish the capacity of natural and agricultural ecosystems [53]. Soil erosion can be triggered in three main pathways. These include (a) the acreage expansion of corn, (b) the removal of residue from the cornfield, and (c) land-use change and the type of land in addition to tillage practice and intensity of land use and the type of bioenergy crops. The rising demand of ethanol may cause the expansion of land use, and this may have adverse consequences in soil retention due to increase corn cropping on the same land and removal of crop residue [2]. The residue of crops left on the field after harvesting the crop can protect water from corrosion [66]. On the other hand, the removal of crop residue can increase the potential risk of erosion in the absence of adequate physical protection of soil surface [67, 68]. These consequences lead to the reduction of the capacity of soil to retain water and other nutrients beneficial to soil quality.

The crop residues are recognized as an important source of bioenergy feedstocks because of their easy availability. It is well informed that the removal of residue may have adverse effects on 'near-surface soil physical, chemical, and biological properties' [69]. The removal of crop residues left in the crop field may significantly affect the soil quality including (a) the loss of sequestering capacity of soil carbon, (b) the damage of the soil structure and the decrease of the water retention capacity, (c) declining the diversity of species of soil fauna, (d) declining the water infiltration rate, and (e) decreasing the retention of water and controlling the water runoff and increasing the risks of erosion of topsoil by water and wind [70]. Moreover, the removal of residue accelerates evaporation, rises soil temperature irregularly, which affects crop production, and diminishes the contribution of organic matter desirable to recover the soils' capability to retain water [70]. It is found that the sloping and erosion-prone land are more adversely affected by residue removal than those on flat lands with proper soil texture and less prone to drainage of water runoff.

One of the key environmental impacts of bioenergy production is the loss or gain of soil organic carbon (SOC), where the key characteristic of SOC is to retain water, soil biodiversity, and harvest yield. The conversion of land for cultivation of bioenergy crops and intensity of cultivation of crops on the same land has negative impacts on SOC [2]. The conversion of marginal or uncultivated land to the production of bioenergy crops may have significant loss of SOC. This is more visible in countries such as Brazil and Australia, where native land is converted to produce sugarcane [71, 72]. In addition, the land-use conversion may worsen erosion or protect soil from erosion. For example, the alteration of the forest into perennial bioenergy crops may enhance the risk of the loss of water and soil [48].

On the other hand, positive impacts on the soil and water retention can be achieved by altering the grain crops into perennial grasses. This is because perennials are able to generate vertical and ribbed stems from sods [73]. Furthermore, the perennial grasses, particularly the switchgrass, may decrease the sediment yield in streamflow in addition to erosion of soil and increase the use of water and infiltration. For example, in Chinese Loess Plateau, which is an erosion-prone area or slope arable land area indicate that there is a benefit of soil and water conservation if perennials are used instead of traditional crops [73]. In this situation, the growth of perennial grasses, particularly in erosion-prone areas or slope arable land, has a greater possibility than corns for e-production of ethanol [53] (Table 1).

TABLE 1 Environmental impacts of bioenergy production

Issues	Positive	Negative	Impacts	Comments
Water quantity and quality		X	The reduction of storage of water, running off surface water may occur at a local and regional scale. Also, an acute shortage of water for agricultural production where water is already under stress	Firming of bioenergy crops in marginal land or erodible areas may reduce the flow of water. The pollution of water may be possible to reduce throughout the selection of appropriate species of bioenergy crops and managing good practice such as production rate
GHGs emission	X		Reduction of the emission of Green House Gases such as SO_2, CO_2, and N_2O and therefore increased air quality and less global warming. Biogas production may reduce the emission of GHGs. The traditional use of bioenergy in developing countries for cooking release suspended particular matters and deteriorates local air quality	The conversion of marginal land for bioenergy crops may reduce the emission of CHGs
Biodiversity		X	Increased clearing of forestlands for growing more grains and oilseeds for biofuels affects biodiversity and ecological balance. The use of fertilizers and pesticides for bioenergy crops can contaminate water and thereby affect aquatic biodiversity	The use of degraded land for bioenergy crops may stimulate the enhancement of the variety of species and thereby may bring benefits for biodiversity

Continued

TABLE 1 Environmental impacts of bioenergy production—cont'd

Issues	Positive	Negative	Impacts	Comments
Soil quality and erosion		X	Expansion of corn cultivation for bioenergy may lead to the reduction of soil capacity to retain water and other nutrients beneficial to soil quality. The removal of crop residues left in the crop field may significantly affect the soil quality, including the loss of sequestering capacity of soil carbon	The conversion from the grain crops to perennial grasses, e.g., siwtchgrass may generate positive effects on the soil and water retention

3 Impacts of bioenergy production: Social aspects

According to [74], impacts of bioenergy production from social point of view include the preservation of livelihoods, access to food, ensuring the continuous supply of energy, and guaranteeing the safety of people, facilities, and regions [74]. Social impacts also include the use of open and transparent participatory processes to engage stakeholders and stakeholders' right to participate in the decision-making process, the establishment of an obligation to respect human rights. Furthermore, the impacts of bioenergy production also include social equity and equality of gender, access to resources and opportunities to be involved in decision-making, and right to have information in addition to equal rights for them [75]. In the developing countries, improved access to local energy sources will have significant social merits, such as increased time for women and children to pursue education and other activities rather than collecting fuel biomass [49]. Moreover, social rights and community access to bioenergy production are essential aspects of resource stewardship [74, 76, 77]. Some of the key social impacts associated with bioenergy resources are as follows.

3.1 Gender and equity

One of the key social impacts relating to bioenergy production is the intragenerational equity, which has received less attention than intergenerational issue [74, 76]. The growing demand for bioenergy and resulting socioeconomic transformation may affect men and women differently irrespective of households headed by male or female [78, 79]. The reasons that women experience socioeconomic impacts differently depends on the lack of adequate opportunities to access to resources including land and forest in addition to other sources of livelihood, their participation in decision-making, income opportunities, access to markets to sell their products at market price, and their bargaining capacity. It is well recognized that the level of participation of women in decision-making particularly in the developing counties is mostly passive or no voice at all [80]. For example, access to and control over land is one of the key impacts that may be felt differently by women and men, where women have far less say in what happens to it even though they are often the primary food producers. This is of particular concern where companies buy up land for energy crops or contract landowners to produce energy crops instead of food [81]. The nature and the extent of impacts associated with bioenergy depend on the technologies for bioenergy production and policy context available in place [74, 78]. Therefore, gender-related issues need to be mainstreamed into planning related to bioenergy production. This will assure women that their concerns and needs are taken into account by the policymakers. The involvement of women in bioenergy planning is imperative because without active women participation, the current plan may omit the key issues such as food security, energy security, and preventing women from benefiting from bioenergy. For example, women are discriminated in some of the sub-Saharan African countries in terms of allocation of land for cultivation. They are allocated low-quality lands by their husbands, and importantly conversion of these lands to biofuel plantations may cause partial or total displacement of women's agricultural activities. These women finally experience negative consequences relating to their ability to produce food or ensuring food security [78, 82].

Biofuel production may have an impact on the security of food because of the competition for the same resources such as land and water for agricultural production such as rice or wheat. This competition may cause the price of food high and affect the availability and accessibility of food particularly for the poor. It is observed that female-head households relatively become more vulnerable than male-headed households due to the high price of food [83]. Another aspect of women is that many poor and marginal women (and men) live in rural areas, where the only source of energy for domestic use is biomass either from forest or agricultural field. In India, for example, firewood, crop waste, and cattle dug form 92% of the domestic energy. Also, women in rural areas in India gather 85% of their cooking wood from forests [84]. When there is any stress on forest ecosystem or the environment, this affects women negatively to collect biomass. The biomass either from forest or crop field becomes scarce and the collection of biomass for women and girls becomes time consuming.

In various situations, often women may have limited informal access to land and resources compared to men [85]. Consequently, the farm labor time and household budgets of women increase unreasonably [86, 87]. In general, women and female-headed households may likely to be excluded from modern bioenergy supply chains. This happens in the developing countries particularly where prevalent and constant gender-based inequalities exist in terms of access to and control over the resources including land, water, and other natural resources; agricultural inputs and equipment; and markets. The exclusion of women and female-headed households from the benefits of modern bioenergy developments may have potential risks to be vulnerable, and their food security can be affected significantly [88].

3.2 Food security

One of the key social issues of bioenergy is food security, which can be affected by the expansion of bioenergy crops that may compete with the production of food using the same resources such as land and water [2, 74]. When agricultural land is used for bioenergy production, the issue of food security inevitably comes forward [61]. The total use of land worldwide for bioethanol and biodiesel together is about 155.47 million acres, which are about 4% of total agricultural land and 3%–4% of its freshwater. The use of agricultural land for bioenergy production may push up food prices [15]. The production of bioenergy may have an indirect effect, for example, in the case of maize production for ethanol in the United States. In this case, food security can be affected indirectly through the high price of grain and other food products. Due to the increasing use of maize where 40% of corn is now produced for the production of ethanol in the United States [89], the price of food grains can be high. The prediction shows that about 50% of food price will increase by about 32% by 2022 due to the increasing demand of biofuel. The rest 50% of the price of food will increase because of the demand-side effects such as population growth and food preferences.

Moreover, the price of food to be increased due to the influence of biofuel demand may affect the expenditures in the developing countries [8]. The economy of the developing countries by and large is dependent on agricultural activities rather than industry. This is because a significant portion of the population is engaged in agricultural production. Given the increasing demand of biofuel and the use of agricultural land for biofuel in some cases (e.g., Cambodia, Brazil, or Thailand), the price for food can be proportionately much higher in the developing countries [15]. In general, however, the net impact on food security in the developing countries either positive or negative depends on the country or the region and eventually the economic position of households or individuals.

Furthermore, the extent of food security that may obstruct or encourage the development of biofuel depends on the overall situation of the country. The situation includes the size of the population, annual production of food, predicted growth, land availability, and importantly whether there is available land for food crops and energy crops. Finally, food security is also influenced by the extent of the end products of bioenergy, contributing to add value that increases the income of farmers.

In addition, to address the possible concern on food security that may be caused by the production of bioenergy, some factors need to be taken into account. A well-functioning market is a precondition for biofuels, food, other bio-products and food security and supply of biofuels [24]. The food security conditions that include availability and access to food with stable supply, the policy for food in place, access to and utilization of land in addition to the availability of land for biofuel production may influence the food security in a region [61], and these issues need to be taken into account to implement a biofuel project.

3.3 Land ownership and tenure

The expansion of bioenergy production is likely to lead to competing for access to land [15]. The farmers and people depending on forests living in areas with no formal land tenure rights suffer from livelihood impacts when the cultivation of agricultural land or any kind of land is converted for bioenergy crops. Although the production of bioenergy and byproducts creates business opportunities for individuals and local people and farmers, the use of the land for cultivation of biofuels depends on the choices, which have rights over land. The right of ownership of landholders and the use of land for grazing or water is established when there is regulations, customary rules, or combinations of any of these [90]. In reality, confusion arises when there are land rights that are in lack of clarity. Also, the lack of clarity exists relating to the access to land in many developing countries. The lack of clarity to access to land in the developing countries tends to create conflict between farmers or landowners and the people or companies who like to acquire the land to deploy bioenergy production [15]. Furthermore, conflict is created when the land of marginal landowners and poor farmers is taken over by companies without prior discussion or reasonable compensation [91].

There are some opportunities for land use for the production of bioenergy when the bioenergy crops are cultivated using unproductive land or the land used as public resources by the local community. Despite that, the acquiring of such land even if not used for any crop production may require to consult with the users of land or local community. Otherwise, this may create conflict between the local community and the companies, which will acquire the land. The users have been using the land generation after generation, and they have significant claims to the land for the collection of fuelwood or for grazing. However, the users of common land may not be able to uphold their claims because the claims are constructed by the common law, which is lacking any formal tenancy arrangements. Therefore, there is a potential threat that the users of common land or unproductive land may be excluded from the use of the land [90]. For example, under the National Mission on Biofuels in India, the government intended to produce Jatropha for biofuel production using marginal land and targeted to acquire 400,000 ha of marginal lands. The marginal land in India is usually treated as common resources. The most part of the 400,000 ha of land was common resources, and this land was the fundamental source of livelihood for the local poor people. The land also provided foodstuff, forage, fuelwood, and materials for household use. For the poor households, the contribution to the income from the land was between 12% and 25% [78]. As the local community had no legal rights to use the marginal land, the government acquired the land for the production of biodiesel and the poor local households were deprived of using the land, the source of livelihood.

The situation of tenure insecurity is a driver of inequality affecting local communities, including indigenous peoples. There are about 1–2 billion people worldwide depend on commonly used land, but they have no legal rights to use the land [92]. The lack of formal rights to their traditional territories often results in a situation of unclear and overlapping claims to lands, especially when there is a growing demand for energy. This, in turn, constitutes one of the major drivers of conflict and instability [92]. Furthermore, involuntary dislocation of poor people from the land used for a food source, woodlands, or forage for livestock, likely to lose customary land rights and access to land [85]. As well, local communities at the local level have limited access or even become excluded from the production of bioenergy and associated benefits if they have no legal land rights [87]. Similarly, rural poor likely to be vulnerable to lose the land where government policy incentives are arranged for large-scale bioenergy production usually through top-down approach and the local poor have no voice due to lack of official land rights [93]. Overall, customary landowners and common resource users who are formerly depended on forest or land resources for their livelihood may undergo restraints when companies use the land with the support of government for biofuel production [86, 94].

Importantly, the income and livelihood opportunities of local communities are affected when the land rights are handed over to large-scale industries and this leads to conflict over land rights [95]. When land tenure is secured, there is less likely to develop any social conflict because the tenure security works as a protection against social conflict in addition to facilitating the production of biofuels [96], where the quantity of land is comparatively abundant and breach on traditional land rights is rare, biofuel production may alter the land with manageable minimum social conflict. For instance, the cultivation of Jatropha in Zambia did not need to compromise with local food crop production [53]. Overall, biofuel production frequently though not continuously needs to negotiate with some groups to secure land tenure. The impact of bioenergy on land-use rights is clearly formed by the production process, political environment in place, and power relations among the potential stakeholders.

3.4 Health concerns

In the developing countries particularly where women use traditional biomass (e.g., cow dung, fuelwood, and straw) for cooking and heating directly, there is a potential health risk for women due to indoor air pollution and smoke. There are potential health concerns associated with bioenergy recourses and production process. The extent of negative impacts on the health of the local community and employees may vary given the types of feedstock, technology to be used for bioenergy production, and the scale of production. Health hazard may arise from GHGs emissions, and longtime exposure to emissions may affect human health that is living around the bioenergy production plants. The degree on negative impacts on health relies on the sources to be used for bioenergy production, for example, straw, wood, and agricultural or bioenergy crops [81]. The burning of biomass produced from sugarcane plantation is often damaging to human health [97] comprising respiratory problems and employees, and local communities may suffer from this disease.

On the other hand, it is recognized that there are also positive health impacts when biofuels considerably decrease the emission of primary fine particle matter [81] as well as of GHGs. The plantation of biofuel can support to improve the public health of local communities when the local communities have a higher income that can be spent on buying drugs and health care. Also, food insecurity has direct impacts on public health and well-being [97]. At the household level, food insecurity varies by gender. The food security is strongly related to health and well-being. In some socioeconomic context, for example, in African and Asian countries, the women are given less food than men, and therefore, women health is susceptible to various diseases including malnutrition [98]. The labor from the fetching and carrying fuelwood makes affect women's and girls' well-being in addition to prenatal mortality and anemia.

3.5 Social acceptability

Social acceptance is a prerequisite to the deployment of bioenergy and technology. Social acceptance depends on cultural values, the attitude of the local community, and adherence to traditional cultivation, aesthetic view, heritage, recreational value, and community perception about the new technologies and production system of bioenergy. The production system will not be sustainable if the local community does not accept it. The social acceptance also trusts, transparency of the activities undertaken by the investors, and how the community people are communicated in addition to community participation in the decision-making process about the deployment of bioenergy. The use of land either own by community unused by the investor may lead to social conflict; particular, the investors tend to avoid community involvement in the bioenergy process. The social acceptance mainly depends on three key issues such as public opinion, transparency, and effective stakeholder participation. Stakeholder participation can contribute to more effective and enduring progress toward other environmental and social issues [99]. The continuous reporting to the community and relevant stakeholders on bioenergy production and benefits improve the confidence of the community on the investors. Equitable access to energy and associated benefits are very important for the community to be ensured by the investors. This will ensure the sustainable operation of the bioenergy plant and thereby energy security [100]. (No similarity in the previous report.)

4 Impacts of bioenergy production: Economic aspects

The bioenergy impacts from economic aspects (both short and long term) had considerable importance [74] in order to understand the sustainable bioenergy production [22]. The economic aspect of bioenergy impacts includes trade, income, energy security, employment generation, and interaction with markets. Some of the key issues of bioenergy from economic aspects are discussed as follows.

4.1 Trade of bioenergy

The share of total primary energy supply today is roughly 10% and the trade of the bioenergy market is progressively expanding. In 2006, the biofuel market started internationally at a very small scale; the trade of biofuels is already established in many countries. The trade market has increased mainly because of the increasing demand of the transport sector, where vehicles use biofuels either in pure form or as mixed form (blended with diesel or gasoline) [101]. In 2018, the production of biofuel globally had increased to 10 billion liters and expected to grow 154 billion liters by 2024. The rapid increase of biofuel is due to expanding market prospects in Brazil, the United States, and particularly in China [12]. Given the future energy demand, the primary consumers have the developed countries and the developing countries are producing and exporting the biofuels [12, 102]. While the countries of production consume most of the bioenergy products locally, some products such as timber, pulpwood, palm oil, and wood pallets are exported that constitute a substantial quantity of total products.

However, biofuel trade often becomes limited due to providing subsidization for domestic producers in industrial countries [103]. The United States, for example, under the American Jobs Creation Act of 2004 and Renewable Fuels Reinvestment Act of 2010 has subsidies domestic producers to protect the jobs of the employees who are working in the bioenergy industries. The European Union has preferential treatments for ethanol imports from many developing countries but excluding some potential exporters such as Brazil and Argentina. Some developing countries in Asia and Africa have the capacity to harvest and export bioenergy products. The trade restrictions imposed by importing countries on bioenergy export are there, and these policies by importing countries may limit the free trade opportunities. Even if this incentive-based policy continues, the energy industries are assumed to be self-reliant now. The stable demand for biofuel along with the overall rate of production of biofuel is due to the growth of self-reliant bioenergy industries [101].

Furthermore, it is observed that the international trade of bioenergy feedstock including forest wood, and residues from agricultural production and ethanol, biodiesel, and wood pellets processed from bioenergy sources will continue to grow because of government policy strategies, for example, countries in European Union and other parts of the world [24]. It has been estimated that the trade of bioenergy is around 1.1 EJ. This is almost 2% of existing biofuel usage; though the bulk of bioenergy trade is growing gradually. Particularly, the market of biofuel trade has been growing fast and will remain to raise overtime [24].

4.2 Employment generation

Bioenergy resources and subsequent production process using suitable technology offer opportunities for job creations, since they consists of many activities such as collection, transformation, storage construction, and distribution. The better opportunities come when feedstocks are produced and supplied from locally [76, 104]. Studies confirm that the labor force prerequisite for the cultivation of biomass resources is almost 5 times higher than that needed for the production of fossil fuels [105]. In the same way, the potential of bioenergy production industries is likely to employ about 100 times more workers per unit of energy produced than the fossil fuel industries [3]. It is noted that the bioenergy sector requires labor force for the preprocess and production of bioenergy sources and for the production of biofuel as the end product [18, 22]. For the operation of a bioenergy plant for the production of bioelectricity, the level of the direct labor force is approximately 4 times higher than that required for the operation of fossil fuel power plants [106, 107].

Furthermore, the use of renewable energy technologies for the production of renewable energy creates job opportunities. It is estimated that about 11 million people are now employed in the renewable energy industry sector [21], which accounts for one-third of employees employed in renewable energy production worldwide. Among the renewable energy technologies, bioenergy is the second largest employer after solar energy. In 2018, there were 3.2 million people employed in the bioenergy industry throughout the entire production process: the production of bioenergy feedstock, transportation, conversion to bio-based products for energy and manufacturing of equipment for bioenergy technologies, and finally the production of energy using the technology. It is important to note that the people engaged in traditional biomass for collecting fuelwood and other cooking materials has not been taken into account here; otherwise, the total renewable jobs would indicate a much higher employment figure for the sector [11].

In the developing countries and rural areas of the developed countries, employment of people for producing bioenergy often offers as an important opportunity [8]. Many governments also have developed policy strategies for biofuels technologies. The intention is to create employment opportunities for establishing industries for biofuel and biofuel feedstock production. These industries export biofuels and promote not only domestic economic growth but also domestic employment. For example, the policy for biofuels in the Philippines has taken an ambitious target that is the country will be one of the leading exporter of biofuel by 2030. Some developing countries such as Indonesia, Brazil, Thailand, and Argentina already have created substantial employment opportunities using bioenergy industries. There are approximately 1.7–3 million Indonesia people working in the palm oil sector [108, 109]. Indonesia has exported 1225 million liters of biodiesel in 2011 and targeted to create an extra 2.5 million jobs through the increase of palm biodiesel production [110].

One of the downsides is that the establishment of bioenergy industries may create employment-related problems given the nature of farms to be operated for biofuel production and type and scale of production and the management quality. The problem arises for the local communities when foreign investors bring their manpower along with them or hire workforce from other parts of the same country or other counters. This situation reduces employment opportunities for local communities. Also, in the developing countries, local employees are exploited by foreign investors when the employees are paid wages lower than the actual wagers they deserve to be paid. Strong governance from the local and national government is necessary to address this issue. In reality, the overall employment associated with bioenergy production depends on the diverse local socioeconomic condition, governance structures, nature of technology, and laws available in the place. These issues are, however, diverse and vary from one country to another [32].

4.3 Interaction with market price

The development of the bioenergy sector may have an influence on markets mainly for agricultural products such as food and feed products and straw as well as forest products such as paper and board. Although all forms of bioenergy are related to agriculture and forest markets through their feedstock demand, it is clear that the impact of first-generation liquid biofuels on food price is still a topic of strong debate. Most studies, however, indicate that there is a relatively limited risk of biofuels significantly affecting the price of food crops [24]. For the efficient function of markets, access to monetary and physical resources is essential to address the pressure of increasing demand. Therefore, policy in place is needed to provide support to an increasingly efficient industry, including the production of crops, the use of residue and wastes, and land use. Also, suitable regulation is needed to evade negative externalities associated with the utilization of biophysical resources. In this situation, effective coordination between energy, agriculture, forestry, trade, and environmental policies is required [24]. In Europe, for example, policymakers recognize that there are added economic benefits from bioenergy, especially in terms of the potential for employment creation and the development of a strong export industry. The renewable energy industry is one of Europe's fastest growing sectors, and is encouraged by the Member States as a locally available and environmentally friendly source of energy [105]. In the European Union, the predicted capacity of bioenergy technology has increased from 180 to 282.88 GW, where most of the technical capacity includes biofuel combustion. The output of bioenergy has increased from 394.71 TWh in 1995 to 876.04 TWh, where the output from biofuel combustion accounts for 367.51 TWh of total output [32].

4.4 Energy security

Energy security means the assurance of uninterrupted supply of energy either as a part of the military policy or foreign policy and national security dimensions. In order to ensure energy security, biofuel may lead to a shift in consumption toward the uninterrupted supply of fuels [9, 22]. The reliable and available supply of resources may play a vital to support the biofuel supply chain, including water nutrients and production process. For the promotion of biofuel-based energy and thereby ensuring energy security, three key factors are essential. These include the stable supply of feedstock for energy, the stable supply of products and coproducts, and the continuous demand and flexibility of feedstock and fuel systems.

There are economic advantages when there is an available and reliable local supply of biofuels. Countries those import fuel, a considerable amount of foreign exchange can be saved using local biomass or sources of bioenergy. For example, in East-Asian countries, it is estimated that the use of renewable energy instead of fossil fuels may save about US$ 2 trillion in fuel costs over the next 23 years, or more than $80 billion annually [12, 21]. The International Energy Agency (IEA) projected that between 2004 and 2030, the investment costs for the establishment of coal or fossil fuel-based new power plants in these countries would total $490 billion. The IEA predicted that the costs of fuel (e.g., coal and oil) for generating electricity using the new power plants would be $6.3 trillion over a 23-year period. Nevertheless, these countries shift to renewable energy, the costs of fuel over the same period would total $4.2 trillion, transforming into savings of $2.1 trillion [21, 111] as mentioned earlier.

Another example is the Philippines which is heavily dependent on imported fuels for electricity generation and other purposes. To avoid the import of fuel, the government has adopted a policy to promote the exploitation of native renewable sources such as coconut methyl ester as diesel enhancer. This policy, if implemented successfully would reduce the import of diesel significantly and save foreign exchange. Philippines can produce about 270,058 MT of coconut methyl ester and can generate US $80 million worth of savings from reduced diesel imports [21]. For an agricultural country like the Philippines, the development of products based on biomass is a great opportunity as the country has huge sources of biomass [12].

5 Discussions

The impacts of bioenergy include environmental, social, and economic aspects. There are both positive and negative environmental, social, and economic impacts of bioenergy production. However, the extent of environmental impacts of bioenergy production seems to be more dominant than social and economic impacts. Tables 1–3 show that some of the social issues particularly

TABLE 2 Social impacts of bioenergy production

Issues	Positive	Negative	Impacts	Comments
Gender inequity	X		Women have relatively less access to and control over land than women and women are deprived of having the equal share when bioenergy crops are produced in common land. When common forests are cleared for bioenergy corps, women suffer from shortage of cooking fuel. Also, they are paid less than men when they are employed by bioenergy companies	
Food security		X	Bioenergy production using agricultural land may lead to food insecurity and the price of food may be higher and go beyond the buying capacity of poor people particularly in developing countries	

TABLE 2 Social impacts of bioenergy production—cont'd

Issues	Positive	Negative	Impacts	Comments
			Women suffer from shortage of food and nutrition when bioenergy crops occupy the common land from where women would collect food	
Land ownership and tenure		X	In the absence of land tenure security, conflicts may arise between local community and bioenergy companies. Local communities can be excluded from the use of land and forests those are the sources of livelihood. The indigenous community depends on common land and forests and they may lose the traditional rights of the use of land and forest when bioenergy company acquires the land	The use of unproductive and public land for bioenergy may bring some benefits for local community
Health issue	X	X	The use of biomass in developing for traditional cooking affect the women's health, increase the emission of GHGs. The burning of bioenergy crops such as sugarcane affects human health. Food insecurity due to bioenergy production using agricultural land and	The solar energy and improved stove for cooking may address some health hazards in developing countries

Continued

TABLE 2 Social impacts of bioenergy production—cont'd

Issues	Positive	Negative	Impacts	Comments
			forests may affect human health due to high price of food	
Social acceptability		X	The cultivation of bioenergy crops and establishment of bioenergy production facilities may cause conflict between if the local community is not consulted at the beginning. The local community may feel insecurity due to the presence of foreign or unknown people in their community. The lack of trust between two parties may create social unrest	

TABLE 3 Social impacts of bioenergy production

Issues	Positive	Negative	Impacts	Comments
Trade of market price	X		The trade of bioenergy contributes to national income particularly for the countries where land is available. Also, trade makes bioenergy available for many countries and thereby the use of fossil fuel is being reduced	Illegal clearance of forests may increase and therefore biodiversity and ecological balance may be negatively affected

TABLE 3 Social impacts of bioenergy production—cont'd

Issues	Positive	Negative	Impacts	Comments
Employment generation	X		Bioenergy has potentials to create more employment opportunities than biofuel production and thereby address the unemployment opportunities	The local community can be deprived of being employed if the bioenergy companies hire manpower from other places. Social tension and conflict may arise
Interaction with market price			The expansion of markets may increase the production of agricultural crops in addition to bioenergy crops	Policies and regulations are needed at place to promote positive impact of marketing
Energy security	X		Uninterrupted supply of bio-based energy. Less dependency on biofuel-based energy	

the women and indigenous communities tend to be negatively affected by bioenergy production in addition to some other issues such as land tenure rights. Importantly, the negative environmental, social, and economic impacts of bioenergy relatively more dominant in the developing countries than the developed countries. This is due to the absence of adequate policies, rules, and regulations and their effective enforcement. For example, land tenure rights in many developing counties are inadequate and not up-to-date. Therefore, the absence of land tenure rights may make the local community more vulnerable in terms of access to their land and the use of land for their livelihood. The absence of proper land tenure rights may prevent the local community from bargaining with bioenergy comparing standing on equal footing. Overall, the tables show that there are more negative impacts of bioenergy than positive impacts despite the fact that the type and extent of impacts depend on many factors including local context, governance, and socioeconomic situations. However, there are some similarities between the developed and developing countries, particularly in terms of availability of land. For example, Brazil, China, Canada, and Australia have available land for the production of bioenergy. Also, the developing countries have available agricultural residues can be used for bioenergy production; similar to that of some developed countries.

While there are more negative impacts of bioenergy production than positive impacts of bioenergy production, particularly in the developing countries, it is possible to address these negative impacts and thereby to harness the benefits of bioenergy production worldwide. To address the negative impacts, proper policy strategies, rules and regulations for protecting women rights, land tenure rights, and rights for indigenous communities need to be available at the place. In addition, institutional arrangements are necessary to enforce these rules and regulations, particularly in the developing countries. These initiatives will promote bioenergy production and enhance energy security both for the developing and developed countries.

6 Conclusions

This chapter aimed to provide an overview of the economic, social, and environmental aspects of bioenergy resources. The environmental aspects of bioenergy resources include the release of less greenhouse gas (GHGs) emissions than biofuels, the deterioration of soil quality, and erosion of top fertile soil due to the removal of biomass from the crop fields. The environmental impacts of bioenergy also include the deterioration of quality and quantity of water due to water runoff from the bioenergy crop field and the use of fertilizer. The use and production of bioenergy affect biodiversity due to the conversion of land for bioenergy production. The impact on biodiversity includes the loss of habitat and ecosystem services and the removal of vegetation.

The social aspects of bioenergy include the generation of job opportunities, impacts on women including their access to job and livelihood, rights of the local community on land, food security, health impacts, and the social acceptance of bioenergy technologies. Women are often more vulnerable than men as the head of the households. Therefore, it is the responsibility of the project proponents of bioenergy project to provide them with equal opportunities for employment to improve their livelihood. The women, particularly in rural areas, are responsible for collecting fuelwood for cooking and heating. The use of marginal land and expansion of bioenergy crop often exclude them or make it difficult for women to collect fuelwood for the purpose of domestic use. These issues need to be addressed during the bioenergy production. Land tenure is another key social aspect that varies from counties to countries, and local communities are affected in many ways. The land tenure, if not defined clearly, may create conflict and also exclude the poor local community from the use of common land they depend for their livelihood. Appropriate arrangements are needed in place for addressing these social aspects.

The economic aspects of bioenergy include the trade of bioenergy and energy security in addition to employment generation and market price. There is a potential for employment opportunities due to the deployment of bioenergy production. This needs, however, proper concentration on local people, including women particularly those who are poor and jobless. Otherwise, there is the potential of conflicts between investors and the local community. The volume of trade of bioenergy is increasing globally, this means the demand for bioenergy is increasing. Like other renewable energies, bioenergy production definitely can play a vital role in ensuring energy security by providing biofuel for the transport sector and fuelwood for household or domestic use. Another aspect of bioenergy is market price. The governments concerned about the market of biofuel are keen to provide incentives so that there is a stable supply of biofuel. Finally, to expand the bioenergy use globally, it is imperative to address the negative socioeconomic and environmental aspects of bioenergy and thereby possible make the utilization of bioenergy more effective as an alternative to fossil fuel.

Acknowledgment

The author is thankful to the reviewer for providing valuable comments for further improvement of the chapter.

References

[1] Padilla-Rivera A, Paredes MG, Güereca LP. A systematic review of the sustainability assessment of bioenergy: the case of gaseous biofuels. Biomass Bioenergy 2019;125:79–94.

[2] Hoekman SK, Broch A, Liu X. Environmental implications of higher ethanol production and use in the US: a literature review. Part I—impacts on water, soil, and air quality. Renew Sust Energ Rev 2018;81:3140–58.

[3] Gabisaa EW, Gheewalaa SH. Potential of bio-energy production in Ethiopia based on available biomass residues. Biomass Bioenergy 2018;111:77–87.

[4] Rooney WL, Blumenthal J, Bean B. Designing a sorghum as a dedicated bioenergy feedstock. Biofuels Bioprod Biorefin 2007;B1:147–57.

[5] Energy Information Administration (EIA). International energy outlook. available at https://www.eia.gov/todayinenergy/detail.php?id=41433; 2018.

[6] Giri S, Nejadhashemi P, Woznicki SA. Regulators' and stakeholders perspectives in a framework for bioenergy development. Land Use Policy 2016;59:143–53.

[7] Engel B, Chaubey I, Thomas M. Biofuels and water quality: challenges and opportunities for simulation modeling. Biofuels 2010;1(3):463–77.

[8] Gheewala S, Damen B, Shi X. Biofuels: economic, environmental and social benefits and costs for developing countries in Asia. Wiley Interdiscip Rev Clim Chang 2013;4(6):497–511 [Q1].

[9] Azad AK, Rasul MG, Khan MMK, Sharma SC, Bhuiya MMK, Mofijur M. A review on socio-economic aspects of sustainable biofuels. Int J Glob Warm 2016;10(1):32–54. https://doi.org/10.1504/IJGW.2016.077903.

[10] Wu Y, Liu S, Young CJ, Dahal D, Sohl TL, Davis B. Projection of corn production and stover-harvesting impacts on soil organic carbon dynamics in the US Temperate Prairies. Sci Rep 2015;5:10830.

[11] World Bioenergy Association. Global bioenergy statistics. available at https://worldbioenergy.org/uploads/191129%20WBA%20GBS%202019_LQ.pdf; 2019. accessed on 12/2/2020.
[12] IEA. Southeast Asia energy outlook 2019: analysis. Paris, available at https://www.iea.org/reports; 2019. accessed on 16/2/2020.
[13] World Bank. Bioenergy development: Issues and impacts for poverty and natural resource management. Washington D.C., USA; 2010.
[14] Gvero P, Petrovic S, Papuga S. Biomass as potential sustainable development driver—case of bosnia and herzegovina; 2013. p. 1–33 [Chapter 1].
[15] Popp J, Laknerb Z, Harangi-Rákosa M. The effect of bioenergy expansion: food, energy, and environment. Renew Sust Energ Rev 2014;32:559–78.
[16] Overend RP. Biomass energy heat provision for cooking and heating in developing countries. In: Kaltschmitt M, editor. Energy from organic materials (biomass). Encyclopedia of sustainability science and technology series. New York, NY: Springer; 2019.
[17] Proskurina S, Heinimö J, Schipfer F, Vakkilainen E. Biomass for industrial applications: the role of torrefaction. Renew Energy 2017;111:265–74.
[18] Azad AK, Rasul MG, Khan MMK, Sharma SC, Bhuiya MMK. Study on Australian energy policy, socio-economic, and environment issues. J Renewable Sustainable Energy 2015;7:063131 [1 – 20] https://doi.org/10.1063/1.4938227.
[19] Barros S., Biofuels in Brazil, annual report, 2019. https://apps.fas.usda.gov/newgainapi/api/report/downloadreportbyfilename?filename=Biofuels%20Annual_Sao%20Paulo%20ATO_Brazil_8-9-2019.pdf, https://doi.org/10.1016/j.indcrop.2019.111507.
[20] Tareen WUK, Dilbar MT, Farhan F. Present status and potential of biomass energy in pakistan based on existing and future renewable resources. Sustainability 2020;12(249):1–40. https://doi.org/10.3390/su12010249.
[21] IRENA. Renewable energy market analysis: Southeast Asia. Dubai, UAE, available at https://www.iena.org; 2018. accessed on 16/3/2020.
[22] Azad AK, Rasul MG, Khan MMK, Sharma SC, Hazrat MA. Prospect of biofuels as an alternative transport fuel in Australia. Renew Sust Energ Rev 2015;43:331–51. https://doi.org/10.1016/j.rser.2014.11.047.
[23] Gielen D, Boshe F, Saygin D. The role of renewable energy in the global energy transformation. Energ Strat Rev 2019;24:38–50.
[24] World Energy Council. World energy resources: bioenergy [Chapter 7]. In: Bioenergy; 2013. https://www.worldenergy.org/assets/images/imported/2013/09/Complete_WER_2013_Survey.pdf.
[25] Welfle A. Balancing growing global bioenergy resource demands Brazil biomass potential and the availability of resource for trade. Biomass Bioenergy 2017;105:83–95.
[26] Szarka N, Eichhorn M, Kittler R. Interpreting long-term energy scenarios and the role of bioenergy in Germany. Renew Sust Energ Rev 2012;68:1222–33.
[27] AEBIOM. Forest sustainability and carbon balance of EU importation of north American Forest. Biomass for bioenergy production. AEBIOM European Biomass Association; 2013.
[28] Kumar A, Adamopoulos S, Jones D. Forest biomass availability and utilization potential in Sweden: a review waste biomass valor; 2020. https://doi.org/10.1007/s12649-020-00947-0.
[29] Kabir Z, Khan I. Environmental impact assessment of waste to energy projects in developing countries: a guideline in the context of Bangladesh. Sustainable Energy Technol Assess 2020;37:1–13.
[30] Khan I, Kabir Z. Sustainability assessment of waste-to-energy (electricity) generation technologies and its prospect in the developing world: a case of Bangladesh. Renew Energy 2020;150:320–33.

[31] International Energy Agency. Bioenergy, https://www.iea.org/fuels-and-technologies/bioenergy; 2018.
[32] Domac J, Segon V. IEA bioenergy update-31. International Energy Agency; 2020. http://www.ieabioenergy.com/wp-content/uploads/2013/10/IEA-Bioenergy-Update-31-Task-29-Technology-Report.pdf.
[33] Keerthi S, Miller SA. Regional differences in impacts to water quality from bioenergy mandate. Bioenergy Biomass 2017;106:115–26.
[34] Demissie Y, Yan E, Wu M. Assessing regional hydrology and water quality implications of large-scale biofuel feedstock production in the Upper Mississippi River Basin. Environ Sci Technol 2012;46:9174–82.
[35] Wu Y, Zhao F, Liu S. Bioenergy production and environmental impacts. Geosci Lett 2018;5(14):1–9. available at https://doi.org/10.1186/s40562-018-0114-y.
[36] Wu M, Zhang Z, Chiu Y. Life-cycle water quantity and water quality implications of biofuels. Curr Sustain Renew Energy Rep 2014;1(1):3–10.
[37] Gasparatos A, Stromberg P, Takeuchi K. Biofuels, ecosystem services and human wellbeing: putting biofuels in the ecosystem services narrative. Agric Ecosyst Environ 2011;142(3–4):111–28.
[38] Zhou X, Clark CD, Nair SS, Hawkins SA, Lambert DM. Environmental and economic analysis of using SWAT to simulate the effects of switchgrass production on water quality in an impaired watershed. Agric Water Manag 2015;160:1–13.
[39] Ashwath N, Kabir Z. Environmental, economic, and social impacts of biofuel production from sugarcane in Australia. In: Khan MT, Khan IA, editors. Sugarcane biofuels: status, potential, and prospects of the sweet crop to fuel the world. USA: Springer International Publishers; 2019. p. 267–84 [Chapter 13].
[40] Cibin R, Trybula E, Chaubey I, Brouder SM, Volenec JJ. Watershed-scale impacts of bioenergy crops on hydrology and water quality using improved SWAT model. GCB Bioenergy 2016;8(4):837–48.
[41] Wu Y, Liu S. Impacts of biofuels production alternatives on water quantity and quality in the Iowa River basin. Biomass Bioenergy 2012;36:182–91.
[42] Guo T, Cibin R, Chaubey I, Gitau M, Arnold JG, Srinivasan R, Kiniry JR, Engel BA. Evaluation of bioenergy crop growth and the impacts of bioenergy crops on streamflow, tile drain flow and nutrient losses in an extensively tile-drained watershed using SWAT. Sci Total Environ 2018;613–614:724–35.
[43] Chen Y, Ale S, Rajan N, Srinivasan R. Modeling the effects of land use change from cotton (Gossypium hirsutum L.) to perennial bioenergy grasses on watershed hydrology and water quality under changing climate. Agric Water Manag 2017;192:198–208.
[44] Dunn JB, Mueller S, Kwon H, Wang MQ. Land-use change and greenhouse gas emissions from corn and cellulosic ethanol. Biotechnol Biofuels 2013;6:51.
[45] Fu J, Jiang D, Huang Y, Zhuang D, Ji W. Evaluating the marginal land resources suitable for developing bioenergy in Asia. Adv Meteorol 2014;4:1–9.
[46] Wang M, Han J, Dunn JB, Cai H, Elgowainy A. Well-to-wheels energy use and greenhouse gas emissions of ethanol from corn, sugarcane and cellulosic biomass for US use. Environ Res Lett 2012;7(4), 045905.
[47] Liu J, Huffman T, Green M. Potential impacts of agricultural land use on soil cover in response to bioenergy production in Canada. Land Use Policy 2017;75:133–42.
[48] Liu T, McConkey BG, Ma Z, Liu Z, Li X, Cheng L. Strengths, weaknesses, opportunities and threats analysis of bioenergy production on marginal land. Energy Procedia 2011;5:2378–86.

[49] Li YL, Khanal SK. Bioenergy: principles and applications. John Wiley & Sons; 2016. http://ebookcentral.proquest.com/lib/newcastle/detail.action?docID=4690034.
[50] Birdsey R, Duffy P, Smyth C. Climate, economic, and environmental impacts of producing wood for bioenergy. Environ Res Lett 2018;13, 050201.
[51] Chum H, Faaij A, Moreira J. Bioenergy. In: IPCC special report on renewable energy sources and climate change mitigation. Cambridge University Press, Cambridge, United Kingdom and New York, NY, USA; 2011.
[52] Lisboa CC, Butterbach-bahl K, Mauder M, Kiese R. Bioethanol production from sugarcane and emissions of greenhouse gases: known and unknowns. GCB Bioenergy 2011;3:277–92. https://doi.org/10.1111/j.1757-1707.2011.01095.
[53] German L, George CS, Pacheco P. Local social and environmental impacts of biofuels global comparative assessment and implications for governance. Ecol Soc 2011;16(4):14–23.
[54] Paolini V, Petracchini F, Segreto M, Tomassetti L. Environmental impact of biogas: a short review of current knowledge. J Environ Sci Health A 2018;53(10):899–906. https://doi.org/10.1080/10934529.2018.1459076.
[55] Wang J, Yang Y, Bentley Y. Sustainability assessment of bioenergy from a global perspective: a review. Sustainability 2018;10(2739):1–19. https://doi.org/10.3390/su10082739.
[56] Qin Z, Zhuang Q, Cai X, He Y, Huang Y, Jiang D, et al. Biomass and biofuels in China: toward bioenergy resource potentials and their impacts on the environment. Renew Sust Energ Rev 2018;82:2387–400.
[57] Sala OE, Sax D, Leslie H. Biodiversity consequences of biofuel production. In: Howarth RW, Bringezu S, editors. Biofuels: environmental consequences and interactions with changing land use. Proceedings of the scientific committee on problems of the environment (SCOPE); 2009. p. 22–5. Pages 127-137. September, Germany.
[58] Dauber J, Bolte AS. Bioenergy: challenge or support for the conservation of biodiversity. GCB Bioenergy 2014;6:180–2. https://doi.org/10.1111/gcbb.12188.
[59] Qin Z, Dunn JB, Kwon H, Mueller S, Wander MM. Influence of spatially dependent, modeled soil carbon emission factors on life-cycle greenhouse gas emissions of corn and cellulosic ethanol. GCB Bioenergy 2016;8(6):1136–49.
[60] Correa DF, Beyer HL, Possingham HP, Thomas-Hall SR, Schenk PM. Biodiversity impacts of bioenergy production: microalgae vs. first generation biofuels. Renew Sust Energ Rev 2017;74:1131–46.
[61] UNEP. Impacts of biofuel production case studies: Mozambique, Argentina and Ukraine. In: Final report; 2013. http://www.unido.org/en/resources/publications/energy-and-environment.html.
[62] Ashworth K, Wild O, Hewitt CN. Impacts of biofuel cultivation on mortality and crop yields. Nat Clim Chang 2013;3:492–6.
[63] Sang T, Zhu W. China's bioenergy potential. GCB Bioenergy 2011;3:79–90. https://doi.org/10.1111/j.1757-1707.2010.01064.x.
[64] McCalmont JP, Hastings A, McNamara NP, Richter GM, Robson P, Donnison IS, Clifton-Brown J. Environmental costs and benefits of growing Miscanthus for bioenergy in the UK. GCB Bioenergy 2017;9(3):489–507.
[65] Pourhashem G, Rasool QZ, Zhang R, Medlock KB, Cohan DS, Masiello CA. Valuing the air quality effects of biochar reductions on soil NO emissions. Environ Sci Technol 2017;51(17):9856–63. https://doi.org/10.1021/acs.est.7b00748.
[66] Blanco-Canqui H, Wortmann C. Crop residue removal and soil erosion by wind. J Soil Water Conserv 2017;72(5):97A–104A.
[67] Turmel M, Speratti A, Baudron F. Crop residue management and soil health: a systems analysis. Agric Syst 2015;134:6–16.

[68] EPA. Biofuels and the environment: first triennial report to congress. National Center for Environmental Assessment EPA; 2011.
[69] Blanco-Canque H, Lal R. Crop residue removal impacts on soil productivity and environmental quality. Crit Rev Plant Sci 2009;28(3):139–63.
[70] Lal R. Soil quality impacts of residue removal for bioethanol production. Soil Tillage Res 2009;102(2):233–41.
[71] Anderson-teixeira KJ, Davisw SC, Masters MD, Delucia EH. Changes in soil organic carbon under biofuel crops. GCB Bioenergy 2009;1:75–96. https://doi.org/10.1111/j.1757-1707.2008.01001.x.
[72] Harris ZM, Spake R, Taylor G. Land use change to bioenergy: a meta-analysis of soil carbon and GHG emissions. Biomass Bioenergy 2015;82:27–39.
[73] Cooney D, Kim H, Quinn L, Lee MS, Guo J, Chen SX, Lee DK. Switchgrass as a bioenergy crop in the Loess Plateau, China: potential lignocellulosic feedstock production and environmental conservation. J Integr Agric 2017;16(6):1211–26.
[74] Dale VH, Efroymson RA, Kline KL. Indicators for assessing socio-economic sustainability of bioenergy systems: a short list of practical measures. Ecol Indic 2013;26:87–102.
[75] Jabareen Y. A new conceptual framework for sustainable development. Environ Dev Sustain 2008;10(2):179–92.
[76] Fedorova E, Pongracz E. Renewable energy, cumulative social effect assessment framework to evaluate the accumulation of social sustainability benefits of regional bioenergy value chains. Renew Energy 2019;131:1073–88.
[77] Elbehri A. Biofuels and the sustainability challenge: a global assessment of sustainability issues. In: Trends and policies for biofuels and related feedstocks. FAO; 2012.
[78] Creutzig F, Corbera E, Bolwing S. Integrating place-specific livelihood and equity outcomes into global assessments of bioenergy deployment. Environ Res Lett 2013;8:1–11.
[79] Edrisi SA, Abhilash PC. Exploring marginal and degraded lands for biomass and bioenergy production: an Indian scenario. Renew Sust Energ Rev 2016;54:1537–51.
[80] Gamisonia N. Challenges and opportunities for equal participation of women and girls in rural development in Georgia. Rome, Italy: Georgian NGO Rural Communities Development Agency; 2017.
[81] Women 2030. Bioenergy in West Africa-impacts on women and forests, forest cover. No.59, Global Forest Coalition (GFC); 2019. https://www.women2030.org/forest-cover-59-bioenergy-in-west-africa-impacts-on-women-and-forests/.
[82] Molony T. Bioenergy policies in Africa: mainstreaming gender amid an increasing focus on biofuels. Biofuels Bioprod Biorefin 2011;5(3):330–41.
[83] FAO. Bioenergy environmental impact analysis (BIAS): analytical framework, working paper, Rome, Italy; 2011.
[84] Anon. Preliminary assessment of bioenergy production in the Caribbean, United Nations development program, Barbados and the OECS; 2009.
[85] Schoneveld GC, German LA, Nutakor E. Land-based investments for rural development: a grounded analysis of the local impacts of biofuel feedstock plantations in Ghana. Ecol Soc 2011;16(4):10.
[86] Julia WB. Gendered experiences of dispossession: oil palm expansion in a Dayak Hibun community in West Kalimantan. J Peasant Stud 2012;39:995–1016.
[87] Skutsch M, De los Rios E, Solis S, Riegelhaupt E, Hinojosa D, Gerfert S, et al. Jatropha in Mexico: environmental and social impacts of an incipient biofuel program. Ecol Soc 2011;64 (4):11. https://doi.org/10.5751/ES-04448-160411.
[88] Beall E, Rossi A. Good socio-economic practices in modern bioenergy production: minimizing risks and increasing opportunities for food security. Rome, Italy: FAO; 2011.

[89] Domac J, Segon V. Forests, bioenergy and people—key elements for energy transition of South-East Europe. In: XIV World Forestry Congress, Durban, South Africa, 7–11 September; 2015.

[90] Miyake S, Renouf M, Peterson A. Land-use and environmental pressures resulting from current and future bioenergy crop expansion. A review. J Rural Stud 2012;28:650–8.

[91] Elder M, Romero J, Bhattacharya A. Socioeconomic impacts of biofuels in East Asia. In: Takeuchi K, et al., editors. Biofuels and sustainability. Science for Sustainable Societies; 2018. p. 87–118. https://doi.org/10.1007/978-4-431-54895-9_8 [Chapter 8].

[92] IFAD. Indigenous peoples' collective rights to lands, territories and natural resources: Lessons from IFAD-supported projects. Rome: IFAD; 2019.

[93] Vermeulen S, Cotula L. Over the heads of local people: consultation, consent, and recompense in large-scale land deals for biofuels projects in Africa. J Peasant Stud 2010;37:899–916.

[94] Obidzinski K, Andriani R, Komarudin H, Andrianto A. Environmental and social impacts of oil palm plantations and their implications for biofuel production in Indonesia. Ecol Soc 2012;17(1):25.

[95] McCarthy JF. Processes of inclusion and adverse incorporation: oil palm and agrarian change in Sumatra, Indonesia. J Peasant Stud 2010;37:821–50.

[96] Smalley R, Corbera E. Large-scale land deals from the inside out: findings from Kenya's Tana Delta. J Peasant Stud 2012;39:1039–75.

[97] Goldemberg J, Coelho ST, Guardabassi P. The sustainability of ethanol production from sugarcane. Energy Policy 2008;36:2086–97.

[98] Hardley C, Lindstrom D, Tessema F. Gender bias in the food insecurity experience of Ethiopian adolescents. Soc Sci Med 2008;66(2):437–8.

[99] McBridea AC, Dalea VH, Baskarana LM, Downinga ME, Eatona LM, Efroymson RA. Indicators to support environmental sustainability of bioenergy systems. Ecol Indic 2011;1277–89.

[100] Kates RW, editor. From the unity of nature to sustainability sciences: ideas and practice. Working paper no. 2018. Center for International Development. Cambridge, MA, USA: Harvard University; 2011.

[101] UNCTAD. The state of the biofuel market: regulatory, trade and development perspectives. Geneva, Switzerland: UNCTAD; 2015.

[102] Meyer SJ, Hurlé B. Global biofuel trade: how uncoordinated biofuel policy fuels resource use and GHG emissions. Issue paper-48. International Center for Trade and Sustainable Development; 2013.

[103] Harmer T. Biofuels subsidies and the law of the World Trade Organization: ICTSD programme on agricultural trade and sustainable development. Issue paper No. 20. Geneva, Switzerland: International Centre for Trade and Sustainable Development; 2009.

[104] Sooriyaarachchi T, Tsai I, Khatib S, Farid A, Mezher T. Job creation potentials and skill requirements in, PV, CSP, wind, water-to-energy and energy efficiency value chains. Renew Sust Energ Rev 2015;52:653–66.

[105] Remedio EM, Domac J. Socio-economic analysis of bioenergy systems: a focus on employment. FAO; 2003.

[106] Jackson RW, Amir B, Ferreira Neto ABF, Erfanian E. Woody biomass processing: potential economic impacts on rural regions. Energy Policy 2018;115:66–77.

[107] Paredes-Sánchez JP, Lopez-Ochoa LM, López-González LM, Xiberta J. Evolution and perspectives of the bioenergy applications in Spain. J Clean Prod 2019;213:553–68.

[108] Nepal S, Tran L. Identifying trade-offs between socio-economic and environmental factors for bioenergy crop production: a case study from northern Kentucky. Renew Energy 2019;142:72–83.

[109] USDA. Foreign agriculture service. In: Global agriculture information network report; 2012. Washington, USA.

[110] Kojima M, Mitchell D, Ward W. Considering trade policies for liquid biofuels. In: Renewable energy special report 004/07; 2007.

[111] Terra Daily. Renewable energy can save East Asia two trillion US dollars in fuel costs. Retrieved from the Terra Daily on-line database on the World Wide Web http://www.terradaily.com; 2007.

Chapter 13

An overview of policy framework and measures promoting bioenergy usage in the EU, the United States, and Canada

Zobaidul Kabir[a,*], M.A. Yusuf[b,c,*], and Imran Khan[d,*]
[a]*School of Environmental and Life Sciences, University of Newcastle, Ourimbah, NSW, Australia,* [b]*Ministry of Finance, Bangladesh Secretariet, Dhaka, Bangladesh,* [c]*Adjunct Faculty, AIUB, Bashundhara, Dhaka, Bangladesh,* [d]*Department of Electrical and Electronic Engineering, Jashore University of Science and Technology, Jashore, Bangladesh*

1 Introduction

Bioenergy is one of the key sources of renewable energy. About one-tenth of the world's total primary energy is supplied from bioenergy [1]. Across all sectors, bioenergy contributes about five times higher than wind and solar, as reported by the International Energy Agency (IEA). This is due to the policy supports across countries in relation to bioenergy use in all possible sectors such as electricity generation, transport. To date, heating is still the dominating sector of final bioenergy consumption where 70% of total bioenergy generated is used for heating [1]. Although this is true, modern bioenergy does not take into account the traditional use of biomasses for cooking and heating in the developing world.

In general, bioenergy is a form of energy that is produced from organic matters, commonly known as biomass, predominantly derived from either plants or animals. Using different technologies, bioenergy can be converted into different form of energy such as electricity, heat, and biofuels. The common sources of bioenergy are shown in Fig. 1.

* Each author contributed equally for the chapter.

Bioenergy Resources and Technologies. https://doi.org/10.1016/B978-0-12-822525-7.00015-9
Copyright © 2021 Elsevier Inc. All rights reserved.

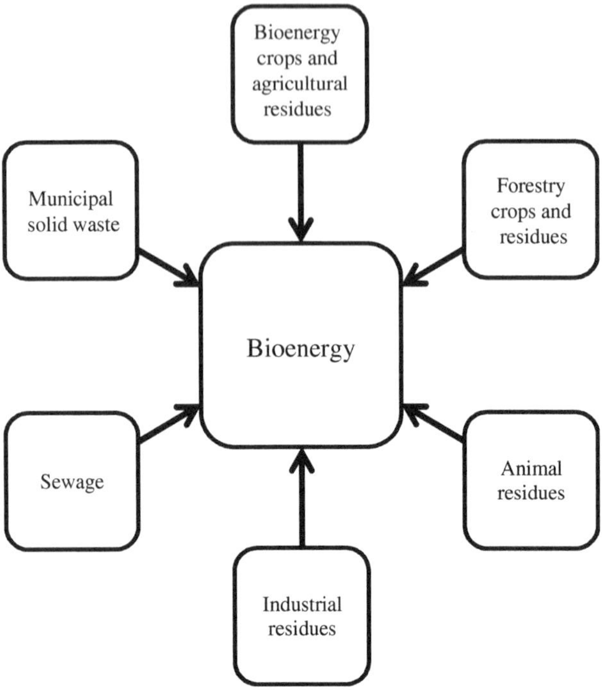

FIG. 1 Common sources of bioenergy/biomass.

Bioenergy can be produced through a number of ways. However, the choice of a specific way depends on the desired energy output and technology required for this conversion. Some conversions are simple in nature, such as burning wood for heat, whereas some are complex, such as biofuel production. Generally, there are five different processes through which bioenergy can be converted to other forms of energy: (i) traditional domestic, (ii) traditional industrial, (iii) modern industrial, (iv) chemical conversion, and (v) biological conversion [2]. The traditional domestic conversion process is used in developing countries for domestic cooking and the conversion efficiency is in between 5% and 15%. The traditional industrial process is used for processing tea, pig iron, bricks, etc. and the conversion efficiency is about 15%. The modern industrial process involves direct combustion such as gasification, pyrolysis, and incineration [3]. The efficiency might vary between 60% and 90% [4]. Chemical conversion involves technologies such as fuel cell, esterification. Biological conversion involves processes like anaerobic digestion, fermentation, for which the efficiency varies between 10% and 15% (e.g., anaerobic digestion) [3]. For instance, bioethanol can be produced through batch fermentation [5]. For details of these conversion processes, see Refs. [2–4].

Being a renewable source of energy, bioenergy offers many advantages. However, there are also some disadvantages of bioenergy. For instance, bioenergy (biomass) has smaller energy content considering its bulk in comparison with fossil fuels. The cost involved for the collection of biomass, that is, labor, transportation, and storage, is higher than for fossil fuels. Therefore, at the initial stage, each country should focus on related policy measures so that maximum bioenergy resources can be utilized toward sustainable development. The focus of this chapter is to shed light on this in the context of developed countries from Europe and North America.

This chapter is divided into six sections. In this chapter, Section 1 outlines the status of bioenergy in the global context followed by the introduction has been described in Section 2. Section 3 outlines the policy measures adopted by the European Union (EU), particularly the key features of Renewable Energy Directive (RED) of 2009 including member states' implementation of the RED. Sections 4 and 5 include a brief description of the US bioenergy scenario and policy measures in addition to policy implications. Section 6 concludes the chapter.

1.1 Recent trends in bioenergy

According to the latest "Global Bioenergy Statistics 2019" published by the World Bioenergy Association (WBA), in 2017, the total primary energy supply was dominated by fossil fuels.

In 2017, the total primary energy supply was dominated by fossil fuels, of which oil contributed about 32%, followed by coal (27%) and natural gas (22%) [6]. On the other hand, the renewable share was only 14% and the rest 5% was from nuclear power. In terms of primary renewable energy supply in 2017, Asia ranked first with 33.2 EJ, followed by the Americas (18.3 EJ) and Africa (16.1). With respect to biomass energy supply, again Asia was in the leading position with 21.6 EJ, followed by Africa (15.4 EJ) [6].

Global continent-wise bioenergy supply in the year of 2017 is shown in Fig. 2. Clearly, Asia was in the top position, followed by Africa and the Americas. Asia ranked first due to the contribution from biomass and primary solid biofuels with amounts of 21.6 and 20.1 EJ, respectively. A similar trend was also found for Africa and the Americas. A majority of the liquid biofuels were produced in the Americas (2.57 EJ), followed by Europe (0.73 EJ). Bioenergy production from municipal waste was about 0.95 EJ in Europe, which was the highest among all the continents. Contrarily, bioenergy from industrial waste was led by Asia with 0.61 EJ. In terms of biogases, this sector was led by Europe with 0.71 EJ in 2017.

Globally, 596 TWh electricity was generated from different bioenergy sources in 2017. Continent-specific electricity generation scenarios are depicted in Fig. 3. Although maximum bioenergy was produced in Asia (see Fig. 2), the highest electricity from bioenergy was generated in Europe (218 TWh), of

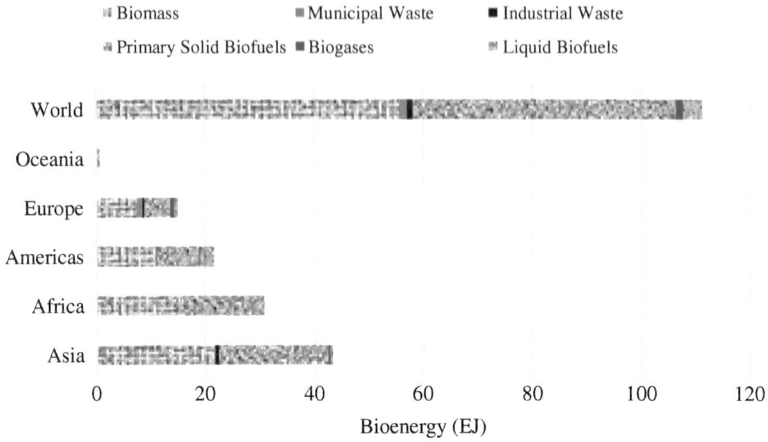

FIG. 2 Global bioenergy supply from different sources in 2017. EJ: Exa Joule (10^{18} J). *(From World Bioenergy Association, Global bioenergy statistics 2019, World Bioenergy Association, Stockholm, Sweden, 2019. https://worldbioenergy.org/uploads/191129 WBA GBS 2019_LQ.pdf, Accessed 6 April 2020).*

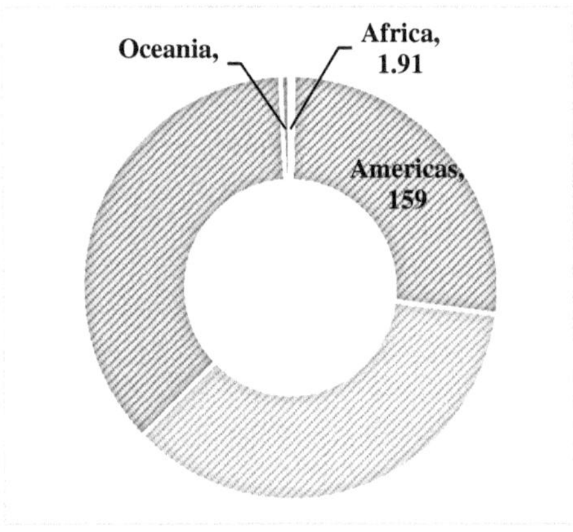

FIG. 3 Continent-specific electricity generation (in TWh) from bioenergy in 2017. *(From World Bioenergy Association, Global bioenergy statistics 2019, World Bioenergy Association, Stockholm, Sweden, 2019. https://worldbioenergy.org/uploads/191129 WBA GBS 2019_LQ.pdf, Accessed 6 April 2020).*

which 95.3 and 64.1 TWh electricity were generated from solid biofuels and biogas, respectively. The second highest electricity generating continent was Asia, followed by the Americas. For Asia and the Americas, the major sectors were solid biofuels (164 TWh) and industrial waste (31.8 TWh), and solid biofuels (125 TWh) and municipal waste (16.3 TWh) [6].

2 Global bioenergy—An analysis

The 'Renewable Capacity Statistics 2020' from the International Renewable Energy Agency (IRENA) will be used here for analysis purposes [7]. The bioenergy trend varies significantly from one region to another in the world. Few regions are achieving high growth; whereas a few are slower, others are in an almost steady state. For instance, if the bioenergy capacity for the last 10 years is considered, it can be seen that the growth in Asia is exponential. In 2019, the capacity was 40,517 MW in Asia and the capacity growth was about 10% compared to the capacity in 2018. The growth was linear in Europe and North America and the capacities were 41,399 and 16,835 MW in 2019, respectively. However, about 2% reduction in capacity occurred for North America in 2019 compared to the previous year. For South America, a curvature nature was observed for the bioenergy capacity; the capacity was increased to about 1.8% in 2019 compared to the previous year. Among all the regions, the maximum growth was found to be 17% in 2019 compared to 2018 for Eurasia. Nevertheless, the overall capacity for Eurasia was only 2398 MW in 2019. In contrast, no growth was observed for the Middle East. These trends are shown in Fig. 4.

Although there was a linear increase in bioenergy capacity development for Africa, the total capacity was only 1713 MW in 2019. A similar trend was also found for Central America and the Caribbean region and the capacity was 2658 MW in 2019. On the other hand, almost no changes in capacity were observed for the Middle East and Oceania.

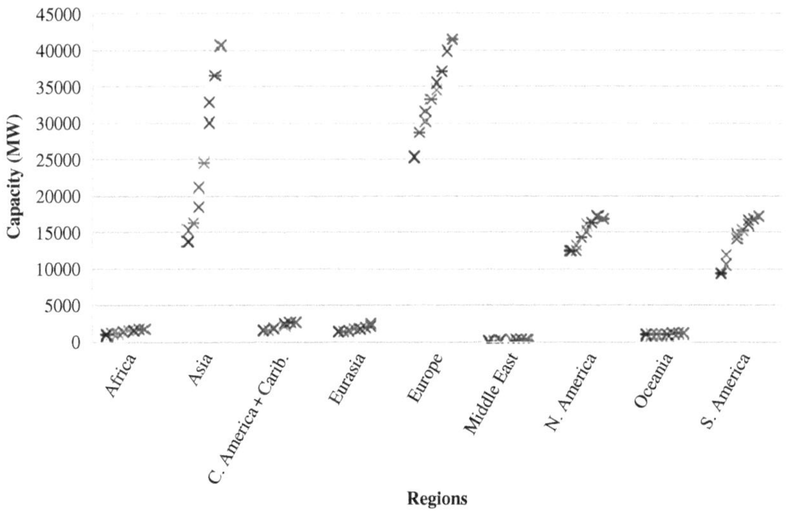

FIG. 4 Bioenergy trends (duration: 2010–2019) in different regions. Note: Carib. stands for the Caribbean. *(From Internatinal Renewable Energy Association (IRENA). Renewable capacity statistics 2020, International Renewable Energy Agency, Abu Dhabi, 2020. Available from: www.irena.org/Publications, Accessed 4/5/2020).*

If the long-term, that is, 10 years (2010–2019), change in capacity is considered, it was found that the maximum capacity development was achieved by the Middle East, which is 75%. However, in actual terms, the capacity was increased from 27 MW in 2010 to 108 MW in 2019. Contrarily, a dramatic increase in bioenergy capacity was achieved by Asia for this period. A growth of about 66% was achieved within this 10-year period. With respect to numerical capacity, it was 13,754 MW in 2010 and improved to 40,417 in 2019, a threefold increase in capacity. For Central America and the Caribbean region, this growth was about 47%, followed by South America and Eurasia with the growth of about 45% for both regions. The lowest growth was achieved by Oceania region with a capacity improvement of 5.39%. Europe and North America achieved 38.78% and 26.33% growth for this 10-year period. For Africa, this growth was about 41% with the capacity increase from 1005 MW in 2010 to 1713 MW in 2019. All these growths are depicted in Fig. 5.

2.1 Renewable energy versus bioenergy: A comparative analysis

In terms of electricity capacity (in MW), a comparative analysis between renewable energy and bioenergy development over a 20-year period (2000–2019) for different continents is conducted in this section.

2.1.1 Africa

In Africa, the renewable energy capacity was 48,443 MW in 2019, which is a twofold increase from 22,040 MW in 2000. In 2019, the countries that

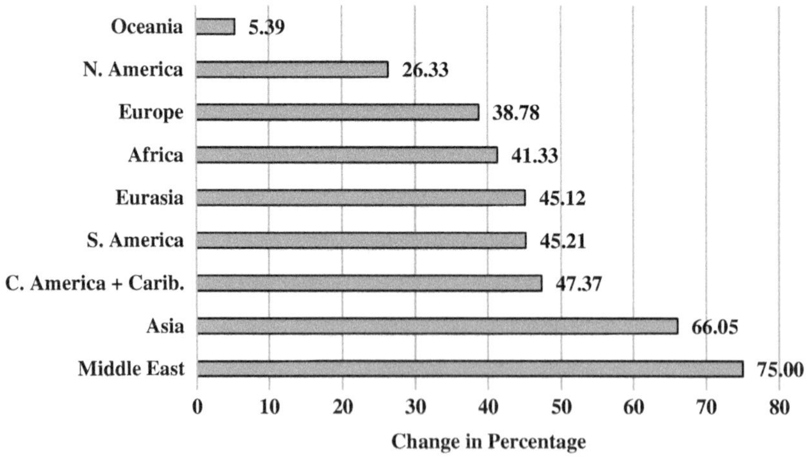

FIG. 5 Change in bioenergy capacity (in %) growth between 2010 and 2019. *(From Internatinal Renewable Energy Association (IRENA). Renewable capacity statistics 2020, International Renewable Energy Agency, Abu Dhabi, 2020. Available from: www.irena.org/Publications, Accessed 4/5/2020).*

predominantly contributed to this total capacity were Angola (2763 MW), Congo DR (2772 MW), Egypt (5972 MW), Ethiopia (4450 MW), Kenya (2178 MW), Morocco (3264 MW), Mozambique (2273 MW), Nigeria (2152 MW), South Africa (6167 MW), Sudan (2146 MW), and Zambia (2538 MW). Most of these capacities are from hydropower. On the other hand, the maximum bioenergy capacity of 290 MW was from Ethiopia, followed by South Africa (265 MW), Sudan (199 MW), Eswatini (106 MW), and Zimbabwe (101 MW). The amount of bioenergy that contributed to the total renewable capacity in Africa was about 4%. This situation is well reflected in Fig. 6. It can be seen that the bioenergy capacity growth ($R^2 = 0.9099$) is more linear in nature than the total renewable capacity growth ($R^2 = 0.8330$) in Africa.

2.1.2 Asia

In Asia, there was a sevenfold increase in electricity generation capacity from renewables in 2019 (1,118,969 MW) compared to the capacity in 2000 (171,445 MW). In this capacity development, almost every country in Asia contributed significantly except a few countries with limited renewable resources or technology limitations such as Brunei (1 MW), Timor Leste (1 MW), Turkmenistan (1 MW), Maldives (15 MW), Mongolia (276 MW), Afghanistan (365 MW), Singapore (467 MW), and Bangladesh (522 MW). Contrarily, a

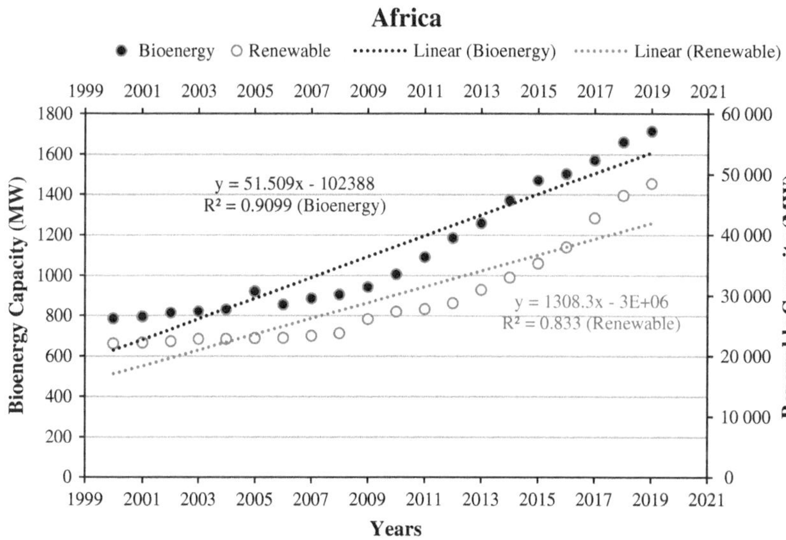

FIG. 6 Comparison between total renewable and bioenergy capacities in Africa. *(From Internatinal Renewable Energy Association (IRENA). Renewable capacity statistics 2020, International Renewable Energy Agency, Abu Dhabi, 2020. Available from: www.irena.org/Publications, Accessed 4/5/2020).*

few giant contributors were China (758,626 MW), India (128,232 MW), Japan (97,461 MW), Vietnam (24,518 MW), and the Republic of Korea (15,653 MW).

Of this total renewable capacity, about 48% was contributed by hydro, 28% from solar, 20% from wind, and 4% from bioenergy. The countries that contributed significantly to enrich the bioenergy capacity of Asia are China (16,537 MW), India (10,227 MW), Thailand (4257 MW), Japan (3196 MW), Indonesia (1840 MW), and the Republic of Korea (1572 MW). The renewable energy development in Asia is exponential in nature, and it is auspicious that bioenergy development is also following a similar trend as shown in Fig. 7.

2.1.3 Central America and Caribbean

A threefold increase in renewable capacity was observed for Central America and the Caribbean (C. America+Carib.) for the 20-years (2000–2019) period. The total renewable capacity in 2019 was 15,572 MW, of which 52%, 14%, and 12% were from hydro, solar, and wind. About 17% renewable capacity was added to the total from bioenergy in the same year. Costa Rica, Guatemala, Panama, Honduras, El Salvador, and the Dominican Republic were the significant contributors to this renewable capacity and they contributed 3141, 2870, 2332, 1700, 1365, 1314 MW, respectively, in 2019. On the other hand, the dominant contributors in bioenergy capacity development were Guatemala (1035 MW),

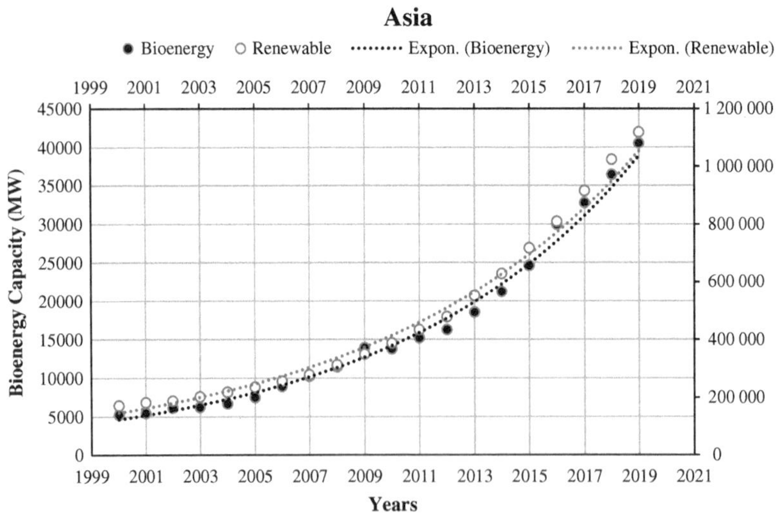

FIG. 7 Comparison between total renewable and bioenergy capacities in Asia. *(From Internatinal Renewable Energy Association (IRENA). Renewable capacity statistics 2020, International Renewable Energy Agency, Abu Dhabi, 2020. Available from: www.irena.org/Publications Accessed 4/5/2020).*

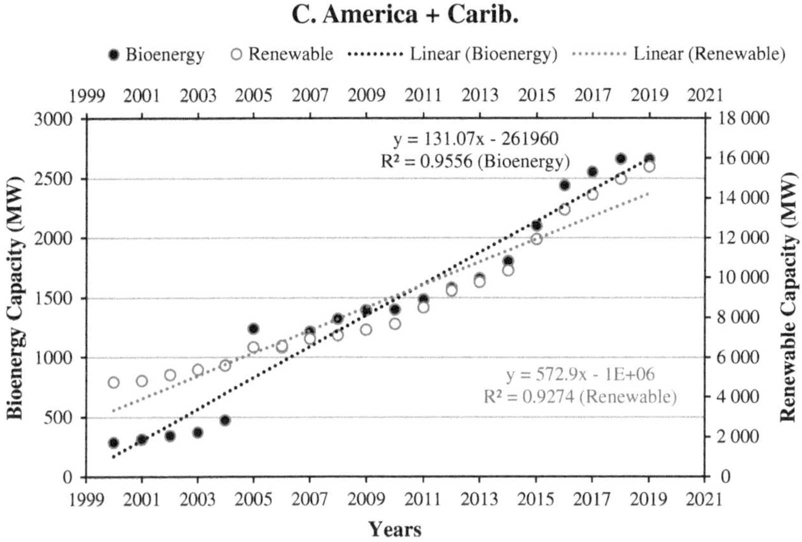

FIG. 8 Comparison between total renewable and bioenergy capacities in Central America and the Caribbean (C. America + Carib.). *(From International Renewable Energy Association (IRENA). Renewable capacity statistics 2020, International Renewable Energy Agency, Abu Dhabi, 2020. Available from: www.irena.org/Publications, Accessed 4/5/2020).*

Cuba (554 MW), El Salvador (316 MW), Nicaragua (218 MW), and Honduras (209 MW). The renewable and bioenergy capacity development trends are both linear in nature (see Fig. 8).

2.1.4 Eurasia

In 2019, the total renewable energy capacity in Eurasia was about 105,799 MW, which is 1.7 times the capacity available in 2000. The most dominating renewable source in this continent is hydro with a contribution of about 83% in the same year, followed by wind and solar with a capacity of 7.36% and 6.76%, respectively. In terms of bioenergy, it contributed about 2.27% to the total renewable capacity in 2019. Predominantly, the Russian Federation contributed more than 50% in this total renewable capacity development, followed by Turkey with a capacity of about 42%. A similar trend was also found for the contribution to the bioenergy capacity development with amounts of 57% and 41% for the former and latter ones, respectively. For renewable capacity development, the continent followed a linear growth ($R^2 = 0.8822$). Whereas bioenergy did not follow this trend, it followed a nonlinear nature as shown in Fig. 9. This nonlinear nature was due to the drop in bioenergy capacity between 2008 and 2013.

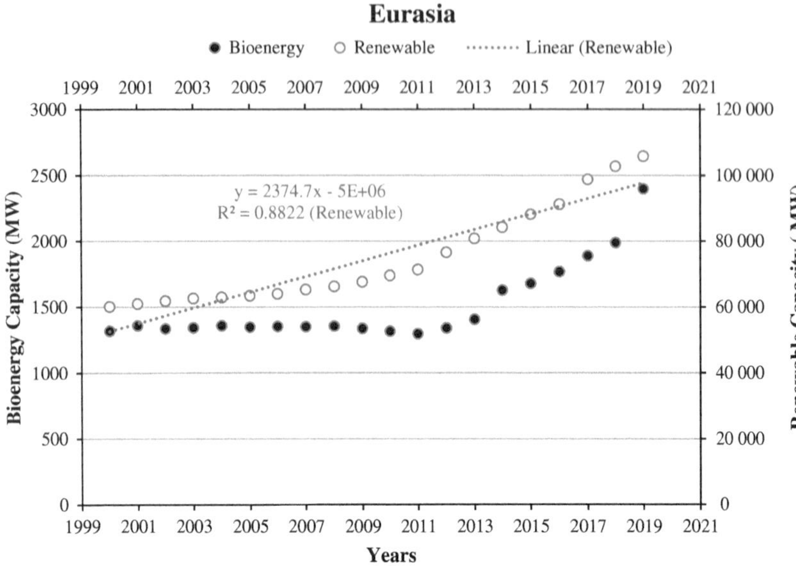

FIG. 9 Comparison between total renewable and bioenergy capacities in Eurasia. *(From Internatinal Renewable Energy Association (IRENA). Renewable capacity statistics 2020, International Renewable Energy Agency, Abu Dhabi, 2020. Available from: www.irena.org/Publications, Accessed 4/5/2020).*

2.1.5 Europe

A threefold increase in total renewable capacity was observed for Europe from 2000 to 2019. The contributions from different renewable sources were: (i) hydro was more than 35%, (ii) wind was more than 30%, (iii) solar was about 25%, and (iv) bioenergy was about 7% in 2019. Germany was the highest contributing country in this continent with a renewable capacity contribution of about 22%. The second highest contributor was Italy with a contribution of about 9.65%, followed by Spain (9.52%), France (9.23%), the United Kingdom (8.15%), and Norway (6.14%).

Similarly, Germany was also found to be the highest contributor to bioenergy capacity development in this continent with a percentage of 24%. The UK was in the second position with a contribution of about 17%. The third place was achieved by Sweden with a contribution of about 12%. The other dominant contributing countries were Italy (9.4%), Finland (5.51%), and Denmark (4.36%). In terms of overall trend of renewable and bioenergy capacity growth in this continent, it was found to be linear for both of them as shown in Fig. 10.

2.1.6 Middle East

Although the total renewable capacity was 22,710 MW in 2019, it was a fivefold increase from 2000. Hydro was the dominant renewable resource in this

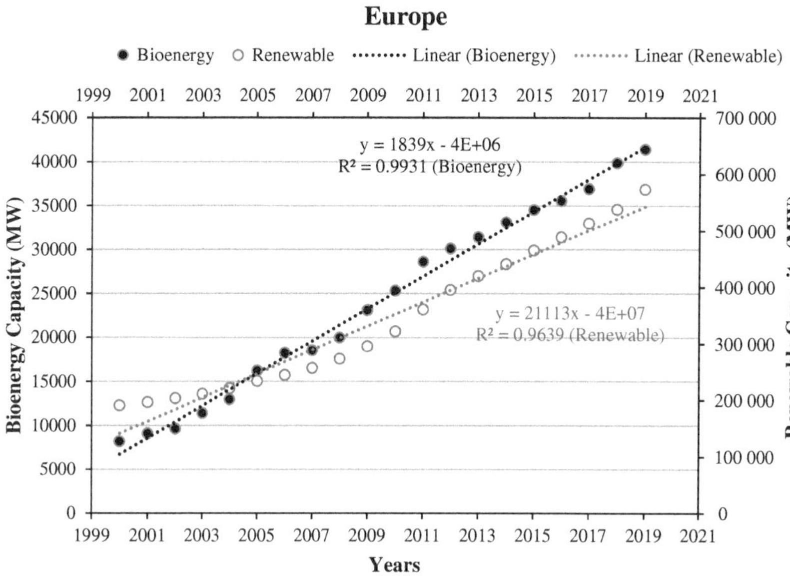

FIG. 10 Comparison between total renewable and bioenergy capacities in Europe. *(From Internatinal Renewable Energy Association (IRENA). Renewable capacity statistics 2020, International Renewable Energy Agency, Abu Dhabi, 2020. Available from: www.irena.org/Publications, Accessed 4/5/2020).*

continent with a contribution of about more than 70%, followed by solar and wind with a capacity of about 24% and 3%, respectively. On the other hand, the contribution from bioenergy is a tiny amount of about 0.5% (108 MW). The highest renewable contribution was from Iran, which was about 57% of the total. The second highest contributing country was Iraq with a contribution of about 10%, followed by the UAE (8.3%), Saudi Arabia (6.62%), Israel (6.61), and Jordan (6.17%).

Out of the 108 MW bioenergy capacity, 38 MW was contributed by Qatar, which was the highest among all the countries in this continent. Israel contributed about 28.3 MW, followed by Jordan (13 MW), Iran (11.6 MW), Lebanon (9 MW), Syria (7 MW), and the UAE (1 MW). With respect to growth trend, renewable capacity followed a linear nature, whereas bioenergy capacity development was nonlinear with a drastic shift in 2011 and 2015 (see Fig. 11).

2.1.7 North America

North America's renewable sector is dominated by hydro with about 50% capacity. The second largest contributing renewable source was wind with a capacity of more than 30%, followed by solar with more than 15% capacity in 2019. In terms of bioenergy, the share was about 4%. The total renewable

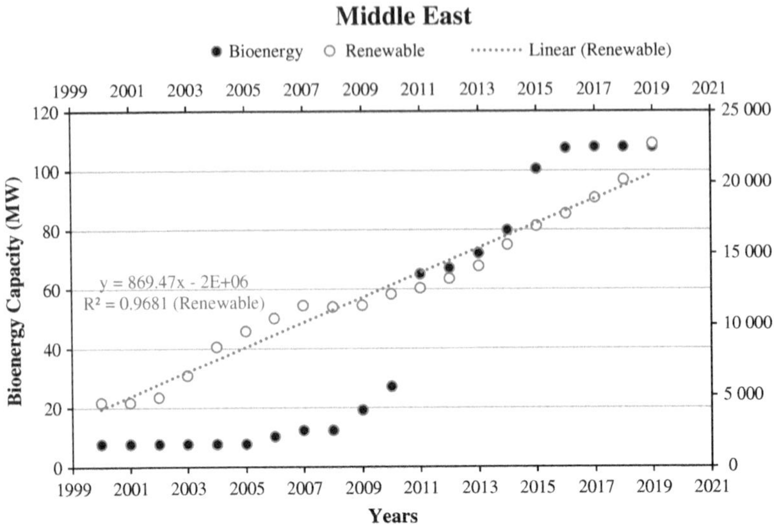

FIG. 11 Comparison between total renewable and bioenergy capacities in the Middle East. *(From Internatinal Renewable Energy Association (IRENA). Renewable capacity statistics 2020, International Renewable Energy Agency, Abu Dhabi, 2020. Available at www.irena.org/Publications, Accessed on 4/5/2020).*

capacity was 173,031 MW in 2000 and increased to 391,240 MW in 2019, which is a twofold increase in capacity.

In 2019, the renewable capacity was dominated by the United States with a share of about 67%, followed by Canada (25%) and Mexico (6%). A similar trend was also found for bioenergy capacity contribution. The United States has the highest capacity of 12,449 MW, followed by Canada (3375 MW) and Mexico (1010 MW). Both renewable and bioenergy capacities followed a linear growth in this region as illustrated in Fig. 12.

2.1.8 Ocenaia

For this region, an almost threefold increase in renewable capacity was observed for the period 2000–2019. Unlike other continents, this region is dominated by solar energy with a share of about 41%. The second highest contributor was hydro with a share of 35%, followed by wind (20%) and bioenergy (2.66%). The leading contributor was Australia with a share of about 80%, followed by New Zealand (18%). The other major contributors were Papua New Guinea (333 MW), Fiji (212 MW), and New Caledonia (175 MW).

In terms of bioenergy capacity, out of 1057 MW, Australia and New Zealand were contributing about 863 and 118 MW, respectively. The other contributors

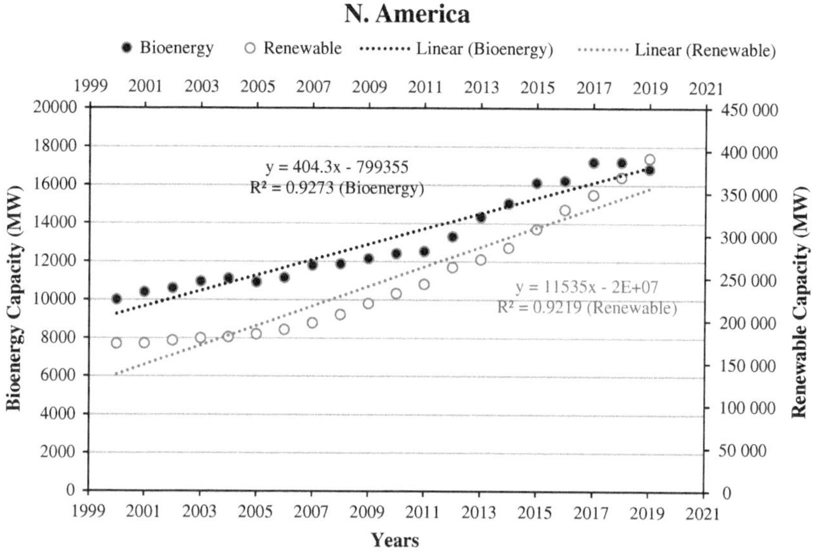

FIG. 12 Comparison between total renewable and bioenergy capacities in North America (N. America). *(From Internatinal Renewable Energy Association (IRENA). Renewable capacity statistics 2020, International Renewable Energy Agency, Abu Dhabi, 2020. Available from: www.irena.org/Publications, acessed on 4/5/2020).*

were Fiji (54 MW), Papua New Guinea (18 MW), Vanuatu (3 MW), and Solomon Island (1 MW). For this region, both renewable and bioenergy capacity growths are following a nonlinear nature as depicted in Fig. 13.

2.1.9 South America

In 2000, renewable capacity was 112,694 MW in this continent, which was increased about twofold and reached 221,130 MW in 2019. Like most other regions, this capacity is also dominated by hydro with a share of about 80%, followed by wind, bioenergy, and solar with shares of about 9%, 8%, and 3%, respectively. In connection with the country's contribution, Brazil ensured a substantial capacity of about 64%. The next country was Venezuela with a share of about 7%. The other countries were Argentina, Colombia, and Chile with shares of 5.78%, 5.60%, and 5.20%, respectively.

In the case of bioenergy, Brazil was the only country that contributed a share of about 87% toward bioenergy capacity in this region. Other major contributors were Chile (501 MW), Uruguay (425 MW), Colombia (340 MW), and Argentina (298 MW). In capacity development over the years, both renewable and bioenergy capacities followed a linear trend as shown in Fig. 14.

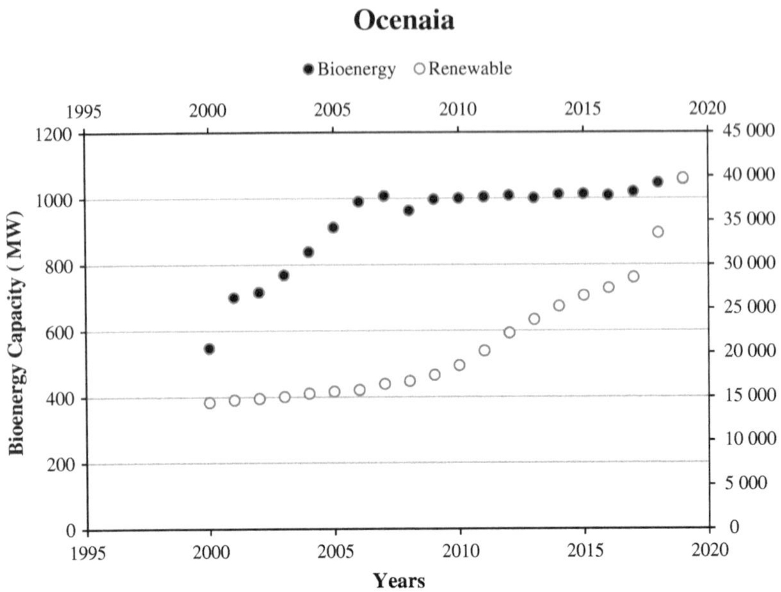

FIG. 13 Comparison between total renewable and bioenergy capacities in Oceania. *(From Internatinal Renewable Energy Association (IRENA). Renewable capacity statistics 2020, International Renewable Energy Agency, Abu Dhabi, 2020. Available from: www.irena.org/Publications, Accessed 4/5/2020).*

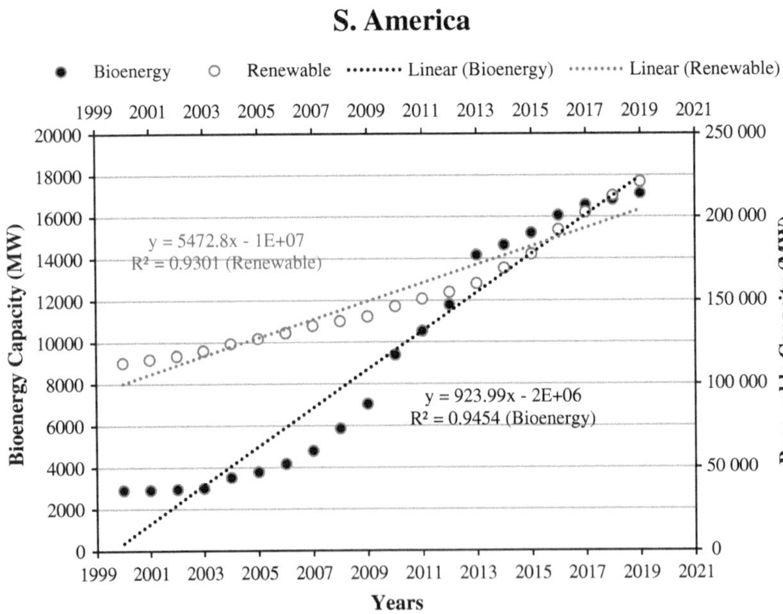

FIG. 14 Comparison between total renewable and bioenergy capacities in South America (S. America). *(From Internatinal Renewable Energy Association (IRENA). Renewable capacity statistics 2020, International Renewable Energy Agency, Abu Dhabi, 2020. Available from: www.irena.org/Publications, Accessed 4/5/2020).*

2.1.10 World

The global and individual renewable electricity capacity growth can be seen from Fig. 15. Clearly, most of the individual growth trends are linear in nature such as hydro, wind. Importantly, the growth of solar capacity follows exponential characteristics. On the other hand, the bioenergy capacity growth is linear. Overall, the global renewable capacity growth is exponential, which is promising to cope with the negative climate change. The nature of bioenergy growth also indicates that extensive policy making at both national and international levels is required to boost the use of this potential source, which is underutilized at the moment compared to other available renewable resources. In contrast, marine resource is still in the infancy stage; thus, it is not suitable to be adopted vastly at present.

The following section provides a brief account of policy measures to promote the development and use of bioenergy resources as renewable energy resource (RES) in the context of the developed world with particular focus on the European Union (EU) countries, the United States, and Canada. The implications of the EU Renewable Energy Directive (RED) 2009 on the member states' (MS) policy regime measures taken to contribute to the achievement of RED have been briefly described. Apart from the EU members, a brief account of the bioenergy sector of the United States and Canada has been discussed.

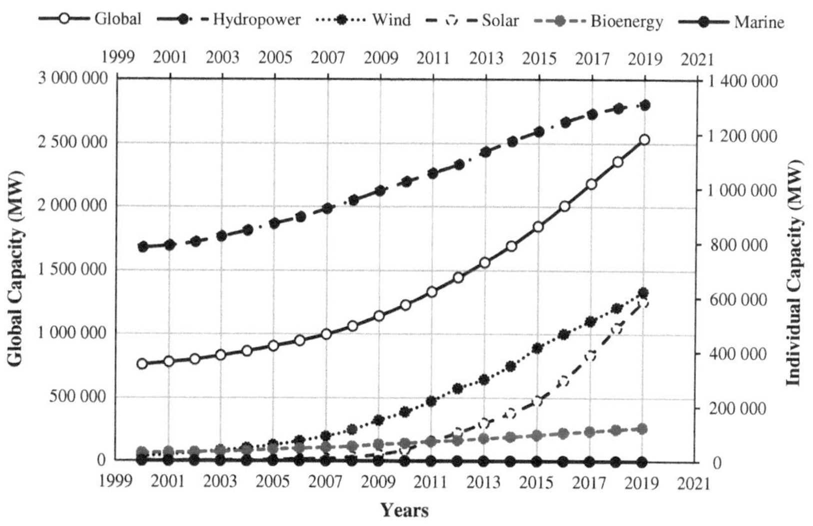

FIG. 15 Global renewable electricity (in MW) growth between 2000 and 2019. *(From Internatinal Renewable Energy Association (IRENA). Renewable capacity statistics 2020, International Renewable Energy Agency, Abu Dhabi, 2020. Available from: www.irena.org/Publications, Accessed 4/5/2020).*

3 Bioenergy resources in the developed world

In the EU, bioenergy is the dominant renewable energy source where it provides almost 70% of the total renewable energy consumption [8]. Bioenergy is expected to be the main contributor to the EU 2020 target of 20% renewable energy of gross final consumption of energy. Further, although there are a variety of sources to generate renewable energy, biofuel is the main source of energy for transport. Bioenergy also has large potential to contribute to the future EU energy services for heat, electricity, and transport. However, notwithstanding its importance, the potential of bioenergy worldwide is inadequate. One of the key reasons is that land is required for bioenergy production. However, the use of land is multifunctional including infrastructural development and urbanization in addition to production of food, conservation of forest resources for timber and nature conservation [9]. Given these issues, the use of bioenergy in the EU is potentially increasing. Bioenergy represents around two-thirds of the total EU primary energy supply of renewable energy sources in 2016, with 5881 PJ (Fig. 16).

In general, developed countries have policy frameworks for the promotion of bioenergy. The EU is a leader in renewable energy development. Its renewable energy policy highlights security of supply, sustainability in the energy sector, and the development of an internal energy market [10]. Renewable Energy Directive (RED) of 2009 is the main policy framework to promote the production and use of renewable energy. Bioenergy is a renewable energy source. As such, a discussion of RED is vital to focus on bioenergy development and its use including challenges in making it widespread as discussed in the following section.

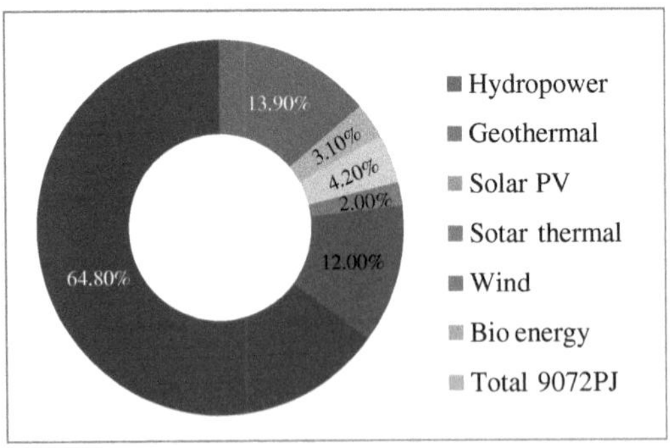

FIG. 16 Total primary energy supply of renewable energy sources in the EU in 2016. *(From Söderberg C, Eckerberg K. Rising policy conflicts in Europe over bioenergy and forestry, Forest Policy Econ, 33:112–119, 2013).*

3.1 The European Union policy framework for bioenergy

In April 2009, the Renewable Energy Directive (2009/28/EC), known as RED-1, was published. **RED 1** established a European framework for the European member states for the promotion of renewable energy. This new legislation (RED) has become the European 20:20:20 Renewable Energy (RE) Strategy including overall targets at the consolidated EU 27 level which are to be achieved by 2020 (Table 1).

The RED sets out binding targets aimed at the promotion of renewable energy. The overall target for the EU is to secure at least 20% of its total energy from RE sources by 2020 and 10% renewable energy in transport by 2020. Another target is to reduce the GHG emissions in the EU by 20% compared to the 1990 emission levels. The directive established RE requirements for each member country (e.g., Italy's target: 17% renewable energy by 2020; for the United Kingdom, the target is 15%) based on a number of factors, including existing renewable generation and GDP [9, 11].

In achieving the binding targets, the levels of effort are differentiated across member states as specified in Annex A of the Directive.[a] The mandatory targets (20:20:20) aim to stimulate investment and innovation in the renewable energy sector to fulfill the target. As part of the RED (2009/28/EC), by the end of June 2010 all the EU MS had put in place binding renewable energy targets. They also had their National Renewable Energy Action Plans (NREAPs) developed (by sector and technology) to meet their 2020 renewable energy targets. The sectoral targets vary in the Gross Final Energy Consumption as presented in Table 2.

In addition to existing measures and expansion targets, it defines a number of instruments and adaptations in support of the development of renewable energies. RED-1 too mandatorily incorporated a set of criteria for ensuring

TABLE 1 European 20:20:20 renewable energy (RE) target.

• 20% less energy consumption
• 20% of all energy must come from RE sources including 10% of total land transport fuel must come from RE
• 20% less Greenhouse Gas (GHG) emissions compared with 1990

From European Commission (EC), 2020 climate & energy package, Available from: https://ec.europa.eu/clima/policies/strategies/2020_en, 3/4/2020.

a. Each EU directive must be implemented through national legislation. Member states may use a variety of policy instruments to support alternative fuels in order to meet the targets in the directives.

TABLE 2 Gross final energy consumption by sector (Italy).

Gross final energy consumption	2005 (%)	2020 (%)
RES—electricity	16.29	28.97
RES—heating and cooling	2.80	15.83
RES—transport	0.42	6.38

From National National Renewable Energy Action Plan (NREAP), Data for member states, Italy, 2010.

sustainability of renewable energy. This is an integral part of the EU sustainability scheme in addition to monitoring and reporting. The sustainability criteria can be summarized as follows:

- Greenhouse gas (GHG) emission savings from the use of biofuels and bioliquids should be at least 35% (50% from 2017 and 60% from 2018).
- Biofuels and liquids produced from raw materials obtained from the following are prohibited:
  - land with high biodiversity (primary forests, highly diverse grassland)
  - Areas of high carbon stock (wetlands, forest, and peatland)

It is to be noted that bioenergy such as biofuels and bioliquids are the key components to reduce GHS emissions. Overall, biofuels[b] and bioliquids are necessary to meet all sustainability criteria in order to achieve the EU objectives, in addition to being suitable for support. To make sure that sustainability criteria are fulfilled, member states involve economic operators[3] to supply proof of conformity [12]. To evade some risks of direct environmental damage and thereby comply with the EU sustainability criteria, the EU states excluded the land with rich biodiversity and carbon stock for the production of biofuels.

In 2015, the EU amended the RED based on the EU Indirect Land Use Change (ILUC) directive [13] by introducing a 7% cap on the crop based (e.g., sugarcane and rapeseed) biofuels (ILUC Directive) to the RES-transport target. The objective of this amendment was to decrease the impact of *indirect land use change* (ILUC) and to get ready to a shift toward advanced biofuels. For the advanced biofuels starting in 2020, Denmark rearranged the EU ILUC Directive in 2016 and introduced a 0.9% mandate. In addition, other member countries have also transposed the amendments into their national laws.

To monitor and decrease the GHG emissions, it was necessary to set a similar demand in the FQD on the specification of fossil fuels such as petrol, diesel, and gas. Biofuels and bioliquids need to comply with the sustainability criteria to ensure quality for counting toward the targets set by the EU. Economic

b. Biodiesel such as methanol is produced from sugarcane, maize, eucalyptus, etc.

operators are under compulsion from their respective governments to provide proof of compliance in support of their fulfillment of sustainability criteria [12]. There is a restriction on the production of biofuels on direct environmental degradation due to biofuel production.

3.1.1 Key features of the EU bioenergy policy

The EU bioenergy policy originated primarily from the EU environmental and climate policy. The open character of the EU bioenergy policy allows actors to fill in the policy and boosts their own translations. Some members have adopted the necessary policy, strategy, and mandates to translate RED directives into action, while a few others are yet to take the necessary political decision, strategy, and action plans to contribute to RED Directive targets. Denmark has not yet made a political decision on how to fulfill the 2020 target of 10% biofuels for transportation, as instructed by the RED. Denmark's biofuels mandate is still fixed at a level of 5.75% (ICCT, 2018). The open character of the EU bioenergy policy creates scope for business and other stakeholders to influence and twist policies to corporations' best interests rather than environmental priorities [14–16]. It is well recognized that bioenergy policy is a multi-sectoral issue. The use of biomass and biofuels extends its reach over other sectoral policy areas, namely land use, forest resources, transport, and agriculture [8].

There are competing views on the widely acclaimed possible benefits of bioenergy. There exists a competition between biomass producers and pulp-and-paper industries. Certainly pulp-and-paper industries will protect the forest resources to sustain their business interests against the proponents who support and produce biomass for energy and heat. In sum, there are conflicts between different policy goals such as between bioenergy, land use change, and food security in attempting the promotion of bioenergy in the energy sector.

3.2 Implications of RED targets for bioenergy promotion and its deployment

The 10% share of renewable sources in land transportation brought enormous potential for the biofuel sector. This mandatory requirement resulted in growing demand for renewable energy sources mainly from first-generation biofuels, next-generation biofuels, and electric vehicles. The 10% target of the RED can also be interpreted as the key strategy instrument for the mitigation of climate change impact in the EU. This is because transport is definitely the only EU economic sector that is contributing to emission GHGs on a large scale and the increase of emission is unimpeded. It is expected that the achievement of the 10% target will reduce the EU's reliance on fossil fuel-based energy and imports of oil.

To achieve the required objectives for the use of renewable energy in the transport sector, the EU states have enhanced the production, promotion, and use of biofuels. The NREAPs projected that the use of total biofuel would reach around 1216 PJ Mtoe in 2020. This will represent more than 90% of the Renewable Energy expected to be used in the transport sector [17, 18]. In 2012, the bioenergy sector contributed to the creation of employment of about 350,000 people in the EU. In 2020, this figure is projected to be about 450,000 people [19].

3.2.1 Institutional structures with governance regulations

Following the RED, the member states enacted individual national bioenergy legislation to promote bioenergy. Through these legislations, typically a government organization is designated to be responsible and authorized to promote investments necessary to produce biofuels and oversee the implementation of national bioenergy programs. Some of the member states also form national technical committees including experts where the technical committee is responsible for setting standards for the use of biofuels. Member states regularly submit their reports on the progress of the promotion and use of energy from renewable sources to comply with article 22 of Directive 2009/28/EC; the EU member states regularly submit their performance reports on the progress of the generation and use of energy from renewable energy sources.

The EU member states have developed their own institutional structures to adopt and promote bioenergy. For instance, the Ministry of Economic Development and the Ministry of the Environment and Protection of Land and Sea in Italy provide coordination and guidance relating to the development and use of bioenergy. The GSE is responsible for measuring and certifying energy savings, with support from ENEA (the Italian national agency for new technologies, energy and sustainable economic development) and the company RSE (Ricerca sul sistema energetico). The UK Government set up the "Office for Renewable Energy Development" (ORED) in 2009 to coordinate actions aimed at achieving its 2020 renewable energy targets. The aim of ORED is to stimulate investment and develop supply chains in all renewable energy technologies (RETs). Its terms of reference also include encouragement and enabling more use of "sustainable bioenergy" [20], as bioenergy is also an important part of the Government's plans to meet the RED objectives in 2020.

The German Federal Government adopted the National Renewable Energy Action Plan (NREAP) introduced in the EU-RED in 2012. The Renewable Energy Act (REA) has proved to be a very strong tool for development and implementation of a biogas market. At the national level, governance systems for sustainability of biogas value chains in Germany work through traditional top-down governmental approaches. Several acts and laws (e.g., Renewable Energy Act-REA, 2000 read with subsequent revisions, and accompanying regulations) for biogas sector development are in force in Germany. In the case of Italy, the State governments enjoy legislative authority, whereas the regions in the states enjoy administrative power. Based on the responsibilities given to the regions, they have taken necessary steps to develop, operate, and manage the energy sector. Importantly, the regions have their own regulatory measures and develop their own regional energy action plans following the regulatory requirements [21].

3.2.2 Response to insufficient ambition and progress (RED Art.27)

Member states have instituted the necessary mechanism to monitor progress of renewable energy including bioenergy. The case of the Netherlands can be mentioned here. Three key agencies of the Netherlands are entrusted with the responsibility of monitoring regulatory authority to meet the targets, as set out in Directive 2009/28/EC. The regulatory agencies are: (a) the Dutch Enterprise Agency (RVO9/Agentschap NL); (b) the Dutch Authority for Consumers and Markets (NMA/ACM10), and (c) the Dutch Emission Authority (NEA). The Dutch Enterprise Agency [RVO] assesses the Stimulation of Sustainable Energy Production (SDE+) subsidy applications and approves of the applications. The annual amount to be given for subsidies is decided on by the Ministry of Economic Affairs. RVO allocates the funds to private firms, households and not-for-profit initiatives within renewable energy. The Dutch competition authority is called ACM. Its main functions are to regulate energy price as well as to monitor and regulate the trading of emission certificates [22]. Further, it monitors and regulates the amount of greenhouse gas emissions in both ETS and non-ETS sectors, such as transport, households, and SMEs [23].

3.2.3 Transposition of the RED-FQD as amended by the ILUC directive

The EU member states have transposed the EU RED and FQD directives into their domestic legislations to facilitate achievement of RED and FQD targets. Appendix A shows a sample of 5 EU countries. Through transposition of the EU laws/framework into own national legislations/policies, as well as promulgating administrative orders, member states have established their own mechanism to regulate the biofuel market and establish standards. For example, in the context of the French biofuels plan, the European target for the inclusion of 5.75% Lower Heating Value (LHV) in 2010 has been raised to 7% LHV in 2010 in the French domestic law 2005–781. Similarly, to promote biofuels in the transport sector, the UK amended the 3.25% renewable fuel use target for 2009/10 in the transport sector a number of times subsequently to comply with the EU targets. The Renewable Transport Fuel Obligations (RTFO) Order is the main legal instrument to encourage use of biofuels in the transport sector. The Renewable Energy Sources Act 2011 of Bulgaria requires Municipal councils to adopt long-term and short-term programs to induce increased use of energy from renewable sources including biofuels. The regulatory roles of MS shifted from a national to a supranational level as they are monitoring renewable energy developments targets which are set in the EU Directive 2009/28/EC to achieve by a given time, namely, 2010 and 2020 [24].

3.2.4 The incentives or support schemes for bioenergy

Member states have devised various incentives including price supports (e.g., feed-in tariffs and premium payments) and tax incentives (exemptions, deductions), investment aid, tax refunds, and renewable energy obligation support schemes (including those using green certificates) that are granted to promote renewable energy. For the placement of bioelectricity, feed-in tariffs and feed-in premiums are the leading 'support schemes.' The policy supports are different across the EU countries. For example, Germany launched a Program in 2017 that provides grants of up to 60% of investment costs for new, innovative heating/cooling networks based on at least 50% renewable heat [25]. Italy's support schemes include green certificates (based on a minimum quota of RES electricity production) for electricity production; white certificates (energy efficiency for large energy suppliers) and tax deductions are granted to incentivize use of biomass for heating and cooling [26]. To reduce the dependency on imported raw materials to produce bioenergy, Italy has allocated more land for growing oil crops [17].

With a view to reducing competition between fossil fuels and biomass in the heating and cooling sectors, Sweden adopted a combination tax to encourage bioenergy. To reduce competition between fossil fuels and biomass in the heating and cooling sectors, it (Sweden) applied a combination of energy and CO_2 taxes. Since 2009, France has introduced VAT with a lower rate for district heating networks that are using over 50% renewable or recovered energy. Furthermore, the French government announced a series of policies to incentivize the biodiesel industry such as governmental funding to help set up a biodiesel company, a tax-reducing policy for the biodiesel industry, and a zero-rate eco-loan for works to improve the overall energy performance of housing (a fiscal measure) (expected result: 15% reduction in energy consumption of buildings by 2023). Moreover, with the approval of the French Ministry of Industry, the B30 biodiesel (i.e., 30% of biodiesel +70% of petroleum diesel) can be used for public vehicles.[c] Similarly, tax incentives (no energy tax on renewable fuels) are in place for renewable fuels in Denmark [27]. Austria also provides incentives to promote bioenergy. Incentives in the form of a feed-in tariff scheme in the electricity sector are given to achieve the bioenergy goals. Austria is a place of "fuel tourism" due to lower fuel taxes relative to neighboring countries and the resulting low price. Neighboring countries fill in fuel containers of their large transport outfits by increasing the purchase of Austrian fuel. For the promotion and use of bioenergy, Bulgaria has offered the highest monetary

c. With such a policy in place, the French automobile manufacturers including Peugeot, Citroen, and Renault have developed new diesel engines which can use B30 biodiesel as fuel.

incentive. Bulgaria provides 248 €/MWh for power plants that are involved in the indirect use of biomass from plants and animal substances [27]. The Netherlands uses the SDE+ premium feed-in scheme as the main supporting instrument for renewable energy. Other incentives include subsidies and tax reduction on investments made toward renewable energy. Investments made for renewable energy facilities were tax deductible [28].

Denmark allocated more than €530 million for biogas expansion for the period up to 2030. Denmark provides new premiums for the use of biogas for transport, processing, and heat (economic regulation). Support[d] for upgrading of biogas is laid down in the Danish Natural Gas Supply Act. In the UK, bioenergy producers are being remunerated for supplying electricity to the distribution grid via an FiT since April 2010, and a similar 'renewable heat initiative' (RHI) from April 2011. Three countries (Germany, Lithuania, and Slovenia) apply a sliding FIP scheme (also known as contract for difference—CfD) to incentivize more demand-oriented production.

Bulgaria, the Czech Republic, Germany, Ireland, Greece, France, Croatia, Italy, Cyprus, Latvia, Luxembourg, Hungary, Austria, Poland, Portugal, Slovenia, Slovakia, and the United Kingdom (18 EU countries) are using a feed-in tariff (FiT) as their main support schemes for biomass in this sector. Similarly, a constant feed-in premium (FIP) scheme is in use in 13 EU countries [27]. 'Biofuels are supported through a minimum quota of biofuels that suppliers of petrol and diesel must include in all fuel sold for transportation' [29]. Grants are also available for establishing plantations of fast-growing trees under Government Regulation No 342/2014 of the Slovak Republic. The quota system in the form of a tradable green certificate (TGC) is applied only in Belgium, Romania, and Sweden.

3.2.5 Policy measures/tools (instruments)

The RED establishes mandatory national targets to facilitate the EU's overall target of 20%. The RED also comprises biennial indicative target paths. The way the member states achieve their national targets is discretional. For example, the UK introduced the Renewable Transportation Fuel Obligation (RTFO) in 2008, setting a biofuel mandate that started at 2.6% by volume in 2009 and increased up to 6% in 2018. The share of renewable energy in transportation was 4.9% in 2016, with double-counting. The UK amended the RTFO in the spring of 2018 to transpose requirements from the EU ILUC directive.

All member states are required (article 4) to prepare National Renewable Energy Action Plans (NREAPs). The NREAP is based on a given template that defines sector targets and specifies policy measures for reaching the targets. Member states have developed action plans. Some members developed a detailed bioenergy strategy before adoption of the action plan. As for instance,

---

d. E.g., electricity generated by burning biomass, a premium of DKK 0.15 per kWh is paid.

the UK Bioenergy Strategy (2012) was developed to deliver a cost-effective bioenergy. This policy strategy supports the Government's approach to obtaining the welfare of bioenergy. Furthermore, the mechanisms for cooperation included in the RED, such as joint support schemes, joint projects, statistical transfers, or even third country cooperation, are set out in Articles (6–11). The intention was to assist member states to achieve their national targets and the EU overall target. To date, however, member states except a few are reluctant to use these cooperation mechanisms. The member countries such as Estonia, Lithuania, Luxembourg, Sweden, and Norway are the only countries that follow the cooperation mechanism through exchanging and reporting statistical information and have joint actions under the framework of the RED [30].

The Guarantees of Origin scheme (Article 15) is an important measure under RED. It provides a mechanism for consumer disclosure of the sources of renewable electricity. It is to be noted that any of the member states has rights not to acknowledge a "guarantee of origin" only when if there are well established confusions about the accuracy, reliability, or veracity of the scheme. It has not yet delivered sufficient transparency with regard to the suppliers fuel mix [31]. The RED also stipulates a number of sustainability criteria to ensure that biofuels placed on the EU market are sustainable.[e] One such sustainability criterion concerns the protection of a "highly biodiverse grassland" (including endangered or vulnerable species) on that land. These criteria are to be respected by economic operators. For access to and operation of the grids, member states are assigned with taking proper steps to improve the grid infrastructure (e.g., smart grids), networks for intelligence, storage facilities for storing energy generated from renewable energy sources, and the overall electricity system for the effective operation of the electricity system. In order to monitor and control the progress toward achieving targets, member states are requested to submit a progress report (according to articles 22–23) every 2 years stating the measures taken and plans made to promote the growth of RES.

3.3 Impact of RED on promotion and further deployment of bioenergy

The EU RED 2009/28/EC has created a compulsion on member states to set time bound goals to increase use of renewable energy. Due to the adoption of RED, use of RE increased significantly in the last two decades. In 2004, where the share of renewable energy of total consumption was 9.6%, the share of renewable energy of total consumption has increased to 18.9% in 2018. The renewable energy share increased between 2004 and 2018 in all member states in the EU. Among the member states, 14 countries were able to double the share of renewable energy of total energy consumption [17]. The total gross energy

e. Only biofuels certified as sustainable can be counted by the member states toward achievement of their 10% transport target.

consumption of the EU had gone down to 1162 Mtoe in 2017 from 1226 Mtoe in 2005. On the other hand, the share of agricultural biomass, forest biomass, and renewable waste combined climbed from 5.9% in 2005 to 10.3% in 2017. In the EU, there has been a considerable increase of bioenergy consumption in Germany, France, Sweden, Italy, and Finland. The total consumption of these Members stood at about 55% in 2017 of the final bioenergy consumption in the EU. By 2017, the biomass installed capacity in the EU surged by almost three times than of the 2005 capacity reaching 32 GW [27]. The setting of mandatory national targets and NREAPs and their proper monitoring played a positive role in enhanced deployment of bioenergy. The NREAPs were monitored with the help of an indicative trajectory defined in (part B, Annex I) of the target during 2010–20 and the submission of biennial progress reports.

In some countries, waste oils are generated from used cooking oils, and tallow as a feedstock for biodiesel is considered more energy efficient. Biodiesel produced from these waste oils is counted twice in meeting energy targets. One of the key aspects of RED is that Article 21(2) of RED allows the double-counting of biofuels from waste and residues. In this case, biofuels in the line with the Article count double toward the target of 10% because of those biofuels are produced from waste and residues. In comparison to single-counting biofuels, only half of the volumes are required to realize the same contribution. The EU average share of double-counting biofuels in total biofuel consumption is 8.8%. The Netherlands is the first country to implement double-counting incentives in its biofuels obligation, and now has a total share of 42% of biofuels from nonfood feedstock such as wastes and residues mainly used cooking oil (data for 2013). Most of the member states in the EU are currently using double-counting (in Germany, double-counting expired at the end of 2014 with transition to a GHG reduction mandate). For instance, the United Kingdom has introduced 'double counting' for the production of biofuels. Biofuels are produced from wastes, residues, nonfood cellulosic material, and lignocellulosic material. Similarly, Italy adopted 'double counting' for promoting biofuel and advancing the production of biofuels in order to achieve the targets at 0.6% in 2018 [27, 30].

However, there is no uniform procedure of double-counting across the EU countries. As such, clarity is needed regarding the use of 'double counting' procedures. While the adoption of double-counting procedures for biodiesel produced from used cooking oil in France and the UK, for example, is well under way, other countries such as Italy and Spain have yet to fully implement the double-counting tool [32]. Also, the effect of this double-counting tool has not been found significant in promoting innovation (e.g., research into advanced biofuel processes) or in bringing new types of biofuels onto the market, such as bioethanol from lignocellulosic feedstock [31].

In addition, the promotion of the use of renewable energy resources under the EU directive has strong potential to create vibrant market opportunities (has been discussed in **detail** in the following section) for the biofuel sector.

The European Biodiesel Board (EBB) shows that the total number of biodiesel facilities in 2011 stood at 254 compared to 185 in 2007 [33]. In EU countries, there are limited areas for productive woodland, and expansion of land for production of biofuel sources is not possible due to competing land use. Therefore, the additional wood required to satisfy the EU bioenergy demand cannot be met by domestic resources. While countries such as France and Germany are mainly depending on domestic supplies, many EU countries such as Italy, the Netherlands, Sweden, and the United Kingdom are importing increasing volumes of wood pellets [34] from other countries such as Brazil and Indonesia to meet their needs.

3.4 Marketing of bioenergy and the EU

The storage facility for bioenergy and availability of proper transport influence the marketing of bioenergy. The development/expansion of infrastructure for storage and transportation of bioenergy, particularly for biofuels, is often supported with government funds. These supports take different forms such as providing infrastructure directly or by proving tax benefits/exemptions and subsidized loans to the private sector to help them develop infrastructures for bioenergy. Some countries provide support for research in energy technology, e.g., Germany [35]. For sustainable use of bioenergy as well as development of a market for biofuel, a market chain is necessary involving all relevant stakeholders (e.g., forest industry). The taxation regime also promotes the RE market. For example, carbon and energy taxes generate the market for biofuels. Price supports such as a Feed-in tariff facilitate new players in the market.

The market for biomass/feedstocks is often dictated by the demand situation. For example, demand for Finnish wood pellets in Sweden helped the Finland pellets market grow [36]. In the EU, initiatives have been taken to promote the RE market. The EU third energy package (2009) focuses on the development of a single European energy market, through adoption of EU-wide network codes (NCs) [27].

The EU has taken projects to promote marketing of RE through promotion of grids/infrastructure. The Integrated Smart GRID CrossFunctional Solutions for Optimized Synergetic Energy Distribution, Utilization Storage Technologies (The inteGRIDy project) project carries out ten pilots, with existing infrastructure, in eight European countries. The objectives of the project are to improve aspects such as demand response, smart distribution grid, energy storage, and smart grid integration of electric vehicles. A total funding of €15,743,171 is allocated for this project, out of which the EU contribution amounts to €12,329,013 [32]. Additionally, under the Connecting Europe Facility (CEF), €5.85 billion are available for boosting energy and transport infrastructure between 2014 and 2020 in the EU to facilitate marketing of RE. Different projects on transmission grids, gas storage, and smart grids will be implemented to develop the trans-European energy infrastructure [37].

In the energy market, there exist a few barriers; the oligopoly energy market is such a barrier that hampers fair competition [25]. The lack of harmonization of requirements with regard to specific testing procedures is another barrier to marketing of renewable energy. For example, specific testing procedures have to be followed in some EU member states to install RE equipment. Such a requirement escalates the cost of RE providers to introduce new technology on the market. In the absence of an EU-wide harmonization of such requirements, trade of renewable energy is inhibited [38].

Despite the significant positive effects of biofuel, recent studies [38] have shown that bioenergy sources also have some drawbacks. The bioenergy sector has potential competition with the food sector either directly or indirectly. There is direct competition between the bioenergy sector and food sector when food commodities are used as the energy source. Problems also arise when bioenergy crops are cultivated on soil that would otherwise be used for food production (indirect competition). These two effects may affect food prices and food security even significantly in some parts of the world. This concern proved valid in 2008 when the competition between food and fuel crops due to the increase of biofuel caused a considerable volatility in the food price and led to riots in several countries across the global South, e.g., for food in Egypt, Haiti [39]. The key challenges are mentioned in Section 3.5.

3.5 Key challenges

(a) The absence of low tariffs on imports of equipment and parts for RE Plants results in increasing the production costs of RE, leading to high costs of RE. Duties and taxes on imports of equipment and parts required for renewable energy plant are reducing the efforts to increase the use of RE.

(b) Many conflicts are seen between different policy goals when attempting to implement bioenergy promotion. Different degrees of conflict have also been identified within the sector with regard to climate change, land use change, and biodiversity [8]. Land use change from bioenergy expansion may increase carbon emissions or reduce biodiversity.[f] Bioenergy has already emerged as a threat to food security and for its potential links to deforestation (e.g., palm biodiesel in Indonesia). The EU's sustainability criteria with regard to ILUC and ruling out biofuels produced from feedstock grown on certain lands have been sensible. But this does not ensure that there are no indirect effects of the policy. For example, higher demand for palm oil (a biodiesel feedstock) could cause producers to shift production of palm oil intended for the EU market to land that meets the sustainability criteria. But production of palm oil (or biofuel) for export to other destinations (not for shipment to the EU) could then be relocated to land

f. **ILUC** may cause the release of CO_2 stored in trees and soil, indirect land use changes risks negating the greenhouse gas savings that result from increased biofuels.

that does not meet the sustainability criteria, causing further deforestation [39]. Realizing the problem, the UK and the EU have introduced new requirements to sustainably source biomass for electricity, focusing on waste and residues rather than pure wood [18].

(c) Other challenges include regulatory and policy uncertainty (e.g., unstable energy policy), and discontinuity in the application of measures [40]. The RED does not specify how the biofuels (or any of the other renewable energy carriers) are to be marketed, but at present biofuels are mostly sold through blending with petrol and diesel. In addition, a major weakness of the EU directive is that RED does not have a specific enforcement (or penalty) mechanism for Member State(s) in case it does not meet its target. The trying or lack of trying to meet national targets is mainly a moral one. However, the EU could initiate an infringement procedure after 2020 (article 258 of the TFEU[g]) [40].

3.6 Beyond 2020

In July 2018, an ambitious political agreement on increasing renewable energy use in Europe was reached between the Commission, the European Parliament, and the Council. This is known as The Renewable Energy Directive 2018/2001 (**RED II**). The new Directive set higher EU-wide targets to be achieved by 2030. Some key elements are:

- A binding renewable energy target for the EU for 2030 of 32% with an upward revision clause by 2023. A binding target of 14% is set for renewables in the transport sector [41].
- Binding targets for the use of advanced, nonfood-based biofuels to 3.5% by 2030.
- Advanced biofuels will be double-counted toward both the 3.5% target and the 14% target.
- Biofuel targets increased to 14% by 2030 with food/feed-based biofuels, a 1% point above 2020 levels but not exceeding 7%.
- The counting of biofuels with a high risk of indirect land use change (ILUC) will be frozen at 2019 levels and gradually phased out from 2023 to 2030 [42].

To address the issue of ILUC in the Clean Energy for All Europeans package, the revised renewable energy directive introduces a new approach. It sets limits on high ILUC-risk biofuels, bioliquids, and biomass fuels with a significant expansion in land with high carbon stock. The directive also introduces an exemption from these limits for biofuels, bioliquids, and biomass fuels **certified as low ILUC-risk**. As the directive is yet to be enacted, how the goals of the new directive are achieved remains to be seen.

g. TFEU: Treaty on the Functioning of the European Union.

4 Bioenergy policy in the United States

As an alternative source of fossil fuel-based energy, the US government has focused on renewable energy from different sources including solar, wind, biomass, for example. The government has adopted policy measures to promote and generate renewable energy to reduce dependency on fossil fuel. In particular, the United States has adopted policies for biofuel or bioenergy generation. The key aspect of the policy is subsidization that has dramatically increased the production of biofuels. The United States has started to produce maize in the early 1980s mainly to revive the agricultural sector when there was an oversupply of agricultural production. The key purpose of the production of maize was to produce ethanol or biofuel. Like other major producing countries, the United States uses targets as part of its biofuel policy. The mandatory target for bioenergy of the United States vis-à-vis other developed countries is presented in Table 3.

4.1 Institutional structures

The Environmental Protection Agency (EPA) administers the Renewable Fuel Standards (RFS) and is responsible for several related tasks. EPA evaluates which renewable fuels are eligible for the RFS program. It also establishes the amount of renewable fuel that will be required for the coming year based on the statutory targets, fuel supply, and other conditions. The EPA Administrator has the authority to waive the RFS requirements [43]. The EPA presently controls the Renewable Fuel Standards (RFS) with particular focus on the production of a minimum volume of biofuel annually on the basis of the estimated total volume of fuels required for the transport sector. The EPA first estimates the total volume of fuel that is expected to be consumed by the transport sector

TABLE 3 Bioenergy targets for transport fuels: United States versus other developed countries.

Country	Mandate (targets) as part of their biofuel policy
EU	In all, 10% share of renewables (including biofuels) in transport fuels by 2020 [M=Mandatory]
United States	Mandatory target of 9 billion gallons of biofuels by 2008, rising to 36 billion by 2022 [M] (of the 36 billion gallons, 21 billion to be from advanced biofuels) [proposed target for 2022]
Canada	In all, 5% renewable content in petrol by 2010; 2% renewables in diesel fuel and heating oil by 2012

From FAO, The State of Food and Agriculture 2008: biofuels: prospects, risks and opportunities, FAO Rome, Italy, 2008.

in the forthcoming year [44] and then calculates the combination requirement to satisfy the total amount of renewable fuels mandated to be used for the upcoming year. Given the biofuel mandate, all relevant companies are blending biofuels to satisfy the renewable volume obligation (RVO) that has been specified by the EPA's standards for the four categories of biofuels from corn, biomass, cellulose, and other bio-sources [45].

Policy tools/measures: (encouragement/punishment): The US Energy Policy Act of 2005 included tax reductions for fuel-ethanol and biodiesel at state and federal levels. Federal tax credit for fuel-ethanol was also provided that was valid until 2010. Other policy instruments to promote bioenergy included loans, loan guarantees, and grants to farmers and biofuel producers at the Federal level and direct payments, grants, and low interest loans at the state level. State governments provided these policy supports including additional fuel tax exemptions and many others to ethanol producers.

4.2 Regulating the biofuel market and establishing standards

The regulation of the biofuel market and establishment of standards is important for promotion of bioenergy in the United States. For instance, the Energy Policy Act of 2005 developed the Renewable Fuel Standards (RFS) and there was a target to increase the production of ethanol from 4 billion gallons in 2006 to 7.5 billion gallons by 2012 [46]. The United States adopted the Energy Independence and Security Act (EISA) in 2007 and this has significantly expanded the biofuel blending mandates. The mandates under the Act emphasized that by 2022, the transportation fuels to be sold in the United States will contain a minimum of 36 billion gallons of biofuels. This act covered four types of biofuels: renewable fuels (including corn-based ethanol), advanced biofuels, biomass-based diesel, and cellulosic biofuels.

Policy interventions in the form of tax reductions for fuel-ethanol and biodiesel at the state and federal levels were important aspects in addition to federal tax credit for fuel-ethanol. These key policy features were valid until 2010. Incentives such as providing loans, loan guarantees, and grants are also provided to farmers and biofuel producers through the Federal Bioenergy Program. Renewable energy usage was underpinned by the passage of the Clean Air Act and the Reformulated Gasoline Program in the early 1990s. The US Energy Policy Act, 42U.S.C. 15,801 adopted in August 2005 amended the Clean Air Act to establish a Renewable Fuel Standard Program [47].

4.3 Policy implications

Although renewable fuel targets are outlined in the statute such as the Energy Policy Act of 2005 and Energy Independence and Security Act 2007, the total renewable fuel statutory target has not been met, since 2014, with the advanced biofuel portion falling below the statutory target by a relatively large margin since 2015. There are adverse impacts of the expansion of biofuels in the United States on the agricultural sector because corn is the main feedstock for ethanol production and the production of corn requires the use of agricultural land. Because of the increasing production of ethanol from corn, this has caused food to become more expensive. The higher cost of food resulted from a structural change in the demand for corn that has pushed up not only corn prices but also prices of wheat and soybean.

5 Bioenergy policy in Canada

Canada has abundant renewable resources with its large landmass and diversified geography. The sources of renewable energy in Canada include water, wind, sunshine, biomass, geothermal, and ocean energy. Despite its huge land area, bioenergy is still at a very small portion of the total primary energy supply. About 75% of the total primary energy supply (TPES) comes from fossil fuels. Renewable energy sources have a share of 17.4% or 2043 PJ—4.5% bioenergy and 12.9% other renewable energy sources. A solid biomass (432 PJ) constitutes the major share in Canadian bioenergy as seen in Fig. 1. Bioethanol, referred to as Biogasoline in the chart below (59 PJ), represents a little over 10% of the bioenergy supply. Biogas (17 PJ), biodiesel (12 PJ), and renewable municipal solid waste (MSW) (5 PJ) reach lower shares (Fig. 17).

On a volume basis, Canada ranked fifth in the world for liquid biofuels production in 2016. In 2016, Canadian production of biofuels reached 1700 million liters of ethanol and 430 million liters of biodiesel. Canada has the third largest total forest area and the sixth largest cultivated land area in the world. The large

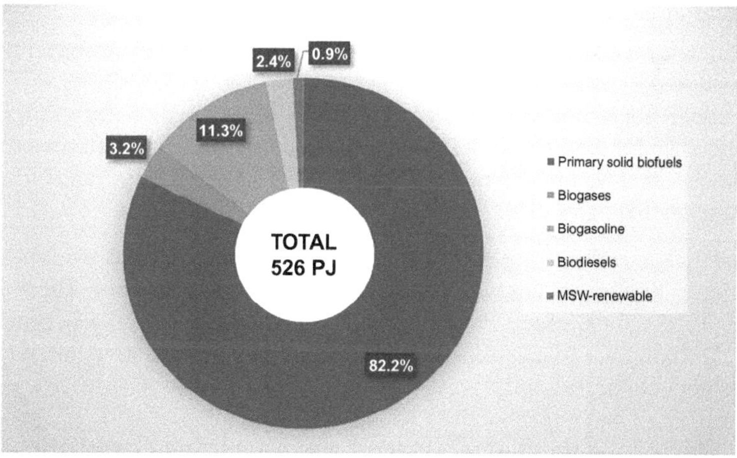

FIG. 17 Total primary energy supply from bioenergy in Canada, 2016. *(From International Energy Association (IEA). World energy balances, OECD 2018. https://webstore.iea.org/world-energy-balances-2018).*

supply of biomass in Canada combined with a strong innovation system and globally leading bioenergy companies indicates that bioenergy could be a large contributor to Canada's energy supply. Despite its relatively small population, Canada was the eighth most important GHG producer in 2012. Bioenergy is an important component to Canada's commitment to reduce its GHG emissions 30% below 2005 levels by 2030 [48].

5.1 Policy features (features of bioenergy laws)

Natural resources fall under provincial jurisdiction (Constitution Act 1982) in Canada. As such, most provinces have established bioenergy policies (or strategies) of their own. For example, BC developed a bioenergy strategy in 2008 and established the Bioenergy Network, a nonprofit organization focusing on organic waste from forestry, agricultural, and municipal sources, with a C$25 million grant to support bioenergy development through investments in capital and technology development/demonstration, targeted capacity building.

The federal Renewable Fuels Regulations mandates 5% renewable content in gasoline (some provinces have a higher target) and 2% in diesel fuel. This regulation is supposed to be replaced by the Canadian Clean Fuels Standard (CFS) to be implemented in 2022 for the liquid fuel stream. The Clean Fuel Standard (CFS) policy, currently under negotiation, aims to reduce GHG emissions from the fuels consumed by introducing Carbon Savings Credits [49].

Government Measures to promote bioenergy: The Canadian government's policy supports to promote renewable energy, among others, are:

- Support for farmer participation in the industry—ecoAgriculture Biofuel Capital Initiative ($200 million over 4 years);
- Support for domestic production through an operating incentive program—ecoENERGY for Biofuels program ($1.5 billion over 9 years);
- Support for next-generation technologies for biofuels from nonconventional feedstocks—NextGen Biofuels Fund ($500 million) [50].
- Support for biomass-based electricity from Alberta's Bioenergy Programs
- Provincial incentives for domestic wood pellets in BC, New Brunswick, and Québec and Natural Resources Canada's IFIT program helps fund "first-in-kind technologies [10].

Canada has measures in place for the quality of biofuel. Five provinces have developed biofuel polices with specific percent volume mandates. The federal ecoENERGY for Biofuels Program (April 1, 2008–March 31, 2017), provided C$1.5 billion over 9 years for operating incentives to stimulate domestic biofuels production.

5.2 Regulating the biofuels market and establishing standards

Different provinces have adopted fuel standards. For example, since January 1, 2007, a Renewable Fuel Standard (RFS) has been in place in Ontario, requiring

5% renewable content in gasoline. For sustainability, Canadian producers primarily use a cascading principle where trees are harvested for an initial purpose, like lumber, and residues from logging and sawmill operations are used to produce pellets. This is in line with current European sustainability priorities [10].

5.3 State of bioenergy in Canada and policy implications

Canada has no binding national or provincial targets with regard to the share of renewable energy in gross final consumption—apart from the transport sector. The share of bioenergy in total primary energy supply (TPES) has not increased in the last 25 years (1990–2016) and remained largely stable (about 5%/Total primary energy supply per capita in 2016 was 323.2 GJ. Of 323.2 GJ, 14.5 GJ are from bioenergy). Rather, a small decrease is seen in recent times [51]. Although its contribution is not big enough, biofuel is the second most important source of renewable energy in Canada [52].

Canadian bioenergy development has not been up to the level of its Scandinavian and Austrian counterparts because of a few barriers. Some of these barriers include high staffing costs, high capital equipment cost, and lack of easy access to bioenergy feedstocks on public lands. Canada requires that any steam installation must have a steam engineer on-site 24 h a day [53]. This results in high staff costs for most projects under 1 MW in Canada.

However, the Canadian government has been providing policy supports to promote renewable energy. The Canadian BioEnergy Centre provides technological support to the bioenergy sector in Research and development and in Product testing and certification. As Canada has a huge abundant biomass for bioenergy resources (including everything from forestry residue to diverted municipal waste), its bioenergy industry has immense potential to be much bigger. Federal policy supports such as the provision for early strategic funding of over $130 million over 5 years for renewable/clean technology, $62.5 million over 2 years to facilitate deployment of infrastructure for alternative transport fuels, increased the scope of eligibility for accelerated Capital Cost Allowance (CCA) rates in electric vehicle charging and electrical energy storage enshrined in the Federal Budget 2016, and the intent to double bioenergy use indicates the considerable likely growth of bioenergy in Canada [54]. Such supports are also creating impacts on renewable energy development as evident in the success story of Bioenergy AE Côte-Nord (Côte-Nord, Quebec). A partnership between Arbec Forest Products and Ensyn, Bioenergy AE has built the world's first commercial facility for the conversion of forest residues to liquid renewable fuel oil. With the $17.5 million funds from the Investments in the Forest Industry Transformation program, a new liquid forest bioenergy product is now available for sale in Canada and North America. This liquid bioenergy reduces greenhouse gas emissions [55].

6 Conclusion

The aim of this chapter was to analyze the policy measures to promote sustainable bioenergy sources in the context of developed countries. In this chapter, the policy measures for bioenergy development were considered from the EU and North American countries. The demand for bioenergy is increasing globally in both developed and developing counties. Policy measures for bioenergy development in the EU countries were analyzed along with the Renewable Energy Directive (RED) framework. The analysis shows that while member states comply with the RED, they have their individual national policy measures for achieving the targets mentioned in the RED. Most of the counties are applying policy measures, most commonly FIT and FIM, in addition to incentives and the tax system. Also, the implications of RED were analyzed with particular focus on challenges. Among the challenges, the competition with land use is most common in addition to environmental degradation. In addition to the EU countries, this chapter also analyzed the policy measures of United States and Canada for bioenergy development including policy implications. Overall, the policy for bioenergy development in development countries shows that while other renewable energy sources such as solar, windmills, and geothermal are important for power generation, bioenergy sources will remain the main contributor. In addition, the policy measures taken by the developed counties are found to be functional to sustainably promote bioenergy.

Appendix A Transposition of EU regulations into MS's legislation

	EU regulations	Transposition into MS legislation
Greece	The RED and the FQD, including the sustainability criteria for biofuels	Fully transposed into the Greek legislation, through several pieces of legislation, including: • Article 1 of L.3851/2010-transposing the RED • L.4062/2012 unit C, OG A'70-transposing the RED and FQD, including the sustainability criteria for biofuels 316/2010 Decision (OG B' 501/2012) of the Chemical Supreme Council-transposing the FQD
Italy	Do	• Legislative Decree No. 28 of March 3, 2011-transposing the RED • Legislative Decree No. 55 of March 31, 2011-transposing the FQD and the sustainability criteria • Decree of October 10, 2014-updating the national biofuel targets until 2022, introducing a definition of "advanced biofuel" • Legislative Decree No. 51 of March 21, 2017-transposing the ILUC Directive • Crop cap: Crop-based biofuels are capped at 7% by 2020

	EU regulations	Transposition into MS legislation
Lithuania	Do	Transposed into the Lithuanian legislation through several pieces of legislation, including: • Act on Use of Biofuels n. 2–83 (2011) • Renewable Energy Law as amended- transposing ILUC provisions • Decree No. 1–348/D1–1014/3–742 as amended-transposing the FQD Crop cap: The renewable energy law sets the crop-based biofuels cap at 7% • Advanced biofuel subtarget: The renewable energy law sets a subtarget for advanced biofuels at 0.5%
Netherlands	Do	Transposed into the Dutch legislation, including • Decision of April 18, 2011 known as "Decree on Renewable Energy in Transport" • Regulation No BJZ2011043268 of 20 April 2011 of the Minister of Infrastructure and the Environment, laying down rules concerning the quality and sulfur content of fuels • Decision of April 8, 2011 laying down the fuels requirements to implement Directive 2009/30/EC (known as the Decree on Fuel Pollution) Crop cap: The Dutch law sets the cap for conventional biofuels of 3% for 2018, 4% for 2019 and 5% for 2020 Advanced biofuel subtarget: The Dutch law sets a mandate for biofuels from Annex IX part A of at least 0.6% for 2018, 0.8% for 2019 and 1% for 2020 after double counting
Poland		Transposed into the Polish legislation through: • Act no. 984/2013 of July 26, 2013 amending the 1997 Energy law and some other acts • Act no. 457/2014 on bio-components and liquid biofuels, amending Act. No 169/2006-transposing sustainability criteria • Act. No 1088/2014 on the system of monitoring and controlling the quality of fuels implementing the provision of FQD in terms of monitoring and reducing GHG of fuels and introducing B7 and the obligation to inform consumers on the content of bio-components on liquid fuels • Act no. 478/2015 on renewable energy

	Transposition into member country legislation
Spain	transposed into the Spanish legislation through: o Sustainable Economy Law 2/2011 of March 4, 2011 • Royal Decree 1088/2010 of September 4, 2010 concerning the technical specifications of petrol, diesel and gas-oil, the use of biofuels and the sulfur content of fuel (this royal decree modifies Royal Decree 61/2006 of January 31, 2006)

Continued

	Transposition into member country legislation
	• Royal Decree 235/2018 of May 1, 2018-establishing calculation methods and information requirements in relation to the intensity of greenhouse gas emissions from fuels and energy in transportation • Royal Decree 1085/2015 and Royal Decree 235/2018-establishing calculation methods and information requirements in relation to the intensity of GHG emissions from fuels and energy in transportation-transposing the ILUC Directive
Sweden	Transposed into the Swedish national legislation by amendments to already existing national legislation, i.e., Swedish Tax law, Green Certificates law, Environmental legislation, etc., and the following: • The law Lag (2010:598) on sustainability criteria for biofuels and liquid fuels, amended by law SFS (2011:1065)-transposing the RED and the sustainability criteria for biofuels and the ILUC Directive • The Fuel Regulation of May 1, 2011-transposing the FQD
United Kingdom	Transposed into the British legislation through several pieces of legislation, including: • Renewable Transport Fuel Obligations (RTFO) (Amendment) Order 2011 No. 2937 of December 7, 2011-transposing the transport elements of the RED, including the sustainability criteria • Renewables Obligation (Amendment) Order 2011, No. 984 of March 29, 2011-transposing the RED • RTFO (Amendment) Order 2013 No. 816 of April 8, 2013-transposing the FQD regarding non-road mobile machinery • Motor Fuel (Road Vehicle and Mobile Machinery) GHG Emissions Reporting Regulations No. 3030/2012-transposing some of the FQD provisions • Statutory Instrument No. 374 of March 13, 2018-amending the RTFO Order and the Motor Fuel that intends to implement the provisions form the ILUC Directive and to implement the provisions of the implementing measures under the FQD

Sector Leiden University MPA: Governing Markets: Competition and Regulation, 2016.
From Webers Z. Institutional Stakeholder Perspective: Corporate Governance in the Dutch Energy

References

[1] International Energy Association (IEA). Bioenergy. International Energy Agency; 2019. Available from: https://www.iea.org/fuels-and-technologies/bioenergy. [Accessed 6 April 2020].

[2] Woods J, Hall DO. Bioenergy conversion technologies. London: FAO; 1994. Available from: http://www.fao.org/3/T1804E/t1804e06.htm#4. [Accessed 7 April 2020].

[3] Khan I, Kabir Z. Waste-to-energy generation technologies and the developing economies: a multi-criteria analysis for sustainability assessment. Renew Energy 2020;150:320–33. https://doi.org/10.1016/j.renene.2019.12.132.

[4] Faaij A. Modern biomass conversion technologies, mitigation and adaptation strategy. Glob Change 2006;11:343–75. https://doi.org/10.1007/s11027-005-9004-7.

[5] Swaraz MA, Satter MA, Rahman MM, Asad MA, Khan I, Amin MZ. Bioethanol production potential in Bangladesh from wild date palm (*Phoenix sylvestris* Roxb.): an experimental proof. Ind Crop Prod 2019;139:1–9. https://doi.org/10.1016/j.indcrop.2019.111507.

[6] World Bioenergy Association. Global bioenergy statistics 2019. Stockholm, Sweden: World Bioenergy Association; 2019. https://worldbioenergy.org/uploads/191129 WBA GBS 2019_LQ.pdf. [Accessed 6 April 2020].
[7] Internatinal Renewable Energy Association (IRENA). Renewable capacity statistics 2020. Abu Dhabi: International Renewable Energy Agency; 2020. Available from: www.irena.org/Publications. [Accessed 4 May 2020].
[8] Söderberg C, Eckerberg K. Rising policy conflicts in Europe over bioenergy and forestry. Forest Policy Econ 2013;33:112–9.
[9] Popp J, Lakner Z, Harangi M, Fari RM. The effect of bioenergy expansion: food, energy, and environment. Renew Sust Energ Rev 2014;32:559–78.
[10] Sawatzkya M. Canadian bioenergy development and EU influences. Alue Ja Ympäristö 2017;46(2):32–46.
[11] Frohne L, Pachter R, Quinlan P. Offshore wind. 2nd ed. Academic Press; 2014.
[12] German L, Schoneveld G. Social sustainability of EU-approved voluntary schemes for biofuels implications for rural livelihoods, working paper no. 75. Center for International Forestry Research; 2011.
[13] European Union (EU). Directive (EU) 2015/1513 of the European Parliament and of the Council of 9 September 2015 amending Directive 98/70/EC relating to the quality of petrol and diesel fuels and amending Directive 2009/28/EC on the promotion of the use of energy from renewable sources. Off J Eur Union 2015;239(1).
[14] Albrecht M, Kortelainen J, Sawatzky M, Lukkarinen J, Rytteri T. Translating bioenergy policy in Europe: mutation, aims and boosterism in EU energy governance. Geoforum 2017;87:73–84.
[15] Kortelainen J, Rytteri T. EU policy on the move - mobility and domestic translation of the European Union's renewable energy policy. J Environ Policy Plan 2017;19(4):360–73.
[16] Hamilinck C. Biofuels and food security risks and opportunities. Netherlands: ECOFYS; 2013.
[17] Anon, https://ec.europa.eu/eurostat/cache/infographs/energy/bloc-4c.html. [Accessed 8 May 2020].
[18] European Commission (EC). Bioenergy in the EU. Amsterdam: European Commission (EC); 2016. Available from: https://www.tni.org/files/publication-downloads/bioenergy_in_the_eu.pdf. [Accessed 22 April 2020].
[19] Scarlat N, Dallemand JF, Monforti-Ferrario F, Nita V. The role of biomass and bioenergy in a future bioeconomy: policies and facts. Environ Dev 2015;15:3–34.
[20] Allen PE, Hammond GP. Bioenergy utilization for a low carbon future in the UK: the evaluation of some alternative scenarios and projections. BMC Energy 2019;1(3).
[21] Cavicchi B, Bryden JM, Matteo V. A comparison of bioenergy policies and institutional frameworks in the rural areas of Emilia Romagna and Norway. Energy Policy 2014;67:355–63.
[22] ACM. Thermogravimetric kinetic modelling of in-situ catalytic pyrolytic conversion of rice husk to bioenergy using rice hull ash catalyst. Bioresour Technol 2020;261:213–22.
[23] NEA. Rapportage hernieuwbare energie 2014. Naleving jaarverplichting hernieuwbare energie vervoer en verplichting brandstoffen luchtverontreiniging. Available from: http://www.emissieautoriteit.nl; 2015. [Accessed 16 April 2020].
[24] IRENA. Bioenergy from boreal forests: Swedish approach to sustainable wood use. Abu Dhabi: IRENA; 2019.
[25] Cavicchi B. A "system dynamics perspective" of bioenergy governance and local, sustainable development. Syst Res Behav Sci 2020;37:315–32.
[26] Ma Y, Liu Y. Biofuels: alternative feedstocks and conversion processes for the production of liquid and gaseous biofuels. 2nd ed. Academic Press; 2019.

[27] Banjaa M, Sikkema R, Jégard M, Motola V, Dallemanda JF. Biomass for energy in the EU: the support framework. Energy Policy 2019;131.
[28] Webers Z. Institutional stakeholder perspective: corporate governance in the Dutch energy sector Leiden University MPA: governing markets: competition and regulation; 2016.
[29] European Union. Amendment directive 2009/28/EC on the promotion of the use of energy from renewable sources. Off J Eur Union 2015;239(1).
[30] Bettina Kampman B, Sina S, Lucha C, Cesbron S, Pato Z, Florcken O, et al. Mid-term evaluation of the renewable energy directive, a study in the context of the REFIT programme. CE Delft; 2015.
[31] Jay M. Emission trading scheme in Rio suffers setback. In: Financial risk management and energy risk; 2012. Available from: http://www.risk.net/energyrisk/news/2184727/emission-trading-scheme-rio-suffers-setback. [Accessed 4 April 2020].
[32] European Commission. Supporting innovative solutions for smart grids and storage, 2020, https://ec.europa.eu/inea/sites/inea/files/h2020_sgs_brochure_2019-010.pdf; 2020. [Accessed 6 May 2020].
[33] British Petrolium (BP). BP's statistical review of world energy. 68th ed. British Petrolium (BP); 2012.
[34] Bentsen NS, Felby C. Biomass for energy in the European Union – a review of bioenergy resource assessments. Biotechnol Biofuels 2012;5:25. http://www.biotechnologyforbiofuels.com/content/5/1/25. [Accessed 5 April 2020].
[35] Schubert R, Schellnhuber H, Buchmann N, Epiney A, Grießhammer R, Kulessa M, et al. Future bioenergy and sustainable land use: flagship report. Berlin: German Advisory Council on Global Change; 2008.
[36] Ericsson K, Huttunen S, Nilsson LJ, Svenningsson P. Bioenergy policy and market development in Finland and Sweden. Energy Policy 2004;3:1707–21.
[37] European Union (EU). EU funding possibilities in the energy sector. n.d. https://ec.europa.eu/energy/funding-and-contracts/eu-funding-possibilities-in-the-energy-sector Accessed 5 May 2020.
[38] Roldán IA. The renewable energy directive and its contribution to the elimination of barriers for the deployment of renewable energy in the EU beyond 2020; 2019.
[39] Searchinger T, Heimlich R, Houghton RA, Dong F, Elobeid A, Fabiosa J, et al. Use of US croplands for biofuels increases greenhouse gases through emissions from land-use change. Science 2008;319:1238–42.
[40] Lendle A, Schaus M. Sustainability criteria in the EU renewable energy directive: consistent with WTo rules? ICTSD Information Note no. 2; 2010.
[41] European Commission (EC). REFIT evaluation of the Directive 2009/28/EC of the European Parliament and of the Council, Commission staff working document; 2016.
[42] Edenhofer O, Pichs-Madruga R, Sokońa Y, Seyboth K, Matschoss P, Kadner S, et al. Renewable energy sources and climate change mitigation. Cambridge: Cambridge University; 2011.
[43] Hoover K, Riddle AA. Forest carbon primer, CRS report; 2020.
[44] EPA. Renewable Fuel Standard Program, USA; 2007.
[45] Guan Z, Oh J. United States biofuel policies: overview and discussion. Accessed from: https://edis.ifas.ufl.edu/pdffiles/FE/FE97400.pdf; 2015.
[46] Schnepf R, Yacobucci BD. Renewable fuel standard (RFS): overview and issues Congressional Research Service report; 2012. p. 27.
[47] Jull C, Redondo PC, Mosoti V, Vapnek J. Recent trends in the law and policy of bioenergy production, promotion and use, FAO Legal Paper; 2007.

[48] Nzihou A. Handbook on characterization of biomass, bio waste and related by-products. Springer; 2020.
[49] FAO. OECD-FAO Agricultural Outlook 2019–2028. Biofuels [Chapter 9]. OECD-FAO; 2019. http://www.fao.org/3/CA4076EN/CA4076EN_Chapter-9_Biofuels.pdf. [Accessed 5 May 2020].
[50] Paré D, Bernier P, Thiffault E, Brian D. Titus the potential of forest biomass as an energy supply for Canada. For Chron 2011;87(1):71–7.
[51] International Energy Association (IEA). IEA bioenergy country report. European Union-2018 Update; 2018.
[52] Meyer JE. The renewable energy transition: realities for Canada and the world. Springer; 2019.
[53] Roger L, Kemp CJ, Stephani M. Global models of urban planning: best practices outside the United States; 2013.
[54] Government of Canada. Budget 2016: a clean gpprowth economy; 2016. [Chapter 4], Canada.ca.
[55] Government of Canada. Budget 2019, building a better Canada. In: Supporting innovation and market diversification in Canada's forest sector: success stories; 2019 [Chapter 2] https://www.budget.gc.ca/2019/docs/plan/chap-02-en.html. [Accessed 21 May 2020].

Index

Note: Page numbers followed by *f* indicate figures and *t* indicate tables.

B

F

I

J

K

L

O

P

R

U

V

W

Y

Z

FSC
www.fsc.org
MIX
Papier | Fördert gute Waldnutzung
FSC® C083411

Kolibri 360 GmbH
Ferdinand-Jühlke-Straße 7
99095 Erfurt, Deutschland
produktsicherheit@kolibri360.de